Ecological Methods

WITH PARTICULAR REFERENCE TO THE
STUDY OF INSECT POPULATIONS

T. R. E. SOUTHWOOD Kt., D.Sc., Ph.D., FRS

Linacre Professor of Zoology in the University of Oxford

CHAPMAN AND HALL

LONDON • NEW YORK • TOKYO • MELBOURNE • MADRAS

UK   Chapman and Hall, 2–6 Boundary Row, London SE1 8HN

USA   Chapman and Hall, 29 West 35th Street, New York NY10001

JAPAN   Chapman and Hall Japan, Thomson Publishing Japan,
            Hirakawacho Nemoto Building, 7F, 1-7-11 Hirakawa-cho,
            Chiyoda-ku, Tokyo 102

AUSTRALIA   Chapman and Hall Australia, Thomas Nelson Australia,
            480 La Trobe Street, PO Box 4725, Melbourne 3000

INDIA   Chapman and Hall India, R. Seshadri, 32 Second Main Road,
            CIT East, Madras 600 035

First edition 1966
Reprinted 1968, 1971, 1975, 1976
Second edition 1978
Reprinted 1980, 1984
First published as a paperback 1987
Reprinted 1989, 1991

© 1966, 1978 T.R.E. Southwood

Printed in Great Britain at the University press, Cambridge

ISBN 0 412 307103

Contents

Most population theories . . . are based on imperfect field data that are not derived from planned population studies in which all the relevant factors were measured simultaneously. . . . When . . . more of such fundamental studies [are] available, we may be able to discuss our theories with more light and less heat.

O. W. RICHARDS, 1961
Annual Review of Entomology p. 147

The ecology of pest populations should be studied to gain understanding of the dynamics of the populations in hope that its mechanisms may be revealed. We hope this will enable us to manipulate the populations and ability to manipulate surely is the aim of all attempts to control animal populations.

D. A. CHANT, 1963
*Memoirs of the Entomological
Society of Canada* No. 32, p. 33

Preface to First Edition

This volume aims to provide a handbook of ecological methods pertinent for the study of animals. Emphasis is placed on those most relevant to work on insects and other non-microscopic invertebrates of terrestrial and aquatic environments, but it is believed that the principles and general techniques will be found of value in studies on vertebrates and marine animals.

The term ecology is now widely used in the field of social, as well as biological, science; whilst the subject of ecology, covering as it does the relationship of the organism to its environment, has many facets. It is, in fact, true to say that the ecologist may have need of recourse to almost all of the methods of the biologist and many of those of the physical scientist: the measurement of the physical factors of the environment may be a particularly important part of an ecologist's work and he will refer to books such as R. B. Platt & J. E. Griffiths' (1964) *Environmental Measurement and Interpretation.*

There are, however, certain methods that are peculiar to the ecologist, those concerning the central themes of his subject, the measurement, description and analysis of both the population and the community. These are *ecological* methods (as opposed to 'methods for ecologists' which would need to cover everything from laboratory workshop practice to information theory); they are the topic of this book.

During the ten years that I have been giving advanced and elementary courses on ecological methods at Imperial College, London, and at various Field Centres, the number and range of techniques available to the ecologist have increased enormously. It has been the comments of past students on the utility of these courses in helping to overcome the difficulties of coping with the scattered and growing literature that have encouraged me to attempt the present compilation. I am grateful to many former students for their criticisms and comments, as I am to the members of classes I was privileged to teach at the University of California, Berkeley, and at the Escuela Nacional de Agricultura, Mexico, whilst writing this book.

Although the general principles of most methods are of wide application, the study of a particular animal in a particular habitat may require certain special modifications. It is clearly impossible to cover all variants and therefore the reader is urged to consult the original papers that appear relevant to his problem. I am grateful to my publishers for agreeing to the publication of the extensive bibliographies, it is hoped that these will provide many leads on specific problems; they are, however, by no means exhaustive.

The present book is designed to be of use to those who teach the practical aspects of animal ecology in schools, training colleges and universities; insects, being numerically the dominant component of the macrofauna of terrestrial and many aquatic habitats, almost invariably come to the forefront of ecological field work. This volume is intended as an aid to all who need to measure and compare populations and communities of animals, not only for the research ecologist, but also for the conservationist and the economic entomologist. Population measurement is as necessary in the assessment of the effects of a pesticide and in the determination of the need for control measures, as it is in intensive ecological studies. It is frequently pointed out that ecological theories have outstripped facts about animal populations and I trust that it is not too presumptuous of me to hope that this collection of methods may encourage more precise studies and more critical analysis of the assembled data so that, in the words of O. W. Richards, we may have 'more light and less heat', in our discussions.

The topics have been arranged on a functional basis, that is, according to the type of information given by a particular method. As a result some techniques are discussed in several places, e.g. radiotracers will be found under marking methods for absolute population estimates (chapter 3), the measurement of predation and dispersal (chapter 9) and the construction of energy budgets (chapter 14). By its very nature ecology cannot be divided into rigid compartments, but frequent cross-references in the text, together with the detailed contents list and index, should enable the reader to find the information he needs. The sequence of chapters parallels, to a large extent, the succession of operations in a piece of intensive research.

It is a pleasure to express my great indebtedness to colleagues who have read and criticized various chapters in draft: Dr N. H. Anderson (ch. 6), Dr R. E. Blackith (ch. 2 & 13), Dr J. P. Dempster (ch. 1, 2 & 3), Mr G. R. Gradwell (ch. 10), Dr C. S. Holling (ch. 12), Mr S. Hubbell (ch. 14), Dr C. B. Huffaker (ch. 2), Dr G. M. Jolly, (section II of ch. 3), Dr C. T. Lewis, (section I of ch. 3), Dr R. F. Morris (ch. 10), Dr O. H. Paris (ch. 9 & 14), Mr L. R. Taylor (ch. 2 & 4), Professor G. C. Varley (ch. 10) and Dr N. Waloff (ch. 1, 2 & 10); frequently these colleagues have also made available unpublished material; they are of course in no way responsible for the views I have expressed or any errors. For access to 'in press' manuscripts, for unpublished data and for advice on specific points I am grateful to: Drs J. R. Anderson, R. Craig and D. J. Cross, Mr R. J. Dalleske, Drs W. Danthanarayana, H. V. Daly, E. A. G. Duffey, P. J. M. Greenslade, M. P. Hassell, P. H. Leslie, J. MacLeod, C. O. Mohr, W. W. Murdoch and F. Sonleitner, Mr W. O. Steel, Drs A. J. Thorsteinson, R. L. Usinger, H. F. van Emden, E. G. White, D. L. Wood and E. C. Young. Ecologists in all parts of the world have greatly helped by sending me reprints of their papers. I have been extremely fortunate too in the assistance I have received in translating; Mrs M. Van Emden has generously made extensive translations of works in German, and with other languages I have been helped by Dr F. Baranyovits (Hungarian), Dr T. Bilewicz-Pawinska (Polish), Mr Guro Kuno (Japanese), Dr P. Stys (Czechoslovakian) and Dr N. Waloff (Russian).

Much of the manuscript was prepared whilst I held a visiting professorship in the Department of Entomology and Parasitology of the University of California, Berkeley; I am indebted to the Chairman of that Department, Dr Ray F. Smith, for his interest and the many kindnesses and facilities extended to me and to the Head of my own Department, Professor O. W. Richards, F.R.S., for his support and advice. I wish to thank Mrs M. P. Candey and Mrs C. A. Lunn for assisting me greatly in the tedious tasks of preparing the bibliographies and checking the manuscript. My wife has encouraged me throughout and helped in many ways, including typing the manuscript.

T. R. E. Southwood

London, October 1965

Preface to Second Edition

In the twelve years since the First Edition was prepared there have been remarkable developments in ecology. The subject has changed its lay image, from a rather recondite branch of biology, to something that is widely considered 'good', but only vaguely understood. The public's focus on environmental problems and the insights into these that ecology can provide are a great challenge to ecologists to develop their subject: they need to be able to provide reliable quantitative inputs for the management of the biosphere. The enormous volume of work that it has been necessary to review for this edition is evidence of the extent to which ecologists are seeking to meet this challenge.

I believe that the theme of the first edition, the need for precise measurement and critical analysis, is equally valid today; although many recent studies show levels of sophistication that were beyond my wildest hopes when I embarked on the preparation of the first edition. In his review of the first edition Dr R. R. Sokal was kind enough to say it was an 'unusual book' for it covered both traps and mathematical formulae, topics that were usually of interest to different people. This, I am glad to say, is now no longer generally true. The computer and the electronic calculator have revolutionized the handling of ecological data, but neither can make a 'silk purse' of sound insight, out of a 'sow's ear' of unreliable raw data or confused analytical procedure. More than ever the ecologist needs to keep his biological assumptions in mind and remember the value of preliminary simple graphical analysis as a means of recognizing new patterns and gaining fresh insights.

It has been a gratifying, though exhausting, experience preparing the new edition! Progress has been so rapid in several areas that some chapters have been completely or largely rewritten (e.g. 11, 12, 13), whilst most have large new sections. It has been necessary to be highly selective in the additions to the bibliographies, even so there are nearly a thousand new entries and only a few older references could be deleted. I hope that, with the advent of *Ecological Abstracts*, the selective nature of the bibliographies will not handicap workers. As the mathematical side of ecology has grown, so have the problems of notation and it is now quite impossible (without extending far beyond the roman and greek alphabets!) to retain a unique notation throughout. Apart from widespread and generally accepted symbols, I have merely aimed to be consistent within a section.

I am most grateful to many ecologists who have helped me in this revision by sending me reprints of their papers or notes on difficulties and errors in the first edition. Detailed criticism, advice and help, including access to unpublished work has been generously given by M. H. Birley, P. F. L. Boreham, M. J. W. Cock, G. R. Conway, M. P. Hassell, R. M. May, A. Milne, S. McNeill, G. Murdie, G. A. Norton, S. Parry, P. M. Reader, G. Seber, N. E. A. Scopes, L. R. Taylor, R. A. J. Taylor, T. W. Tinsley, J. M. Webster, R. G. Wiegert and I. J. Wyatt. I am deeply indebted to them, and to others, especially Margaret Clements and my wife, who have assisted with patience and forbearance in the essential tasks associated with the preparation of the manuscript of the new edition.

T. R. E. Southwood

Imperial College, Silwood Park, Ascot.
October 1977

Acknowledgements

Grateful acknowledgement is hereby made to authors and publishers of the original material that has been modified to give the figures, tables, and formulae used in this book; full citations are given in the appropriate places in the text, together with the relevant entry in the bibliography. Gratitude is expressed to the Editor of the Journal of Ecology (Fig. 13.5) and the Controller of H.M. Stationery Office (Fig. 4.6) for permission to reproduce the figures indicated, and to Dr F. Winsor and Messrs Simon and Schuster Ltd, for permission to reprint part of *The Theory that Jack Built* that appears on page xxiv.

For permission to reproduce re-drawn illustrations, modify published tables or to make short quotations, thanks are due to the Editors of the Annals of Applied Biology, Annals of Limnology, Annals of the Entomological Society of America, Archives Néerlandaises de Zoologie, Biological Reviews, Biometrics, Biometrika, Bulletin of Entomological Research, Canadian Entomologist, Canadian Journal of Zoology, Ecological Monographs, Ecology, Entomologia Experimentia et Applicata, Entomologist's Monthly Magazine, Indian Journal of Entomology, Journal of Animal Ecology, Journal of Economic Entomology, Journal of Theoretical Biology, Nature, Oikos, Pedobiologia, Proceedings of the Ceylon Association for the Advancement of Science, Researches in Population Ecology, Statistica Neerlandica and Zeitschrift für Angewandte Entomologie, and to Academic Press, Annual Reviews Inc., the Director of the Anti-Locust Research Centre, Belknap Press, Blackwells Scientific Publications, Blakiston Co., B. Bishop Museum, British Entomological Society, Butterworths, Finnish State Agricultural Research Board, Holt, Rinehart & Winston, Reinhold Inc., Royal Entomological Society of London, F. Warne & Co., and Dr R. L. Usinger. For additional permissions in respect of the new figures in the Second Edition we are grateful to several of the above and to Gustav Fischer Verlag, Harvard University Press, Hydrobiologica, Lepidopterists' Society and Taxon.

A cautionary rhyme

This is the Cybernetics and Stuff
That covered Chaotic Confusion and Bluff
That hung on the Turn of a Plausible Phrase
And thickened the Erudite Verbal Haze
Cloaking Constant K
That saved the Summary
Based on the Mummery
Hiding the Flaw
That lay in the Theory Jack built.

F. WINSOR: *The Space Child's Mother Goose*

Simon & Schuster

1

Introduction to the Study of Animal Populations

Information about animal populations is sought for a variety of purposes; but the *object* of a study will largely determine the methods used and thus this must be clearly defined at the outset. Very broadly studies may be divided into *extensive* and *intensive* (Morris, 1960). Extensive studies are carried out over a large area and are normally concerned with the distribution of insect species or with the relation of insect pest population to crop damage or with the prediction of damage and the application of control measures (e.g. Kaelin & Auer, 1954; Strickland, 1961; Chiang *et al.*, 1961; National Academy of Sciences, 1969). A particular area will be sampled once or at the most a few times during the season, and emphasis will normally be placed on a particular developmental stage of the insect. The timing of such sampling is obviously of critical importance: it must be appropriate in relation to the phenology of the chosen stage (Morris & Bennett, 1967). Such studies will produce considerable information about the pattern of population level over a large area or in successive years, and it is often possible to relate the level of the population to certain edaphic or climatic factors (Kaelin & Auer, 1954; Chiang *et al.*, 1961).

Intensive studies involve the continual observation of the population of an animal in the same area. Usually information is required on the sizes of the populations of successive developmental stages so that a life-table or budget may be constructed and an attempt made at determining the factors that cause the major fluctuations in population size (key factors) and those that govern or regulate it (Morris, 1960; Richards, 1961; Varley & Gradwell, 1963). It is important to consider at the start the type of analysis (see Chapter 10) that will be applied and so ensure that the necessary data is collected in the best manner. Intensive studies may have more limited objectives, such as the determination of the level of parasitism, the amount of dispersal or the overall rate of population change.

The census of populations and the stages at which mortality factors operate are necessary first stages in the estimation of the productivity (Chapter 14) of ecosystems. In survey and conservation work, the species make-up of the population and changes in its diversity associated with man's activities are most frequently the features it is desired to measure. Special methods of analysis need to be used (Chapter 13), but difficulties usually arise because of the virtual impossibility of extracting the many different species from a habitat with equal efficiency by a single method (e.g. Nef, 1960).

1.1 Population estimates

Population estimates can be classified into a number of different types; the most convenient classification is that adopted by Morris (1955), although he used the terms somewhat differently in a later paper (1960).

1.1.1 Absolute and related estimates
The animal numbers may be expressed as a density per unit area of the ground of the habitat. Such estimates are given by nearest neighbour and related techniques (Chapter 2), marking and recapture (Chapter 3), by sampling a known fraction of the habitat (Chapter 4–6) and by removal sampling and random walk techniques (Chapter 7).

Absolute population
The number of animals per unit area (e.g. hectare, acre). It is almost impossible to construct a budget or to study mortality factors without the conversion of population estimates to absolute figures, for not only do insects often move from the plant to the soil at different developmental stages, but the amount of plant material is itself always changing. The importance of obtaining absolute estimates cannot be overemphasized.

Population intensity
The number of animals per unit of habitat, e.g. per leaf, per shoot, per plant, per host. Such a measure is often, from the nature of the sampling, the type first obtained (see also p. 138) and when the level of the insect population is being related to plant or host damage it is more meaningful than an estimate in absolute terms. It is also valuable in comparing the densities of natural enemies and their prey. However, the number of habitat units/area should be assessed, for differences in plant density can easily lead to the most intense population being the least dense in absolute terms (Pimentel, 1961). When dealing with different varieties of plants differences in leaf area may account for apparently denser populations, in absolute terms, on certain varieties (Bradley, 1952), and the actual choice of the leaf or of the plant as the unit for expressing population intensity can affect the relative population levels (Broadbent, 1948) (Fig. 1.1). With litter fauna owing to the effects of seasonal leaf fall the intensity measure (on animals/weight of litter) will give a different seasonal picture from an absolute estimate per square metre (Gabbutt, 1958). These examples also underline the importance of absolute estimates where one's interest lies primarily in the animal population.

Basic population
In some habitats, especially forests and orchards, it is often convenient to have

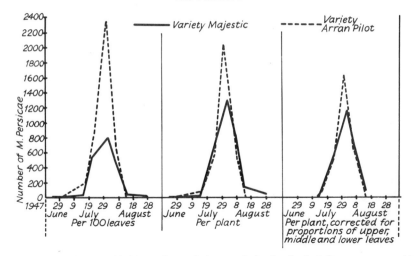

Fig. 1.1. The influence of habitat unit on relative population levels when these are measured in terms of population intensity; the populations of *Myzus persicae* on different varieties of potato (after Broadbent, 1948).

an intermediate unit between that used for measuring intensity and absolute measures of ground area, e.g. 10 sq ft of branch surface (Morris, 1955), or branches of apple trees (Lord, 1968).

1.1.2 Relative estimates
These estimates, in which the population is measured in unknown units, allow only comparisons in space or time; they are especially useful in extensive work, in studies on animal activity or in the investigation of the constitution of a polymorphic population. The methods employed are either the catch per unit effort type or various forms of trapping, the results of which depend on a number of factors besides population (Chapter 7). There is no hard and fast line between relative and absolute methods, for absolute methods of sampling are seldom 100 % efficient and relative methods can sometimes be corrected in various ways to give density estimates.

1.1.3 Population indices
The animals themselves are not counted, but their products (e.g. frass, webs, exuviae, nests) or effects (especially plant damage) are recorded.

Both population indices and relative estimates of population can sometimes be related to absolute population (if this is measured at the same time) by regression analysis, and if such a study has been based on sufficient data subsequent estimates from relative methods or indices can be converted to absolute terms using various correction factors; such an approach is common in fisheries research (e.g. Beverton & Holt, 1957).

1.2 Errors and confidence

The statistical errors of various estimates can usually be calculated and are referred to as the fiducial limits (the estimate (x) being expressed as $x \pm y$, where $y =$ fiducial limits). These are sometimes incorrectly referred to as 'confidence' limits, but the distinction between the two terms is in practice unimportant. The fiducial limits are calculated for a given probability level, normally the 0.05 level, which means that there are only 5 chances in 100 that the range given by the fiducial limits does not include the true value (hence the espressions 5 % probability level and 95 % fiducial limits). If more samples are taken the limits will be narrower; but the biologist is often worried about some of the assumptions (e.g. does the 'knockdown' method really collect all the weevils on a tree?) and intuitively believes, correctly, that the first estimate should be compared with the second made by a *different* method that has different assumptions. If the estimates are of the same order of magnitude, then the investigator can have much greater confidence in his biological assumptions being true. It is therefore sound practice for the ecologist to *estimate the population or other variables by more than one method simultaneously* (see Richards & Waloff, 1954; Lamb & Wellington, 1974).

When two estimates have been obtained from different sampling procedures in this way, provided they are internally consistent (e.g. a t-test shows that the means are not significantly different) they may be combined to give a weighted mean, weighting each estimate inversely as its variance (Cochran, 1954). Under some circumstances Bayes' Theorem could be used (see p. 98) to give the combined estimate. Laughlin (1976) has suggested that the ecologist may be satisfied with a higher probability level (say 0.2) and thus narrower fiducial limits for estimates based on more than one method, because such estimates have a qualitative, biological assurance, additional to that from the consistancy of the data, that the true mean lies near to them.

Most population studies are based on sampling and the values obtained are considered to have a generality that scales with the area from which the samples were drawn. However, all these estimates have fiducial limits and the level of accuracy that should be aimed at is a difficult problem for the ecologist; Morris (1960) has aptly said that 'we are not likely to learn what precision is required by pessimistic contemplation of individual' fiducial limits. As the amount of time and labour that can be put into any problem is invariably limited it should always be borne in mind that the law of diminishing returns applies as one attempts to reduce the statistical errors of sampling and, in the long run, more knowledge of the ecology of the animal may be gained by studying other areas or making other estimates or even by just taking further samples than by straining for a very high level of accuracy in each operation. Against this must be set the fact that when animals are being extracted from samples the errors will all lie on one side of the true value (i.e. too few will be found). A number of very carefully conducted control samples may allow a correction factor to be applied, but the percentage of animals missed may vary with density; sometimes more are overlooked at the lowest densities (Morris, 1955).

An alternative to sampling is the continuous, or regularly repeated, study of a restricted cohort, e.g. the population of an aphid on a particular leaf or leaf-miners on a bough. Such studies have a very high level of accuracy, but they sacrifice generality. A combination of some cohort studies with larger scale sampling often provides valuable insights.

REFERENCES

BEVERTON, R. J. H. and HOLT, S. J., 1957. On the dynamics of exploited fish populations. *Fisheries investigations, ser. 2.* **19,** 533 pp. Min. Agric. Fish. Food Gt Britain, London, H.M.S.O.

BRADLEY, R. H. E., 1952. Methods of recording aphid (Homoptera: Aphididae) populations on potatoes and the distribution of species on the plant. *Can. Ent.* **84,** 93–102.

BROADBENT, L., 1948. Methods of recording aphid populations for use in research on potato virus diseases. *Ann. appl. Biol.* **35,** 551–66.

CHIANG, H.C., *et al.*, 1961. Populations of european corn borer, *Ostrinia nubilalis* (Hbn.) in field corn, *Zea mays* (L.) *Univ. Missouri Res. Bull.* **776,** 96 pp.

COCHRAN, W. G., 1954. The combination of estimates from different experiments. *Biometrics* **10,** 101–29.

GABBUTT, P. D., 1958. The seasonal abundance of some arthropods collected from oak leaf litter in S. E. Devon. *Proc. X int. Congr. Ent.* **2,** 717.

KAELIN, A. and AUER, C., 1954. Statistische Methoden zur Untersuchung von Insektenpopulationen dargestellt am Beispiel des Grauen Lärchen-Wicklers (*Eucosoma grisea* Hb., *Semasia diniana* Gm.). *Z. angew. Ent.* **36,** 241–83.

LAMB, R. J. & WELLINGTON, W. G., 1974. Techniques for studying the behaviour and ecology of the European earwig, *Forficula duricularia* (Dermaptera: Forficulidae). *Can. Ent.* **106,** 881–8.

LAUGHLIN, R., 1976. Counting the flowers in the forest: combining two population estimates. *Aust. J. Ecol.* **1,** 97–101.

LORD, F. T., 1968. An appraisal of methods of sampling apple trees and results of some tests using a sampling unit common to insect predators and their prey. *Can. Ent.* **100,** 23–33.

MORRIS, R. F., 1955. The development of sampling techniques for forest insect defoliators, with particular reference to the spruce budworm. *Can. J. Zool.* **33,** 225–94.

MORRIS, R. F., 1960. Sampling insect populations. *A. Rev. Ent.* **5,** 243–64.

MORRIS, R. F. and BENNETT, C. W., 1967. Seasonal population trends and extensive census methods for *Hyphantria cunea. Can. Ent.* **99,** 9–17.

NATIONAL ACADEMY OF SCIENCES, 1969. *Principles of plant and animal pest control,* vol. 3, Insect-Pest Management and Control. Ch. 3, Insect Surveys. National Academy of Sciences, Washington D.C.

NEF, L., 1960. Comparaison de l'efficacité de différentes variantes de l'appareil de Berlese Tullgren. *Z. angew. Ent.* **46,** 178–99.

PIMENTEL, D., 1961. The influence of plant spatial patterns on insect populations. *Ann. ent. Soc. Am.* **54,** 61–9.

RICHARDS, O. W., 1961. The theoretical and practical study of natural insect populations. *A. Rev. Ent.* **6**, 147–62.

RICHARDS, O. W. and WALOFF, N., 1954. Studies on the biology and population dynamics of British grasshoppers. *Anti-Locust Bull.* **17**, 1–182.

STRICKLAND, A. H., 1961. Sampling crop pests and their hosts. *A. Rev. Ent.* **6**, 201–20.

VARLEY, G. C. and GRADWELL, G. R., 1963. The interpretation of insect population changes. *Proc. Ceylon Assoc. Adv. Sci. 18* (D), 142–56.

FURTHER BIBLIOGRAPHY

ALLEN, S. E., GRIMSHAW, H. M., PARKINSON, J. A. and QUARMBY, C., 1975. *Chemical Analysis of Ecological Material*, 565 pp., Oxford.

CANCELA DA FONSECA, J. P., 1965–9. L'outil statistique en Biologie du sol. *Revue Ecol. Biol. Sol.* **2**, 299–332, 475–88; **3**, 283–91, 381–407; **5**, 41–54; **6**, 1–30, 533–55.

DAVIES, R. G., 1971. *Computer Programming in Quantitative Biology.* 492 pp., Academic Press, London and New York.

GOODALL, D. W., 1962. Bibliography of statistical plant sociology. *Excerpta Bot. B.* **4**, 253–322.

MOSBY, H. S. (ed.), 1963. *Wildlife Investigational Techniques* (2nd ed.). Wildlife Society, U.S.A.

NICHOLLS, C. F., 1963. Some entomological equipment. *Res. Inst. Can. Dept. Agric. Belleville, Inf. Bull.* **2.**

NISHIDA, T. and TORII, T., 1970. *A Handbook of Field Methods for Research on Rice Stem-Borers and their Natural Enemies.* I.B.P. Handbook 14, 132 pp., Blackwells, Oxford.

PETERSON, A., 1934. *A manual of entomological equipment and methods.* Pt 1. Edwards Bros. Inc., Ann Arbor.

SCHULTZ, V., 1961. An annotated bibliography on the uses of statistics in ecology – search of 31 periodicals. *U.S. Atom. Energy Comm. Off. tech. Inf. TID* **3908.**

SERVICE, M., 1976. *Mosquito Ecology: Field Sampling Techniques.* Applied Science Publishers, London.

2

The Sampling Programme and the Measurement and Description of Dispersion

2.1 Preliminary sampling

2.1.1 Planning and field work

As it is normally impossible to count all the invertebrates in a habitat, it is necessary to estimate the population by sampling; naturally the estimates should have the highest accuracy commensurate with the amount of work expended, and if this is to be so a sampling programme which lays down the distribution, size and number of samples will need to be drawn up. There is no universal sampling method and although the statistical principles are given in Cochran (1963), Hansen, Hurwitz & Madow (1953), Stuart (1962), Yates (1953), Elliott (1971), Elliott & Decamps (1973), and Seber (1973), 'the sampling of a particular insect population must be resolved about the distribution and life-cycle of the insect involved' (Graham & Stark, 1954). Assuming that the life-cycle is known, preliminary work will be necessary to gain some knowledge of the distribution of the insect and the cost (work involved) of sampling; the worker will also need to be quite clear as to the exact problem he is proposing to investigate (Lamb, 1958; Morris, 1960; Strickland, 1961). The importance of careful formulation of the hypothesis for test cannot be overstressed.

The first decision concerns the universe to be sampled; whether this is to be a single habitat (e.g. field, woodland) or representatives of the habitat type from a wide geographical area will depend on whether an intensive or an extensive study is planned (p. 1), and the second decision must determine the magnitude of population change it is desired to record. Many species of insect pest exhibit ten- or even hundred-fold population changes in a single season (Southwood & Jepson, 1961) and therefore an estimate of population density with a standard error of about 25 % of the mean, which will enable a doubling or halving of the population to be detected, is sufficiently accurate for damage assessment and control studies on such species (Church & Strickland, 1954). For life-table studies, more especially on natural populations, a higher level of accuracy will be necessary; the level is frequently set at 10 %.

In extensive work the amount of sampling in a particular locality will be limited and therefore a further decision concerns the best stage for sampling (Burrage & Gyrisco, 1954); it may be desirable that this is the stage most closely correlated with the amount of damage, or if the purpose of the survey is to assess the necessity for control the timing should be such that it will give advanced information of an outbreak (Tunstall & Matthews, 1961; Gonzalez, 1970). Within these limits two other factors need consideration; the stage should preferably be present in the field for a long period, at least sufficiently long to allow the survey to be completed before an appreciable number have developed to a later stage, and the easier the stage is to sample and count the better. The reliability of samples in an extensive survey may be particularly sensitive to current weather conditions (Harris, Collis & Magar, 1972).

Although the preliminary sampling and the analysis of the assembled data will provide a measure of many of the variables the actual decisions must still, in many cases, be a matter of judgement. Furthermore, as the density changes, so will many of the statistical parameters, and a method that is suitable at a higher density may be found inadequate if the population level drops. Shaw (1955) found that Thomas & Jacob's (1943) recommendation for sampling potato aphids, one upper, middle and lower leaf from each of fifty plants, was unsatisfactory in Scotland, in certain years, because of the lower densities.

Details of the development of sampling programmes for various insects are given by Morris (1955), LeRoux & Reimer (1959), Harcourt (1961a, 1964) Lyons (1964) Mukerji (1973) and Coulson *et al.* (1975, 1976), amongst others, and these papers may be used as models.

Assuming that the study is planned in one field, this should be divided up into a number of plots, say 10–20. The habitat must now be considered from the biological angle and a decision made as to whether it might need further division; if it is woodland the various levels of the tree, upper, middle and lower canopy and probably the tips and bases of the branches, would on *a priori* grounds be considered as potentially different divisions; the aspect of the tree might also be important. In herbage or grassland, if leaves or other small sampling units are being taken, the upper and lower parts of the plants should be treated separately.

It is also of value to take at least two different sized sampling units (Waters & Henson, 1959), one should be towards the smallest possible limit, e.g. a leaf blade or half a leaf, for as a general principle a higher level of reproducibility is obtained (for the same cost) by taking more smaller units than by taking fewer large ones.* Small sampling units may also enable precision to be increased by distinguishing between favourable and unfavourable microhabitats; Condrashoff (1964) found with a leaf miner that the upper and lower leaf surfaces should be considered as separate units. Two examples of any one size sampling unit should be taken within each sampling plot or subsection. For

---

* The only disadvantage of sampling by small units is the number of zeros that may result at low densities; this truncation may make analysis difficult and has led to the suggestion that larger sized samples should be taken (Pradhan & Menon, 1945; Spiller, 1948, 1952). The decision must be related to the density of the animal, although moderate truncation can be overcome by suitable transformation; in other cases it may be necessary to increase the size of the unit (Andersen, 1965).

example for a field crop a preliminary plan could be:

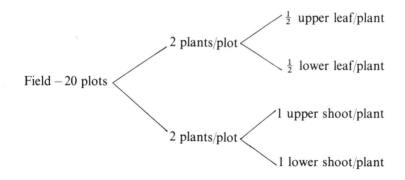

giving a total of 160 samples.

At the same time a record should be kept of the *cost* of each part of the sampling routine; this is normally measured in man hours and will be made up of the time required to select and take the sample and count the animals in it, together with that spent moving from one sampling site to the next.

2.1.2 Statistical aspects
Before the data gathered in the preliminary samples can be analysed some aspects of statistics need to be considered.

THE NORMAL DISTRIBUTION AND TRANSFORMATIONS
The normal or Gaussian distribution is the term applied to a continuous variable which when frequency is plotted against magnitude gives a symmetrical bell-shaped curve. Although the properties of animals – for example the heights of man – often describe a normal curve, the dispersion pattern of the individuals of a population is seldom, if ever, normal. The dispersion of a field population could approach towards the normal only if the dispersion was random, but the population was so dense or the size of sampling unit so large that considerable numbers were present in each sample. Therefore, in contrast to other distributions described below, the normal distribution is not of interest to ecologists as a means of describing dispersion; its importance arises solely from the fact that for most statistical methods to be applied to a set of data the frequency distribution must be normal and it should possess the associated properties that the variance is independent of the mean (or, more strictly, the variance is homogeneous, all errors coming from the same frequency curve), and its components additive. Although the analysis of variance is more robust in some respects than the χ^2 (Reimer, 1959; Abrahamsen & Strand, 1970), data whose frequency distribution is considerably skewed and with the variance closely related to the mean cannot be analysed without the risk of errors (Beall, 1942).

In order to overcome these problems the data are transformed; that is the actual numbers are replaced by a function whose distribution is such that it normalizes the data or stabilizes the variance. For example, if the square root transformation were applied to 9, 16 and 64 they would become 3, 4 and 8, and it will be observed that this tends to reduce the spread of the larger values, the interval between the second and third observations (16 and 64) is on the first scale nearly 7 times that between the first and second observations; when transformed the interval between the second and third observations is only 4 times that between the first and second. It is thus easy to visualize that a transformation of this type would tend to 'push' the long tail of a skew distribution in, so that the curve becomes more symmetrically bell-shaped. It must be stressed that transformation does however lead to difficulties, particularly in the consideration of the mean and other estimates (see below). It should not be undertaken routinely, but only when the conditions for statistical tests are grossly violated (Le Roux & Reimer, 1959; Finney, 1973). Before proceeding to discuss the correct transformation, the relationship of the mean and variance must be considered further.

Taylor's power law

As mentioned above the distribution of individuals in natural populations is such that the variance is not independent of the mean.* Now if the mean and variance of a series of samples are plotted, they tend to increase together (Fracker & Brischle, 1944; Kleczkowski, 1949; Bancroft & Brindley, 1958; Waters & Henson, 1959; Harcourt, 1961b, 1963). This relationship has been shown by Taylor (1961, 1965, 1971) to obey a power law. It holds in a continuous series of distributions from regular through random to highly aggregated and is expressed by:

$$s^2 = a\bar{x}^b \tag{2.1}$$

where a and b are constants, a is largely a sampling factor, while b appears to be an index of aggregation characteristic of the species. The same relationship between the mean ($\bar{x}$) and the variance (s^2) has been discovered independently by Fracker & Brischle (1944) for quadrat counts of the currant, *Ribes*, and by Hayman & Lowe (1961) for counts of the cabbage aphid, *Brevicoryne brassicae*.

The series of means and variances necessary to calculate a and b may be obtained from several sets of samples from different areas, from sets of samples of different sizes or by combining samples to form different sized sampling units. The mean and the variance are calculated from the raw data by the usual methods or by the use of probability paper (p. 13) for a less exact

---

* The mean, $(\bar{x}) = \dfrac{\Sigma x}{N}$ and the estimate of the variance, $s^2 = \dfrac{\Sigma(x^2) - (\Sigma x)^2/N}{N-1}$ where x = no. of animals/sample; N = total number of samples of that size and Σ = sum of . . .

approach. The values of $\bar{x}$ and s^2 are plotted on a log/log scale (Fig. 2.1) and the value of a read off on the s^2 axis at the value of $\bar{x} = 1$. The value of b can then be found from the equation:

$$\log s^2 = \log a + b \log \bar{x} \qquad (2.2)$$

It can be shown that as the variance varies with the mean in this way the appropriate variance stabilizing transformation function, $f(\bar{x})$, is of the form:

$$f(x) = Q \int \bar{x}^{-b/2} d\bar{x} \qquad (2.3)$$

Therefore, to transform one finds the value z in the expression:

$$z = x^p \qquad (2.4)$$

where $x =$ the original ('raw') number, $z =$ the transformed value and $p = 1 - \frac{1}{2}b$.

Choosing the transformation
As Taylor's power law holds so widely, the value of b may be found as above and hence that of p. If $p = 0$ a logarithmic transformation should be used, $p = 0.5$ square roots, $p = -0.5$ reciprocal square roots, -1.0 reciprocals. Healy & Taylor (1962) give tables for $p = 0.2, 0.4, 0.6, 0.8$ and for the negative powers.

These are precise transformations, but in practice where sampling and other errors are fairly large it will usually be found adequate to transform the data from a regular population by using squares, that from a slightly contagious one by using square roots and that from distinctly aggregated or contagious populations by using logarithms.

In order to overcome difficulties with zero counts in log transformations a constant (normally one) is generally added to the original count (x); this is expressed as 'log $(x + 1)$'. Anscombe (1948) has suggested that a better transformation would be obtained by taking log $(x + k/2)$ where k is the dispersion parameter of the negative binomial (see below). As k is frequently in the region of 2, in many cases this refinement would have little effect. Andersen (1965) has shown that if the mean and k are very small (less than 3 and approaching zero, respectively) then the variance will not be stabilized by $k^{1/2}$ or any of the common transformations, but if independent samples are pooled or the size of the sampling unit increased the data may be satisfactorily transformed.

It is customary to transform percentages to angles (arcsin), but Reimer (1959) has suggested this is only of value when the probability of finding the individual bearing the attribute (i.e. the leaf having a gall) is uniform within each area (for which a % infestation has been calculated), but varies considerably between the different classes (whose various % infestation one wishes to analyse). Even in such cases it is necessary only when a number of the

% points lie outside the 20–80 range. When the various percentages are based on grossly unequal numbers of individuals, they will need to be weighted before the analysis of variance can be applied (Reimer, 1959).

The use of transformations can lead to problems and as stressed above transformation should not be routinely undertaken. If the fiducial limits are calculated from the transformed mean this may be erroneous (Abrahamsen, 1969). The biological interpretation of estimates based on transformed data is often difficult and may lead to unsurmountable problems in the construction of life-budgets. There is indeed much to commend the use of the arithmetic mean (i.e. that based on the untransformed data) in population studies (van Emden, Jepson & Southwood, 1961; Lyons, 1964) and if the distribution of the animal is random (see below) the fiducial limits are available in tables (Pearson & Hartley, 1958). If data have been transformed, the means of the untransformed and transformed values should be provided; back transforms of, say, geometric means from logarithmic transformations are more difficult to interpret and contain biases unless the variances are small (Finney, 1973). Beauchamp & Olson (1973) suggest that the bias will be corrected by using the log normal distribution for detransforming; they developed a computer programme for this procedure with regression estimates.

Checking the adequacy of the transformation

An adequate transformation should eliminate or considerably reduce two attributes of the data that are easily tested for. These are:

(*i*) the skewness of the frequency curve, which is shown on arithmetic probability paper (Fig. 2.2) as well as by plotting the frequencies of different sample sizes:

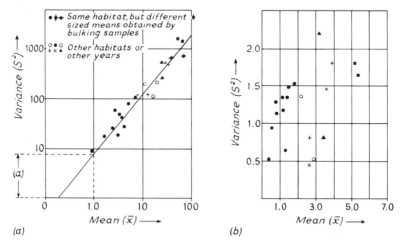

Fig. 2.1. *a*. The plot of the variance against the mean on a log/log scale to obtain the constant *a* of Taylor's power law, data from samples of olive scales (*Parlatoria oleae*) per twig; *b*. the same data transformed to $x^{0.4}$, showing the relative independence of the variance from the mean.

(*ii*) the dependence of the variance on the mean (instability or non-homogeneity of the variance) which may be shown graphically (Fig. 2.1).

Thus the value of any transformation in normalizing the data (eliminating skewness) may be crudely tested by the use of probability paper (see below), but as Hayman & Lowe (1961) have pointed out 'as non-normality must be extreme to invalidate the analysis of variance it is better to concentrate on stabilizing the variance of the samples'. A correct transformation for this property will also ensure the third property necessary for the analysis of variance, the additivity of the variance (Bliss & Owen, 1958); indeed all three properties are related and for practical purposes the distinction between transformation for normality and that for stabilizing the variance need not be emphasized. The adequacy of a transformation in stabilizing the variance may be tested for graphically (Fig. 2.1b) or by calculating the correlation coefficient for the two terms (Harcourt, 1961b, 1965).

Alternatively more precise methods may be used as described by Forsythe & Gyrisco (1961), who used a non-parametric rank correlation coefficient to test for non-normality. When normality had been established the heterogeneity of the variance could be tested for by Hartley's (1950) test and its non-additivity by Tukey's (1949).

THE USE OF PROBABILITY PAPER

Probability paper is often a valuable time-saving tool in the preliminary stages of statistical analysis. Its use enables the following to be assessed graphically: the normality of the data, the mean and the variance; it also tests the uniformity of the samples, discloses any polymodality and may allow the separation of the components responsible for the polymodality (Harding, 1949). Estimates made from probability paper may not be as accurate as those that are computed because of the difficulty of fitting a line by eye and reading off the graph.

There are two methods of using probability paper; first with a large number of samples, say over 20, they can be grouped in frequency classes and the cumulative frequencies plotted. If for example one had 100 observations on the numbers of galls on a leaf and the smallest was 5 and the largest 85, the vertical scale of the probability would then be marked out from 5 to 85; the smallest sample of which there was only one therefore accounts for 1 % of the samples and thus a point is inserted on intersection of the 1 % and 5 lines. The next smallest number of galls per leaf are 10, 13, 15 (two leaves), then our next points are at 2 % and 10, 3 % and 13, and 5 % and 15 (because 5 % of the leaves sampled have 5 or fewer galls). This process is continued until the size of the penultimate sample provides the 99 % point and the last sample the 99.99 % point.

The second method which is more applicable to the preliminary type of analysis, for which probability paper is useful, is valid for a smaller number of samples. Here the samples are ranked, but *not* grouped, and points corresponding to their size are placed at equal intervals along the frequency scale. The smallest sample is inserted at the % value corresponding to $50/N$ (where $N =$ the number of samples). Subsequent samples are inserted at intervals of $100/N\%$. An example is given in Table 2.1 and Fig. 2.2.

Table 2.1 The numbers of eggs of the whitefly, *Aleurotrachelus jelinekii*, per leaf of its host plant, *Viburnum tinus*; data used for Fig. 2.2. (By calculation log mean $= 1.528$; log standard deviation $= 0.555$; read off Fig. 2.2, line drawn by eye, log mean $= 1.52$, log standard deviation $= 0.59$.)

Leaf No.	% point (for plotting on probability paper)	No. of eggs/leaf (ranked)	Log No. of eggs/leaf
1	5	6	0.77
2	15	10	1.00
3	25	11	1.04
4	35	13	1.11
5	45	29	1.46
6	55	34	1.53
7	65	77	1.88
8	75	94	1.97
9	85	106	2.02
10	95	320	2.50

If a normally distributed set of frequency data are plotted on probability paper they will give a straight line, and the mean will lie at the 50% mark. The standard deviation (s) is found by reading off the values of x at the 15.87% mark (x_1) and at the 84.13% mark (x_2), then

$$s = \frac{x_2 - x_1}{2} \tag{2.5}$$

However, if the data are not normally distributed they will form a curve (Fig. 2.2). Shallow curves can be made to approximate to the straight line by transforming them in square roots, deeper curves by transformation in logarithms. The transformed values are then plotted on the x axis or logarithmic probability paper may be used (Table 2.1, Fig. 2.2).

If, however, the distribution is not of the same type throughout the field (or area sampled) a straight line will not be obtained, the plot will have a kink in it (Fig. 2.3) (Harding, 1949; Cassie, 1954; Harris, 1968; Southwood & Cross, 1969). An algorithm to separate the normal components of a compound distribution has been described by Gregor (1969). Such a change in distribution (polymodality) is almost certainly of considerable biological significance, and quite apart from statistical considerations it is important to recognize it at the outset. If it is found, then further preliminary sampling will be necessary to delimit the areas of differing distributions and each must, in subsequent work, be treated as a separate universe.

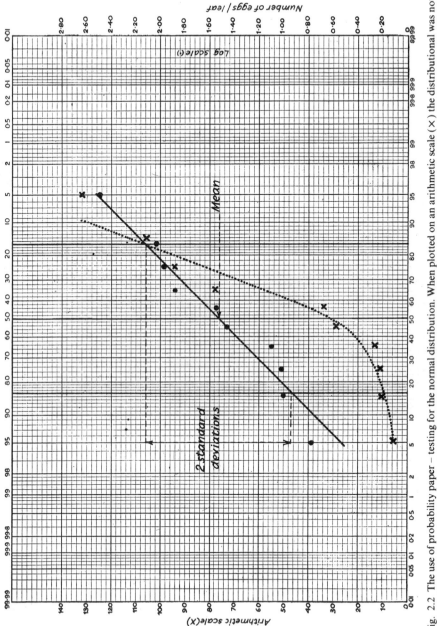

Fig. 2.2 The use of probability paper – testing for the normal distribution. When plotted on an arithmetic scale (×) the distributional was not normal, but when transformed to logs (●) a straight line can be drawn through the points showing that the distribution has become normalized; the mean and standard deviation may then be read off (data from Table 2.1).

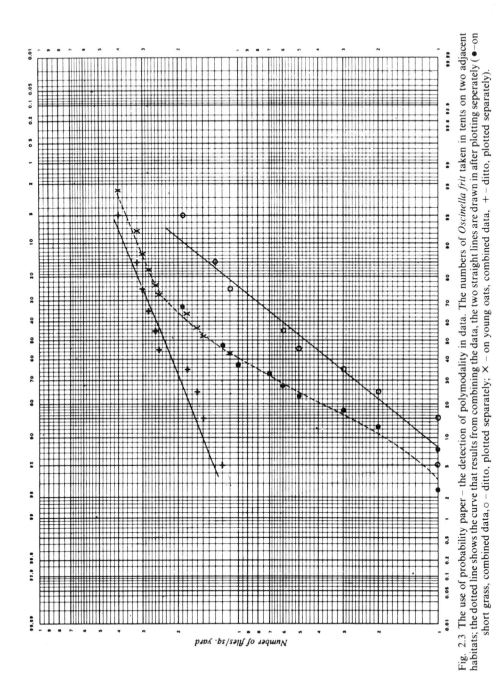

Fig. 2.3 The use of probability paper – the detection of polymodality in data. The numbers of *Oscinella frit* taken in tents on two adjacent habitats; the dotted line shows the curve that results from combining the data, the two straight lines are drawn in after plotting seperately (●–on short grass, combined data, ○ – ditto, plotted separately; × – on young oats, combined data, + – ditto, plotted separately).

Besides arithmetic and logarithmic probability paper described above, Poisson and binomial probability papers are available and may be of considerable assistance with certain problems (Mosteller & Tukey, 1949; Ferguson, 1957).

In any complex situation, e.g. where different parts of the plant have been sampled, it is necessary to carry out an analysis of the variance (see standard statistical textbooks) of the whole data. The amount of the variance due to within-plant and within-plots variation should be compared with that for between plants and between plots.

2.2 The sampling programme

2.2.1 The number of samples per habitat unit (e.g. plant)
There are two aspects, firstly whether different regions of the unit need to be sampled separately and secondly the number of samples within each unit or subunit (if these are necessary) that should be taken for maximum efficiency. Although the habitat unit could, for example, be the fleece of a sheep, a bag of grain or a rock in a stream, for convenience the word plant will, in general, be used in its place in the discussion below.

SUBDIVISION OF THE HABITAT
If the distribution of the population throughout the habitat is biased towards certain subdivisions, but the samples are taken randomly, what LeRoux and Reimer (1959) aptly term *systematic errors* will arise. This can be overcome either by sampling so that the differential number of samples from each subdivision reproduces in the samples the gradient in the habitat, or by regarding each part separately and correcting at the end. The question of the estimation of the area or volume of the plant is discussed in Chapter 4. The amount of subdivision of the plant that various workers have found necessary varies greatly. On apple the eggs, larvae and pupae of the tortricid moth, *Archips argyrospilus*, were found for most of the year to be randomly distributed over the tree so that only one level (the lower for ease) needed to be sampled (Paradis & LeRoux, 1962). In contrast on the same trees and in the same years the immature stages of two other moths showed marked differences between levels at all seasons (LeRoux & Reimer, 1959). With the spruce budworm (*Choristoneura fumiferana*), Morris (1955) found that there were 'substantial and significant differences from one crown level to another' and that there was a tendency for eggs and larvae to be more abundant at the top levels, but there was no significant difference associated with different sides of the same tree. A similar variation with height was found with the eggs of the larch sawfly (*Pristiphora erichsonii*), although here it was concluded that in view of the cost and mechanical difficulties of stratified sampling at different

heights a reasonable index of the population would be obtained by sampling the mid-crown only (Ives, 1955). Studying all the organisms on aspen (*Populus tremuloides*) Henson (1954) found it is necessary to sample at three different levels of the crown and even with field crops height often needs to be considered: Broadbent (1948) recommended that potato aphids be estimated by picking three leaves – lower, middle and upper – from each plant and when estimating the population of the european corn borer (*Ostrinia nubilalis*), Hudson & LeRoux (1961) showed that the lower and the upper halves of the maize stem needed to be considered separately, the former containing the majority of the larvae. The distribution of the eggs of various cabbage-feeding Lepidoptera was found by Hirata (1962) to depend on the age of the plant.

Aspect is sometimes important; in Nova Scotia in the early part of the season the codling moth lays mostly on the south-east of the apple trees, but later this bias disappears (MacLellan, 1962). Aspect has also been found to influence the distribution on citrus of the long-tailed mealy bug, *Pseudococcus adonidum* (Browning, 1959), the eggs of the oak leafroller moth, *Archips* (Ellenberger & Cameron, 1977), and of three species of mite, each of which was most prevalent on a different side (Dean, 1959), but not that of the pine beetle, *Dendroctonus* (Dudley, 1971). Variations in the spatial distribution of similar species in the same habitat, which complicates a sampling programme designed to record both, has also been recorded for two potato aphids by Helson (1958). Some insects are distributed without bias on either side of the mid vein of leaves, so that they may be conveniently subsampled, and Nelson, Slen & Banky (1957) record that when estimating populations of sheep keds, the fleece of only one side need be sampled.

Occasionally it may be found that such a large and constant proportion of the population occur on a part of the plant that sampling may be restricted to this: Wilson (1959) showed that in Minnesota 84% of the eggs of the spruce budworm (*Choristoneura fumiferana*) are laid on the tips of the branches and if sampling is confined to these, rather than entire branches, sampling time may be reduced by up to 40%

The taking of a certain number of samples randomly within a site which is itself selected randomly from within a larger area, e.g. a field, is often referred to as *nested sampling*, and may be on two, three or more levels (Bancroft & Brindley, 1958).

THE NUMBER OF SAMPLES PER SUBDIVISION

To determine the optimum number of samples per plant (or part of it) (n), the variance of within-plant samples (s_s^2) must be compared with the variance of the between-plant samples (s_p^2) and set against the cost of sampling within the same plant (c_s) or of moving to another plant and sampling within it (c_p):

$$n_s = \sqrt{\frac{s_s^2}{s_p^2} \times \frac{C_p}{C_s}} \qquad (2.6)$$

If the interplant variance (s_p^2) is the major source of variance and unless the cost of moving from plant to plant is very high n will be in the order of one or less (which means one in practice). Interplant variance has been found to be much greater than within-plant variance for the spruce sawfly (*Diprion hercyniae*) (Prebble, 1943), the lodgepole needle miner (*Recurvaria starki*) (Stark, 1952*b*), the cabbage aphid (*Brevicoryne brassicae*) (Church & Strickland, 1954), the spruce budworm (*Choristoneura fumiferana*) (Morris, 1955), the winter moth (*Operophtera brumata*) (Morris & Reeks, 1954), the diamondback moth (*Plutella maculipennis*) (Harcourt, 1961*a*), the cabbage butterfly (*Pieris rapae*) (Harcourt, 1962), the Western pine beetle (*Dendroctonus brevicomis*) (Dudley, 1971), the pine chermid (*Pineus pinifoliae,*) (Ford & Dimond, 1973) and the spider mite (*Panonychus ulmi,*) (Herbert & Butler, 1973). In most of these examples the within-plant variance was small so that only one sample was taken per plant or per stratum of that plant, although of course when this is done, within and between tree variances cannot be separated. With some apple insects the within-tree variance (s_s^2) becomes larger, especially at certain seasons, and then as many as seven samples may be taken from a single tree (LeRoux & Reimer, 1959; LeRoux, 1961; Paradis & LeRoux, 1962).

Often a considerable saving in cost without loss of accuracy in the estimation of the population, but with loss of information on the sampling error, may be obtained by taking randomly a number of subsamples which are bulked before sorting and counting. This is especially true where the extraction process is complex as with soil samples; Jepson & Southwood (1958) bulked four random, 3-in row samples of young oat plants and soil to make a single 1-ft row sample that was then washed and the eggs of the frit fly (*Oscinella frit*) extracted. Such a process gave a mean as accurate as that obtained by washing all the 3-in samples separately (greater cost). Paradis and LeRoux (1962) sampled the eggs of a tortricid moth, *Archips argyrospilus*, on apple by bulking 25 cluster samples.

2.2.2 The sampling unit, its selection, size and shape

The criteria for the sample unit are broadly (Morris, 1955):

(1) It must be such that all units of the universe have an equal chance of selection.

(2) It must have stability (or if not its changes should be easily and continuously measured — as with the number of shoots in a cereal crop).

(3) The *proportion* of the insect population using the sample unit as a habitat must remain constant.

(4) The sampling unit must lend itself to conversion to unit areas.

(5) The sampling unit must be easily delineated in the field.

(6) The sampling unit should be of such a size as to provide a reasonable balance between the variance and the cost.

(7) The sampling unit must not be too small in relation to the animal's size as

this will increase edge effect errors.

(8) The sampling unit for mobile animals should approximate to the average ambit of an individual. This 'condition' is particularly significant in studies on dispersion involving contiguous sampling units, Lloyd (1967) has suggested that a test of the appropriate size would be provided by several series of counts of animals in contiguous quadrats conforming to a Poisson series.

A sampling unit defined in relation to the animal's ambit or territory (e.g. gallery of a bark-beetle), will give different information from one defined in terms of the habitat (e.g. Cole, 1970). This re-emphasizes the need to be very clear as to objectives and hypothesis before commencing a sampling programme.

To compare various sampling units in respect to variance and cost it is generally convenient to keep one or other constant. The same method of sampling must, of course, be used throughout. From preliminary sampling the variances of each of the different units (s_u^2) can be calculated; these should then be computed to a common basis, which is often conveniently the size of the smallest unit. For example if the smallest unit is 1 ft of row, then the variance of 2 ft row unit will be divided by 2 and those of 4 ft by 4. The costs will similarly be reduced to a common basis (C_u). Then the relative net cost for the same precision for each unit will be proportional to:

$$C_u s_u^2$$

where C_u = cost per unit on a common basis and s_u^2 = variance per unit on a common basis. Alternatively the relative net precision of each will be proportional to $1/C_u s_u^2$. The higher this value, the greater the precision for the same cost.

A full treatment of the methods of selecting the optimum size sampling unit is given in Cochran (1963) and other textbooks, but as population density, and hence variance, is always fluctuating, too much stress should not be placed on a precise determination of optimum size of the sampling unit.

Even with soil animals, where such a procedure might be of most value, Yates & Finney (1942) showed that although 4- and 6-in diameter cores are equally efficient at low densities, at high densities the comparative efficiency of the 6-in sample falls off. With insects on plants the nature of the plant usually restricts the possible sizes to, for example, half leaf, single leaf, or shoot (see p. 141).

The shape of the sampling unit when this is of the quadrat type, rather than a biological unit, is theoretically of importance because of the bias introduced by edge effects. These are minimal with circles, maximal with squares and rectangles and intermediate with hexagonals (Seber, 1973), because they are proportional to the ratio of sample unit boundary length to sampling unit area. If the total habitat is to be divided into numbered sampling units (for random number selection), then circular units are impractical because of the gaps between and it is doubtful if the reduction of error from the use of hexagons normally justifies the difficulties of lay-out. Clearly the larger the sampling unit, proportionally the less the boundary edge effect. The size of the organism will also influence this effect: the larger it is in relation to the sample size, the greater the chance of an individual lying across a boundary (Safranyik & Graham, 1971). This problem has been investigated for sub-cortical insects where the damage to the edge individuals by the punch, and the curved nature of the sampled substrate pose special problems (Safranyik & Graham, 1971). In general edge effects can be minimized by a convention (e.g. of the animals crossing the boundaries only those on the top and left-hand boundaries are counted).

2.2.3 The number of samples

The total number of samples depends on the degree of precision required. This may be expressed either in terms of achieving a standard error of a predetermined size, or in probability terms, getting confidence limits of a predetermined half-width, a percentage of the mean (Karandinos, 1976).

For many purposes a standard error of 5% of the mean is satisfactory. Within a homogeneous habitat the number of samples (n) required is given by:-

$$n = \left(\frac{s}{E\bar{x}} \right)^2 \tag{2.7}$$

where $s =$ standard deviation, $\bar{x} =$ mean and $E =$ is the predetermined standard error as a decimal of the mean (i.e. normally 0.05). This expression compares the standard deviation (s) of the observations with the standard error ($E\bar{x}$) acceptable for the contrasts we need to make; it will be noted from this equation that in any given situation the value of the standard error will change with the square root of the number of samples: thus a large increase in n is necessary to bring about a small improvement in s.

Where sampling is necessary at two levels, e.g. a number of clusters per tree, the number of units (n_t) that need to be sampled at the higher level, e.g. trees (LeRoux & Reimer, 1959; Harcourt, 1961a) is given by:

$$n_t = \frac{(s_s^2/n_s) + S_p^2}{(\bar{x} \times E)^2} \tag{2.8}$$

where $n_s =$ the number of samples within the habitat unit (calculated as above), $s_s^2 =$ variance within the habitat unit, $s_p^2 =$ variance between the habitat unit ($=$ interplant variance), $\bar{x} =$ mean per sample (calculated from the transformed data and given in this form and E as above.

Rojas (1964) has shown that if the dispersion of the population has been found to be well described by the negative binomial the desired number of samples is given by:

$$N = \frac{1/\bar{x} + 1/k}{E^2} \tag{2.9}$$

where $k =$ the dispersion parameter of the negative binomial (see below).

When the confidence limits are used as the predetermined standard, the required half-width is usually set at 10% of the mean. The general formula (2.7) then becomes:-

$$n' = \left(\frac{ts}{D\bar{x}}\right)^2 \tag{2.10}$$

where $t =$ 'Students t' of standard statistical tables, it depends on the number of samples and approximates to 2 for more than 10 samples at the 5% level, and $D =$ the predetermined half-width of the confidence limits as a decimal (usually 0.1). It will be seen that normally this gives a similar estimate to equation 2.7, provided the values of E and D are adjusted to accord with their meanings. The procedure is perforce somewhat approximate, depending on the preliminary estimates of the mean and standard deviation, and the inclusion of t does perhaps give it a slightly bogus air of precision! Additionally there is normally a biological approximation, for as population characters change with time, so will the optimal number of samples (e.g. Bryant, 1976; Kapatos *et al.*, 1977).

Another type of sampling programme concerns the measurement of the frequency of occurrence of a particular organism or event, for example the frequency of occurrence of galls on a leaf or of a certain genotype in the population (Cornfield, 1951; Cochran, 1963; Oakland, 1953; Henson, 1954). Before an estimate can be made of the total number of samples required, an approximate value of the probability of occurrence must be obtained. For example, if it is found in a preliminary survey that 25% of the leaves of oak trees bear galls the probability is 0.25. The number of samples (N) is given by:

$$N = \frac{t^2 pq}{D^2} \tag{2.11}$$

where $p =$ the probability of occurrence (i.e. 0.25 in the above example), $q = 1 - p$; t and D are as above.

If it is found that the leaves (or other units) are distributed differently in the different parts of the habitat, they should be sampled in proportion to the variances. For example, Henson (1954) found from an analysis of variance of the distribution of the leaf-bunches of aspen that the level of the crown from which the leaves had been drawn caused a significant variation and when this variance was portioned into levels the values were: lower 112993, middle 68012, upper 39436. Therefore leaf-bunches were sampled in the ratio of $3:2:1$ from these three levels of the crown.

2.2.4 The pattern of sampling

Once again it is important to consider the object of the programme carefully. If the aim is to obtain estimates of the mean density for use in, for example, life-tables, then it is desirable to minimize variance. But if the dispersion

(= distribution = pattern) of the animal is of prime interest then there is no virtue in a small variance.

In order to obtain an unbiased estimate of the population the sampling data should be collected at *random*, that is so that every sampling unit in the universe has an equal chance of selection. In the simplest form – the *unrestricted random sample* – the samples are selected by the use of random numbers from the whole area (universe) being studied (random number tables are in many statistical works, or the last two figures in the columns of numbers in most telephone books provide a substitute). The position of the sample site is selected on the basis of two random numbers giving the distances along two co-ordinates, the point of intersection is taken as the centre or a specified corner of the sample. If the size of the sample is large compared with the total area then the area should be divided in plots which will be numbered and selected using a single random number (e.g. Lloyd, 1967). Such a method eliminates any personal choice by the worker whose bias in selecting sampling sites may lead to large errors (Handford, 1956).

However, just because it is absolutely random this method is not very efficient for minimizing the variance, since the majority of the samples may turn out to come from one area of the field. The method of *stratified random* sampling is therefore to be preferred for most ecological work (Yates & Finney, 1942; Healy, 1962; Abrahamsen, 1969); here the area is divided up into a number of equal sized subdivisions or strata and one sample is randomly selected from each strata. Alternatively if the strata are unequal in size the number of units taken in each part is proportional to the size of the part; this is referred to as self-weighting (Wadley, 1952). Such an approach maximizes the accuracy of the estimate of the population, but an exact estimate of sampling error can only be obtained if additional samples are taken from one (or two) strata (Yates & Finney, 1942). The taking of one sample randomly and the other a fixed distance from it has been recommended by Hughes (1962) as a method of mapping aggregations. The fixed distance must be less than the diameter of the aggregations and assumes these are circular; the standard error cannot be calculated. However the method has been found useful for soil and benthnic faunas (Gardefors & Orrhage, 1968).

When the habitat is stratified, biological knowledge can often be used to eliminate strata in which few insects would be found. Such a restricted universe will give a greater level of precision for the calculation of a mean than an unrestricted and completely random sample with a wide variance. Prebble (1943) found with a pine sawfly: satisfactory estimates of the pupae were only obtained if sampling was limited to the areas around the bases of the trees, the variance of completely random sampling throughout the whole forest was too great, as many areas were included that were unsuitable pupation sites (see also Stark & Dahlsten, 1961).

The other approach is the *systematic sample*, taken at a fixed interval in space (or time). In general such data cannot be analysed statistically, but Milne (1959) has shown that if the *centric systematic area-sample* is analysed as if it were a random sample, the resulting statistics are 'at least as good, if not rather better', than those obtained from random sampling. The centric systematic sample is the one drawn from the exact centre of each area or stratum and its theoretical weakness is that it might coincide with some unsuspected systematic distribution pattern. As Milne points out, the biologist should, and probably would, always watch for any systematic pattern, either disclosing itself as the samples are recorded on the sampling plan or apparent from other knowledge. Such a sampling programme may be carried out more quickly than the random method and so has a distinct advantage from the aspect of cost (see also p. 58).

An example of an unbiased systematic method is given by Anscombe (1948). All the units (e.g. leaves) are counted systematically (e.g. from top to bottom and each stem in turn), then every time a certain number (say 50) is reached that unit is sampled and the numbering is commenced again from 1; only one allocation of a random number is needed and that is the number (say somewhere between 1 and 20) allotted to the first unit.

Biologists often use methods for random sampling that are less precise than the use of random numbers, such as throwing a stick or quadrant or the haphazard selection of sites. Such methods are not strictly random; their most serious objection is that they allow the intrusion of a personal bias, quite frequently marginal areas tend to be under sampled.

It may be worthwhile doing an extensive trial comparing a simple haphazard method with a fully randomized or systematic one, especially if the cost of the latter is high when compared with the former. Spiller (1952) found that scale insects on citrus leaves could be satisfactorily sampled by walking round the tree, clockwise and then anticlockwise, with the eyes shut and picking leaves haphazardly. For assessing the level of red bollworm eggs (*Diparopsis castanea*) to determine the application of control measures Tunstall & Matthews (1961) recommended two diagonal traverses across the field counting the eggs at regular intervals.

Bias may intrude due to causes other than personal selection by the worker: grains of wheat that contain the older larvae or pupae of the grain weevil (*Sitophilus granarius*) are lighter than uninfested grains. The most widespread method of sampling is to spread the grains over the bottom of a glass dish and then scoop up samples of a certain volume; as the lighter infected grains tend to be at the top this method can easily overestimate the population of these stages (Howe, 1963). In contrast, for the earlier larval instars before they have appreciably altered the weight of the grain such a simple method gives reliable results (Howe, 1963); it was undoubtedly this difference that led to Krause & Pedersen (1960) stressing the need for samples to contain a relatively high proportion of the same stage if good replication was to be obtained.

2.2.5 The timing of sampling

The seasonal timing of sampling will be determined by the life-cycle of the insect (Morris, 1955). In extensive work when only a single stage is being sampled, it is obviously most important that this operation should coincide with peak numbers (e.g. Edwards, 1962). This can sometimes be determined by phenological considerations (Unterstenhöfer, 1957), but the possibility of a control population in an outdoor cage (Harcourt, 1961*a*) to act as an indicator should be borne in mind. The faster the development rate, the more critical the timing. With intensive studies that are designed to provide a life-table regular sampling will be needed throughout the season.

It is not always realized that the time of day at which the samples are taken may also considerably affect them. The diurnal rhythms of the insects may cause them to move from one part of the habitat to another as Dempster (1957) found with the Moroccan locust (*Dociostaurus maroccanus*) (Fig. 2.4).

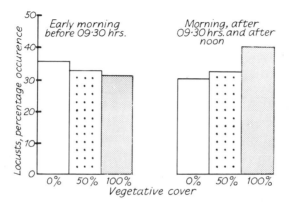

Fig. 2.4. The variation in the distribution of adults of the Moroccan locust *Dociostaurus maroccanus* at different times of the day; the histograms show the relative numbers on bare ground and areas with moderate and dense vegetation. (After Dempster, 1957.)

Many grassland insects move up and down the vegetation not only in response to weather changes, but also at certain times of the day or night (p. 242) and during the day quite a proportion of active insects may be airborne (cf. the observations of Southwood, Jepson & Van Emden (1961) on the numbers of adults of the frit fly (*Oscinella frit*) on oats). There is a marked periodicity of host-seeking behaviour in many blood-sucking invertebrates (e.g. Camin *et al.*, 1971; Corbet & Smith, 1974). The ecologist may find that some of his sampling problems can be overcome, or at least additional information gained, if he works at night or at dusk and dawn, rather than during conventional working hours.

2.3 Dispersion

The dispersion of a population, the description of the pattern of the distribution or disposition of the animals in space, is of considerable ecological significance. Not only does it affect the sampling programme (Rojas, 1964) and the method of analysis of the data, but it may be used to give a measure of population size (nearest neighbour and related techniques) and, in its own right, is a description of the condition of the population. Changes in the dispersion pattern should be considered alongside changes in size when interpreting population dynamics. For example, if a mortality factor reduces the clumping of a sessile organism this is an indication that it acts most severely on the highest densities, or if the dispersion of a population becomes more regular then intensification of competition should be suspected (Iwao, 1970c). An understanding of dispersion is vital in the analysis of predator – prey and host – parasite relationships (Crofton, 1971; Murdie & Hassell, 1973; Hassell & May, 1974; Anderson, 1974).

2.3.1 Mathematical distributions that serve as models

It is necessary to outline some of the mathematical models that have been proposed to describe the distribution of organisms in space; for a fuller treatment reference should be made to Anscombe (1950), Wadley (1950), Cassie (1962) and Katti (1966), to other papers cited in the text and to textbooks, e.g. Bliss & Calhoun (1954), and Patil & Joshi's (1968) dictionary of distributions.

BINOMIAL FAMILY

The central place in this family is occupied by the *Poisson series* which describes a *random distribution* (Fig. 2.5). It is important to realize that this does not mean an even or uniform distribution (Fig. 2.5), but that there is an equal probability of an organism occupying any point in space and that the presence of one individual does not influence the distribution of another. When plotted, the Poisson series gives a curve which is described completely by one parameter, for the variance (s^2) is equal to the mean $(\bar{x})$. A full discussion of variance and its calculation will be found in most statistical textbooks (e.g. Bailey, 1959; Bliss & Calhoun, 1954; Goulden, 1952), but the observed, variance (s^2) of a distribution may be calculated:

$$s^2 = \frac{\Sigma(fx^2) - \left[(\Sigma fx)^2/N\right]}{N-1} \tag{2.12}$$

where Σ = the sum of . . .
f = frequency of . . .
x = various values of the number of animal/sample
N = number of samples

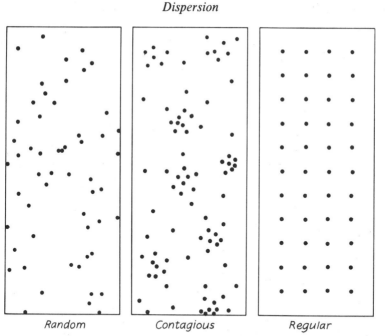

Random   Contagious   Regular

Fig. 2.5. Different types of distribution

The probability (p) of finding a certain number (x) of animals in a sample from a population with a given mean ($\bar{x}$) and a Poisson distribution is given by:

$$p_x = e^{-\bar{x}} \frac{\bar{x}^x}{x!} \tag{2.13}$$

where e = base of natural (Napierian) logarithms, so that $e^{-\bar{x}}$ may be found by using a table of these logs 'backwards'. The goodness of fit of a set of data to the Poisson distribution may be tested by a χ^2 on the observed and expected values or by the index of dispersion (p.39).

Occasionally it may be found that the variance is less than the mean; this implies a more regular (or uniform or even) distribution than is described by a Poisson series (Fig. 2.5).

Most commonly in ecological studies the variance will be found to be larger than the mean, that is, the distribution is contagious* (Fig. 2.5), the

* The term 'contagious' is a mathematical one coined in connection with work on epidemiology and has certain implications that to some extent make its use in ecology inappropriate (Waters & Henson, 1959). An alternative is the term 'over-dispersion' first introduced into ecology by Romell (1930), with its opposite – for more uniform spacing – 'under-dispersion'; unfortunately, however, the use of these terms has been reversed by some ecologists, therefore the terms used here are contagious and regular, which are also those commonly used in plant ecology (Greig-Smith, 1964).

population is clumped or aggregated. Many contagious insect populations that have been studied can adequately be expressed by the *negative binomial* (or *Pascal*) *distribution* (Bliss & Owen, 1958; Rojas, 1964; Lyons, 1964; Harcourt, 1965; Ibarra *et al.*, 1965). This distribution is described by two parameters, the mean and the exponent k, which is a measure of the amount of clumping and is often referred to as the dispersion parameter.

Generally values of k are in the region of 2; as they become larger the distribution approaches and is eventually (at infinity) identical with that of the Poisson, whilst fractional values of k indicate a distribution tending towards the logarithmic series, which occurs when k is zero. The value of k is not a constant for a population, but often increases wih the mean (Anscombe, 1949; Bliss & Owen, 1958; Waters & Henson, 1959 (see p. 40).

The truncated Poisson may be of value where the distribution is non-random, but the data are too limited to allow the fitting of the negative binomial (Finney & Varley, 1955).

Calculating k *of the negative binomial*
The value of k may be computed by several methods (Anscombe, 1949, 1950; Bliss & Fisher, 1953; Debauche, 1962; Legay, 1963). Three are presented here, two approximate and one (No. 3) more accurate. Another method is described by Katti and Gurland (1962).

$$(1) \qquad\qquad\qquad k = \frac{\bar{x}^2}{s^2 - \bar{x}} \qquad\qquad\qquad (2.14)$$

The variance is calculated by the formula given above. The efficiency of this estimate in relation to various values of k and $\bar{x}$ can be seen in Fig. 2.6, from which it is clear that unless the mean is low it is not reliable for use with populations that show a moderate degree of clumping (i.e. a k of 3 or less). This virtually limits its use to *low density populations*.

$$(2) \qquad\qquad \log\left(\frac{N}{n_0}\right) = k \log\left(1 + \frac{\bar{x}}{k}\right) \qquad\qquad (2.15)$$

where N=total number of samples, n_0=number of samples containing no animals and others as before.

The easiest way to find the unknown 'k' is by the method of trial and error (iterative solution); that is, various values of k are substituted into the equations until the two sides are equal. A reasonable value of k to start with is 2.

It will be seen from Fig. 2.6 that this method is efficient for most populations with very small means, but for large ones only where there is extensive clumping. This can be expressed another way by saying that about one-third of the samples must be blank (empty) if the mean is below 10; with larger

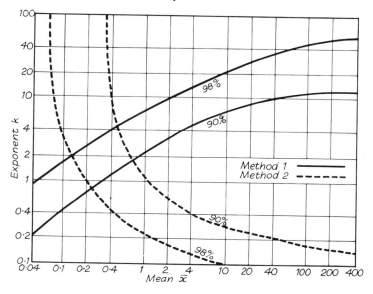

Fig. 2.6. Large sample efficiencies of the estimation of k of the negative binomial distribution by methods 1 and 2 (redrawn from Anscombe, 1950).

means more blank samples must have been found for the method to be reasonably efficient.

$$(3) \qquad N \log_e \left(1 + \frac{\bar{x}}{k}\right) = \sum \left(\frac{Ax}{k+x}\right) \qquad (2.16)$$

where $\log_e$ = natural logs, Ax = the sum of all frequencies of sampling units containing more than x individuals (e.g. $A_6 = \Sigma f_7 + f_8 + f_9$).

The two sides of this equation must again be made to balance by iteration (trial and error); correctly it is solved by maximum likelihood. As a first step the value of k should be obtained from one of the approximate methods above. If the left-hand side of the equation is found to be too large, then the approximate estimate of k is too large; conversely if the right-hand side is the larger the estimate of k is too low. When two estimates of k have been tried and have given slight excesses in the left- and right-hand sides respectively, a final k can be found by proportion, but this solution should be checked back in the equation. Shenton & Wallington (1962) have shown that even this maximum likelihood estimator of k will have a bias if the mean is small and k large.

Testing the fit of the negative binomial
When the value of k has been found the agreement between the negative binomial series as a model and the actual distribution can be tested in three ways.

The expected frequencies of each value may be calculated and these compared by a χ^2 with the actual values (Bliss & Fisher, 1953). The expected values are given by:

$$p_x = \frac{\Gamma(k+x)}{x!\,\Gamma(k)} \times \left(\frac{\bar{x}}{\bar{x}+k}\right)^x \times \left(\frac{k}{k+\bar{x}}\right)^k \tag{2.17}$$

where p_x = the probability of a sample containing x animals and the values of $x!$ and $\Gamma(k)$ can be found from tables of factorials and of log gamma functions respectively. χ^2 has three fewer degrees of freedom than the number of comparisons that are made between expected and actual frequencies; those with small expectations generally pooled; but Pahl (1969) considers that this is not necessary. A computer programme (No. 41) to estimate k by method 3 and test its fit by this method is given by Davies (1971).

Such a comparison between actual and expected frequencies may be distorted by irregularities due to chance; there are two alternative tests based on the difference between the actual and expected moments (mean, variance or skewness) compared with their standard errors (Anscombe, 1950; Bliss & Fisher 1953). Which of these tests is most efficient will depend on the sizes of $\bar{x}$ and k. Evans (1953) has shown that with small means ($\bar{x}$), unless k is large, the most efficient test is that based on the second moment (variance); for other values of $\bar{x}$ and k the test based on the third moment (skewness) should be used. The actual dividing line between the two tests (E–E) is given in Fig. 2.7.

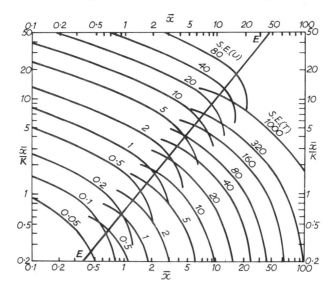

Fig. 2.7. Testing the fit of the negative binomial distribution – the standard errors of T and U for $N = 100$. For other values of N multiply the standard error by $10/\sqrt{N}$. The line E–E represents the contour along which the efficiencies of the methods are equal, the use of the second moment test is preferable in the area to the left of the line and of the third moment test in the area to the right (after Evans, 1953).

The test based on the second moment involves the calculation of the statistic U which is the difference between the actual variance and the expected variance given by:

$$U = s^2 - \left(\bar{x} + \frac{\bar{x}^2}{k} \right)$$ (2.18)

the value of k being derived by the second method above. If U is significantly less than its standard error, (S.E. (U)), most conveniently calculated from Fig. 2.7 (the exact formula is given by Evans, 1953), then the negative binomial may be taken as a satisfactory model.

The third moment test involves the calculation of T, the difference between the actual third moment (skewness) of the data and its value predicted from the first two moments (mean and variance) of the same sample:

$$T = \left(\frac{\Sigma fx^3 - 3\bar{x} \Sigma fx^2 + 2\bar{x}^2 \Sigma fx}{N} \right) - s^2 \left(\frac{2s^2}{\bar{x}} - 1 \right)$$ (2.19)

This should be compared with its standard error (the square root of its large variance) and if the negative binomial is a satisfactory model, T will be significantly smaller than the S.E. The standard error (S.E. (T)) may be approximately read off from Fig. 2.7 or calculated:

$$\text{S.E.}(T) = \sqrt{\frac{2\bar{x}(k+1)\frac{\bar{x}^2}{k^2}\left(1+\frac{\bar{x}}{k}\right)^2\left[2\left(3+5\frac{\bar{x}}{k}\right)+3k\left(1-\frac{\bar{x}}{k}\right)\right]}{N}}$$ (2.20)

where the symbols are as above and the value of k derived by method 3 above. A large positive value of U or T means that the actual distribution is more skew than that described by the negative binomial, a large negative value that the actual distribution is less skew.

Calculating a common k

Samples may be taken from various fields or other units and each will have a separate k. The comparison of these and the calculation of a common k (if there is one) will be of value in transforming the data for the analysis of variance (p. 17) and for sequential sampling (p. 51). The simplest method is the moment or regression method (Bliss & Owen, 1958; Bliss, 1958). Two statistics are calculated for each unit

$$x^1 = \bar{x}^2 - \left(\frac{s^2}{N} \right)$$ (2.21)

$$y^1 = s^2 - \bar{x}$$ (2.22)

where $\bar{x}$ = the mean, s^2 = variance and N = number of individual counts on which $\bar{x}$ is based. When y^1 is plotted against x^1 (Fig. 2.8) (including occasional negative or zero values of y^1) the regression line of y^1 on x^1 passes through the

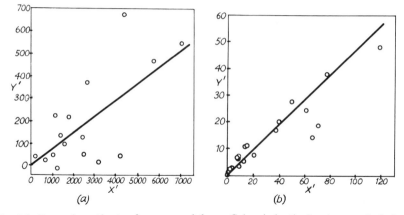

Fig. 2.8. Regression estimate of a common k for: a. Colorado beetle, *Leptinotarsa*, in 8 plots within each of 16 blocks, b. wireworms (Col. Elateridae) in 175 sampling units in each of 24 irrigated fields (after Bliss & Owen, 1958).

origins and has the slope $1/k$. An approximate estimate of the common k (k_c) is given by

$$\frac{1}{k_c} = \frac{\Sigma y^1}{\Sigma x^1} \tag{2.23}$$

It may be apparent from the plotting of y^1 on x^1 that a few points lie completely outside the main trend, and therefore although their exclusion will mean that the resultant k is not common to the whole series of samples, it is doubtful if the k_c derived by including them would really be meaningful.

A further graphical test of the homogeneity of the samples is obtained by plotting $(1/k)(=y^1/x^1)$ against the mean ($\bar{x}$) for each sub-area or group of samples. If there is neither trend nor clustering (Fig. 2.9a) we may regard the fitting of a common k as justified, but if a trend (Fig. 2.9b) or clustering occurs it is doubtful if the calculation of a common k is justified.

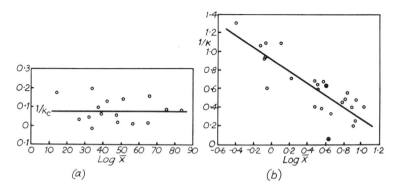

Fig. 2.9. The relation of $1/k$ to the mean for: a. Colorado beetle, b. wireworms based on the same data as Fig. 2.8 (after Bliss & Owen, 1958).

Fairly rough estimates of k_c are usually adequate, but in critical cases a weighted estimate of k_c should be obtained and whether or not this lies within the sampling error can be tested by the computation of χ^2. The method is given by Bliss (1958) and Bliss & Owen (1958). The latter authors describe another method of calculating a common k especially suitable for field experiments arranged in restricted designs (e.g. randomized blocks).

LOGARITHMIC AND OTHER CONTAGIOUS MODELS

A number of other mathematical models have been developed to describe various non-random distributions, and several of these may have more than one mode. Anscombe (1950) and Evans (1953) have reviewed these and they include the Thomas, Neyman's types A, B and C and the Polya-Aeppli. The Thomas (1949) type is based on the assumption of randomly distributed colonies whose individual populations are values plus one, from a Poisson series. Neyman's (1939) distributions are similar and are intended to describe conditions found soon after insect larvae hatched from egg batches; the modes are equally spaced. Skellam (1958) has shown that Neyman's type A is particularly applicable where the organisms occur in compact clusters, but that it can be used as an approximation in certain conditions when the clustering is less compact. The Polya-Aeppli distribution describes the situation when an initial wave of simultaneous invaders have settled and produced clusters of offspring; it may have one or two modes.

As mentioned above, the logarithmic model describes situations for which the negative binomial would give a very small value of k. The logarithmic series (Fisher, Corbet & Williams, 1943), which is derived from the negative binomial with k tending to zero and with the zero readings neglected, and the *discrete* and *truncated* (and censored) *log normal* distributions (Preston, 1948; Grundy, 1952) have been found of most value in the description of the relationship between numbers of species and numbers of individuals and are discussed later (p. 423). Also they have been found to give a reasonable description of the distribution of the individuals of some insects (e.g. the citrus scale insect, *Aonidiella ornatum* (Spiller, 1952), eggs of the larch sawfly, *Pristiphora erichsonii* (Ives, 1955)).

Working with plankton, with large mean values, Cassie (1962) suggested that the action of a series of environmental factors led to the population being distributed in a succession of Poisson series, the means of the series being themselves distributed according to the log normal model. He called this the *Poisson log normal*; it differs from the negative binomial mainly in the left-hand flank, as ordinarily plotted, where it allows for fewer zero values.

IMPLICATIONS OF THE DISTRIBUTION MODELS AND OF CHANGES IN THE TYPE OF DISTRIBUTION

Neyman's distributions are based on precise models, and as Upholt & Craig (1940) found, if the biological assumptions underlying the distribution are not fulfilled it will not adequately describe the disperson of the population. It is perhaps for this reason that Neyman's distributions have been used relatively little by entomologists, most preferring the other distributions that can be derived from a number of different hypotheses. The negative binomial can arise in at least five different ways (Anscombe, 1950; Waters & Henson, 1959):

(1) Inverse binomial sampling: if a proportion of individuals in a population possess a certain character, the number of samples, in excess of k, that have to be taken to obtain k individuals with this character will have a negative binomial distribution with exponent k.

(2) Heterogeneous Poisson sampling: if the mean of a Poisson varies randomly from occasion, under certain conditions a negative binomial results (Pielou, 1969). A biological example of this is the observation of Itô *et al.* (1962) that a series of counts of a gall-wasp on chestnut trees were distributed as a Poisson for each single tree, but when the counts from all trees were combined they were described by a negative binomial. The distribution of oribatid mites in the soil (Berthet & Gerard, 1965) and of a tapeworm in fish (Anderson, 1974) also appear to arise according to this model.

(3) Compounding of Poisson and logarithmic distributions: if a number of colonies are distributed as a Poisson, but the number of individuals per colony follows a logarithmic distribution, the resulting distribution per unit area (i.e. independent of colonies) will be a negative binomial. Counts of bacteria (Quenouille, 1949), and the dispersion of eggs of the cabbage butterfly, *Pieris rapae* (Kobayashi, 1966), have been shown to satisfy this model.

(4) Constant birth−death−immigration rates, the former two expressed per individual and the immigration rate per unit of time, will lead to a population whose size will form a negative binomial series.

(5) True contagion: where the presence of one individual in a unit increases the chance that another will occur there also.

The logarithmic series (Fisher, Corbet & Williams, 1943) can also be derived from several modes of population growth (Kendall, 1948; Shinozaki & Urata, 1953).

It is clear, therefore, that from mathematical considerations alone it is unsound to attempt to analyse the details of the biological processes involved in generating a distribution from the mathematical model it can be shown to fit or, more often, approximately fit (Waters, 1959). There is value in expressing the various possible mechanisms in biological terms: this has been done for parasite − host interactions by Crofton (1971) and Anderson (1974): these then provide alternative hypotheses that may be tested by other observations.

The extent of clumping and the changes in it provide important evidence about the population. The uses and interpretation of the parameters will be discussed later (p. 39), here changes in the actual type of distribution will be discussed. The main distinction lies between regular, random and contagious distributions, with the respective implications that the animals compete (or at least tend to keep apart), have no effect on each other, or are aggregated or clumped, but one must be careful to define the area over which the dispersion is described. The behavioural significance of the spacing pattern within the colony area (see below) is different from that between colonies.

The first difficulty is that the sampling method chosen by the experimenter may effect the apparent distribution (Waters & Henson, 1959), as is shown by the example from the work of Itô *et al.* (1962) quoted above and that of Shibuya & Ouchi (1955), who found that the distribution of the gall-midge, *Asphondylia*, on soya bean was contagious if the plant was taken as the sampling unit, but random if the numbers per pod were considered. The contagion appeared to be due to there being more eggs on those plants with more pods. The effect of the size of the sampling unit on the apparent distribution has also been demonstrated for chafer beetle larvae (Burrage & Gyrisco, 1954) and grasshopper egg pods (Putnam & Shklov, 1956). Careful testing of the size and pattern of sampling as suggested above will help to discover whether it is an artifact or a reflection of the mixture of dispersion within 'colony areas' and the pattern of 'colony' distribution. Kennedy & Crawley (1967) aptly describe the dispersion of the sycamore aphid, *Drephanosiphon platanoides*, as 'spaced out gregarious'; they point out that at close proximity, how close depending on the size of their 'reactive envelopes', animals may repel each other (e.g. caddis larvae, Glass & Boubjerg, 1969), but on a larger scale will often show grouping or contagion. This may arise from patchiness of the habitat, including differential predation (Waters, 1959) or from the behaviour of the animals themselves, or a combination of both. The behaviour leading to aggregation in the absence of special attractive areas in the habitat, may be of two types; inter-individual attraction or the laying of eggs (or young) in groups (Cole, 1946).

The dispersion of the initial insect invaders of a crop is often random; this randomness may be real or an artifact due to the low density relative to the sample size (see below). The distribution of aphids in a field during the initial phase of infestation is random, becoming contagious as each aphid reproduces (Sylvester & Cox, 1961; Shiyomi & Nakamura, 1964), although there may be differences between species (Kieckhefer, 1975). The egg masses of many insects are randomly distributed (e.g. Chiang & Hodson, 1959) and the individual eggs and young larvae are clumped; however, the dispersion of the larvae may become random, or approach it, in later instars. Such changes in distribution with the age of the population have been observed in wireworms (Salt & Hollick, 1946), the rice stem borer (*Chilo simplex*) (Kono, 1953), a chafer beetle, *Amphimallon majalis* (Burrage & Gyrisco, 1954), the cabbage butterfly (*Pieris rapae*) (Itô, Gotoh & Miyashita, 1960), the diamond-back moth (*Plutella maculipennis*) (Harcourt, 1961*a*), white grubs (*Phyllophaga*) (Guppy & Harcourt, 1970) and others. It is tempting to infer from such changes that either mortality or emigration or both are density-dependent; however, as Morisita (1962) has shown, this conclusion may not be justified with non-sedentary animals, and such changes could result from the alteration of the size of the area occupied by the colony relative to that of the sample or from the decrease in population density (see below). In contrast, the tendency to aggregate may be such that density-dependent mortality is masked, as in the pear lace-bug (*Stephanitis nashi*), where even after the bugs have been artificially removed so as to produce a random distribution, it returns to non-randomness in a few days (Takeda & Hukusima, 1961); and in the cabbage root fly (*Erioschia brassicae*) k is similar for all immature stages (Mukerji & Harcourt, 1970).

Changes in the density of the insect often lead to changes, or at least apparent changes, in the distribution. When the population is very sparse the chances of individuals occurring in any sampling unit is so low that their distribution is effectively random. The random distribution of low pop-ulations and the contagious distributions of higher ones have often been observed; for example, with wireworms (Finney, 1941), grasshopper egg pods (Davis & Wadley, 1949), a ladybird beetle, *Epilachna 28-maculata* (Iwao, 1956), the cabbage butterfly (*Pieris rapae*) (Harcourt, 1961*b*), the pea aphid (*Acyrthosiphon pisum*) (Forsythe & Gyrisco, 1963) and cotton insects (Kuehl & Fye, 1972). However, with the Nantucket pine tip moth (*Rhyacionia frustrana*), as the population density increases still further the distribution tends towards the Poisson again, i.e. the k value becomes high (Waters, 1959). Populations of several other forest insects (Waters, 1959) and of the periodical cicadas (Dybas & Davis, 1962) also show a tendency to become more random at higher densities (p. 26). Indeed one can envisage that at even higher densities the distribution would pass beyond the Poisson and become regular. Dispersions approaching a regular distribution have been but rarely observed: in the bivalve, *Tellina* (Holme, 1950), ants (Waloff & Blackith, 1962), aphids (Kennedy & Crawley, 1967) and a coral (Stimson, 1974).

In different habitats the type of distribution may change, as Yoshihara (1953) found with populations of the winkle, *Tectarius granularis* (Mollusca) on rough and smooth rocks, although it is difficult to separate this habitat effect from associated changes in density.

Within the same habitat different species will usually show different dispersion patterns as has been found with leaf hoppers (Homoptera: Auchenorrhyncha) on rice plants (Kuno, 1963). These differences can arise from several biological causes: one species may aggregate more than another in the same habitat because it disperses less or because it reproduces more or because only certain parts of the habitat are suitable for it. However, Abrahamsen & Strand (1971) suggest that for many soil organisms there may be a relationship between k and $\bar{x}$ that can be expressed by the regression:

$$k = 0.27 + 0.334 \sqrt{\bar{x}}$$

COMPARISON OF AGGREGATION INDICES

When a population is sampled three basic bits of information are available:

(i) the estimate ($\bar{x}$) of the true mean, m (ii) the estimate (s^2) of the true variance (σ^2) and (iii) the size (unit). The indices used for the description of animal populations are derived from various arrangements of this information. The simplest approach is the variance/mean ratio: $s^2/\bar{x}$ which as described above is unity for a Poisson (random) distribution. The Index of Clumping of David & Moore (1954) is simply:

$$I_{\text{DM}} = \frac{s^2}{\bar{x}} - 1 \tag{2.24}$$

which gives a value of zero for a random (non-clumped) population.

A more general approach to the variance – mean relationship is given by Taylor's Power Law (equation 2.1):

$$s^2 = a + \bar{x}^b$$

This is an empirical relationship that holds for most species (Taylor *et al.*, 1978).

Pattern analysis in plant communities was developed by P. Greig-Smith (see Greig-Smith, 1964). The method is based on increasing quadrat size; where a peak arises in the plot of a measure of variance against quadrat size this is considered as 'clump area'. This method can only be applied to contiguous quadrats, Morisita (1959, 1962, 1964, 1971) developed an index I_{δ} that provided similar information but could be used on scattered quadrats. Zahl (1974) has suggested that Greig-Smith's approach can be made more robust by the use of the S-method of Scheffé.

With mobile animals, the degree of crowding experienced by an individual interested Lloyd (1967), who devised an Index of Mean Crowding ($\dot{m}$):

$$\dot{m} = m + \left(\frac{\sigma^2}{m} - 1 \right) \tag{2.25}$$

where m and σ^2 are the true mean and variance respectively. Thus $\dot{m}$ is the amount by which the variance mean ratio exceeds unity added to the mean. If the distribution conforms to a Poisson (random) then $\dfrac{\sigma^2}{m} = 1$ and $\dot{m} = m$. With estimates based on samples the Index of Mean Crowding ($\dot{x}$) is given by:

$$\dot{x} \approx \bar{x} + \left(\frac{s^2}{\bar{x}} - 1 \right) \tag{2.26}$$

which for contagious dispersions described by the negative binomial as (eq. 2.14):

$$s^2 = \bar{x} + \frac{\bar{x}^2}{k}$$

may be written

$$\overset{*}{x} = \bar{x} + \frac{\bar{x}}{k} \tag{2.27}$$

If the ratio of mean crowding ($\overset{*}{x}$) to mean density ($\bar{x}$) is compared:—

$$\frac{\overset{*}{x}}{\bar{x}} = 1 + \frac{1}{k} \tag{2.28}$$

one sees that the reciprocal of k is the proportion by which mean crowding exceeds mean density. Lloyd (1967) termed $\overset{*}{x}/\bar{x}$ the 'patchiness'.

A clarification and unification of these various approaches has been achieved by Iwao (1968, 1970 *a b c*, 1972) who initially demonstrated that the relationship of mean crowding ($\overset{*}{x}$) to mean density ($\bar{x}$) for a species could be expressed over a range of densities by a linear regression:

$$\overset{*}{x} = \alpha + \beta\bar{x} \tag{2.29}$$

(Iwao used the expression '$\overset{*}{m} - m$ method', but in reality one is normally working with estimates ($\overset{*}{x}$ and $\bar{x}$), rather than the true mean and its derivatives. α and β are characteristic for the species and a particular habitat (Fig. 2.11) and their interpretation will be discussed later. Subsequently (Iwao, 1972) he developed the ρ-index:

$$\rho_i = \frac{\overset{*}{x}_1 - \overset{*}{x}_{i-1}}{\bar{x}_i - \bar{x}_{i-1}} \tag{2.30}$$

where $\overset{*}{x}_1$ and $\bar{x}_i$ are the mean crowding index and mean density for the ith sized quadrat sample. Thus with a series of different sized quadrats values of ρ may be plotted for the second and subsequent quadrat sizes (Fig. 2.12). Changes in the value of the ρ-index will indicate clump or territory size in a manner analagous, but theoretically preferable, to Greig-Smith's method.

$\overset{*}{x}$ expresses 'crowding', Lloyd derives it by estimating the mean number in addition to one present in the area. Working with mites on leaves, Tanigoshi, Browne & Hoyt (1975) restored the one to give an index they have called 'mean colony size' ($\overset{*}{C}$)

$$\overset{*}{C} = \overset{*}{x} + 1 \tag{2.31}$$

Lloyd (1967) termed this 'mean demand'–the mean number of individuals per quadrat per individual–and used it as an expression of trophic demand. The concept of mean colony size is useful, but only likely to be meaningful as formulated when the sampling unit, as with spider mites on a leaf, contains an entire colony.

Thus all these indices are related to each other and to the most simple concept, the comparison of the estimate of the variance (s^2) with that for the mean ($\bar{x}$) [see also Patil & Stiteler, 1974]:

$$1 + \frac{s^2 - \bar{x}}{\bar{x}^2} = 1 + \frac{1}{k} = \frac{\overset{*}{x}}{\bar{x}} = \frac{\overset{*}{C} - 1}{\bar{x}} \approx I_\delta \qquad (2.32)$$

Iwao's patchiness regression expressed as:

$$\left(\frac{s^2}{\bar{x}} - 1 \right) + \bar{x} = \alpha + \beta \bar{x} \qquad (2.33)$$

may be seen as a parallel to Taylor's power law (2.1):

$$\log s^2 = \log a + b \log \bar{x}$$

Bringing in the third piece of information, quadrat size, the ρ-index is approximately the same as plotting the changes in I with quadrat size; it is the proportion by which mean crowding changes against mean density with increased sample area. Essentially in Greig-Smith's method variance is plotted against sample area.

2.3.2. Biological interpretation of dispersion parameters

INDEX OF DISPERSION – THE DEPARTURE OF THE DISTRIBUTION FROM RANDOMNESS

If the dispersion follows the Poisson distribution (p. 26) the mean ($\bar{x}$) will equal the variance and therefore departures of the coefficient of dispersion (or variation) $s^2/\bar{x}$ from unity will be a measure of the departure from the Poisson (from randomness). This is tested by calculating the index:

$$I_D = \frac{s^2(n-1)}{\bar{x}} \qquad (2.34)$$

where n = the number of samples. I_D is approximately distributed as χ^2 with $n-1$ degrees of freedom, so that if the distribution is in fact Poisson the value of I_D will *not* lie outside the limits (taken as 0.95 and 0.05) of χ^2 for $n-1$ as given in standard tables.* The index is therefore used as a test criterion for the null hypothesis that the pattern is random: we may or may not confirm this. The coefficient of variation or mean/variance ratio is a descriptive sample statistic that will approach zero for regularly distributed organisms, whilst a large value implies aggregation. Davies (1971) provides a computer programme (No. 39) for fitting a Poisson and testing the goodness of fit (see also p. 27).

As Naylor (1959) has pointed out, the indices may be added if they are from the same sized samples. Examples of the use of this index are in Salt & Hollick's (1946) studies on wireworms, Naylor's (1959) on *Tribolium*,

* For values of n^* larger than those given in tables χ^2 may be calculated (see p. 96)

Nielsen's (1963) on *Culicoides* and Milne's (1964) on the chafer beetle, *Phyllopertha*. In contagious populations the value of this index is influenced by the density and size of the sampling unit (i.e. by the size of the mean). Greig-Smith (1964), Green (1966), Pielou (1969) and Seber (1973) discuss this and other indices in detail.

'k' OF THE NEGATIVE BINOMIAL– AN INDEX OF AGGREGATION IN THE POPULATION

If the negative binomial (p. 28) can be fitted to the data the value of k gives a measure of dispersion; the smaller the value of k the greater the extent of aggregation, whereas a large value (over about 8) indicates that the distribution is approaching a Poisson, i.e. is virtually random. This may be appreciated from the relationship of k to the coefficient of variation (above):

$$c.v. = \frac{\sqrt{s^2}}{\bar{x}} = \left[\frac{1}{k} + \frac{1}{\bar{x}} \right]^{\frac{1}{2}} \tag{2.35}$$

Clearly the smaller k, the larger the $c.v.$ Unfortunately the value of k is often influenced by the size of the sampling unit (Cole, 1946; Morris, 1954; Waters & Henson, 1959; Harcourt, 1961a) and therefore comparisons can only be made using the same sized unit. But within this restriction it does provide a useful measure of the degree of aggregation of the particular population, varying with the habitat and the developmental stage (Hairston, 1959; Waters, 1959; Harcourt, 1961a; Dybas & Davis, 1962). Examples of these variations are given in Table 2.2. As Waters (1959) has pointed out, the degree of aggregation of a population, which k expresses, could well affect the influence of predators and parasites.

The aggregation recognized by the negative binomial may be due either to active aggregation by the insects or to some heterogeneity of the environment at large (microclimate, soil, plant, natural enemies (p. 321). Dr. R. E. Blackith has suggested (see also in Richards & Waloff, 1961) that if mean size of a clump is calculated using Arbous & Kerrich's (1951) formula and this is found to be *less than 2* then the 'aggregation' would seem to be due to some environmental effect and not to an active process. Aggregations of 2 or more insects could be caused by either factor. The mean number of individuals in the aggregation is calculated by:

$$\lambda = \frac{\bar{x}}{2k} v \tag{2.36}$$

where $\bar{x}$ = the mean, v is a function with a χ^2 distribution with $2k$ degrees of freedom and λ = the number of individuals in the aggregation for the probability level allocated to v. To find the mean size of the 'aggregate' the value at the 0.5 probability level is used. $2k$ degrees of freedom will usually be fractional but an adequate χ^2 can be calculated by reference to graphs or by proportionality.

To take two examples from Table 2.2d, the eggs and pupae of *Pieris rapae* —

Eggs:
$$\lambda_m = \frac{9.5}{3.1 \times 2} \times 5.55 = 8.5$$

Pupae:
$$\lambda_m = \frac{1.7}{2.3 \times 2} \times 3.54 = 1.3$$

Table 2.2 The variation k of the negative binomial with some factors. a, b & c with sampling unit. d with developmental stage and change in density. a. *Eriophyes* leaf galls on *Populus tremuloides*. b. Nantucket pine tip moth, *Rhyacionia frustrana* (from Waters & Henson, 1959); c & d cabbage white butterfly, *Pieris rapae* (from Harcourt, 1961b)

Sampling unit	Mean number per unit	k
a. Single leaf	0.21 ± 0.01	0.0611 ± 0.0013
1 leaf-bunch	0.88 ± 0.04	0.8830 ± 0.0024
2 leaf-bunches	1.65 ± 0.08	0.1150 ± 0.0035
5 leaf-bunches	3.78 ± 0.26	0.1740 ± 0.0064
Branch	14.23 ± 1.68	0.2000 ± 0.0119
b. Tip of shoot	0.49 ± 0.03	0.449 ± 0.021
Branch whorl	2.36 ± 0.36	0.214 ± 0.035
Tree	13.13 ± 4.21	0.253 ± 0.074

Sampling unit	k for 1st instar larvae	k for 4th instar larvae
c. Quadrant of cabbage	1.38	1.96
Half cabbage	2.28	4.24
Whole cabbage	2.32	4.28

Stage	Mean density	k
d. Egg	9.5	3.1
1st instar	5.6	2.8
2nd instar	4.4	2.8
3rd instar	4.0	4.6
4th instar	3.6	5.1
5th instar	2.6	7.8
Pupa	1.7	2.3

From this it can be concluded that the 'clumping' of the eggs could be due to a behavioural cause — in this instance the females tending to lay a number of eggs in proximity to one another or to the heterogeneity of the environment — only certain areas being suitable for oviposition, but 'clumping' of the pupae is due to environmental causes.

The values of the k and the mean for $\lambda = 2$ are plotted in Fig. 2.10, from which it can be ascertained for a particular population whose k and mean are

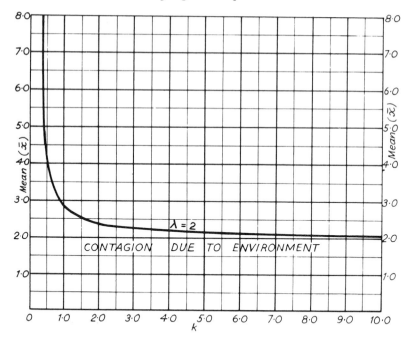

Fig. 2.10. The cause of contagion in the data – the plot of the 'mean aggregation' (λ) of two individuals for various values of the mean and k of the negative binomial. Populations whose value for the mean plotted against k lie below the line may be considered to exhibit contagion because of some environmental factor; in those populations whose value lies above the line contagion could be due either to an active behavioural process or to the environment.

known whether the mean 'aggregation' size is above or below 2. 'Aggregations' of less than 2 in sedentary animals could be due to behavioural causes if the majority of the population had been killed between settling and sampling.

Recently the negative binomial has been found useful in describing the pattern of predator attacks on patches of prey (Hassell, 1978). May (1978) has shown that the k of the attack pattern is the reciprocal of the coefficient of dispersion (or variation) ($\sqrt{s/\bar{x}}$) of the numbers of predator per patch.

'b' OF TAYLOR'S POWER LAW – AN INDEX OF AGGREGATION FOR THE SPECIES

The parameter b of Taylor's power law (p. 10) is a measure of aggregation that is generally considered characteristic and constant for the species (Taylor, Woiwod & Perry, 1978): although Banerjee (1976) disputes this.

Bliss (1971) has pointed out that special techniques should be applied for the calculation of this regression as the independent variate is not error-free: the same criticisms apply even more forcefully to Iwao's patchiness regression (see below), but in practice the more accurate approach is seldom required. As shown by Iwao (1970), Taylor's Power Law will not fit certain theoretical distributions, although the biological frequency (and significance) of these is uncertain. It remains a simple and useful description of species distribution (Healey, 1964; Youdeowei, 1965; Green, 1970; Taylor, 1970; Usher, 1971; Bardner & Lofty, 1971; Evans, 1972; Kieckheffer, 1975; Egwuatu & Taylor, 1976; Croft *et al.*, 1976). High values of b show strong contagion, but if the regression line crosses the Poisson line (slope $b = 1$) this shows a change to random distribution at lower densities, hence, unless $a = 1$, the value of b in itself cannot be taken as a test of randomness as shown by George (1974) for zooplankton. The constant a is largely a scaling factor related to sample size, but if sample size is larger than the colony size yet a biologically meaningful entity (e.g. a leaf for spider mites) then a is also related to colony size. It would seem that the amount of scatter of the points around the line (Fig. 2.1) which can be approximately estimated as the variance around the appropriate regression line (p. 10) is a measure of the effect of habitat variation on the extent of aggregation.

LLOYD'S MEAN CROWDING AND IWAO'S PATCHINESS REGRESSION – INDICES FOR THE POPULATION AND SPECIES

The index of mean crowding ($\overset{*}{x}$) is given (Lloyd, 1967) approximately by:

$$\overset{*}{x} \approx \bar{x} + \left(\frac{s^2}{\bar{x}} - 1 \right) \tag{2.37}$$

If the underlying distribution is known to fit the negative binomial:

$$\overset{*}{x} = \bar{x} + \frac{\bar{x}}{k} \tag{2.38}$$

This is an efficient estimate provided the number of samples (N) is large. The most accurate method (3 above) for k is desirable. A general formula is:

$$\overset{*}{x} = \bar{x} + \left(\frac{s^2}{\bar{x}} - 1 \right) \left(1 + \frac{s^2}{N\bar{x}^2} \right) \tag{2.39}$$

the bracket term correcting for sampling bias and k being calculated by moments (method 1 above). The appropriate large-sample variance of the estimate is:

$$\text{var}(\overset{*}{x}) \approx \left[\frac{1}{N} \left(\frac{\overset{*}{x}}{\bar{x}} \right) \left(\frac{s^2}{\bar{x}} \right) \left(\overset{*}{x} + \frac{2s^2}{\bar{x}} \right) \right] \tag{2.40}$$

Lloyd (1967) discusses the problems of applying these to rare species and the statistical requirement, difficult for the biologist to meet, that one is *a priori* assured of encountering the species. The index $\overset{*}{x}$ is the mean number of other individuals per sample unit per individual and is thus an expression of the intensity of interaction between individuals.

As will be seen from the first formula for the calculation of $\overset{*}{x}$ when the population is randomly distributed the mean crowding index equals the mean density. Many examples have been given in the previous section of increased contagion with increased density. Iwao (1968) showed that $\overset{*}{x}$ is related to the mean ($\bar{x}$) over a series of densities:

$$\overset{*}{x} = \alpha + \beta\bar{x} \qquad (2.41)$$

This is based on the changing relationship of Lloyd's Index of Patchiness ($\overset{*}{x}/\bar{x}$) and may be termed Iwao's Patchiness Regression. The constant α indicates the tendency to crowding ($+$ve) or repulsion ($-$ve) and Iwao (1970b) termed it the 'Index of Basic Contagion'. It is a property of the species. The coefficient β is related to the pattern in which the organism utilises its habitat and Iwao named it the 'Density contagiousness coefficient'. The theoretical effects of various underlying mathematical distributions on α and β and their application to many sets of field data are described by Iwao (1968, 1970a & b, 1972) and Iwao & Kuno (1971) and provides a measure of the numbers in the minimum clump in the sample unit, e.g. 38 for the individual eggs of *Epilachna* (Fig. 2.11a) or 21.5 for *Macrosiphon* (excluding 1st instar) (Fig. 2.11). A negative value of α shows there is a tendency for the animals to repel each other, although, of course, there will not be actual points in this region (see Iwao & Kuno, 1971, Fig. 3). Coefficient β expresses the extent to which the colonies (as defined by β) are contagious at higher densities e.g. in *Choristoneura* (Fig. 2.11b) or *Macrosiphum* (Fig. 2.11c). Although a single 'patchiness regression' can often be fitted for data from different areas, significantly different β's could theoretically arise where the pattern of spatial heterogeneity differed between the two areas Likewise strong density dependent mortality would influence the value of β. Density

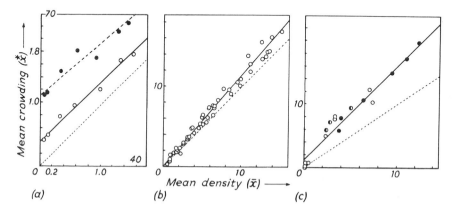

Fig. 2.11. Iwao's patchiness regression: *a.* Potato beetle, *Epilachna,* number of egg masses per plant (○) and of individual eggs (●); *b.* Spruce budworm, *Choristoneura,* numbers of larvae/twig; *c.* Pea aphid, *Macrosiphum,* total aphids (excluding 1st instar) per shoot (after Iwao, 1968).

independent mortality should alter the value of α but leave β unaffected (Iwao, 1970).

The area occupied by a colony may be determined by the use of Iwao's (1972) index:

$$\rho = \frac{\dot{x}_i - \dot{x}_{i-1}}{x_i - x_{i-1}} \qquad (2.42)$$

where subscripts 1, 2, 3 ---- i stand for successively increasing sizes of quadrats. Clearly this approach is only possible with animals living in habitats where sample size can be regularly increased. The index shows a sharp change in magnitude at the quadrat size corresponding to colony area (Fig. 2.12).

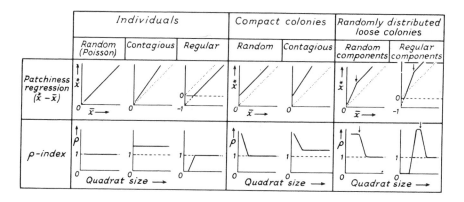

Fig. 2.12. The expected patchiness regressions and ρ-index plots for various patterns of dispersion (after Iwao, 1972).

BREDER'S EQUATIONS – A MEASURE OF THE COHÉSION OF AGGREGATIONS
Working on fish schools Breder (1954) found that the distance between individuals varied only slightly, and such uniform or regular distribution is indeed characteristic of the distribution of 'crowded' individuals (p. 26). With such schools or social swarms, when the individuals are their normal distance (d) apart the attractive force (a) between individuals can be said to equal the repulsive force (r), so that c, a measure of the cohesion in the equation $c = (a - r)/d^2$, is zero. Negative values of c will show 'overcrowding', there will be a tendency for the individuals to move apart; positive values that normal aggregation has not been obtained – these will approach $+1$ asymptotically if the attractive force is taken as 1. If r_1 and r_2 are the repulsive forces of two

animals on each other, then the equation becomes:

$$c = \frac{1 - r_1 r_2}{d^2} \qquad (2.43)$$

or

$$r_1 r_2 = d_s^2 \qquad (2.44)$$

where $d_s =$ the mean normal distance between individuals in the aggregation, because by definition $c = 0$ under these conditions.

The mean distance between individuals in a moving school may be measured from photographs and so the value for r found. It is then possible to plot c against d for a series of values (see Fig. 2.13). The angle at which the curves cross the 0 value is of interest: the steeper the line the tighter the aggregation, angles of intersection of $40° - 50°$ being characteristic of fish that keep apart by a distance considerably less than their own length; smaller angles indicate looser aggregations, both spatially and probably, Breder suggests, in their tendency to split up. The values of r given in Fig. 2.13 represent $\sqrt{r_1 . r_2}$.

Some insects, such as locusts and army worms, form swarms many of the characters of which would seem to parallel fish schools, and therefore the equations developed by Breder might well be used for comparisons with these insects. From Breder's equations it would seem that the smaller the mean distance, in terms of their total lengths, between individuals the more stable the aggregation.

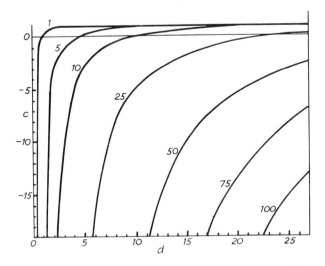

Fig. 2.13. Breder's curves for the cohesion of aggregations. The mean normal distance between individuals as a decimal fraction of the animal total length gives the value of d for $c = 0$, the appropriate curve for r ($= r_1 r_2$) may then be read off (after Breder, 1954).

DEEVEY'S COEFFICIENT OF CROWDING

This measure was used by Deevey (1947) to express the extent of crowding in barnacle populations, where the individuals actually impinge on one another as they grow and is given by:

$$C_c = 2\pi r^2 N^2 \tag{2.45}$$

where r = the radius of the fully grown animal and N = the density of the animal per unit area. The units of r and N must of course be comparable (e.g. cm and cm^2) and then C_c will define the number of contacts per unit area. This coefficient could be used with any sedentary, approximately circular animal or with plant galls on leaves; when a realistic value can be given to r, for example the radius of a circular home range (p. 338), it could also be used with mobile forms.

2.3.3 Nearest neighbour and related techniques – measures of population size or of the departure from randomness of the distribution

There are basically two separate approaches: one may either select an individual at random and measure the distance between it and its nearest neighbour (true nearest neighbour techniques) or one may select a point and measure the distance between this point and the nearest or nth nearest individuals (sometimes referred to as closest individual techniques). From each of these, conclusions may either be drawn about the departure of the distribution from random, if the population density is known from some other method, or alternatively, if one can assume the distribution to be random, its density can be estimated (Morisita, 1954). Clearly tests of randomness must be carried out. If the population has been sampled with quadrats and the mean is greater than one, the Poisson Index of Dispersion (above) is a satisfactory test. Other test procedures are given by Seber (1973).

These techniques are most easily used with stationary, discrete, easily mapped organisms, e.g. trees, and have been used extensively by botanists (Greig-Smith, 1964). With animals, their mobility and the risk that one will fail to find the nearest neighbour limit their application (Turner, 1960). However, such methods have been applied to studies on grasshoppers (Blackith, 1958) and frogs (Turner, 1960) and, with more success, to studies of fairly conspicuous and relatively stationary animals (e.g. snails; Keuls, Over & de Wit, 1963) or of well-marked colonies (e.g. ant mounds; Waloff & Blackith, 1962; Blackith *et al.*, 1963). Nearest neighbour and related techniques could also be used with other relatively sedentary animals such as barnacles, limpets (*Patella*), scale insects, tube-building animals and gall formers (especially on leaves). The use of photography in mapping natural distributions of snails has been described by Heywood & Edwards (1961), and this could provide a valuable aid to the application of these techniques, not only to fairly sedentary organisms, but also where the substratum is suitable, to far more mobile forms. The various dolichopodid and other flies on mud are an example of a community that might be studied in this way. There are, furthermore, a few other techniques, such as the 'squashing' or 'imprinting' method for estimating mite numbers on leaves (p. 145), or the deadheart method for estimating stem borers (p. 293) that do, or could easily, provide maps of the distribution of the individuals to which nearest neighbour or closest individual methods could be applied. Henson (1961) has used a combination of photography and nearest neighbour methods to study the aggregations of a scolytid beetle in the laboratory. (See also p. 141.)

NEAREST NEIGHBOUR METHODS

In this method a point is selected at random, and then one searches around in tight concentric rings until an animal is found; the searching is then continued until its nearest neighbour is found and the distance between these two animals is measured. Strictly speaking, this is not a random method (Pielou, 1969); a truly random method would involve numbering all the animals in advance. The extent to which the method is robust in respect of this violation is not clear and thus in general the closest individual method is to be preferred (Seber, 1973). There are basically two expressions; the simplest is due to Clark & Evans (1954):

$$m = \frac{1}{4\bar{r}^2} \tag{2.46}$$

where m = density per unit area; $\bar{r}$ = mean distance between nearest neighbours. As has been pointed out by Blackith (1958), this formula is based on the mean r squared, whereas Craig's (1953) formula is based more correctly on the sum of r squared; the latter formula is, however, less easy to use and describes the density in a quarter segment. Turner (1960) has concluded that for randomly distributed populations Clark & Evans's formula is fairly satisfactory.

The importance of a number of population estimates by different methods has been emphasized elsewhere (p. 4) and therefore in fairly homogeneous habitats (e.g. field crops), unless the species obviously aggregates, these methods would provide a useful 'order of magnitude check' on other estimates obtained by marking and recapture or by sampling methods, and like marking and recapture they are independent of sample size.

If, however, the density has been measured by one of these other methods, the departure of the distribution from randomness will be given by the value of the numerator in the expressions (Clark & Evans, 1954; Waloff & Blackith, 1962).

$$\bar{r}^2 = \frac{1.154}{m} \qquad \text{Uniform, hexagonal spacing-competition}$$

$$\bar{r}^2 = \frac{0.2500}{m} \qquad \text{Random}$$

where $\bar{r}$ = mean distance between nearest neighbours, m = number of individuals per unit area.

Smaller values of the numerator will indicate aggregation or clumping due to environmental or behavioural factors. These estimates should be based on large samples and, of course, the density and distance units of measurement must be the same.

The use of the second, third, ... nth nearest neighbours, discussed by Morisita (1954) and Thompson (1956), not only enables more accurate density determinations to be made, but also the dispersion pattern to be detected over a larger area. However, as Waloff & Blackith (1962) have pointed out, competition effects detectable between nearest neighbours may be masked at greater distances because of heterogeneity of the habitat. Thompson shows that the statistic* :

$$2\pi m \Sigma r_n{}^2 \qquad (2.47)$$

where m = density and r = distance, is distributed as a χ^2 with $2N_n$ degrees of freedom (N = number of observations). This value may be calculated and compared with the expected value under the hypothesis of randomness (the χ^2 value for $2N_n$ is found from tables). A probability of χ^2 greater than 0.95 indicates significant aggregation, whilst a probability of less than 0.05 indicates significant regularity.

A simpler but – according to Thompson – less reliable comparison is given by the proportionality constants in the equation:

$$\bar{r}_n{}^2 = \frac{p}{m} \qquad (2.48)$$

where p = proportionality constant and other symbols as above. For a random distribution for $n=1$ (i.e. the nearest neighbour) as indicated above $p=0.25$; for $n=2$, $p=0.87$; for $n=3$, $p=0.97$; for $n=4$, $p=1.05$.

CLOSEST INDIVIDUAL OR DISTANCE METHOD

In this method a point is selected at random and the distance between this and the nth nearest individual measured. The main theoretical difficulty is the possibility of the random point being closer to the boundary than to an individual. Practically, it is more time consuming than the nearest neighbour, because of the need to find the 3rd or more distant individual. However, in spite of these problems, because of the difficulties of true random selection of an animal, referred to above (difficulties that apply less to botanical studies), this method is to be preferred.

The method has also been derived, apparently independently, by Keuls, Over & de Wit (1963) and used for population estimations of the snail *Limnaea* (= *Galba*) *truncatula*. It depends on the population being randomly

* See also SMALL, R. *et al.*, 1985. *Oikos*, **44**, 511–2.

distributed, although if the discontinuities are small, the errors that are made will tend to cancel each other out. The basic equation is:

$$m = \frac{n-1}{\pi} \times \frac{1}{a_n{}^2} \qquad (2.49)$$

where n = the rank of the individual in distance from the randomly selected point, e.g. for the nearest neighbour $n=1$, for the second nearest $2 \ldots$; a = the distance between the randomly selected point and the individual and other symbols are as above. The values of $(n-1)/\pi$ for a series of values of n are given in Table 2.3; clearly, therefore, in order to estimate population density the reciprocal of the square of the distance of the nth animal from the random point $(1/a^2)$ need only be multiplied by the factor in the middle column of this table. The 95 % fiducial limits and the coefficient of variation are also given.

Table 2.3 Multiplication factors for estimating the density of a population by the distance method with 95% fiducial limits and the coefficient of variance of the estimate of density (after Keuls *et al.*, 1963).

| | Multiply $1/\bar{a}^2$ for nth nearest animal from point by value in table for: | | | |
| | 95% | Estimate | 95% | Coefficient |
n	lower limit	of density	upper limit	of variation
1	0.016	—	0.95	—
2	0.113	0.32	1.51	—
3	0.26	0.64	2.01	1.00
4	0.44	0.95	2.47	0.71
5	0.63	1.27	2.91	0.58
6	0.83	1.59	3.35	0.50
7	1.05	1.91	3.77	0.45
8	1.27	2.23	4.19	0.41
9	1.50	2.55	4.60	0.38
10	1.73	2.86	5.00	0.35

In the original paper values of n up to 100 are given, these are not reproduced here as the labour of assessing even the 10th nearest animal is considerable. However, the fiducial limits become narrower with higher values and the third to fifth nearest individuals are probably the most practical. The nearest animal cannot be used for it is without the estimator $(n-1)/\pi$ in the equation for density, neither can the second nearest, as an unbiased estimate of the variance is not available.

Keuls *et al.* suggest that the method could be applied to very mobile animals by recording the nearest individual at s moments (perhaps photographically, p. 47) when the formula becomes:

$$m = \frac{sn - 1}{\pi} \times \frac{1}{\sum_{1 \to s} a_n^{\,2}} \qquad (2.50)$$

that is one simply adds up, say, four measurements of the nearest animal and regards the sum as the measurement of the fourth nearest.

Various indices of non-randomness have been derived [for a fuller discussion see Pielou (1967) and Seber (1973)]. Two are:

(i) Pielou's Index (I_α):

$$I_\alpha = \bar{a}_i^{\,2}\, m\pi \qquad (2.51)$$

that requires a knowledge of the density per unit area (m) as well as the mean distance from a random point to the closest individual ($\bar{a}_i$). $I_\alpha = 1$ for random distributions, > 1 for contagious and < 1 for regular. It has been used by Underwood (1976) with intertidal molluscs.

(ii) Eberhardt's (1967) Index (I_E):

$$I_E = \frac{\Sigma(\bar{a}_i^{\,2})}{(\Sigma(\bar{a}_i))^2} \qquad (2.52)$$

This has the advantage that density need not be known, but the departure from randomness cannot easily be determined. $I_E = 1.27$ for a random distribution, being greater or less for contagious and regular respectively.

2.4 Sequential sampling

In this type of sampling the total number of samples taken is variable and depends on whether or not the results so far obtained give a definite answer to the question posed about the frequency of occurrence of an event (i.e. abundance of an insect). It is of particular value for the assessment of pest density in relation to control measures, when these are applied only if the pest density has reached a certain level. Extensive preliminary work is necessary to establish the type of distribution (the relation of the mean and the variance) and the density levels that are permissible and those that we associated with extensive damage. From such data it would be possible to lay down a fixed number of samples that would enable one to be sure in all instances what was the population level; however, this would frequently lead to unnecessary sampling and a sequential plan usually allows one to stop sampling as soon as enough data have been gathered. For example, to take extremes, if 80 insects per shoot, or more, caused severe damage and no insects were found in the first 5 samples, common sense would tempt one not to continue for the additional, say 15, samples. Sequential sampling gives an exact measure (based on the known variance) so that with extremely high and extremely low populations very few samples need be taken and the expenditure of time and effort (cost) is minimal. It may also be used to obtain population estimates with a fixed level of precision (Kuno, 1969; Green, 1970).

As the distribution of most insect species can be fitted to the negative binomial, or to Iwao's Patchiness Regression, formulae will be given only for these types of distribution; of course the principles are the same with other distributions. The method is described by Wald (1948), Goulden (1952) and Waters (1955). Stark (1952*b*) has applied a sequential plan to normalized data for a needle miner (Lepidoptera).

The first decision must be to fix the insect population levels related to the infestation classes; we wish to distinguish between the hypothesis (H_1) that there are e.g. 200 or more egg masses per branch, sufficient to cause heavy damage, and the hypothesis (H_0) that there are e.g. 100 or fewer egg masses per branch, insufficient to cause damage.

The second decision concerns the level of probability of incorrect assessment one is prepared to tolerate; there are two types of error:

ψ = the probability of accepting H_1 when H_0 is the true situation
ω = the probability of accepting H_0 when H_1 is the true situation

Let us say that the same level is accepted for both, 1 false assessment in 20, i.e. a probability level of 0.05 for ψ and ω. Different levels may be necessary because the costs of failing to apply control measures when an outbreak occurs, may be much higher than those arising from the application of the treatment that is proved not to have been required.

For an animal whose distribution corresponds to the negative binomial the following values need to be calculated; the common k having been found already (p. 3) and the means being fixed as above (Oakland, 1950; Morris, 1954):

	Infestation level	
	H_0	H_1
Mean $= kp$	kp_0 (e.g. 100)	kp_1 (e.g. 200)
$p = kp/k$	p_0	p_1
$q = 1 + p$	q_0	q_1
Variance $= kpq$	kp_0q_0	kp_1q_1

The next aim is to plot the two lines (Fig. 2.14) that mark the 'acceptance' and 'rejection' areas (these terms have come from quality control work where sequential sampling was originally developed).

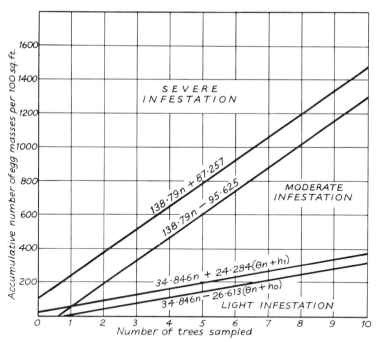

Fig. 2.14. A sequential sampling chart with two sets of acceptance and rejection lines (after Morris, 1954).

The formulae for the lines are:

$$d_0 = \theta n + h_0 \qquad (2.51)$$

$$d_1 = \theta n + h_1 \qquad (2.52)$$

where d = cumulative number of insects, n = number of samples taken, θ = the slope of the lines given by

$$\theta = k\frac{\log(q_1/q_0)}{\log(p_1 q_0/p_0 q_1)} \qquad (2.53)$$

and h_0 and h_1 the lower and upper intercepts given by

$$h_0 = \frac{\log[\omega/(1-\psi)]}{\log(p_1 q_0/p_0 q_1)} \qquad (2.54)$$

$$h_1 = \frac{\log[(1-\omega)/\psi]}{\log(p_1 q_0/p_0 q_1)} \qquad (2.55)$$

Once these calculations are made a graph similar to Fig. 2.14 can be drawn up, and it will be clear that occasionally (especially when the true population lies between the population levels chosen, i.e. 100 and 200 in this example) the cumulative total continues to lie in the uncertain zone and therefore an arbitrary upper limit must be set to the number of samples taken. This limit (x) will be decided on considerations of cost, and normally it will be laid down that if after x samples the result still lies in the uncertain zone treatment will be applied.

If Iwao's Patchiness Regression (p. 43) is used the lines can be best calculated for successive values of n (number of samples taken) by (Iwao, 1975):

$$d_1 = nx_c + t \sqrt{n[(\alpha + 1)x_c + (\beta - 1)x_c^2]} \tag{2.56}$$

$$d_o = nx_c - t \sqrt{n[(\alpha - 1)x_c + (\beta - 1)x_c^2]} \tag{2.57}$$

where $x_c = $ the critical density, α and β are the parameters from the patchiness regression and t is taken from statistical tables for the acceptable error levels for ψ (for d_o equation) and ω (for the d_1 equation).

It is possible to draw up operating characteristic curves and average sample number curves which give a measure of the probability of accepting the two hypotheses and the average number of samples necessary, respectively, at different population levels (Oakland, 1950; Morris, 1954).

The actual field work may be carried out based on the graph, or a sequential table may be prepared from it. This gives the uncertainty band for various numbers of samples; e.g. (after Morris, 1954):

Sample tree	Moderate vs severe infestation uncertainty band (cumulative total of insects)
1	42–225
2	183–366
3	323–506
4	460–643
5	599–782
6	738–921

It is often considered desirable to set up three classes: severe, moderate and light, and then two pairs of lines are calculated.

A general account of the utility of sequential sampling plans for forest insects is given by Ives (1954) and, in addition to the papers already referred to, plans have been developed for the winter moth (*Operophtera brumata*) (Reeks, 1956), the forest tent caterpillar (*Malacosoma disstria*) (Connola, Waters & Smith, 1957), the larch sawfly (*Pristiphora erichsonii*) (Ives & Prentice, 1958), the red-pine sawfly (*Neodiprion nanulus*) (Connola, Waters & Nason, 1959), the coffee shield bug, *Antestiopsis* (Rennison, 1962), lodgepole needle miner (*Evagora milleri*) (Stevens & Stark, 1962), the aphid, *Myzus persicae* (Sylvester & Cox, 1961), white grubs (Ives & Warren, 1965), *Pieris rapae* (Harcourt, 1966), wireworms (Onsager, Landis & Fox, 1975), cotton fleahopper (*Pseudatomoscelis*) (Pieters & Sterling, 1974), cotton arthropods

(Sterling & Pieters, 1974, 1975), and others. It must, however, be remembered that these plans are based on particular sampling methods in a certain area with a given developmental stage and, as has been shown (p. 41), the value of k and hence the validity of the plan may alter if any of these are changed.

2.5 Presence or absence sampling

It is sometimes very costly to estimate the actual numbers in a sample, but presence or absence can be easily assessed. If the dispersion of such an animal in a particular habitat, e.g. forest, can be described by the negative binomial distribution with a known k, the probability of a particular mean population per tree (N_y) being reached can be estimated (Gerrard & Chiang, 1970; Wilson & Gerrard, 1971; Gerrard & Cook, 1972):

$$(N_y) = k\left[\left(\frac{1}{1-p'}\right)^{1/k} - 1\right] \tag{2.58}$$

where p' = probability of the animal being present in the sampling unit – determined on the basis of presence or absence sampling. The form of the relationship is shown in Fig. 2.15. This method provides an alternative to

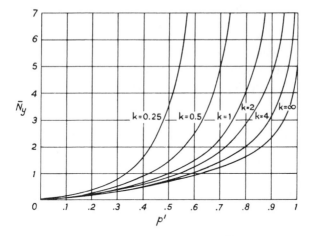

Fig. 2.15. Relationship between mean population density $(\bar{N}_y)$ and the probability of cell occupancy (p') for selected values of k (after Gerrard and Cook, 1972).

sequential sampling as a basis for pest control decisions. It is particularly useful with highly clumped animals; the saving on counting time is greater, as is the sensitivity (Fig. 2.15) and is reliable only if the critical density levels are related to values of p' less than about 0.8; above this the uncertainty associated with the predictions is too great. Obviously the sampling unit can be varied to achieve this relationship but for it to retain its value for rapid assessment the new unit must be easily recognized. Furthermore any major changes in the value of k with density would invalidate this approach.

2.6 Sampling a fauna

Emphasis has been placed on the problems involved in sampling the population of a single species and these are formidable enough. But with faunal surveys the problem of ensuring the detection of an adequate proportion of the species present must be considered. When sampling vegetation, Gleason (1922) pointed out as the area sampled increased the numbers of hitherto unrecorded species added decreased, a concept referred to as the species-area curve, and one that has been often used and disputed in plant ecology (Goodall, 1952; Evans, Clark & Brand, 1955; Greig-Smith, 1978; Watt, 1964). The logarithmic series and log normal distributions (p. 423) express the same basic assumptions another way.

One approach to this problem is therefore to take a number of preliminary samples and calculate the index of diversity based on the log series (p. 429). If one can make an estimate of the total number of individuals in the fauna then the theoretical total number of species present can be arrived at from the index of diversity, and the total number of individuals that should be collected to find $y \%$ of species present can be determined. The value of y will, of course, depend on the aim of the study. Unfortunately, as pointed out later (Chapter 13), this approach has a number of practical and theoretical limitations. The practical ones stem from the mosaic nature of most habitats (Hairston, 1964), and therefore, unless sampling is by a light trap or similar device that is independent of microhabitats, more new species are added by further samples than would be expected. The theoretical problems arise from the discovery that the log normal distribution provides a better description of the phenomenon than the simpler logarithmic series (see p. 421 for a further discussion). An approach which makes no assumptions about the type of distribution is given by Good & Toulmin (1956).

Sometimes the mosaic nature of the habitat can be utilized to facilitate the planning of an optimal sampling programme. This is especially true where the effects of pollution are being studied in streams, but might be extended to other situations where a factor is acting simultaneously over a wide range of special habitats within a major habitat or community type, for example the effect of aerial spraying on the fauna of a mixed woodland. If a fixed number of samples of the fauna are taken – say from three different zones of a stream: riffles, pools and marginal areas – it will be found that the first few samples will provide a more complete picture of the total observed fauna in one zone than in the other (Gaufin, Harris & Walter, 1956). In other words because of the different diversities and microdistributions of the fauna in the different zones, and also possibly because of the differing efficiencies of the collecting methods under these conditions, the rates of accumulation of new species vary from zone to zone. A zone in which a high percentage of the total species taken in a large number of samples (n) were recorded in the first few samples, can have a smaller number of samples drawn from it than a zone in which the rate of accumulation is less rapid.

A method of calculating the average number of new species contributed by the Kth successive sample (when K can equal any number from 1 upwards) has been devised by Gaufin, Harris & Walter (1956), who also provide a table of the coefficients necessary for up to a total of ten samples. Their formula is:

$$S_K = \frac{1}{S} \sum_{i=1}^{n-K+1} a_i K S_i \qquad (2.59)$$

where S_K = the number of new species added by the Kth sample, n = the total number of samples taken in the preliminary survey, S = the total number of species taken in the n samples, S_i = the number of different species appearing in i out of n samples and $a_i K$ are coefficients by which successive values of S_i are multiplied in calculating the summation item. By summing a series of values of S_K (for $K = 1$ to a predetermined number K') it is possible to compute the average number of species found in K' samples. A method of calculating the standard errors of such estimates has been developed by Harris (1957).

2.7 Biological and other qualitative aspects of sampling

Hitherto stress has been laid on the statistical aspects of sampling; however, there are certain biological problems that are of cardinal importance. The ecologist should always remember that the computation of fiducial limits of estimates only tells him the consistency of the samples he has collected, and the statistical techniques cannot be blamed if these have been consistently excluding a major part of the population. This is indeed the most serious biological problem in all sampling work, namely to ensure that there is not a part of the population with a behavioural pattern or habitat preference such that it is never sampled (Macleod, 1958).

Less dangerous because the phenomenon will be recorded, although it may well be misinterpreted, is the tendency for the behaviour of the animal to change and affect its sampling properties. The reactions of the fly *Meromyza* to weather conditions (Hughes, 1955) and of the larvae of different species of corixid water-bug to light (Teyrovský, 1956) lead to errors when net sampling is carried out under different conditions (p. 240). In many, perhaps almost all, insects the behaviour alters with age, and this may lead to a confusion between a change in behaviour and death. For example older females of the mirid grass bug, *Leptopterna dolobrata*, spend more time on the base of the grass shoots and less on the tops than do males. If a marking and recapture experiment is carried out and the bugs collected by sweeping, the females will appear to be much shorter-lived than the males. But this is because as they age, the females enter the sampling zone less frequently; they are actually longer-lived than the males.

The variations in the distribution of the insect may well be linked with some character of the environment or its host plant, and it is important to distinguish this variation from sampling variance otherwise the fiducial limits of the estimates may become so wide that no conclusions can be drawn (cf. Prebble, 1943), that is systematic errors will arise (p. 17). When population trend is eliminated, approximately 50% of the variance of the counts of populations of the sugar cane froghopper (*Aeneolamia varia*), on sugar cane stools could be accounted for in terms of the size of the stools (Fewkes, 1961); hence more 'accurate' estimates would be obtained by taking stool size into consideration and adjusting population estimates, or by allowing for it in the analysis. This is particularly important when comparisons of froghopper numbers are being made between insecticide-treated fields; the fields are of different ages and hence with stools of different sizes. Many other examples could be given of environmental factors influencing distribution; they are commonly microclimatic in nature, e.g. soil moisture affects the distribution of the cocoons of a sawfly (Ives, 1955). Even when the environment appears uniform distribution may be systematically non-random; for example, the most dense populations of the beetle, *Tribolium*, in flour are always adjacent to the walls of the container (Cox & Smith, 1957). Covariance analyses are often helpful in such situations.

In most habitats a random sample can be selected by numbering the habitat on a grid system and using a table of random numbers; an acceptable approximation is to move a certain number of units determined by a random number along one side of a plot and then turn at right angles and move a second number of units into the plot. Such a method can be used for soil samples, crops and herbage, aquatic samples and, with some modifications, with trees (p. 24).

The studies of Howe (1963) on stored grain insects illustrate some of the problems of random sampling. The bag of grain may be selected randomly or the suction probe (p. 157) may be inserted at a random point (the depth of sampling is usually an important biological component of the variability and hence each depth band tends to be sampled separately). The quantity of grain obtained in this way is often more than can conveniently be examined for insects, and thus a subsample must be taken. Howe (1963) has compared four ways of subsampling grain and he has shown that two mechanical separators, a machine devised for the separation of fine granular fuels (Anon, 1955) and the standard apparatus for splitting grain samples for grading, and one hand method gave reliable randomized subsamples. The hand method consisted of pouring the grain over the spherical bottom of a short-necked flask upturned in a glass dish; the grain becomes more or less evenly distributed round the dish and samples of a fixed volume are scooped up radially, the dish being turned through a random angle between each scoop. Although this method was found reliable by Howe, it does depend, as he stresses, on the personal skill and avoidance of bias by the operator.

REFERENCES

ABRAHAMSEN, G., 1969. Sampling design in studies of population densities in Enchytraeidae (Oligochaeta). *Oikos* **20**, 54–66.

ABRAHAMSEN, G. and STRAND, L., 1970. Statistical analysis of population density data of soil animals, with particular reference to Enchytraeidae (Oligochaeta). *Oikos* **21**, 276–84.

ANDERSEN, F. S., 1965. The negative binomial distribution and the sampling of insect populations. *Proc. XII int. Congr. Ent.* 395.

ANDERSON, R. M., 1974. Population dynamics of the cestode *Caryophyllaeus laticeps* (Pallas, 1781) in the bream (*Abramis brama* L.). *J. Anim. Ecol.* **43**, 305–21.

ANON., 1955. British Instrument Industries' exhibition. *Engineering* **180** (22 July 1955). Particle sampling p. 116.

ANSCOMBE, F. J., 1948. On estimating the population of aphids in a potato field. *Ann. appl. Biol.* **35**, 567–71

ANSCOMBE, F. J., 1949. The statistical analysis of insect counts based on the negative binomial distribution. *Biometrics* **5**, 165–73.

ANSCOMBE, F. J., 1950. Sampling theory of the negative binomial and logarithmic series distributions, *Biometrika* **37**, 358–82.

ARBOUS, A. G. and KERRICH, J. E., 1951. Accident statistics and the concept of accident-proneness. *Biometrics* **7**, 340–432.

BAILEY, N. T. J., 1959. *Statistical methods in biology.* 200 pp., English Universities Press, London.

BANCROFT, T. A. and BRINDLEY, T. A., 1958. Methods for estimation of size of corn borer populations. *Proc. X int. Congr. Ent.* **2**, 1003–14.

BANERJEE, B., 1976. Variance to mean ratio and the spatial distribution of animals. *Experientia* **32**, 993–4.

BARDNER, R. and LOFTY, J. R., 1971. The distribution of eggs, larvae and plants with crops attacked by wheat bulb fly, *Leptohylemyia coarctata* (Fall.). *J. appl. Ecol.* **8**, 683–6.

BEALL, G., 1942. The transformation of data from entomological field experiments so that the analysis of variance becomes applicable. *Biometrika* **32**, 243–362.

BEAUCHAMP, J. J. and OLSON, J. S., 1973. Corrections for bias in regression estimates after logarithmic transformation. *Ecology* **54**, 1403–7.

BERTHET P. and GERARD, G., 1965. A statistical study of microdistribution of Oribatei (Acari). Part I. The distribution pattern. *Oikos* **16**, 214–27.

BLACKITH, R. E., 1958. Nearest-neighbour distance measurements for the estimation of animal populations. *Ecology* **39**, 147–50.

BLACKITH, R. E., SIDDORN, J. W., WALOFF, N. and EMDEN, H. F. VAN, 1963. Mound nests of the yellow ant, *Lasius flavus* L., on water-logged pasture in Devonshire. *Ent. mon. Mag.* **99**, 48–9.

BLISS C. I., 1958. The analysis of insect counts as negative binomial distributions. *Proc. X int. Congr. Ent.* **2**, 1015–32.

BLISS, C. I., 1971. The aggregation of species within spatial units. *In* Patil, G. P., Pielou, E. C. & Waters, W. E. (eds.), *Statistical Ecology 1* (Spatial Patterns and Statistical distributions), pp. 311–35, Penn State University Press, Philadelphia.

BLISS, C. I. and CALHOUN, D. W., 1954. *An outline of biometry.* Yale Co-operative Corp., New Haven.

BLISS, C. I. and FISHER, R. A., 1953. Fitting the negative binomial distribution to biological data and note on the efficient fitting of the negative binmial. *Biometrics* **9**, 176–200.

BLISS, C. I. and OWEN, A. R. G., 1958. Negative binomial distributions with a common *k*. *Biometrika* **45**, 37–58.

BREDER, C. M., 1954. Equations descriptive of fish schools and other animal aggregations. *Ecology* **35**, 361–70.

BROADBENT, L., 1948. Methods of recording aphid populations for use in research on potato virus diseases. *Ann. appl. Biol.* **35**, 551–66.

BROWNING, T. O., 1959. The long-tailed mealybug, *Pseudococcus adonidum* L. in South Australia. *Aust. J. Agric. Res.* **10**, 322–37.

BRYANT, D. G., 1976. Sampling populations of *Adelges piceae* (Homoptera: Phylloxeridae) on Balsam fir, *Abies balsamea*. *Can. Ent.* **108**, 1113–24.

BURRAGE R. H. and GYRISCO, G. G., 1954. Estimates of populations and sampling variance of European chafer larvae from samples taken during the first, second and third instar. *J. econ. Ent.* **47**, 811–17.

CAMIN, J. H., GEORGE, J. E., and NELSON, V. E., 1971. An automatic tick collector for studying the rhythmicity of "drop off" Ixodidae. *J. med. Ent.* **8**, 394–8.

CASSIE, R. M., 1954. Some uses of probability paper in the analysis of size frequency distribution. *Aust. J. Mar. Freshw. Res.* **5**, 513–22.

CASSIE, R. M., 1962. Frequency distribution models in the ecology of plankton and other organisms. *J. Anim. Ecol.* **31**, 65–92.

CHIANG, H. C. and HODSON, A. C., 1959. Distribution of the first-generation egg masses of the European corn borer in corn fields. *J. econ. Ent.* **52**, 295–9.

CHURCH, B. M. and STRICKLAND, A. H., 1954. Sampling cabbage aphid populations on brussels sprouts. *Plant Path.* **3**, 76–80.

CLARK, P. J. and EVANS, F. C., 1954. Distance to nearest neighbor as a measure of spatial relationships in populations. *Ecology* **35**, 445–53.

COCHRAN, W. G., 1963. *Sampling techniques*. 2nd ed. 413 pp., Wiley, New York.

COLE, L. C., 1946. A theory for analyzing contagiously distributed populations. *Ecology* **27**, 329–41.

COLE, W. E., 1970. The statistical and biological implications of sampling units for mountain pine beetle populations in lodgepole pine. *Res. Popul. Ecol.* **12**, 243–48.

CONDRASHOFF, S. F., 1964. Bionomics of the aspen leaf miner, *Phyllocnistis populiella* Cham. (Lepidoptera: Gracillariidae). *Can. Ent.* **96**, 857–74.

CONNOLA, D. P., WATERS, W. E. and NASON, E. R., 1959. A sequential sampling plan for Red-pine sawfly *Neodiprion nanulus* Schedl. *J. econ. Ent.* **52**, 600–2.

CONNOLA, D. P., WATERS, W. E. and SMITH, W. E., 1957. The development and application of a sequential sampling plan for forest tent caterpillar in New York. *Bull. N.Y. St. Mus. Sci. Serv.* **366**, 22 pp.

CORBET, P. S. and SMITH, S. M., 1974. Diel periodicities of landing of multiparous and parous *Aedes aegypti* (L.) at Dar es Salaam, Tanzania (Diptera, Culicidae). *Bull. ent. Res.* **64**, 111–21.

CORNFIELD, J., 1951. The determination of sample size. *Am. J. Pub. Health* **41**, 654–61.

COULSON, R. N., HAIN, F. P., FOLTZ. J. L. and MAYYASI, A. M., 1975. Techniques for sampling the dynamics of Southern Pine beetle populations. *Texas Agric. Exp. Stn. Misc. Pub.* **1185**, 3–18.

COULSON, R. N., PULLEY, P. E., FOLTZ, J. L. and MARTIN, W. C., 1976. Procedural guide for quantitatively sampling within-tree populations of *Dendroctonus frontails*. *Texas Agric. Exp. Stn. Misc. Pub.* **1267**, 3–26.

COX, D. R. and SMITH, W. L., 1957. On the distribution of *Tribolium confusum* in a container. *Biometrika* **44**, 328–35.

CRAIG, C. C., 1953. On a method of estimating biological populations in the field. *Biometrika* **40**, 216–18.

CROFT, B. A., WELCH, S. M. and DOVER, M. J., 1976. Dispersion statistics and sample size estimates for populations of the mite species *Panonychus ulmi* and *Amblyseius fallacis* on Apple. *Environmental Ent.* **5**, 227–34.

CROFTON, H. D., 1971. A quantitative approach to parasitism. *Parasitology* **62**, 179–93.

DAVID, F. N. and MOORE, P. G., 1954. Notes on contagious distributions in plant populations. *Ann. Bot. Lond. N. S.* **18**, 47–53.

DAVIES, R. G., 1971. *Computer Programming in Quantitative Biology.* 492 pp., Academic Press, London and New York.

DAVIS, E. G. and WADLEY, F. M., 1949. Grasshopper egg-pod distribution in the northern Great Plains and its relation to egg-survey methods. *U.S.D.A. Circ.* **816**, 16 pp.

DEAN, H. A., 1959. Quadrant distribution of mites on leaves of Texas grapefruit. *J. econ. Ent.* **52**, 725–7.

DEBAUCHE, H. R., 1962. The structural analysis of animal communities of the soil. *In* Murphy, P. W. (ed.), *Progress in soil zoology*, 10–25.

DEEVEY, E. S., 1947. Life tables for natural populations of animals. *Quart. Rev. Biol.* **22**, 283–314.

DEMPSTER, J. P., 1957. The population dynamics of the Moroccan locust (*Dociostaurus maroccanus* Thunberg.) in Cyprus. *Anti-locust Bull.* **27**, 60 pp.

DUDLEY, C. O., 1971. A sampling design for the egg and first instar larval populations of the western pine beetle, *Dendroctonus brevicomis* (Coleoptera: Scolytidae). *Can. Ent.* **103**, 1291–313.

DYBAS, H. S. and DAVIS, D. D., 1962. A population census of seventeen-year periodical cicadas (Homoptera: Cicadidae: *Magicicada*). *Ecology* **43**, 432–44.

EBERHARDT, L. L., 1967. Some developments in 'distance sampling'. *Biometrics* **23**, 207–16.

EDWARDS, R. L., 1962. The importance of timing in adult grasshopper surveys. *J. econ. Ent.* **55**, 263–4.

EGWUATU, R. I. and TAYLOR, T. A., 1975. Aspects of the spatial distribution of *Acanthomia tomentosicollis* Stal. (Heteroptera, Coreidae) in *Cajanus cajan* (Pigeon Pea). *J. econ. Ent.* **69**, 591–4.

ELLENBERGER, J. S. and CAMERON, E. A., 1977. The spatial distribution of oak leaf roller egg masses on primary host trees. *Environmental Ent.* **6**, 101–6.

ELLIOTT, J. M., 1971. Some methods for the statistical analysis of samples of benthic invertebrates. *Sci. Publ. Freshw. Biol. Assoc.* **25**, 148 pp.

ELLIOTT, J. M. and DÉCAMPS, H., 1973. Guide pour l'analyse statistique des échantillons d'invertébrés benthiques. *Annls. limnol.* **9**, 79–120.

EMDEN, H. F. VAN., JEPSON, W. F. and SOUTHWOOD, T. R. E., 1961. The occurrence of a partial fourth generation of *Oscinella frit* L. (Diptera: Chloropidae) in southern England. *Entomologia exp. appl.* **4**, 220–5.

EVANS, D. A., 1953. Experimental evidence concerning contagious distributions in ecology. *Biometrika* **40**, 186–211.

EVANS, D. E., 1972. The spatial distribution and sampling of *Aeneolamis varia saccharina* and *A. postien fugata* (Homoptera: Cercopidae). *Entomologia exp. appl.* **15**, 305–18.

EVANS, F. C., CLARK, P. J. and BRAND, R. H., 1955. Estimation of the number of species present in a given area. *Ecology* **36**, 342–3.

FERGUSON, J. H. A., 1957. Some applications of binomial probability paper in genetic analyses. *Euphytica* **5**, 329–38.

FEWKES, D. W., 1961. Stool size as a factor in the sampling of sugarcane froghopper nymph populations. *J. econ. Ent.* **54**, 771–2.

FINNEY, D. J., 1941. Wireworm populations and their effect on crops. *Ann. appl. Biol.* **28**, 282–95.

FINNEY, D. J., 1973. Transformation of observations for statistical analysis *Cott. Gr. Rev.* **50**, 1–14.

FINNEY, D. J. and VARLEY, G. C., 1955. An example of the truncated Poisson distribution. *Biometrics* **11**, 387–94.

FISHER, R. A., CORBET, A. S. and WILLIAMS, C. B., 1943. The relation between the number of species and the number of individuals in a random sample of an animal population. *J. Anim. Ecol.* **12**, 42–58.

FORD, R. P. and DIMOND, J. B., 1973. Sampling populations of pine leaf chermid *Pirieus pinifoliae* (Homoptera: Chermidae) II. Adult Gallicolae on the secondary host. *Can. Ent.* **105**, 1265–74.

FORSYTHE, H. Y. and GYRISCO, G. G., 1961. Determining the appropriate transformation of data from insect control experiments for use in the analysis of variance. *J. econ. Ent.* **54**, 859–61.

FORSYTHE, H. Y. and GYRISCO, G. G., 1963. The spatial pattern of the pea aphid in alfalfa fields. *J. econ. Ent.*, **56**, 104–7.

FRACKER, S. B. and BRISCHLE, H. A., 1944. Measuring the local distribution of *Ribes*. *Ecology* **25**, 283–303.

GARDEFORS, D. and ORRHAGE, L., 1968. Patchiness of some marine bottom animals. A methodological study. *Oikos* **19**, 311–21.

GAUFIN, A. R., HARRIS, E. K. and WALTER, H. J., 1956. A statistical evaluation of stream bottom sampling data obtained from three standard samples. *Ecology* **37**, 643–8.

GEORGE, D. G., 1974. Dispersion patterns in the zooplankton populations of a eutrophic reservoir. *J. Anim. Ecol.* **43**, 537–51.

GERRARD, D. J. and CHIANG, H. C., 1970. Density estimation of corn rootworm egg populations based upon frequency of occurrence. *Ecology* **51**, 237–45.

GERRARD, D. J. and COOK, R. D., 1972. Inverse binomial sampling as a basis for estimating negative binomial population densities. *Biometrics* **28**, 971–80.

GLASS, L. W. and BOUBJERG, R. V., 1969. Density and dispersion in laboratory populations of caddis fly larvae (Cheumatopsyche, Hydropsychidae). *Ecology* **50**, 1082–4.

GLEASON, H. A., 1922. On the relation between species and area. *Ecology* **3**, 158–62.

GONZALEZ, D., 1970. Sampling as a basis for pest management strategies. *Tall Timbers Confr. Ecological Animal Control by Habitat Management* **2**, 83–101.

GOOD, I. J. and TOULMIN, G. H., 1956. The number of new species, and the increase in population coverage, when a sample is increased. *Biometrika* **43**, 45–63.

GOODALL, D. W., 1952. Quantitative aspects of plant distribution. *Biol. Rev.* **27**, 194–245.

GOULDEN, C. H., 1952. *Methods of statistical analysis* (2nd ed.), Wiley, New York, 467 pp.

GRAHAM, K. and STARK, R. W., 1954. Insect population sampling. *Proc. ent. Soc. B. C.* **51**, 15–20.

GREEN, R. H., 1966. Measurement of non-randomness in spatial distributions. *Res. Popul. Ecol.* **8**, 1–7.

GREEN, R. H., 1970. On fixed precision level sequential sampling. *Res. Popul. Ecol.* **12**, 249–51.

GREGOR, J., 1969. An algorithm for the decomposition of a distribution into Gaussian components. *Biometrics* **25**, 79–93.

GREIG-SMITH, P., 1964. *Quantitative plant ecology* (2nd ed.). 256 pp., Butterworths, London.

GRUNDY, P. M., 1952. The fitting of grouped truncated and grouped censored normal distributions. *Biometrika* **39**, 252–9.

GUPPY, J. C. and HARCOURT, D. G., 1970. Spatial pattern of the immature stages and teneral adults of *Phyllophaga* spp. (Coleoptera: Scarabaeidae) in a permanent meadow. *Can. Ent.* **102**, 1354–9.

HAIRSTON, N. G., 1959. Species abundance and community organisation. *Ecology.* **40**, 404–16.

HAIRSTON, N. G., 1964. Studies on the organization of animal communities. *J. Anim. Ecol.* **33** (suppl.), 227–39.

HANDFORD, R. H., 1956. Grasshopper population sampling. *Proc. ent. Soc. B. C.* (1955) **52**, 3–7.

HANSEN, M. H., HURWITZ, W. N. and MADOW, W.G., 1953. *Sample survey methods and theory.* Vol. 1 Wiley, New York.

HARCOURT, D. G., 1961*a*. Design of a sampling plan for studies on the population dynamics of the diamond back moth, *Plutella maculipennis* (Curt.) (Lepidoptera: Plutellidae). *Can. Ent.* **93**, 820–31.

HARCOURT, D. G., 1961*b*. Spatial pattern of the imported cabbageworm, *Pieris rapae* (L.) (Lepidoptera: Pieridae), on cultivated Cruciferae. *Can. Ent.* **93**, 945–52.

HARCOURT, D. G., 1962. Design of a sampling plan for studies on the population dynamics of the imported cabbageworm, *Pieris rapae* (L.) (Lepidoptera: Pieridae). *Can. Ent.* **94**, 849–59.

HARCOURT, D. G., 1963. Population dynamics of *Leptinotarsa decemlineata* (Say) in Eastern Ontario. I. Spatial pattern and transformation of field counts. *Can. Ent.* **95**, 813–20.

HARCOURT, D. G., 1964. Population dynamics of *Leptinotarsa decemlineata* (Say) in Eastern Ontario. II. Population and mortality estimation during six age intervals. *Can. Ent.* **96**, 1190–8.

HARCOURT, D. G., 1965. Spatial pattern of the cabbage looper, *Trichoplusia ni*, on Crucifers. *Ann. ent. Soc. Am.* **58**, 89–94.

HARCOURT, D. G., 1966. Sequential sampling for the imported cabbageworm, *Pieris rapae* (L.). *Can. Ent.* **98**, 741–6.

HARDING, J. P., 1949. The use of probability paper for the graphical analysis of polymodal frequency distributions. *J. mar. biol. Soc.* **28**, 141–53.

HARRIS, D., 1968. A method of separating two superimposed normal distributions using arithmetic probability paper. *J. Anim. Ecol.* **37**, 315–9.

HARRIS, E. K., 1957. Further results in the statistical analysis of stream sampling. *Ecology* **38**, 463–8.

HARRIS, J. W. E., COLLIS, D. G. and MAGAR, K. M., 1972. Evaluation of the tree-beating method for sampling defoliating forest insects. *Can. Ent.* **104**, 723–9.

HARTLEY, H. O., 1950. The maximum F-ratio as a short-cut test for heterogeneity of variance. *Biometrika* **37**, 308–12.

HASSELL, M. P., 1978. *The Dynamics of Arthropod Predator-Prey Systems.* Princeton Monographs in Population Biology Princeton, New Jersey (in press).

HASSELL, M. P. and MAY, R. M., 1974. Aggregation of predators and insect parasites and its effect on stability. *J. Anim. Ecol.* **43**, 567–94.

HAYMAN, B. I. and LOWE, A. D., 1961. The transformation of counts of the cabbage aphid (*Brevicoryne brassicae* (L.)). *N. Z. J. Sci.* **4**, 271–8.

HEALEY, V., 1964. The density and distribution of two species of *Aptinothrips* (Thysanoptera) in the grass of a woodland. *Entomologist* **97**, 258–63.

HEALY, M. J. R., 1962. Some basic statistical techniques in soil zoology.*In* Murphy, P. W. (ed.), *Progress in soil zoology* pp. 3–9, Butterworths, London.

HEALY, M. J. R. and TAYLOR, L. R., 1962. Tables for power-law transformations. *Biometrika* **49**, 557–9.

HELSON, G. A. H., 1958. Aphid populations: Ecology and methods of sampling aphids *Myzus persicae* (Sulz.) and *Aulacorthum solani* (Kltb.). *N. Z. Entomologist* **2**, 20–3.

HENSON, W. R., 1954. A sampling system for Poplar insects. *Can. J. Zool.* **32**, 421–33.

HENSON, W. R., 1961. Laboratory studies on the adult behaviour of *Conopthorus coniperda* (Schwarz) (Coleoptera: Scolytidae). II. Thigmotropic aggregation. *Ann. ent. Soc. Am.* **54**, 810–19.

HERBERT, H. J. and BUTLER, K. P., 1973. Sampling systems for European red mite, *Panonychus ulmi* (Acarina: Tetranychidae) eggs on apple in Nova Scotia. *Can. Ent.* **105**, 1519–23.

HEYWOOD, J. and EDWARDS, R. W., 1961. Some aspects of the ecology of *Potamopyrgus jenkinsi* Smith. *J. Anim. Ecol.* **31**, 239–50.

HIRATA, S., 1962. Comparative studies on the population dynamics of important Lepidopterous pests on cabbage. 2. On the habits of oviposition of *Pieris rapae crucivora, Plusia nigrisigna* and *Manestra (Barathra) brassicae* on cabbage plants. *Jap. J. appl. Ent. Zool.* **6**, 200–7.

HOLME, N. A., 1950. Population dispersion in *Tellina tenuis* Da Costa. *J. mar. biol. Ass. U.K.* **29**, 267–80.

HOWE, R. W., 1963. The random sampling of cultures of grain weevils. *Bull. ent. Res.* **54**, 135–46.

HUDSON, M. and LEROUX, E. J., 1961. Variation between samples of immature stages, and of mortalities from some factors, of the european corn borer, *Ostrinia nubilalis* (Hübner) (Lepidoptera: Pyralidae) on sweet corn in Quebec. *Can. Ent.* **93**, 867–88.

HUGHES, R. D., 1955. The influence of the prevailing weather on the numbers of *Meromyza variegata* Meigen (Diptera, Chloropidae) caught with a sweepnet. *J. Anim. Ecol.* **24**, 324–35.

HUGHES, R. D., 1962. The study of aggregated populations. *In* Murphy, P. W. (ed.), *Progress in soil zoology*, 51–5, Butterworths, London.

IBARRA, E. L., WALLWORK, J. A. and RODRIGUEZ, J. G., 1965. Ecological studies of mites found in sheep and cattle pastures. 1. Distribution patterns of Oribatid mites. *Ann. ent. Soc. Am.* **58**, 153–9.

ITÔ, Y., GOTOH, A. and MIYASHITA, K., 1960. On the spatial distribution of *Pieris rapae crucivora* population. *Jap. J. appl. Ent. Zool.* **4**, 141–5.

ITÔ, Y., NAKAMURA, M., KONDO, M., MIYASHITA, K. and NAKAMURA, K., 1962. Population dynamics of the chestnut gall-wasp, *Dryocosmus kuriphilus* Yasumatsu (Hymenoptera: Cynipidae). II. Distribution of individuals in bud of chestnut tree. *Res. Popul. Ecol.* **4**, 35–46.

IVES, W. G. H., 1954. Sequential sampling of insect populations. *Forestry Chron.* **30**, 287–91.

IVES, W. G. H., 1955. Effect of moisture on the selection of cocooning sites by the larch sawfly, *Pristiphora erichsonii* (Hartig). *Can. Ent.* **87**, 301–11.

IVES, W. G. H. and PRENTICE, R. M., 1958. A sequential sampling technique for surveys of the larch sawfly. *Can. Ent.* **90**, 331–8.

IVES, W. G. H. and WARREN, G. L., 1965. Sequential sampling for white grubs. *Can. Ent.* **97**, 596–604.

IWAO, S., 1956. The relation between the distribution pattern and the population density of the large twenty-eight-spotted lady beetle, *Epilachna 28-maculata* Motschulsky, in egg-plant field. Pattern of the spatial distribution of insect 6. *Jap. J. Ecol.* **5**, 130–5.

IWAO, S., 1968. A new regression method for analyzing the aggregation pattern of animal populations. *Res. Pop. Ecol.* **10**, 1–20.

IWAO, S., 1970a. Problems of spatial distribution in animal population ecology. *In* Patil, G. P. (ed.), *Random Counts in Biomedical and Social Sciences.* pp. 117–49, Penn State University Press, Philadelphia.

IWAO, S., 1970b. Analysis of spatial patterns in animal populations: progress of research in Japan. *Rev. Pl. Protec. Res.* **3**, 41–54.

IWAO, S., 1970c. Analysis of contagiousness in the action of mortality factors on the western tent caterpillar population by using the $\overset{*}{m}$–m relationship. *Res. Popul. Ecol.* **12**, 100–10.

IWAO, S., 1972. Application of the $\overset{*}{m}$–m method to the analysis of spatial patterns by changing the quadrat size. *Res. Popul. Ecol.* **14**, 97–128.

IWAO, S., 1975. A new method of sequential sampling to classify populations relative to a critical density. *Res. Popul. Ecol.* **16**, 281–8.

IWAO, S. and KUNO, E., 1971. An approach to the analysis of aggregation pattern in biological populations. *In* Patil, G. P., Pielou, E. C. & Waters, W. E. (Eds.) *Statistical Ecology* 1: Spatial Patterns & Statistical Distributions. Penn State University Press, Philadelphia.

JEPSON, W. F. and SOUTHWOOD, T. R. E., 1958. Population studies on *Oscinella frit* L. *Ann. appl. Biol.* **46**, 465–74.

KAPATOS, E., McFADDEN, M. W. and PAPPAS, S., 1977. Sampling techniques and preparation of partial life tables for the olive fly, *Dacus oleae* (Diptera: Trypetidae) in Corfu. *Ecol. Ent.* **2**, 193–6.

KARANDINOS, M. G., 1976. Optimum sample size and comments on some published formulae. *Bull. Ent. Soc. Am.* **22**, 417–21.

KATTI, S. K., 1966. Interrelations among generalized distributions and their components. *Biometrics* **22**, 44–52.

KATTI, S. K. and GURLAND, J., 1962. Efficiency of certain methods of estimation for the negative binomial and the Neyman type A distributions. *Biometrika* **49**, 215–26.

KENDALL, D. G., 1948. On some modes of population growth leading to R. A. Fisher's logarithmic series distribution. *Biometrika* **35**, 6–15.

KENNEDY, J. S. and CRAWLEY, L., 1967. Spaced-out gregariousness in sycamore aphids *Drepanosiphum platanoides* (Schrank) (Hemiptera Callaphididae). *J. Anim. Ecol.* **36**, 147–70.

KEULS, M., OVER, H. J. and DE WIT, C. T., 1963. The distance method for estimating densities. *Statistica Neerlandica* **17**, 71–91.

KIECKHEFER, R. W., 1975. Field populations of cereal aphids in South Dakota spring grains. *J. econ. Ent.* **68**, 161–4.

KLECZKOWSKI, A., 1949. The transformation of local lesion counts for statistical analysis. *Ann. appl. Biol.* **36**, 139–52.

KOBAYASHI, S., 1966. Process generating the distribution pattern of eggs of the common cabbage butterfly, *Pieris rapae crucivora. Res. popul. Ecol.* **8**, 51–60.

KONO, T., 1953. Basic unit of population observed in the distribution of the rice-stem borer, *Chilo simplex*, in a paddy field. *Res. Popul. Ecol.* **2**, 95–105.

KRAUSE, G. F. and PEDERSEN, J. R., 1960. Estimating immature populations of rice weevils in wheat by using subsamples. *J. econ. Ent.* **53**, 215–6.

KUEHL, R. O. and FYE, R. E., 1972. An analysis of the sampling distributions of cotton insects in Arizona. *J. econ. Ent.* **65**, 855–60.

KUNO, E., 1963. A comparative analysis on the distribution of nymphal populations of some leaf- and planthoppers on rice plant. *Res. Popul. Ecol.* **5**, 31–43.

KUNO, E., 1969. A new method of sequential sampling to obtain the population estimates with a fixed level of precision. *Res. Popul. Ecol.* **11**, 127–36.

LAMB, K. P., 1958. Aphid sampling. *N.Z. Entomologist* **2**, 6–11.

LEGAY, J. M., 1963. A propos de la répartition de la cecidomyie du Hêtre, *Mikiola fagi*. Un exemple de distribution binomiale négative. *Ann. Epiphyt. C* **14**, 49–56.

LEROUX, E. J., 1961. Variations between samples of fruit, and of fruit damage mainly from insect pests, on apple in Quebec. *Can. Ent.* **93**, 680–94.

LEROUX, E. J. and REIMER, C., 1959. Variation between samples of immature stages and of mortalities from some factors, of the Eye-spotted Bud Moth, *Spilonota ocellana* (D. & S.) (Lepidoptera: Olethreutidae), and the Pistol Casebearer, *Coleophora serratella* (L.) (Lepidoptera: Coleophoridae), on apple in Quebec. *Can. Ent.* **91**, 428–49.

LLOYD, M., 1967. Mean crowding. *J. Anim. Ecol.* **36**, 1–30.

LYONS, L. A., 1964. The spatial distribution of two pine sawflies and methods of sampling for the study of population dynamics. *Can. Ent.* **96**, 1373–407.

MACLELLAN, C. R., 1962. Mortality of codling moth eggs and young larvae in an integrated control orchard. *Can. Ent.* **94**, 655–66.

MACLEOD, J., 1958 The estimation of numbers of mobile insects from low incidence recapture data. *Trans. R. ent. Soc. Lond.* **110**, 363–92.

MAY, R. M., 1978. Host–parasitoid systems in a patchy environment: a phenomenological study. *J. Anim. Ecol.* **47** (in press).

MILNE, A., 1959. The centric systematic area-sample treated as a random sample. *Biometrics* **15**, 270–97.

MILNE, A., 1964. Biology and ecology of the garden chafer, *Phyllopertha horticola* (L.). IX. Spatial distribution. *Bull. ent. Res.* **54**, 761–95.

MORISITA, M., 1954. Estimation of population density by spacing method. *Mem. Fac. Sci. Kyushu Univ. E* **1**, 187–97.

MORISITA, M., 1959. Measuring of the dispersion of individuals and analysis of the distributional patterns. *Mem. Fac. Sci. Kyushu Univ. E (Biol.)* **2**, 215–35.

MORISITA, M., 1962. I_δ-index, a measure of dispersion of individuals. *Res. Popul. Ecol.* **4**, 1–7.

MORISITA, M., 1964. Application of I_δ-index to sampling techniques. *Res. Popul. Ecol.* **6**, 43–53.

MORISITA, M., 1971. Composition of the I_δ-index. *Res. Popul. Ecol.* **13**, 1–27.

MORRIS, R. F., 1954. A sequential sampling technique for spruce budworm egg surveys. *Can. J. Zool.* **32**, 302–13.

MORRIS, R. F., 1955. The development of sampling techniques for forest insect defoliators, with particular reference to the spruce budworm. *Can. J. Zool.* **33**, 225–94.

MORRIS, R. F., 1960. Sampling insect populations. *A. Rev. Ent.* **5**, 243–64.

MORRIS, R. F. and REEKS, W. A., 1954. A larval population technique for the winter moth, *Operophtera brumata* (Linn.) (Lepidoptera: Geometridae). *Can. Ent.* **86**, 433–8.

MOSTELLER, F. and TUKEY, J. W., 1949. The uses and usefulness of binomial probability paper. *J. Am. Stats. Assoc.* **44**, 174–212.

MUKERJI, M. K., 1973. The development of sampling techniques for populations of the tarnished plant bug, *Lygus lineolaris* (Hemiptera: Miridae.) *Res. Popul. Ecol.* **15**, 50–63.

MUKERJI, M. K. and HARCOURT, D. G., 1970. Spatial pattern of the immature stages of *Hylemya brassicae* on cabbage. *Can. Ent.* **102**, 1216–22.

MURDIE, G. and HASSELL, M. P., 1973. Food distribution, searching success and predator-prey models. *In* Bartlett, M. S. & Hiorns, R. W. (eds). *The Mathematical Theory of the Dynamics of Biological Populations*, pp. 87–101. Academic Press, London.

NAYLOR, A. F., 1959. An experimental analysis of dispersal in the flour beetle, *Tribolium confusum. Ecology* **40**, 453–65.

NELSON, W. A., SLEN, S. B. and BANKY, E. C., 1957. Evaluation of methods of estimating populations of the sheep ked, *Melophagus ovinus* (L.) (Diptera: Hippoboscidae), on mature ewes and young lambs. *Can. J. Anim. Sci.* **37**, 8–13.

NEYMAN, J., 1939. On a new class of 'contagious' distributions, applicable in entomology and bacteriology. *Ann. Math. Stat.* **10**, 35–57.

NIELSEN, B. OVERGAARD, 1963. The biting midges of Lyngby Aamose (Culicoides: Ceratopogonidae). *Natura Jutlandica* **10**, 46 pp.

OAKLAND, G. B., 1950. An application of sequential analysis to whitefish sampling. *Biometrics* **6**, 59–67.

OAKLAND, G. B., 1953. Determining sample size. *Can. Ent.* **85,** 108–13.

ONSAGER, J. A., LANDIS, B. J. and FOX, L., 1975. Efficacy of fonofos band treatments and a sampling plan for estimating wireworm populations on potatoes. *J. econ. Ent.* **68,** 199–202.

PAHL, P. J., 1969. On testing for goodness-of-fit of the negative binomial distribution when expectations are small. *Biometrics* **25,** 143–51.

PARADIS, R. O. and LEROUX, E. J., 1962. A sampling technique for population and mortality factors of the fruit-tree leaf roller, *Archips argyrospilus* (Wlk.) (Lepidoptera: Tortricidae), on apple in Quebec. *Can. Ent.* **94,** 561–73.

PATIL, G. P. and JOSHI, S. W., 1968. *A dictionary and bibliography of discrete distributions.* Oliver & Boyd, Edinburgh.

PATIL, G. P. and STITELER, W. M., 1974. Concepts of aggregation and their quantification: a critical review with some new results and applications. *Res. Popul. Ecol.* **15,** 238–54.

PEARSON, E. S. and HARTLEY, H. O., 1958. *Biometrika tables for statisticians.* Vol. 1. Cambridge University Press, Cambridge, 240 pp.

PIELOU, E. C., 1969. *An Introduction to Mathematical Ecology.* 286 pp, Wiley-Interscience, New York and London.

PIETERS, E. P. and STERLING, W., 1974. A sequential sampling plan for the cotton fleahopper, *Pseudatomoscelis seriatus. Environ. Ent.* **3,** 102–6.

PRADHAN, S. and MENON, R., 1945. Insect population studies. I. Distribution and sampling of spotted bollworm of cotton. *Proc. Nat. Inst. Sci. India* **6** (2), 61–73.

PREBBLE, M. L., 1943. Sampling methods in population studies of the European spruce sawfly, *Gilpinia hercyniae* (Hartig.), in Eastern Canada. *Trans. R. Soc. Canada* III, V, **37,** 93–126.

PRESTON, F. W., 1948. The commonness, and rarity, of species. *Ecology* **29,** 254–83.

PUTNAM, L. G. and SHKLOV, N. 1956. Observations on the distribution of grasshopper egg-pods in Western Canadian stubble fields. *Can. Ent.* **88,** 110–17.

QUENOUILLE, M. H., 1949. A relation between the logarithmic, Poisson, and negative binomial series. *Biometrics* **5,** 162–4.

REEKS, W. A., 1956. Sequential sampling of the winter moth, *Operophtera brumata* (Linn.). *Can. Ent.* **88,** 241–6.

REIMER, C., 1959. Statistical analysis of percentages based on unequal numbers, with examples from entomological research. *Can. Ent.* **91,** 88–92.

RENNISON, B. D., 1962. A method of sampling *Antestiopsis* in arabsia coffee in chemical control schemes. *E. Afr. agric. For. J.* **27,** 197–200.

RICHARDS, O. W. and WALOFF, N., 1961. A study of a natural population of *Phytodecta olivacea* (Forster) (Coleoptera, Chrysomeloidea). *Phil. Trans.* B. **244,** 205–57.

ROJAS, B. A., 1964. La binomial negativa y la estimación de intensidad de plagas en el suelo. *Fitotecnia Latinamer.* **1,** (1), 27–36.

ROMELL, L. G., 1930. Comments on Raunkiaer's and similar methods of vegetation analysis and the 'law of frequency'. *Ecology* **11,** 589–96.

SAFRANYIK, L. and GRAHAM, K., 1971. Edge-effect bias in the sampling of sub-cortical insects. *Can. Ent.* **103,** 240–55.

SALT, G. and HOLLICK, F. S., 1946. Studies of wireworm populations. II. Spatial distribution, *J. exp. Biol.* **23,** 1–46.

SEBER, G. A. F., 1973. *The Estimation of Animal Abundance and Related Parameters.* 506 pp, Griffin, London.

SHAW, M. W., 1955. Preliminary studies on potato aphids in north and north-east Scotland. *Ann. appl. Biol.* **43,** 37–50.

SHENTON, L. R. and WALLINGTON, P. A., 1962. The bias of moment estimators with an application to the negative binomial distribution. *Biometrika* **49,** 193–204.

SHIBUYA, M. and OUCHI, Y., 1955. Pattern of spatial distribution of soy bean pod gall midge in a soy bean field. *Ôyô-Kontya* **11**, 91–7.

SHINOZAKI, K. and URATA, N., 1953. Apparent abundance of different species and heterogeneity. *Res. Popul. Ecol.* **2**, 8–21.

SHIYOMI, M. and NAKAMURA, K., 1964. Experimental studies on the distribution of the aphid counts. *Res. Popul. Ecol.* **6**, 79–87.

SKELLAM, J. G., 1958. On the derivation and applicability of Neyman's type A distribution. *Biometrika* **45**, 32–6.

SOUTHWOOD, T. R. E. and CROSS, D. J., 1969. The ecology of the partridge. III. Breeding success and the abundance of insects in natural habitats. *J. Anim. Ecol.* **38**, 497–509.

SOUTHWOOD, T. R. E. and JEPSON, W. F., 1961. The frit fly – a denizen of grassland and a pest of oats. *Ann. appl. Biol.* **49**, 556.

SOUTHWOOD, T. R. E., JEPSON, W. F. and EMDEN, H. F. VAN, 1961. Studies on the behaviour of *Oscinella frit* L. (Diptera) adults of the panicle generation. *Entomologia exp. appl.* **4**, 196–210.

SPILLER, D., 1948. Truncated log-normal and root-normal frequency distributions of insect populations. *Nature* **162**, 530.

SPILLER, D., 1952. Truncated log-normal distribution of red scale (*Aonidiella aurantii* Mask.) on citrus leaves. *N.Z. J. Sci. Tech. (B)* **33**, 483–7.

STARK, R. W., 1952a. Analysis of a population sampling method for the lodgepole needle miner in Canadian Rocky Mountain Parks. *Can. Ent.* **84**, 316–21.

STARK, R. W., 1952b. Sequential sampling of the lodgepole needle miner. *Forestry Chron.* **28**, 57–60.

STARK, R. W. and DAHLSTEN, D. L., 1961. Distribution of cocoons of a *Neodiprion* sawfly under open-grown conditions. *Can. Ent.* **93**, 443–50.

STERLING, W. L. and PIETERS, E. P., 1974. A sequential sampling package for key cotton arthropods in Texas. *Texas Agric. Exp. Stn. Dept. Tech. Rep.* **74–32**, 1–28.

STERLING, W. L. and PIETERS, E. P., 1975. Sequential sampling for key arthropods of cotton. *Texas Agric. Exp. Stn. Dept. Tech. Rep.* **75–124** 1–21.

STEVENS, R. E. and STARK, R. W., 1962. Sequential sampling for the lodgepole needle miner, *Evagora milleri*. *J. econ. Ent.* **55**, 491–4.

STIMSON, J., 1974. An analysis of the pattern of dispersion of the hermatypic coral, *Pocillopora meandrina* var *nobilis* Verrill. *Ecology*, **55**, 445–9.

STRICKLAND, A. H., 1961. Sampling crop pests and their hosts. *A. Rev. Ent.* **6**, 201–20.

STUART, A., 1962. *Basic ideas of scientific sampling*, 99 pp. Griffin, London.

SYLVESTER, E. S. and COX, E. L., 1961. Sequential plans for sampling aphids on sugar beets in Kern County, California. *J. econ. Ent.* **54**, 1080–5.

TAKEDA, S. and HUKUSIMA, S., 1961. Spatial distribution of the pear lace bugs, *Stephanitis naski* Esaki et Takeya (Hemiptera: Tingitidae) in an apple tree and an attempt for estimating their populations. *Res. Bull. Fac. Agric., Gifu Univ.* **14**, 68–77.

TANIGOSHI, L. K., BROWNE, R. W. and HOYT, S. C., 1975. A study on the dispersion pattern and foliage injury by *Tetranychus medanieli* (Acarina: Tetranychidae) in simple apple ecosystems. *Can. Ent.* **107**, 439–46.

TAYLOR, L. R., 1961. Aggregation, variance and the mean. *Nature* **189**, 732–5.

TAYLOR, L. R., 1965. A natural law for the spatial disposition of insects. *Proc. XII int. Congr. Ent.* 396–7.

TAYLOR, L. R., 1970. Aggregation and the transformation of counts of *Aphis fabae* Scop. on beans. *Ann. appl. Biol.* **65**, 181–9.

TAYLOR, L. R., 1971. Aggregation as a species characteristic. *In* Patil G. P., Pielou, E. C. & Waters, W. E. (eds) *Statistical Ecology* **1**, pp. 357–77. Penn State University Press, Philadelphia.

TAYLOR, L. R., WOIWOD, I. P. and PERRY, J. N., 1978. The density-dependence of spatial behaviour and the variety of randomness, *J. Anim. Ecol.* **47**, 383–406.

TEYROVSKY, V., 1956. Fotopathie larey klestánek (Corixinae). *Acta Univ. agric. silv. Brunn.* **2**, 147–77.

THOMAS, I. and JACOB, F. H., 1943. Ecology of potato aphids in north Wales. *Ann. appl. Biol.* **30**, 97–101.

THOMAS, M., 1949. A generalization of Poisson's binomial limit for use in ecology. *Biometrika* **36**, 18–25.

THOMPSON, H. R., 1956. Distribution of distance to *n*th neighbour in a population of randomly distributed individuals. *Ecology* **37**, 391–4.

TUKEY, J. W., 1949. One degree of freedom for non-additivity. *Biometrics* **5**, 232–42.

TUNSTALL, J. P. and MATTHEWS, G. A., 1961. Cotton insect control recommendations for 1961–2 in the Federation of Rhodesia and Nyasaland. *Rhodesia Agric. J.* **58** (5), 289–99.

TURNER, F. B., 1960. Size and dispersion of a Louisiana population of the cricket frog, *Acris gryllus. Ecology* **41**, 258–68.

UNDERWOOD, A. J., 1976. Nearest neighbour analysis of spatial dispersion of intertidal Prosobrunel Gastropods within two substrata. *Oecologia* **26**, 257–66.

UNTERSTENHÖFER, G., 1957. The basic principles of plant protection field tests. *Höfchen Briefe* **10** (4) (English ed.), 173–236.

UPHOLT, W. M. and CRAIG, R., 1940. A note on the frequency distribution of black scale insects. *J. econ. Ent.* **33** (1), 113–14.

USHER, M. B., 1971. Properties of the aggregations of soil arthropods, particularly Mesostigmata (Acarina). *Oikos* **22**, 43–9.

WADLEY, F. M., 1950. Notes on the form of distribution of insect and plant populations. *Ann. ent. Soc. Am.* **43**, 581–6.

WADLEY, F M., 1952. Elementary sampling principles in entomology. *U.S.D.A. Pl. Quar. Bur. Ent. E.T.* **302**, 17 pp.

WALD, A., 1948. *Sequential sampling*. Wiley & Sons, New York.

WALOFF, N. and BLACKITH, R. E., 1962. The growth and distribution of the mounds of *Lasius flavus* (Fabricius) (Hym.: Formicidae) in Silwood Park, Berkshire, *J. Anim. Ecol.* **31**, 421–37.

WATERS, W. E., 1955. Sequential sampling in forest insect surveys. *For. Sci.* **1**, 68–79.

WATERS, W. E., 1959. A quantitative measure of aggregation in insects. *J. econ. Ent.* **52**, 1180–4.

WATERS, W. E. and HENSON, W. R., 1959. Some sampling attributes of the negative binomial distribution with special reference to forest insects. *For. Sci.* **5**, (4), 397–412.

WATT, A. S., 1964. The community and the individual. *J. Ecol.* **52** (suppl.), 203–11.

WILSON, L. F., 1959. Branch 'tip' sampling for determining abundance of spruce budworm egg masses. *J. econ. Ent.* **52**, 618–21.

WILSON, L. F. and GERRARD, D. J., 1971. A new procedure for rapidly estimating european pine sawfly (Hymenoptera: Diprionidae) population levels in young pine plantations. *Can. Ent.* **103**, 1315–22.

YATES, F., 1953. *Sampling methods for censuses and surveys*. Griffin, London.

YATES, F. and FINNEY, D. J., 1942. Statistical problems in field sampling for wireworms. *Ann. appl. Biol.* **29**, 156–67.

YOSHIHARA, T., 1953. On the distribution of *Tectarius granularis. Res. Popul. Ecol.* **2**, 112–22.

YOUDEOWEI, A., 1965. A note on the spatial distribution of the cocoa mirid *Sahlbergella singularis* Hagl. in a cocoa farm in Western Nigeria. *J. Agric. Soc. Nigeria* **2**, 66–7.

ZAHL, S., 1974. Application of the S-method to the analysis of spatial pattern. *Biometrics* **30**, 513–24.

3

Absolute Population Estimates using Marking Techniques

Studying plaice and waterfowl populations respectively, Petersen and Lincoln independently developed a marking method from which the total population may be estimated (Le Cren, 1965). This was based on the principle that if a proportion of the population was marked in some way, returned to the original population and then, after complete mixing, a second sample was taken, the number of marked individuals in the second sample would have the same ratio to the total numbers in the second sample as the total of marked individuals originally released would have to the total population. As the first three quantities were known the latter could easily be calculated. This method has been extensively developed and provides the major alternative absolute method to those based on the count of animals within a fixed unit of the habitat; it has the advantage that its accuracy does not depend on an assessment of the number of sampling units in the habitat and, as has already been stressed (Chapter 1), it is a wise practice to use more than one method simultaneously. There are also certain other methods of estimating a population that depend on the presence of marked individuals, but use a different principle to the Lincoln Index. A comprehensive survey of these methods is provided by Seber (1973).

A basic prerequisite to the use of these methods of population estimation is a technique for marking the animals so that they can be released unharmed and unaffected into the wild and recognized again on recapture. Such techniques may also be used in studies on behaviour, e.g. dispersal, longevity and growth; but for convenience all aspects of marking insects and other invertebrates will be discussed here.

3.1 Methods of marking animals

A fundamental requirement of any marking technique is that it shall not affect the longevity or behaviour of the animals. An attempt should always be made to confirm that this is true in the particular case under investigation because, for example, although the pigments used in most markers may be non-toxic, the solvents are often toxic. This may be checked in the laboratory or field cage by keeping samples of living marked and unmarked individuals and comparing longevity (e.g. Pal, 1947; Crumpacker, 1974; LaBrecque, Bailey,

70

Meifert & Weidhaas, 1975) or in the field by comparing the longevity of individuals bearing differing numbers of marks (e.g. Richards & Waloff, 1954; Dobson & Morris, 1961). Newly emerged insects may be more sensitive to the toxic substances used in markers than older insects (Jackson, 1948; Dobson, Stephenson & Lofty, 1958) and the attachment of labels to their wings, which has no effect on old insects, may cause distortion due to interference with the blood circulation (Waloff, 1963).

It should also be borne in mind that the presence of conspicuous marks may well destroy an animal's natural camouflage and make it more liable or, as Hartley (1954) observed with marked snails, less liable to predation. This effect is difficult to assess; it can to some extent be avoided by marking in inconspicuous places or by the use of fluorescent powders (Pal, 1947), dyes in powder form (Quarterman, Mathis & Kilpatrick, 1954), phenolphthalein solution (Peffly & Labrecque, 1956) or radioactive tracers whose presence is only detectable by the use of a special technique after recapture. The effect on predation of a conspicuous, but convenient, marking method could be measured by marking further individuals with one of these invisible methods and comparing longevity. The effect of a conspicuous mark can be checked through choice experiments in the laboratory: Buckner (1968) confirmed that small mammals are not influenced if their prey was stained with vital dyes.

Another aspect of the conspicuous mark is that, where the animals are sampled by a method that relies on the sight of the collector, then the marked individuals may tend to be collected more than unmarked ones (Edwards, 1958).

A third problem concerns the durability of the mark; some paints, particularly cellulose lacquers, may flake off leaving the animal apparently unmarked; student's oil paints and powdered dyes may wash off; some fluorescent powders may lose this property on exposure to sunlight (Polivka, 1949) or be abraded during collection (Dow, 1971) and a radioactive isotope could decay and/or be excreted by the animal. Immature animals will lose marks on their cuticle when they moult. Laboratory tests of durability are not always reliable; Blinn (1963) found that cellulose lacquers would remain on the shells of land snails for two years in the laboratory, but they only lasted about one in the field, and the rate at which a radioactive isotope is lost may depend upon the animal's diet (p. 317) and other factors (p. 466).

The handling and release of marked individuals may also affect their subsequent life expectancy and behaviour; these problems are discussed below.

The amount of effort that can be put into a marking programme (i.e. its cost) will be related to the percentage of recoveries that can be expected; a high cost per individual marked will be justified where the recovery rate is high.

3.1.1 Group marking methods
These methods enable a large number of animals to be marked in the same

way and are perfectly adequate for most capture–recapture population estimations and in dispersal studies. Almost all the methods are capable of one or two variants so that two or three groups may be marked differently, and thus the distinction made by Dobson (1962) between group methods and common marking methods in which all individuals are marked in the same way is not followed. Marking methods have also been reviewed by Dobson (1962) and by Gangwere, Chavin & Evans (1964), who give further references.

PAINTS AND SOLUTIONS OF DYES

Materials

Artist's oil paint is perhaps the most extensively used marking material; it can, of course, be obtained in a variety of colours and has been used successfully for marking moths (Collins & Potts, 1932), tsetse flies (Scott, 1931; Jackson, 1933b), bed bugs (Mellanby, 1939), locusts and grasshoppers (Richards, 1953; Richards & Waloff, 1954), mirids (Muir, 1958), flies (Cragg & Hobart, 1955; Dobson, Stephenson & Lofty, 1958), beetles (Mitchell, 1963) and others. However, Davey (1956) found them toxic to certain locusts, although this may have been the effect of the dilutant and they are, of course, slow drying. Artist's poster paints have been used to mark mosquitoes (Gillies, 1961).

Nitrocellulose lacquers or paints (e.g. model aircraft dope) and alkyl vinyl resin paints are quick drying and have been used by a number of workers; on snails, where they were applied to a small area on the underside of the shell, from which the periostracum had been scraped (Sheppard, 1951), and on grass hoppers (Richards & Waloff, 1954; Clark, 1962), although the first-named authors found them less satisfactory than artist's oil paints. They have also been applied to ants (Holt, 1955), lace bugs (Southwood & Scudder, 1956), dragonflies (Corbet, 1952; Pajunen, 1962), mites (Hunter, 1960), tipulid flies (Freeman, 1964) and carabid beetles (Greenslade, 1964) and mosquitoes (Sheppard, Macdonald, Tonn & Grab, 1969). Fluorescent lacquer enamels (e.g. 'Glo-craft' or 'Dayglo') or fluorescent pigments with gum arabic glue plus a trace of detergent have been used to mark tsetse flies (Jewell, 1956, 1958; McDonald, 1960), chafer beetles (Evans & Gyrisco, 1960) and lepidopterous caterpillars (Wood, 1963). Animals marked in these ways may be spotted after dark in the field at distances of up to 25–30 ft (8–10 m) by the use of a beam of ultraviolet light; such a beam may be produced by a battery-powered lamp (McDonald, 1960).

Reflecting paints may also be used to mark animals for detection at night and have the advantage that they can be seen with a small (12-V) hand torch for up to 30 ft (10 m); Rennison, Lumsden & Webb (1958) used this method with tsetse flies and found that a mixture of the minute glass beads in a thin varnish of shellac or in a gum solution with a trace of detergent was preferable to the commercial aerosol form of the paint (e.g. Codit, 7211, reflecting paint).

Aluminium paint was found to adhere well to scraped areas of the elytra of carabid beetles (Murdoch, 1963).

Aniline dyes dissolved in a mixture of alcohol and shellac or in alcohol alone have been used, respectively, to mark cucumber beetles, *Diabrotica* (Dudley & Searles, 1923) and gypsy moths, *Lymantria* (Collins & Potts, 1932). Solutions of various stains in alcohol, such as eosin, orange G and Congo red, have been used to mark adult Lepidoptera (Meder, 1926; Yetter & Steiner, 1932; Leeuwen, 1940; Nielsen, 1961); petroleum based inks (e.g. Easterbrook Flowmaster) have been found particularly useful, staining the integument below the scales (Wolf & Stimmann, 1972). Fluorescent dye solutions, in alcohol or acetone, principally rhodamine B, have been used to mark mosquitoes (Chang, 1946) and *Drosophila* (Wave, Henneberry & Mason, 1963) and the tick, *Argus* (Medley & Ahrens, 1968). Working on house flies Peffly & Labrecque (1956) used a 6 % solution of phenolphthalein in acetone; the marked flies were identified on recapture by placing them in 1 % sodium hydroxide solution, whereupon they became purple. Fales *et al.* (1964) used waterproof inks to mark face flies.

Application

When the paints or solutions are in their most concentrated form they are most conveniently applied by the use of an entomological pin, a sharpened match-stick, a single bristle, or even a fine dry grass stem (Jackson, 1933*b*; Mellanby, 1939; Corbet, 1952; Muir, 1958; Hunter, 1960). With quick-drying cellulose lacquer it may be necessary to dilute them slightly with acetone or another solvent; if this is not done a fine skin may form over the droplet on the pin and it will not adhere firmly to the animal. With artist's oil paints, dyes in solution or diluted lacquers a camel-hair brush may be used (Dudley & Searles, 1923; Wood, 1963), although generally this method has no advantage over the use of a pin and frequently leads to the application of too large a mark. If the mark covers any of the sense organs or joints the specimen will have to be discarded. Freeman (1964) found a fine syringe was suitable for applying cellulose lacquers to tipulid flies. Felt tip pens provide a convenient way for marking some large insects (Iwao, Kiritani & Hokyo, 1966).

The paints or solutions may be further diluted with acetone, dilute alcohol or other solvents and sprayed on. This may be done with a hand atomizer (e.g. a nasal spray) while the insects are contained in a small wire cage (Dudley & Searles, 1923; Leeuwen, 1940; Davey & O'Rourke, 1951; Evans & Gyrisco, 1960); mortality during marking by this method can be reduced if, immediately after spraying, the insects are quickly dried in the draught from an electric fan (Yetter & Steiner, 1932; Leeuwen, 1940). Large numbers of moths may be rapidly marked, the number marked being recorded by a cyclone transfer machine and associated photoelectric cell described by Wolf & Stimman, (1972). This technique has been extended to field marking of locusts (Davey, 1956) and butterflies (Nielsen, 1961) by the use of a spray gun and an oil can (e.g. 'Plews Oiler') respectively; it was found that individuals could be marked at a distance of 15 feet (5 m) or more. Davey showed that for the same cost (labour and time) nearly ten times as many locusts could be marked by this method than by that involving the capture and handling of each individual, and Davey & O'Rourke (1951) found that handling itself could have fatal effects on tabanid flies, *Chrysops*.

DYES AND FLUORESCENT SUBSTANCES IN POWDER FORM

Hairy insects may be marked by dusting them with various dyes in powder form; this is most easily done by applying the dusts from a powder dispenser (insufflator) or by producing a dust storm in a cage with a jet of air [e.g. produced by a bicycle pump (Frankie, 1973)]. Apparatus that could be used for this purpose is described by Dunn & Mechalas (1963) and Frankie (1973). Only a very small quantity of powder is necessary.

Non-fluorescent dyes that have been found useful are the rotor and waxoline group. The marked insects are recognized by laying them on a piece of white filter paper and dropping acetone on to them, when a coloured spot or ring forms beneath those that have been marked. As the testing involves the killing of insects this method is not suitable for extensive recapture work, and laboratory tests showed that with blowflies the mark may only last for one week and seldom for more than two (MacLeod & Donnelly, 1957). However, dyes of two different colours may be applied to the same insect and distinguished in the spotting. This method has been used for calypterate flies by Schoof & Mail (1953), Quarterman, Mathis & Kilpatrick (1954) and MacLeod & Donnelly (1957) and for the frit fly (*Oscinella frit*) (Southwood & Jepson, unpublished) and various mirids (J. P. Dempster, unpublished). It is possible that under certain circumstances it could be used for marking large aggregated populations in the field.

Fluorescent substances (e.g. zinc sulphide powders, especially 'Helecon' nos. 1757, 1953, 2200, 2225, 2267 & 3206), whose presence is detected by placing the animals under an UV lamp, have also been used extensively for marking (Zukel, 1945; Pal, 1947; Polivka, 1949; Taft & Agee, 1962 Stern & Mueller, 1968; Foott, 1976). Multiple marking is possible, although combinations sometimes produce distinct fluorescences (Moth & Baker, 1975), and because they may be detected without killing the insect, they are suitable for use in capture-recapture population estimation (Crumpacker, 1974). Although with the stable fly, *Stomoxys*, field cage experiments suggested that they could adversely affect longevity (LaBrecque *et al.*, 1975), this was not detected in trails with *Drosophila* (Moth & Barker, 1975). Lists of the various materials that may be used are given by Staniland (1959) and Bailey, Eliason & Iltis (1962). They may be applied directly, but better adhesion can be obtained by mixing one part of the dye with six parts of gum arabic, adding water until a paste is formed, then drying the paste and pulverizing it in a mortar. This powder is applied to the insects which are then placed in a high humidity; the gum arabic particles absorb sufficient moisture to make them adhere to the insect. This method has been used for marking mosquitoes by Reeves, Brookman & Hammon (1948). The movements of foraging bees have been studied by marking them with a fluorescent powder as they leave the hive; this is conveniently done by forcing them to walk between two strips of velveteen liberally dusted with the marker (Smith & Townsend, 1951); tabanids caught in a canopy trap (p. 260) have been self-marked a similar way (Sheppard *et al.*, al., 1973). Bees have also been marked when visiting flowers by dusting these with a mixture of the fluorescent powder and a carrier such as talc or lycopodium dust. The bees leave a trail of powder, which can be detected after dark with an UV lamp, on the other flowers they have visited (Musgrave, 1949, 1950; Smith, 1958; Johansson, 1959). Fluorescein and rhodamine B have been founded useful in this work; all the bees leaving the hive are marked and the marks last for weeks (Smith & Townsend, 1951). The possibility that unmarked individuals may bear a few particles that will fluoresce under UV light should be remembered when using these markers. Wild caught mosquitoes have been found with fluorescent blue, purple, green, white, yellow and orange spots (Reeves *et al.*, 1948), therefore the use of rhodamine B, which fluoresces red, was recommended.

An ingenious self-marking method for newly emerged calypterate flies has been devised by Norris (1957). The principle is that the soil or medium which contains the fly puparia is covered with a mixture of about forty parts to one of fine sand and fluorescent powder or the puparia are coated with dye; as the flies emerge a small quantity of the dust adheres to the ptilinum; and when after emergence the ptilinum is retracted the dust becomes lodged in the ptilinal suture. Sometimes in an examination of the faces of such marked flies in UV light this suture will be seen to shine vividly; but a more reliable technique is to crush the whole head on a filter paper, at the same time adding a small amount of the appropriate solvent: the mark may be seen on the paper, using UV if necessary (Steiner, 1965). Many types of fluorescent powder have been found satisfactory: 'Lumogen' (Norris, 1957), 'Tinopal' (Schroeder *et al.*, 1972) and 'Day-glo' (van Dinther, 1972; Sheppard *et al.* 1973). Oil soluble dyes, e.g. Calco red, have also been used with fruit flies (Steiner, 1965).

LABELS

Bands and rings are used extensively in work on birds and mammals (Cottam, 1956; Taber, 1956), but the small size of most insects usually precludes these convenient methods. Butterflies and locusts have, however, been marked by attaching small labels with a word or a code written in waterproof black ink to part of their wings; in the Lepidoptera the area should first be denuded of scales. Earlier workers used paper or cellophane stuck on with an adhesive, such as 'Durofix', but recently 'Sellotape' has been found satisfactory (Fletcher, 1936; Williams *et al.*, 1942; Urquart, 1958; Wojtusiak, 1958; Roer, 1959, 1962; Waloff, 1963). Klock, Pimentel & Stenburg (1953) devised a machine that glued lengths of coloured thread to anaesthetized flies. Punched ferrous labels have been used on bees; they may be magnetically removed (Gary, 1971). It is possible that large-bodied insects might be tagged internally using the methods of fishery workers (Lindroth, 1953). Great progress has also been made with vertebrates through the use of radio-telemetry (e.g. Bartholomew, 1967); developments in electronics may make it possible for the largest insects to carry a transmitter so this technique could be used.

MUTILATION

This method is also more widely used with vertebrates, especially fish, amphibians and reptiles (Ricker, 1956; Woodbury, 1956), than with the smaller insects, for a mark in order to be easily visible may be proportionally so large as to affect the insect's behaviour. Lepidoptera have been marked by clipping their wings (Querci, 1936), carabid beetles by damaging their elytra in various ways: incising the edges (Grüm, 1959), punching or burning small holes (Skuhravý, 1957; Schjøtz-Christensen, 1961) or by scraping away the surface of the elytra between certain striae (Murdoch, 1963); the latter method can be used with large chrysomelids (Southwood, unpublished). Crabs have been marked by cutting some of the teeth on the carapace (Edwards, 1958) and orthopteroids by notching the pronotum and amputating tegmina (Gangwere *et al.*, 1964).

MARKING INTERNALLY BY INJECTION

If the tissues of the animal can be marked in some way this has the great advantage with an arthropod that the mark is not lost during moulting. It has been found possible to mark crayfish by injecting a small amount of 'Bates numbering machine ink' into the venter of the abdomen; the black, blue and red inks were found to be non-poisonous (Slack, 1955; Black, 1963); similarly indian ink may be used in the fish, *Gambusia* (O'Grady & Hoy, 1972). This method might be applied to other large arthropods that have an area of almost transparent cuticle. Attempts have been made to mark adult mosquitoes by feeding the larvae with dyes; these have mostly been unsatisfactory, often leading to high mortalities (Weathersbee & Hasell, 1938; Chang, 1946; Reeves *et al.*, 1948; Bailey *et al.*, 1962).

MARKING BY FEEDING WITH DYES

The marking of invertebrates with a vital dye incorporated in the food, is a valuable tool: its significance is enhanced if the mark is retained from larval to adult life (see section 8 below) and if it can be detected without killing the animal. Many workers have screened a wide range of dyes, but few have been found satisfactory: the majority are either rapidly excreted or prove toxic (Reeves *et al.* 1948; Zacharuk, 1963; Daum *et al.*, 1969; Hendricks, 1971). Calco oil red [N−1700 American Cyanamid Co.] if fed to immature stages, has been found to mark adults and sometimes the resultant eggs (but not the F_1

larvae) of several cotton insects (*Heliothis*, *Platyedra*, *Anthonomus*) (Daum *et al.* 1969; Hendricks & Graham, 1970; Graham & Mangum, 1971). The dye is fed in a natural oil (e.g. cotton seed) with the larval diet at a concentration of about 0.01 % dye/unit diet: it is detected in the adult by crushing the abdomen. Similar results have been obtained with oil soluble 'Deep Black BB' and 'Blue II' (BASF Corp., J.N., U.S.), (Hendricks, 1971).

Sawfly larvae have been marked by feeding them, in the last instar, on foliage treated with solutions of rhodamine B (series 4; 3.7 g/l) and Nile Blue Sulphate (series 5; 0.4g/l) (Heron, 1968). The cocoons, adults and eggs were all marked and the dyes were visible externally; rhodamine B fluorescing bright yellow – orange under UV. Mosquitoes and the eyes of a small fish (O'Grady & Hoy, 1972) have been marked by feeding them on a 0.01 % solution of rhodamine B in a sugar solution (Reeves *et al.*, 1948); the same dye has also been used to stain the gut of *Drosophila* (Wave *et al.*, 1963). House flies have been marked with thiorescin (Shura-Bura & Grageau, 1956) and fluorescein (Zaidenov, 1960); in some cases only a small proportion of the insects could be induced to feed on the solution. South (1965) marked slugs by feeding them on agar jelly containing 0.2 % neutral red; the digestive gland became deeply stained and the colour was easily visible through the foot. Bloodsucking Diptera have been marked by allowing them to feed on a cow which had had 200 ml of an aqueous solution, containing 4 g of trypan blue, intravenously administered over 20 minutes. The dyestuff can be detected by a paper chromatographic technique, in which the gut contents of the fly is mixed with 0.1 N sodium hydroxide solution and applied to a narrow strip of Whatman No. 1 chromatographic filter paper, which is developed in 0.1 N sodium hydroxide solution; the trypan blue remains at the origin, other marks due to the gut contents move away (Knight & Southon, 1963). Haematophagous animals may also be marked with specific agglutinins (Cunningham, Harley & Grainge, 1963). These dyes or stains may often be detected in the faeces of predators that have fed on marked animals (Hawkes, R. B., 1972).

GENES, MUTANT AND NORMAL

Dispersal may be studied by the use of mutant genes (Peer, 1957, Levin, 1961; Hausermann, Fay & Hacker, 1971) or when various genotypes are clearly distinct, by their different proportions in adjacent colonies (e.g. Sheppard, 1951; Richards & Waloff, 1954; Goodhart, 1962); however, it should be remembered that selection may operate differentially in the different colonies, perhaps on young stages before the genotype becomes identifiable. Different sexes and age classes also provide naturally marked groups.

RARE ELEMENTS

Many elements, when exposed to a source of neutrons, become radioactive and emit a characteristic spectrum of gamma-rays. The process is known as neutron activation, further details are given in text books such as Bowen &

Gibbons (1963). Various workers have marked animals by incorporating rare elements in them and subsequently recognizing the mark by neutron activation and gamma spectroscopy. The great advantage of the method is that, theoretically at least, the mark may be retained from larval to adult life and self-marking is possible. For example if a small quantity of a rare element was mixed into a mosquito's breeding pool any adults emerging should be detectable any time in their life: the contributions of different breeding sites to a population could be determined. The disadvantages are that the equipment is extremely expensive (but is often available in nuclear physics or engineering centres); the procedure for gamma spectroscopy for rare elements is fatal for the animals (so it cannot be used in recapture studies) and there are often difficulties, undoubtedly of a basic physiological nature, in getting an adequate quantity of the rare element absorbed into the body (rather than just held in the gut) and retained in the tissues.

Dysprosium, first used by Riebartsch (1963) has been found to be a life-long marker for some Lepidoptera (Jahn *et al.* 1966) and *Drosophila* (Richardson *et al.*, 1969); rubidium for the cabbage looper, *Trichoplusia ni* larva (Berry, Stimmann & Wolf, 1972), europium for Lepidoptera (Jahn *et al.*, 1966; Ito, 1970) and manganese for fruit fly, *Ceratitis* (Monro, 1968). Rubidium when sprayed on host plants is readily absorbed; Berry *et al.* (1972) found a straight line relationship between the concentrations of rubidium in foliar spray and in the male moths that developed from the larvae that had been reared on the sprayed plants; but with the pea aphid, *Acyrthosiphon pisum*, the biological half-life was only about one day and detectable quantities only remained for four days after the aphid had left the plant (Frazer & Raworth, 1974). Ito (1970) showed that incorporation of europium (44 or 80 ppm) into the diets of moth larvae, *Hyphantria* and *Spodoptera*, did not affect growth and survival.

Alternatively the rare elements may be used as labels. Cerium is absorbed into the cuticle of insects; Rahalkar *et al.* (1971) used it (50 ug $CeCl_3$/insect in an alcoholic solution) to mark weevils, *Rhynchophorus*, by topical application on the elytra, and Bate *et al.* (1974) used a similar method with gold for the elm bark beetle. Jahn *et al.* (1966) obtained marking by spraying dysprosium and europium salts, with a little detergent, on to the insects.

Emerging calypterate diptera may self-mark by contamination of the ptilinal suture. Various dyes are often used for this purpose (p. 75), rare elements may be even more effective. Haisch, Stark and Forster (1975) marked emerging *Rhagoletis* with dysprosium and samarium, mixed with fine sand and silica gel (1 % by wt.) at concentrations of 0.1 % by weight. Provided the pupa were 3 cm deep the error in self marking was about 2 %.

RADIOACTIVE ISOTOPES

It is impossible to provide a full bibliography for this extensively used method; one is however given by Anon. (1963) and many papers are listed in reviews by Jenkins & Hassett (1950), Lindquist (1952), Hinton (1954), Pendleton (1956),

Dahms (1957) and Jenkins (1963), and Russian work by Anon. (1961) and by Andreev (1963).

The radioactive isotopes of elements are unstable and they disintegrate emitting radiations and forming other, usually non-radioactive, isotopes; the rate of disintegration is characteristic for each isotope and is described in terms of its half-life. Of the three types of radiation, entitled alpha, beta and gamma, the isotopes used in ecological work produce only the two last named. Beta particles have less power of penetration than gamma rays; however, the actual energies (expressed as MeV = million electron volts) of the radiations differ for the different isotopes. Radioactivity is measured in units of curies (1 curie = 3.7×10^{10} disintegrations/second) and the actual mass of chemical element constituting one curie will vary depending on the half-life of the isotope. The specific activity of any material or solution gives the relationship of the amount of radio-isotope to the total element content and is expressed in terms such as c/g. In biological work smaller units are usually required and these are: millicurie (mc) = 10^{-3}c and microcurie (μc) = 10^{-6}c. Further basic information on isotopes may be found in textbooks, e.g. Comar (1955), Francis, Mulligan & Wormall (1959), Faires & Parks (1958), O'Brien & Wolfe (1964) Thornburn (1972), or in tables, e.g. Hollander, Perlman & Seaborg (1953), Radio Chemical Manual (1966). Table 3.1 gives the characters of the principal isotopes used in entomological work. The use of isotopes in ecology is by no means confined to marking; they may also be used in studies on predation and energy flow (see Chapters 9 and 14). Great care should be exercised in all work with radioactive isotopes and precautions taken not to contaminate the environment (and the operators!). For this reason long half-life isotopes should normally be avoided in fieldwork.

Table 3.1 Some characters of the principal isotopes used in ecological research (* used as labels)

Isotope	Symbol	Half-life (approximate)	Radiations and energies (MeV) Beta (maximum)	Gamma
^{14}Carbon	C	5760 years	0.16	
^{137}Cesium	Cs	30 years	0.31	0.66
* ^{60}Cobalt	Co	5.27 years	0.31	1.2
* ^{65}Zinc	Zn	245 days	0.33	1.1
* ^{195}Gold	Au	185 days		0.1
^{45}Calcium	Ca	165 days	0.25	
* ^{182}Tantalum	Ta	115 days	0.51	range, max. 1.2
^{35}Sulphur	S	87 days	0.17	
* ^{46}Scandium	Sc	84 days	0.36	0.89
* ^{192}Iridium	Ir	74 days	0.67	0.6
^{89}Strontium	Sr	55 days	1.50	
^{59}Iron	Fe	45 days	0.46	1.2
^{32}Phosphorus	P	14.2 days	1.71	
^{131}Iodine	I	8 days	0.61	0.36

There are basically two methods of marking animals with isotopes, although the dividing line is not sharp. The isotopes may be used as a label outside the animal or alternatively they may be fed to the animal and incorporated in its tissues. Which method and which isotope will depend on the precise nature of the work. If it is desired to trace the animal's movements, in the soil or in wood, or to locate it from a distance, then it is the penetrating gamma radiations that will have to be detected.

If the object is to mark part of the population in a way that can be recognized after the animal has been recaptured, then a wider range of isotopes are available, including those of carbon, calcium, phosphorus and sulphur, which, if fed to the animal, are readily incorporated into its tissues. This marking method is preferable to the use of radioactive labels and, indeed, to many other marking methods in that animals can be tagged with the minimum amount of manipulation and the mark is invisible to predators. It is also almost unique (see above) as the mark will not be shed when the insect moults; this character gives the method great potential for the estimation of population size and mortality in immature stages (Cook & Kettlewell, 1960) and may be passed on to the next generation (Beard, 1965). It is necessary, however, to choose an isotope (e.g. sulphur) with a reasonably long half-life (unless the insect's life-cycle is very short) and to carry out laboratory tests on the extent to which the radioactive isotope atoms are excreted by the animal and replaced by normal ones (see below and p. 466).

Ants pass food to other members of their colonies by trophallaxis, whilst some gregarious insects (e.g. *Triatoma*) feel intra-integumentally on the gut contents of others. With these the risk of secondary tagging becomes significant, thereby invalidating the use of marking for Lincoln-Petersen Index type estimates (Odum & Pontin, 1961; Erickson, 1972); they may however be used, with ants, to determine information on foraging areas (p. 94). However, Stradling (1970) has shown that if the ants are starved for four days after marking, before return to the nest, ^{32}P becomes incorporated into the tissues and secondary tagging does not occur.

Labels

Besides acting as a suitable gamma source, radioactive labels need to be non-toxic and durable. They may be attached externally, when corrosion or loss are major problems, or inserted into the body, when toxicity or disturbed behaviour present difficulties. Among the elements used are cobalt (^{60}Co), tantalum (^{182}Ta), iron (^{59}Fe) and iridium (^{192}Ir). For protection, iron may be used in stainless steel wire, iridium in a platinum-iridium wire (Baldwin & Cowper, 1969) and cobalt can be gold plated (Fredericksen & Lilly, 1955). A benzene soluble DeKhontinsky type cement has been found a good glue for attaching labels (Sanders & Baldwin, 1969). Alternatively, cobalt in the form of the nitrate can be made up with a water-resistant resin glue (e.g. 'Bond Fast') at a strength of abour 1.6 mc/ml and applied to the surface of the insect

(Sullivan, 1953; Green, Baldwin & Sullivan, 1957); likewise ^{192}Ir, as the chloride, mixed with some cellulose paint and rubber solution (van der Meijden, 1973) or [65]Zn, as the oxide, mixed with cellulose paint and applied with a special applicator (Mitchell *et al.* 1973). Minute labels (about 0.05 mm $\times$ 0.16 $\times$ 0.23 $-$ 0.46 mm) of metallic tantalum have been used to mark coccinellid larvae and pentatomoid bugs, being attached by seccotine glue or cellulose paint (Banks, 1955; Banks, Brown & Dezfulian, 1961), and to mark earthworms, introduced into the coelomic cavity (Gerard, 1963). The initial specific acitivity of the tantalum labels used by Banks *et al.* on the bugs was about 8c; with the much smaller young coccinellid larva, labels with an average activity of 7.5 μc produced marked adverse effects and those used in the field work had a lower specific activity (1.3 μc). Workers using cobalt have applied labels with much higher specific activities, and Sullivan (1961) has shown that if this is greater than 50 μc the survival of the weevil, *Pissodes strobi*, is affected; this does not, however, become apparent for several months. But I am doubtful of the validity of the implied claim of some workers that, if the dosage is just small enough for the mortality to be negligible, then the animals' behaviour is normal. The problem of tagging and maximum total radiation dose is discussed fully by Griffin (1952); for the ecologist it is a good principle to use the minimum radiation dose consistent with detection, as was done by van der Meijden (1973), who showed that the pupation success of moth larvae was not affected by labels containing 3–4 μc of [192]Ir: these labels could be detected from about 60 cm.

Submerging the animals in a solution containing the radioactive isotope is another method that may be classified as labelling, although it is possible that a small amount of the isotope could become incorporated in the tissues. This method was first developed by Roth & Hoffman (1952), who marked house flies, wasps, leafhoppers, grasshoppers and coleoptera by dipping them for one minute into a solution containing ^{32}P at the concentration of 5 μc/ml together with a wetting agent. The marker did not penetrate the body of the animal, yet it could be removed only by prolonged washing. Large numbers of insects may be marked quickly by this technique and it has frequently been used; some examples are:

Animal	Isotope	Strength of solution applied	Authors
Boll weevi, *Anthonomus*	^{60}Co	0.5 μc/ml	Barbers *et al.*, 1954
Bug, *Eurygaster*	^{60}Co	12 mc/ml	Rakitin, 1963
Pine weevil, *Pissodes*	^{46}Sc	*c.* 7.5 mc/ml	Godwin *et al.*, 1957
Ant, *Lasius*	^{32}P	*c.* 1 μc/ml	Odum & Pontin, 1961
Tick, *Amblyomma*	^{32}P	10 μc/ml	Knapp *et.*, 1956
Tick, *Amblyomma*	^{59}Fe	10 μc/ml	Smittle *et al.*, 1967

The strengths of the solution applied varied greatly; it is possible that in some instances activity levels well above the minimum necessary were used. Roth & Hoffman found that the isotopes on their insects could be detected for about eighteen days and Godwin *et al.* (1957) could recognize their marked weevils for five months; the difference largely stemming from the half-lives of the isotopes. The marking can conveniently be done in a covered funnel, with a constricted neck or straining plate to prevent the insects being washed out (Davis & Nagel, 1956; Knapp *et al.*, 1956). More delicate animals cannot be marked by this method and because of the 'violence' involved it is doubtful if it should be used even with robust species where an alternative technique is available.

A few instances are recorded of marking insects by spraying them with radioactive isotopes; this is, however, extremely hazardous for the operator and cannot be recommended.

Incorporation in tissues

Some of the advantages of this method have been alluded to above. The natural food of either the larva or the adult may be made radioactive or, more conveniently, the animal is offered a sugar solution containing the isotope – a large number of species of animal will take up fluid from such solutions. The actual distribution of the isotope in the body will depend on the distribution of the appropriate element and not on the method of feeding (cf. Hoffman, Lindquist & Butts, 1951).

Plants may be made radioactive in a variety of ways (Sudia & Linck, 1963; Wiegert & Lindeborg, 1964; Krall & Simmons, 1977); two of the simplest are to grow them in a culture solution containing an isotope or to soak the seeds for a day in isotope-containing solution (Monk, 1967). All the parts of a large coniferous tree, including the pollen, may be tagged by introducing ^{32}P into the branches and trunk (Graham, 1957). Examples of phytophagous insects that have been marked through their host plant are turnip fly with ^{32}P (Oughton, 1951), plum weevil with ^{32}P and ^{89}Sr (Rings & Layne, 1953), mealy bugs with ^{32}P (Cornwell, 1955, 1956), lepidopterous larvae with ^{35}S (Cook & Kettlewell, 1960), a pentatomoid, *Eurygaster*, with ^{32}P (Quraishi, 1963*a* and *b*) and a mirid, *Orthotylus*, with ^{32}P (Lewis and Waloff, 1964). The specific activity of the solution depends on the method of detection it is planned to use (see below), on the isotope, on the animal and on the plant. For example, with ^{32}P Cornwell (1956) used 15–43 $\mu c/ml$ where the marked mealy bugs were detected by a Geiger counter, and Lewis & Waloff (1964) 0.5 $\mu c/ml$ where the marked mirid bugs were detected by autoradiography. Hubbell *et al.* (1965) found that the quantity of radio-isotope assimilated from a food source would vary under different conditions (see p. 466). With the gall-midge, *Dasyneura brassicae* marked by treating its galls with ^{32}P, concentrations of over 6 $\mu c/ml$ were necessary to provide a clear auto-radiographic image; concentrations greater than 25 $\mu c/ml$. (1.6 $\mu c/gm$ host plant) reduced midge size and fecundity, but emergence was not affected by strengths up to 100 $\mu c/ml$. (Sylven & Lonsjo, 1970). In the flea beetle *Altica*, the optimum body burden of ^{32}P, for the production of labelled eggs, has been found to be 0.03 μl per female (Riordan & Peschken, 1970). ^{65}Zn is an effective label for fungal host plants (Coleman, 1968).

Bloodsucking animals may be marked through their hosts (normally the isotope is injected interperitoneally into the host); this method has been used for example for mosquitoes marked with ^{32}P (Yates, Gjullin & Lindquist, 1951), ticks marked with ^{14}C (Babenko, 1960), fleas with ^{32}P and ^{89}Sr (Shura-Bura & Kharlamov *in* Anon., 1961) and blackflies with ^{32}P (Bennett, 1963). Engorged female ticks and their eggs can be marked by subcuticular injection of ^{14}C, ^{144}Ce and other isotopes (Sonenshine & Yunker, 1968; Sonenshine, 1970).

Aquatic larvae and the resulting adults become marked if reared in solutions containing isotopes (e.g. the experiments with mosquitoes and blackflies by Bugher & Taylor, 1949; Fredeen *et al.*, 1953; Gillies, 1961; Shemanchuk, Spinks & Fredeen, 1953, Garby, Yasuno & Phurwethaya, 1966).

The sensitivity of the insect to radiation clearly needs to be taken into account in deciding the dose rate. Many workers (e.g. Baldwin *et al.*, 1955, 1966; Abdel-Malek, 1961; James, 1961; Sonenshine, 1970; Dow, 1971; Hawkes, C., 1972) have recorded dosage levels that did, or apparently did not, affect the arthropods they studied. The problem has been thoroughly studied by Robertson (1973) in a scale insect and she has related dose to body size. The effect of the level of β-radiation from an isotope in the issues is cumulative, and corresponds to the effects of a short-term exposure to β-radiations of about one-third the dose. The reproductive female is most sensitive (L. D.$_{50}$ of 4200 rads), high sensitivity is also shown in the period just prior to moulting. She recommends a level of 20 $\mu c/gm$ (of insect body) for insects to survive and reproduce and 150 $\mu c/gm$ of insect body may be permissible for short-term work: these results are probably a general guide for small insects.

The incorporation of an isotope in an artificial food, bait or 'drinking water' is a technique that has been widely used (Jensen & Fay, 1951; Radeleff *et al.*, 1952; Schoof *et al.*, 1952; Macleod & Donnelly, 1957; Baldwin *et al.*, 1958; Barnes, 1959; Dow, 1959; Khudadov, 1959a; Shura-Bura *et al.*, 1962; Eddy *et al.*, 1962; Pelekassis *et al.*, 1962; Fay *et al.*, 1963; Orphanidis *et al.*, 1963).

A convenient method for the mass marking of large numbers of small, delicate insects has been developed by Lewis & Waloff (1964) for the mirid, *Orthotylus*, and has been used extensively with small Diptera, *Oscinella* and *Drosophila* (Southwood, unpublished); it would seem to be of very wide application. The basic principle is that the insects drink an isotope-containing sugar solution which is presented to them on a piece of saturated lens paper in a polystyrene Petri dish (Fig. 3.1). This has three metal studs (rivets) stuck to the base for legs, and holes are made in the upper and lower portions with a

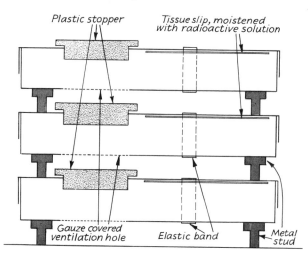

Fig. 3.1. Stacked polystyrene dishes modified for the presentation of radioactive solutions to small insects (after Lewis & Waloff, 1964).

hot metal tube. A plastic stopper fits into the top hole and the lower one is covered with gauze (stuck on with acetone or a similar solvent). Most small insects soon die in the laboratory from desiccation and, as marking takes about 24 hours, the dishes are stacked under bell-jars whose atmosphere is kept humid by a tube of an appropriate glycerine solution with a filter-paper wick (Johnson, 1940); 10 % solution was used in this work giving 98 % R.H.

The procedure is as follows: the lens tissue paper slip is placed on the lid and moistened with the radioactive solution (about 0.2ml) which should also contain the appropriate quantity (i.e. 10 % in the work mentioned above) of glycerine; the paper will then adhere to the lid by surface tension. The two halves of the dish should then be put together and held by an elastic band. A known number of insects (50–100) can then be tipped straight from a specimen tube in which they have been collected into the dish through the hole in the lid and the plastic stopper inserted. The dishes are then stacked under a bell-jar; after about a day they may be taken into the field at an appropriate time (see below) and released by removing the lids.

When working with very low concentrations of an isotope a significant proportion of the ions will become adsorbed on to the fibres of the lens tissue paper and become unavailable to the insects (Lewis & Waloff, 1964). This can be overcome by raising the concentration of non-radioactive ions, i.e. of the 'carrier', normally an orthophosphate or sulphate, so that the final concentration of this salt in the solutions applied to the dishes was not less than 0.1 %. This is conveniently done by diluting the stock isotope solution with an omnibus, 5 % sugar, 0.1 % carrier and 10 % glycerine solution. The specific activity of the ^{32}P solution used by Lewis & Waloff (1964) was 0.5 μc/ml.

Because of the low energy of ^{35}S radiations higher concentrations of this isotope must be used; Lewis & Waloff found a specific activity of 20 $\mu c/ml$ adequate.

Detection

There are two simple methods of detection adequate for mark and recapture studies: by a Geiger–Müller tube and by autoradiography. The former is indicated if the animal is to be traced in the field (e.g. Godfrey, 1954; van der Meijden, 1973); the latter demands that the animals be recaptured and killed, but has the advantage that very low levels of radioactivity can be used for marking so one can discount the possibility that, although the animal may appear normal, its behaviour is affected by the radiations. A method for continuously locating small mammals in a limited arena is described by Graham & Ambrose (1967): it could be used with large carabid beetles or similar insects.

The most convenient autoradiographic method for small insects is to stick them between two strips of 'Sellotape'* (Gillies, 1958); these should be labelled. They may, if necessary, be examined under the microscope and retained for permanent reference. However, in the first place the strips must be applied to the surface of an X-ray film (e.g. 'Kodirex', 'Ilfrex'); the exposure time will vary with specific activity of the insects and the speed of the film; it is usually between 2 and 10 days. The matching of the Sellotape strip to the subsequent print, and hence the recognition of the actual insects that were marked, will be facilitated if the exact position of the strip during exposure is recorded on the film by, for example, pin scratches marking the outline of the corners or by the use of a radioactive ink on the strip. Gardiner (1963) gives the following formula for such an ink: 1 mc ^{147}promethium (half-life 2.6 years) in 0.25 ml N hydrochloric acid, diluted with 1.5 ml water and add N ammonia solution until solution neutral to phenol red; add drop by drop mixing continuously to 56 ml 'Mandarin' indian ink (Windsor & Newton Ltd). Autoradiographic methods have been used by Gillies (1958, 1961), Khudadov (1959*b*), Abdel-Malek & Abdel-Wahab (1961*a*) and Lewis & Waloff (1964) and are discussed in general by Fitzgerald (1958).

In studies on energy or nutrient flow using isotopes (see chapter 14) it is often necessary to use another method, the scintillation counter. This may entail repeatedly confining the living animal while it is radioassayed: a convenient cage is described by Hayes (1969). The quantification of the level of radioactivity provided by the scintillation counter may be useful; Smittle *et al.* (1973) showed that successive egg rafts of mosquitoes tagged as larvae, had significantly different levels of counts.

The length of time a radioactively marked animal remains detectable will depend on the initial dose and on the effective half-life; the latter is the

---

* A transparent adhesive tape – 'Scotch tape' is similar.

resultant of the half-life of the isotope and the rate of replacement of the isotope atoms in the tissues by inactive ones, i.e. on the biological half-life. Jensen & Fay (1951) found with various calypterate flies marked with ^{32}P that the rate of replacement was much faster when they were fed on a diet rich in phosphorus (milk) than if on one poor in this element (honey solution). If maintained on sugar after feeding on solutions containing 25–400 μc/ml ^{32}P, eye gnats, *Hippelates*, remained detectable by a Geiger counter for just over fifteen days (Dow, 1959); Jensen & Fay's results were similar. Autoradiography, of course, will detect lower levels, and Gillies (1961) records that *Anopheles* marked with ^{32}P as larvae were readily detectable eight weeks after emerging as adults. In two acridids (*Schistocerca* and *Anacridium*) it was two weeks before ^{32}P was accumulated in the brain and the gonads, but the rate of disappearance from the gut was very different in the two species (Abdel-Malek & Abdel-Wahab, 1961*b*). McAllan & Neilson (1965) found that, even after a six-day exposure to feeding pads containing ^{89}Sr, different individuals took up different amounts of the isotope; therefore the detection time was determined by the individuals that ingested and assimilated the least quantity of isotope. Although any particular insect should be tested, present evidence would suggest that ^{32}P will remain detectable for at least two to three weeks (Anon., 1961; Bennett, 1963) and ^{35}S considerably longer (Kettlewell, 1952; Cook & Kettlewell, 1960), the half-life of the isotope being the most important factor.

AUTORADIOGRAPHIC DISCRIMINATION BETWEEN ^{32}P AND ^{35}S MARKED INSECTS

Owing to the different energies of these two isotopes the images they produce on X-ray film are distinct; those due to ^{35}S give a clear silhouette of the marked insect; those due to ^{32}P are stronger but the shape of the insect is indistinct. However, when these isotopes have been used as two different marks it would be unwise to rely on the different form of the image. The short half-life of ^{32}P can be used to identify it; if the test strip containing the insects was exposed to film twice, at about a fourteen-day interval, the ^{32}P images could be much weaker on the second than on the first film – those from ^{35}S would alter but little. It is, however, possible to distinguish the two isotopes simultaneously; Duncombe (1959) has pointed out that if X-ray film, coated with emulsion on both sides, is used an image due to ^{35}S will appear only on the side nearest the source due to the weak penetration of the ^{35}S radiation. In practice this is not easy to see and two other ingenious methods, based on the same principle (the difference in MeV of the radiations), have been described by Gillies (1958) and Lewis & Waloff (1964) and are illustrated diagrammatically in Fig 3.2.

3.1.2 Individual marking methods

If each individual can be separately marked additional information can be

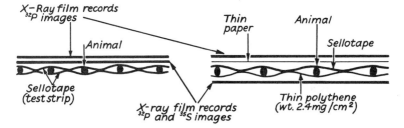

Fig. 3.2. Double exposure autoradiographic methods of discriminating between animals marked with ^{32}P and ^{35}S – diagrammatic sections of arrangement of test strip and films in cassette: *a*. method of Gillies (1958); *b*. method of Lewis & Waloff (1964).

obtained on longevity and dispersal and, if they can be aged and sexed initially, survival can be related to these and other characters. Birth- and death-rates may be more easily calculated and the excessive handling of animals, recaptured more than once, with its attendant problems, may be avoided (Dobson & Morris, 1961). It may also, for example, be possible by weighing individual females to assess the rate of oviposition in the field (Richards & Waloff, 1954), and it provides a method for assessing the randomness of recapture. The extent to which the higher cost, in terms of effort, of marking animals individually is justified will depend on the percentage of the marked individuals that are recovered; high recovery rates justify elaborate marking programmes; with low recovery rates individual marking is seldom justified.

If small labels can be attached to the insects' wings (see above) there is no problem in marking each individual, and Nielsen (1961) records that in the laboratory he marked butterflies individually on the wing with a rubber stamp. With most small invertebrates individuals marks have to be obtained by a combination of spots in various positions, the numerical range of the coding often being increased by the use of various colours. It is generally wise to follow the policy of Michener *et al.* (1955) and ensure that all individuals bear the same number of marks, so that if one is lost number 121 doesn't become say 21, but is immediately recognised as an 'unreadable' mark. It is also desirable to use the minimum number of colours in marking any individual insect, as the change in colours increases the handling time. Obviously a change in colour between say the ranges 1–99 and 100–199 is less of a practical problem, than the use of two colours in a single mark. The actual pattern will depend on the size and shape of the insect, the number of colours available and the number of individual marks required. Patterns for bodies and wings have been devised for dragonflies (Borror, 1934; Parr, Gaskell & George, 1968), bed bugs (Mellanby, 1939), tsetse flies (Jackson, 1953), bees (von Frisch, 1950), grasshoppers (Richards & Waloff, 1954; Nakamura *et al.*, 1971; White, 1970), snails (Blinn, 1963), craneflies (Freeman, 1964) , mosquitoes (Sheppard *et al.* 1969; Conway *et al.* 1974) and lepidoptera

(Ehrlich & Davidson, 1960; Brussard, 1971; Dempster, 1971). The system of Richards & Waloff is logical and versatile, enabling up to 999 individuals to be marked with continuous numbering (Fig. 3.3). On the right of the thorax is a spot that represents the unit, on the left one that represents the tens, the head spot the hundreds; 1–5 and 10–50 are white spots, 6–0 and 60–00 are red; ten different colours are used for the head mark including the zero class (for 1–99). With this system the addition of a single further mark would allow another 10,000 individuals to be numbered. With smaller insects the number of spots can be reduced by increasing the number of colours used, as Richards & Waloff did with the hundred mark; however, the practical problem of switching rapidly between more than three colours is serious. The rapid reading of the mark is another advantage and Richards & Waloff's code is particularly clear in this respect because the same colours and corresponding positions represent the same figures in units and tens.

Richards & Waloff's code may be considered a decimal code, based on tens. Ehrlich & Davidson (1960) developed a 1–2–4–7 marking system, modified by Brussard (1971) to mark up to 1000 individuals with one colour (Fig. 3.3b). A binomial system for marking mosquitoes was developed by Sheppard *et al.* (1969), up to 255 individuals can be marked with a single colour (Fig. 3.3c) but as with Brussard's method the number of marks is variable and the quick and accurate reading of the mark a matter of experience.

A more elaborate code of shapes has been devised by White (1970) (Fig. 3.3d). The system is self-checking (if marks are lost), a single colour may be used up to 160, but as White points out rapid reading becomes more difficult beyond 120.

If the body and/or wings are unsuitable for spotting the legs may be marked, but the removal of such marks by the animals during cleaning movements is a real risk. Kuenzler (1958) was able to number lycosid spiders by marking their legs with white enamel paint and Corbet (1956) hippoboscid flies by marking both the body and the legs.

Mutlilation may be used as an alternative to colour marks. By allocating a code of numbers to the teeth on a crab's carapace it is possible to number up to about forty individuals (depending, of course, on the age and species of crab) by cutting off one or more teeth (Edwards, 1958). In the same way by numbering the interstrial spaces of a carabid, a patch in any one of which could be scraped, approximately 999 individuals could be numbered (Murdoch, 1963) (Fig. 3.4a). Gangwere *et al.* (1964) developed a method of individually marking orthopteroids by notching the pronotal margin and amputating the tegmina.

3.1.3 Handling techniques

If the animals are to be marked by spraying or dusting this can be done whilst they are still active, either in the field or in small cages (see above); but for marking with paints, especially precise spotting, and for most methods of

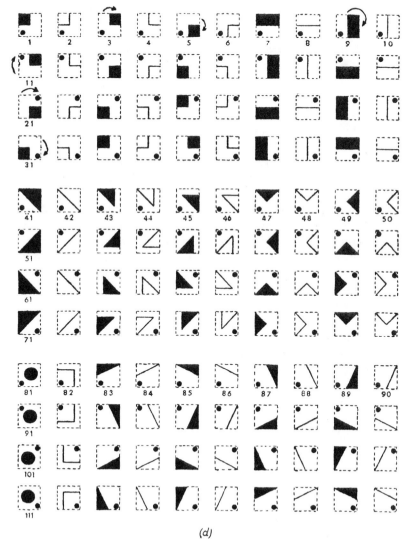

Fig. 3.3 Systems for marking insects individually using colour and position codes; *a*. Richards and Waloff's (1954) 'decimal system'; *b*. Brussard's (1971) modification of Ehrlich and Davidson's '1–2–4–7 system'; *c*. Sheppard *et al.*'s (1969) 'binomial system'; *d*. White's (1970) 'shape code' (see text for further explanations).

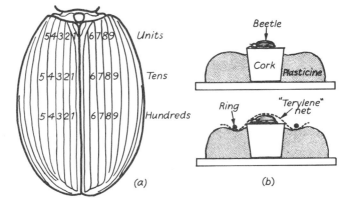

Fig. 3.4. *a*, A 'mutilation system' for marking carabid beetles individually by scraping certain positions on the elytra, applied to *Agonum fuliginosum*. *b*. A device for holding a hard-bodied animal whilst it is marked (modified from Murdoch, 1963).

labelling and mutilation, it is necessary for the animal to be still. Often it can be held between the fingers; where this is impossible the animal must be held by some other methods, anaesthetized or chilled.

The animals may be held still under a net or with a hair. One end of the hair is fixed (e.g. in seccotine) and the other has a piece of Plasticine on it; the animal is placed underneath and the hair may be tightened under a stereoscopic microscope by pressure on the Plasticine (Banks, 1955). More robust insects may be held immobile on top of a cork by a piece of Terylene net which is kept taut by a ring (of about 2 in. diameter, but of course depending on the size of the animal); the ring is pushed down into the surrounding Plasticine (Fig. 3.4*b*) and the animal can be marked through the holes of the net (Murdoch, 1963). Conway, Tripis & McClelland (1974) held mosquitoes for spot marking between diaphragms of nylon mesh and stockinet. These devices may be placed under a microscope.

Insects can also be held by suction (Hewlett, 1954); Muir (1958) found that a mirid bug was conveniently held by the weak suction of a pipette, formed from the ground-down point of a coarse hypodermic needle, connected to a water pump. The picking up of the bugs by this weak suction was facilitated by coating the inside of the glass dish, in which they were held, by an unsintered dispersion of 'Fluon' GPI, polytetrafluoroethylene (PTFE), which presented an almost frictionless surface (Muir, 1958; Radinovsky & Krantz, 1962). Butterflies may be held in a clamp (Horn, 1976).

If none of these methods can be used or if the insects are so active that they cannot even be handled and counted they may have to be immobilized. Chilling at a temperature between 1 and 5°C is probably the best method and may be done in a lagged tank surrounded by an ice-water mixture (MacLeod & Donnelly, 1957). Alternatively an anaesthetic may be employed: carbon dioxide is often used and is easily produced from 'dry ice' (Dalmat, 1950; Roth & Hoffman, 1952; Fredericksen & Lilly, 1955; Green, Baldwin & Sullivan, 1957; Caldwell, 1956). The last mentioned author gives the following table of recovery times of house flies from carbon dioxide anaesthesia at $21°C$ ($70°F$):

Exposure time	Recovery time
Up to 5 min	1–2 min
5–30 min	3–5 min
30–60 min	5–10 min

Working on the earwig, *Forficula*, Lamb & Wellington (1974) found that although activity may be quickly resumed after carbon dioxide anaesthesia of less than 1 minute, normal behaviour could be affected for many hours; they recommend a 24-hour recovery period. Dalmat (1950) found that female blackflies (Simulidae) laid more eggs than normal after being subjected to carbon dioxide, but Edgar (1971) found this did not affect the development of the spider, *Lycosa lugubris*. Other anaesthetics are ether, chloroform, nitrogen and nitrous oxide; honey bees are prematurely aged by these (Ribbands, 1950; Simpson, 1954). Therefore, the use of anaesthetics should, wherever possible, be avoided in ecological and behavioural studies.

3.1.4 Release

The release of the animals after marking is an operation that is too often casually undertaken. A few methods (e.g. Davey, 1956) allow the animals to be marked without capture, others for it to be marked in the field and immediately released again (Jackson, 1933*b*; Richards, 1953). However, frequently after a period of incarceration, handling and disturbance, the animals are released into the field and it should not be surprising if they show a high level of activity immediately after release; indeed Greenslade (1964) recorded with individually marked ground beetles that there was far greater movement on the day after release than subsequently. Two approaches can be used to minimize this effect.

If the animal has a marked periodicity of movement (i.e. is strictly diurnal or nocturnal) then it should be released during its inactive period; for example I released radioactively tagged individuals of the frit fly (*Oscinella frit*) at dusk (its period of activity is from dawn to late afternoon) (Southwood, 1965). Animals that are active at most times of the day may be restrained from flying immediately after release by covering them with small cages (Evans & Gyrisco, 1960). The release sites should be chosen carefully. It is especially important to avoid the release of small flying insects in the middle of the day when their escape flights may carry them beyond the shelter of the habitat into winds or thermals that can transport them for miles. Of course, only apparently healthy, unharmed individuals should be released.

The release points should be scattered throughout the habitat, as it is essential that the marked animals mix freely with the remainder of the population; e.g. Muir (1958) returned the arboreal mirids he had marked to all parts of the tree. Very sedentary animals may indeed invalidate the use of the capture–recature method if they do not move sufficiently to re-mix after marking, as Edwards (1961) found, surprisingly, with a population of the grasshopper, *Melanoplus*. The extent of the re-mixing may be checked, to some degree, by a comparison of the ratio of marked to unmarked individuals in samples from various parts of the habitat; the significance of the difference may be tested by a χ^2 (Iwao *et al.*, 1963).

3.2 Capture–recapture methods of estimating population parameters

3.2.1 Lincoln Index type methods

ASSUMPTIONS

Various assumptions underlie all methods of capture–recapture analysis. If the particular animal does not fulfil one or more conditions it might be possible to allow for this to some extent, but a method of analysis should not be applied without ensuring, as far as is practicable, that its inherent assumptions are satisfied.

The following assumptions underlie most methods of analysis:

(1) The marked animals are not affected (neither in behaviour nor life expectancy) by being marked and the marks will not be lost.

(2) The marked animals become completely mixed in the population.
Cold and inclement weather soon after release may seriously restrict the mixing of marked individuals in the population (Nakamura, Ito, Miyashita & Takai, 1967).

(3) The probability of capturing a marked animal is the same as that of capturing any member of the population; that is, the population is sampled randomly with respect to its mark status, age and sex. Termed 'equal catchability', this assumption has two aspects: firstly, that all individuals of the different age groups and of both sexes are sampled in the proportion in which they occur; secondly, that all the individuals are equally available for capture irrespective of their position in the habitat.

(4) Sampling must be at discrete time intervals and the actual time involved in taking the samples must be small in relation to the total time.

When using the simple Lincoln Index it is also assumed that:

(5) The population is a closed one or, if not, immigration and emigration can be measured or calculated.

(6) There are no births or deaths in the period between sampling or, if there are, allowance must be made for them.

Most of the other more complex methods of estimating population by capture – recapture may be applied in situations where either migration, natality or mortality or all three are occurring. These methods, however, require a series (at least two) of occasions on which animals are marked, on the second and subsequent occasions the recaptured animals are remarked and released again. Hence these methods make the further assumption:

(7) Being captured one or more times does not affect an animal's subsequent chance of capture. This is a further extension of assumption (3) – equal catchability.

Several of these methods estimate the number of marked animals available for recapture from their subsequent survival and that of other cohorts. These approaches therefore make the additional assumption:

(8) Every marked animal has the same probability of surviving through the sampling period. This means that if mark status is in any way related to age then mortality must be independent of age; it must act randomly on the population and some individuals (e.g. of greater age and different mark status) must not be at greater risk.

THE VALIDITY OF THE ASSUMPTIONS

Marking has no effect
The experimental assessment of the effects of marking, the precautions in this respect and the re-mixing of the marked animals in the population have already been discussed above.

Manly (1971a) has described a test to determine if the survival of individuals immediately after marking and release is different from that of animals that have been marked for some while. It is thus a test of the effect of the marking process and not of the influence of bearing a mark. These losses are referred to by fisheries workers as Type 1 losses. It depends on each animal bearing a unique mark and may be used in conjunction with the actual data derived from a series of, at least four, marking occasions. The difference (y) between the survival of animals newly exposed to marking and others is given by:

$$y = \log_e\left(\frac{r'_i}{r_i} \times \frac{u_i}{R_i}\right) \qquad (3.1)$$

where r'_i = the number of marked animals in the ith sample that are recaptured again later, r_i = the total marked animals in the ith sample, u_i = the total unmarked animals in the ith sample and R_i = the total animals marked on the ith occasion and subsequently recaptured (see Table 3.2). Now if the process of marking has no effect on the animal, y will have an expected value of zero. The significance of any value of y can be determined by the statistic g, derived by dividing y by its variance:

$$g = y \bigg/ \left(\frac{1}{r'_i} + \frac{1}{R_i} - \frac{1}{r_i} - \frac{1}{u_i}\right)^{\frac{1}{2}} \qquad (3.2)$$

g will behave as a random normal variate with mean and variance zero. The one-tail test is normally justified because an increase in survival due to the marking process can be ruled out and thus at the 5 % level $g < 1.64$ or at 10 % level $g < 1.29$ may be taken as showing that the marking process has no effect on survival. A similar test has been developed by Robson (1969).

Equal catchability

There may be several different causes for the violation of this assumption; underestimates of the population result (Cormack, 1972; Bohlen & Sundström, 1977). Roff (1973b) has reviewed the statistical tests available for the purpose of testing for equal catchability but, after simulation, has concluded, that under practical conditions they may be unreliable. Such a warning is useful and means that the ecologist must also rely on other biological indications and on other methods for confirmation that his population is homogeneous: the computer cannot replace biological insight. It is to be hoped, however, that further refinements of the mathematical aspect will be achieved, for the excessive pessimism of Roff's concluding sentence does not advance the subject! Phillips & Campbell's (1970) studies on the whelk, suggest that 'frequency of capture' methods may be more robust and, as stressed above, the ecologist should always endeavour to select homogeneous categories (White, 1975).

(1) Sub-categories of the population are differently sampled because by the nature of the habitat only part of the population is available. This problem is particularly acute with subcortical, wood-boring and subterranean insects. It is well exemplified by ants: the number and composition of the foragers of an ant colony are so variable that unless one has a considerable knowledge of these aspects, or can sample from within the nest (Stradling, 1970), Lincoln Index and related methods cannot be used for population estimation (Ayre, 1962; Erickson, 1972). The marking and recapture of foraging ants may however be used to estimate the foraging area (Kloft, Holldobler & Haisch, 1965) and, over a short period of time, the number of foragers (Holt, 1955):

$$F = R_f \times (T_o + T_i) \qquad (3.3)$$

where $F =$ the number of foragers; $R_f =$ rate of flow of foragers per unit time, i.e. the average number of ants passing fixed points on all routes from a nest per minute; $T_o =$ average time (in the same units as RF) spent outside the nest (found by marking individuals on the outward journey and timing their return); and $T_i =$ average time spent inside the nest (found by marking ants as they enter and timing their reappearance). In *Pogonomyrmex badius* only about 10 % of the total population forage above ground in any two-week period (Golley & Gentry, 1964), whilst in the wood ant, *Formica*, a constant group of individuals forage such that if their number ($\hat{F}$) is estimated, colony size ($\hat{C}$) is given by (Kruk-de-Bruin *et al.* 1977):

$$\log \hat{C} = (\log. \hat{F} + 0.75)/1.01 \qquad (3.4)$$

Foragers of this species may remain away from the nest for up to two days.

(2) Sub-categories of the population are differentially sampled because differences due to sex and/or age or other causes. In many animals, especially when trapping techniques are used, differences will be associated with differences in sex or age. It is indeed a good principle to enumerate and estimate males and females separately, at least in the first instance; many workers have found striking differences in their behaviour (e.g. Jackson, 1933a; Cragg & Hobart, 1955; MacLeod, 1958; Muir, 1958). Initially different stages or age classes should also be considered separately and their survival rates determined using one of the formulae given below; if their survival rates differ significantly, one must continue to consider them separately. Indeed every attempt should be made to ensure that the population estimated is as homogeneous as possible.

If there is no mortality, a χ^2 goodness-of-fit test may be applied to a table of the releases, recaptures and non-recaptures of the different classes (Seber, 1973; White, 1975). Lomnicki (1969) found that land snails are very non-random in their movements, some individuals being more inclined to recapture than others.

(3) There is a periodicity in the availability of sub-categories. This problem has been particularly encountered with mosquitoes, where an initial period of emigration (Sheppard *et al.* 1969) and the gonotrophic and linked feeding-cycles invalidate the equal catchability assumption and may lead to five fold over-estimates (Conway *et al.* 1974). Other animals show periodicities in their behaviour and therefore, particularly when sampling is by a behavioural method (e.g. trapping) and the periodicity cycle is long compared with the sampling interval, workers should consider this problem. With the Fisher & Ford method (see below) it may be allowed for by introducing additional terms into the model (Sheppard *et al.* 1969; Conway *et al.* 1974) (see below). The modification of the Fisher & Ford method in this way depends on a knowledge of the periodicities gained independently. The extent to which the incorporation of these additional terms improves the relationship of the expected and observed recaptures gives a measure of their value, but a precise statistical test is not available.

(4) The processes of capturing and marking affect catchability. This can have two origins: the initial experience may alter catchability or it may be a cumulative effect. In the former case (more relevant to vertebrates) a regression method analogous to removal trapping (p. 230) may be used to estimate total population (Marten, 1970) (see p. 113). The effect of repeated capture may be tested for by Leslie's test for random recaptures (Orians & Leslie, 1958; Turner, 1960). A comparison is made of the actual and expected variances of a series of recaptures of individuals known to be alive throughout the sampling period; individual marks must have been used. This test is best illustrated by a worked example taken from Leslie's appendix, based on the recaptures of shearwaters; with insects, of course, the recapture periods would be days or weeks rather than years.

EXAMPLE (after Leslie): 32 individuals marked in 1946 were recovered for the last time in 1952, therefore they were available for recapture in the years 1947–51 inclusive and the following two tables can be prepared:

Year by year analysis		All years analysis	
Year	No. of recaptures in each year (n_i)	No. of recaptures for each individual x	Frequency of x $f(x)$
1947	7	0	15
1948	7	1	7
1949	6	2	7
1950	4	3	2
1951	7	4	1
n_i = 31		5	0
		$N = \Sigma f(x) = 32$	

The actual sum of squares:

$$\Sigma(x - \bar{x})^2 = \Sigma x^2 f(x) - \frac{[\Sigma x f(x)]^2}{N} \tag{3.5}$$

$$= 69 - 30.03 = \underline{38.97}$$

The expected (= theoretical) variance:

$$\sigma^2 = \frac{\Sigma n_1}{N} - \frac{\Sigma(n_i^2)}{N^2} \tag{3.6}$$

$$= \frac{31}{32} - \frac{199}{32^2} = 0.9688 - 0.1943 = \underline{0.7745}$$

Then

$$X^2 = \frac{38.97}{0.7745} = \underline{50.32}$$

X^2 may be treated as equivalent to a χ^2 and for degrees of freedom $(N - 1)$ of between 20 and 30 the probability of a value as great or greater than this can be assessed from χ^2 tables; probabilities of less than 0.50 imply that capture is not random. With values over 30, use can be made of the fact that $\sqrt{2\chi^2} - \sqrt{2\mathrm{d.f.} - 1}$ is approximately normally distributed about a mean of zero without standard deviation; in other words one calculates the value for this expression and looks up its probability in the table of the normal deviate. In the present case:

$$\sqrt{2 \times 50.32} - \sqrt{(2 \times 31) - 1} = 10.03 - 7.81 = +2.22$$

The probability of a deviate as great as this is somewhere between 0.025 and 0.020 and therefore the shearwaters were not collected randomly, Leslie has suggested that the test should only be used when the number of individuals is 20 or more and the number of occasions on which recapture was possible is at least 3. This test will not distinguish whether the higher catchability of some individuals is due to catching effects or, as Lomnicki (1969) presumed with land snails, inherent individual differences. Carothers (1971) has developed an extension of Leslie's test more appropriate when the number of occasions on which recapturing occurs is small.

METHODS OF CALCULATION

Although simple Lincoln Index estimates are not permissible unless conditions 5 and 6 are satisfied, the variants of the methods have all been derived to allow for loss (emigration and death) from or gains (immigration and birth) to the population. Migration rate can also be measured by other means (pp. 244, 335) or it may be artificially prevented by the use of a field cage (Dobson *et al.*, 1958; Dobson & Morris, 1961). Birth and death can also be measured by other methods (Chapter 9). Sometimes mortality rate may be so high that the interval between the release of the marked individuals and the recapture sampling is extremely short, a matter of hours (e.g. Coulson, 1962); this is only possible with a mobile insect and under such conditions Craig's method (p. 113) might be used as well as one based on the Lincoln Index.

The Lincoln Index

Provided the first six conditions listed above are satisfied it is legitimate to estimate the total population from the simple index used by Lincoln (1930):

Total population ÷ Original number marked = Total second sample ÷ Total recaptured

or
$$\hat{N} = \frac{an}{r} \tag{3.7}$$

where $\hat{N}$ = the estimate of the number of individuals in the population (N), n = total number of individuals in the second sample, a = total number marked and r = total recaptures.

Strictly the size of n should be predetermined and normally is approximately equal to a, then the variance of this estimate is given by (Bailey, 1952):

$$\text{var } \hat{N} = \frac{a^2 n(n-r)}{r^3} \tag{3.8}$$

If the second sample (n) consists of a series of subsamples and a large proportion of the population have been marked then it is possible to utilize the recovery ratios ($= r/n$) in each to calculate the standard error of the estimated population (Welch, 1960):

$$\hat{N} = \frac{a}{R_T} \tag{3.9}$$

where R_T = the recovery ratio (r/n) based on the total animals in all of the samples; the variance is approximately:

$$\text{var } \hat{N} = \left(\frac{a}{R_T^2}\right)^2 \times R_T \frac{(1-R_T)}{y} \tag{3.10}$$

where y = total animals in subsamples. This approach is only valid if the marked individuals are distributed randomly in the subsamples, which may be tested by comparing the observed and theoretical variance (see p. 96).

The above methods are applicable to large samples where the value of r is fairly large (say over 20); Bailey (1951, 1952) has suggested that with small samples a less biased estimate is given if 1 is added to n and r, i.e.:

$$\hat{N} = \frac{a(n+1)}{r+1} \tag{3.11}$$

An approximate estimate of the variance of this is given by:

$$\text{var } \hat{N} = \frac{a^2(n+1)(n-r)}{(r+1)^2(r+2)} \tag{3.12}$$

These methods are based on what is referred to as *direct sampling* in which the size of n is predetermined. If it is possible, *inverse sampling*, where the number of marked animals to be captured (i.e. r) is predetermined. has the advantages of giving an unbiased population estimate and variance (Bailey, 1952):

$$\hat{N} = \frac{n(a+1)}{r} - 1 \tag{3.13}$$

$$\text{var } \hat{N} = \frac{(a-r+1)(a+1)n(n-r)}{r^2(r+1)} \tag{3.14}$$

A modification of the Lincoln Index has been developed by Gaskell & George (1972). It is based on Bayes' Theorem that links the probability distribution of a new estimate with the likelihood of the sample and the previous estimate The incorporation of the previous estimate is of considerable practical significance: the value as a biological check of independent estimates by alternative methods is stressed throughout this book. This Bayesian modification now allows their incorporation into the final population estimate. Gaskell & George's formula is essentially:

$$\hat{N} = \frac{an + 2N'}{r+2} \tag{3.15}$$

where $N' =$ the previous estimate obtained, for example, from quadrat sampling or nearest neighbour type techniques.

Other single mark methods
This approach is extensively used in fisheries where the ecologist can mark a number of fish, but must rely on commercial fishing to provide the recapture sample; one of the major problems in this work is to distinguish mortality due to man from that due to other causes (Gulland, 1955).

If a group marking method is used, recaptured individuals must be removed and the successive removals recorded (Ito, 1973); with individual marks the number of previously un-recaptured individuals must be recorded (Marten, 1970): the fall off in these values, determined by regression, may be related to total population (p. 230). The methods used will not be discussed further here; some are described by Beverton & Holt (1957; p. 184 et seq.) and the whole approach lucidly reviewed by Seber (1973).

Review of methods for a series of marking occasions
If the animals are marked on a series of two or more occasions, then an allowance may be made for the loss of marked individuals between the time of the initial release and the time when the population is estimated. These methods have recently been reviewed by Cormack (1968), Parr, Gaskell & George (1968), Robson (1969) and Seber (1973). Their history and development, with particular reference to the entomological literature are given below. A simple, non-algebraic development of the main formulations is provided by Cormack (1973), who shows how the estimates are based on the largest 'known groups'.

Jackson (1933a, 1937, 1939, 1940, 1944, 1948) and Dowdeswell, Fisher & Ford (1940; and Fisher & Ford, 1947) were the first to devise methods for estimating the population using the data from a series of marking and recapture occasions. Besides assumptions 1–4 and 7 listed above (p. 92), they also assumed a constant survival rate over a period of time; even though this was known to be an approximation it was considered necessary for an algebraic solution.

Jackson developed two methods in his work on the tsetse fly where he calculated a theoretical recapture immediately after release by either the 'positive method', where the loss ratio was calculated for that day over all release groups, and the 'negative method', where the loss ratio was calculated from the subsequent daily percentage losses for the two release groups. Richards & Waloff (1954) used Jackson's negative method and give worked examples. It is possible that the negative method might still be found useful in a situation, as provided by the use of radioactive isotopes, where marking is carried out on a limited number of occasions followed by a long series of recaptures.

Fisher's method is often referred to as the 'trellis method', as the data are initially set out in a trellis diagram; details of its working are given by Dowdeswell (1959). The survival rate has to be determined by trial and error, but with the advent of access to computers this comparatively robust method has enjoyed a revival. A useful discussion and summary of these early methods is given by MacLeod (1958).

A slightly different method of analysis, but with the same principles, is given by Leslie & Chitty (1951) and Leslie (1952) and referred to as method A. It has been used with carabids by Grüm (1959) and trypetid flies by Sonleitner & Bateman (1963). Although these methods are of considerable historical interest and have been widely used, it is difficult to envisage a situation where their use would still be recommended in preference to the more recent methods. A possible exception to this statement is the use of Jackson's negative method under the circumstances outlined above.

An important advance was made by Bailey (1951, 1952), who introduced maximum likelihood techniques into capture–recapture analysis and so was able to calculate the variances of his estimates. The equations in Bailey's triple-catch method are simple and may be solved directly to provide estimates of various population parameters.

Another significant step was made at about the same time by Leslie (1952; Leslie & Chitty, 1951), who compared three different methods of classifying the animals according to their marks. In Leslie's method A, mentioned above, the animals were classified according to the occasion on which they were marked and thus an animal bearing several marks would give several entries in the recapture table. Jackson, Fisher and Bailey all also used this method of classification in drawing up their recapture tables, but Leslie & Chitty (1951) showed that it leads to loss of information. This loss is only slight when there are but a small number of multiple recaptures (Sonleitner & Bateman, 1963). His method B, in which the animals were classified according to the date on which they were last marked (ignoring all earlier marks), was completely efficient under the assumptions he made. Leslie (1952) gives a full account of his method and a worked example, but his formulae require solution by iterative methods when more than three sampling occasions are considered and hence the calculations are rather laborious. Jolly (1963) simplified the calculations by providing formulae for explicit solutions (i.e. direct, not 'trial and error'). This method is particularly applicable if a fairly large number of individuals have been recaptured several times (multiple recaptures) in a long series of samples, a situation that often arises with work on mammals. However, Sonleitner & Bateman (1963) used Leslie's formulae to obtain estimates of the population of a trypetid fly and Muir (1958) used Jolly's method with a mirid bug.

Working on blowflies, where the incidence of recapture was very low, MacLeod (1958) found that none of the then known methods were usable and he derived two formulae (based on the Lincoln Index) which gave estimates, albeit rather approximate, of the total population. For both methods the mortality rate had to be ascertained independently, in his case in the laboratory, and for one method the percentage of flies not immigrating over a given time was measured by a separate experiment.

All the above methods are based on deterministic models that assume that the survival rate over an interval is an exact value, whereas it would be more correct to state that in nature an animal has a probability of surviving over the interval. This probability is well expressed by a stochastic model, but initially it was thought that the computations arising from a stochastic model would be too complex. Darroch (1958, 1959) showed that under certain conditions a fully stochastic model, giving explicit solutions for the estimation of population parameters, was possible, and Seber (1965) and Jolly (1965) have, independently, extended this method to cover situations in which there is both loss (death and emigration) and dilution (births and immigration). Their methods give similar solutions, except that Jolly's makes allowance for any animals killed after capture and hence not released again, a common occurrence in entomological experiments.

The Jolly-Seber method efficiently groups the data and is fully stochastic; however, its reliability strictly depends on assumption (8), effectively the probability of any animal surviving through any period is not affected by its age at the start of the period. Manly & Parr (1968) pointed out that with short-lived adult insects (e.g. many Lepidoptera) studied over their emergence period the animals marked on the ith occasion are likely, at that time, to be younger than those originally marked on the first. This in itself would not invalidate the Jolly-Seber method, if all deaths were independent of age: this cannot be assumed. They therefore devised a method free of assumption (8), based on intensity of sampling, but depending on a relatively high frequency of multiple recapture. The term 'multiple recapture' has been used extensively and is appropriate with mark-grouping methods (e.g. Bailey's) where the number of marks an animal bears (or the number of occasions on which it has been captured) affects its classification. However, it will be noted that in the date-grouping methods no significance is attached to any mark other than the last and hence, as Jolly (1965) points out, the term multiple recapture should not be used in connection with these methods.

Choice of method for a series of marking occasions
Comparative studies on actual or simulated data have been made by several workers including Parr (1965), Sheppard *et al.* (1969), Manly (1970), Roff (1973a) and Bishop & Sheppard (1973). Insights may often be gained if different methods are used and their discrepancies carefully considered. There are, however, certain general indications: the use and comparison of more than one method of estimation can be instructive.

As mentioned above the advent of ready access to computers has increased the utility of Fisher & Ford's trellis method: it is a robust method for small samples and, provided the survival rate remains relatively constant, gives population estimates not dissimilar to those of the Jolly-Seber method. It has the additional advantage that periodicity of availability for sampling, known from biological knowledge, may be incorporated in the trellis model (see Conway *et al.* 1974).

Bailey's Triple Catch and the Jolly-Seber method both provide estimates of the variance, however, these have been shown to be related to the population estimate and are particularly unreliable for small samples. The Jolly-Seber method population estimate is usually reliable when 9% or more of the population is sampled and the survival rate is not less than 0.5 (Bishop & Sheppard, 1973); it is less sensitive to age-dependent variations in the mortality rate than Fisher & Ford's method, but is not of course independent of them. Except when there are only a very limited number of sampling occasions the Jolly-Seber method is superior to Bailey's.

The Jolly-Seber method may seriously overestimate the survival rate (Bishop & Sheppard, 1973): it remains however probably the most useful method.

Manly & Parr's method is not affected by age-dependent mortality but required the sampling of a relatively high proportion of the population (>25 % for populations under 250, >10 % for larger populations) (Manly, 1970). When these conditions can be achieved it should be used if mortality is thought to be related to age, particularly if it is high early in the life span studied.

The Fisher — Ford Method

The data from a series of recaptures are set out in a 'trellis diagram'; a constant survival rate that best fits the data is found by a trial and error process, and one essentially makes a series of Lincoln Index estimates working backwards in time from day to day from when the releases were originally made. Then the population estimate ($\hat{N}t$) is:

$$\hat{N}_t = \frac{n_t a_i \phi_{i-t}}{r_{ti}} \tag{3.16}$$

where n_t = total sample at time t, a_i = total marked animals released at time i, ϕ_{i-t} = the survival rate over the period $i - t$ and r_{ti} = the recaptures at time t of animals marked at time i. Details of the method are given by Fisher & Ford (1947) and in an expository form by Dowdeswell (1959) and Parr (1965). Bishop & Sheppard (1973) provide a computer programme in ALGOL.

As already mentioned, one of the strengths of this robust model is that additional variables such as the periodicity of availability may be incorporated in the model and the computer used to find the best fit to the various parameters (Sheppard *et al.* 1969; Conway *et al.* 1974). A model for the mosquito, *Aedes aegypti*, assuming a four-day feeding (and hence availability) cycle with some secondary feeding on the second and third days, is shown in Table 3.2. The total recaptures for any day will have components bearing the marks of different days. For example, r_4 may contain recaptures from the first second and third marking occasions, and the values of these will, if the model is appropriate fit the expressions: $p_4 a_1 \phi^3 \beta_3$, $p_4 a_2 \phi^2 \beta_2$ and $p_4 a_3 \phi \beta_1$ respectively. The sampling intensity (catch rate), p, will vary from occasion to occasion; here it is the value for the fourth occasion. The marked animals (from a_i) available for recapture are modified by the survival rate, ϕ and availability proportion β. As it is a four day cycle β itself will have four values and each represents the different proportions available (in this case feeding) on each day of the cycle: the value of β_0 is 1 (all survivors are available on the fourth day), the other β-values, given by Conway *et al* (1974) are somewhat complex algebraically because in their situation the periodicity of availability was complicated by some mosquitoes feeding on the second and third days of the cycle, as well as on the first.

Table 3.2 Population model for mark and recapture with constant survival (ϕ), but cycling proportionate availability (β) and varying sampling intensity (catch rate) (p): the modification of the Fisher & Ford model by Conway et al. (1974)

1	$n_1\ a_1$							
2	$n_2\ a_2$	$p_2\ a_1\ \phi\ \beta_1$						
3	$n_3\ a_3$	$p_3\ a_1\ \phi^2\ \beta_2$	$p_3\ a_2\ \phi\ \beta_1$					
4	$n_4\ a_4$	$p_4\ a_1\ \phi^3\ \beta_3$	$p_4\ a_2\ \phi^2\ \beta_2$	$p_4\ a_3\ \phi\ \beta_1$				
5	$n_5\ a_5$	$p_5\ a_1\ \phi^4\ \beta_0$	$p_5\ a_2\ \phi^3\ \beta_3$	$p_5\ a_3\ \phi^2\ \beta_2$	$p_5\ a_4\ \phi\ \beta_1$			
6	$n_6\ a_6$	$p_6\ a_1\ \phi^5\ \beta_1$	$p_6\ a_2\ \phi^4\ \beta_0$	$p_6\ a_3\ \phi^3\ \beta_3$	$p_6\ a_4\ \phi^2\ \beta_2$	$p_6\ a_5\ \phi\ \beta_1$		
7	$n_7\ a_7$	$p_7\ a_1\ \phi^6\ \beta_2$	$p_7\ a_2\ \phi^5\ \beta_1$	$p_7\ a_3\ \phi^4\ \beta_0$	$p_7\ a_4\ \phi^3\ \beta_3$	$p_7\ a_5\ \phi^2\ \beta_2$	$p_7\ a_6\ \phi\ \beta_1$	
8	$n_8\ a_8$	$p_8\ a_1\ \phi^7\ \beta_3$	$p_8\ a_2\ \phi^6\ \beta_2$	$p_8\ a_3\ \phi^5\ \beta_1$	$p_8\ a_4\ \phi^4\ \beta_0$	$p_8\ a_5\ \phi^3\ \beta_3$	$p_8\ a_6\ \phi^2\ \beta_2$	$p_8\ a_7\ \phi\ \beta_1$

However if we simplify and generalize the model and assume that any cohort of the animals in question is only available for sampling every fourth day, $\beta_0 = 1$ and β_1, β_2 and $\beta_3 = 0$. Certain terms will then disappear from the trellis (Table 3.2). The survival rate and β-values for the trellis may be determined by the normal least squares method and then the sampling intensities for each catch date determined.

In the simplified general case, with synchronized availability, the total population for the ith occasion is then estimated by

$$\hat{N}_i = \frac{n_i(1 - \phi^x)}{p_i \phi^y (1 - \phi)} \tag{3.17}$$

where x = the period of the cycle of availability (e.g. four days) and y = the period between birth and becoming available for capture (e.g. for mosquitoes one day). If y is zero, then the denominator becomes $p_i(1 - \phi)$.

Full details for the special case of *Aedes aegypti* are given by Conway *et al.* (1974); but the availability of a number of insects (including the rhinoceros beetle) to trapping methods shows a marked periodicity related to age, and the method is therefore of wider application.

Bailey's triple-catch method
As indicated above this method is based on a deterministic model of survival and uses an inefficient method of grouping the data, although the latter is not a serious fault if very few animals are recaptured more than once (Sonleitner & Bateman, 1963). The formulae for the calculation of the various parameters are relatively simple. De Lury (1954) has suggested that a series of only three samples is unlikely to measure the magnitude of biological variation encountered. However birth-and death-rates may well be more constant over a short period than over a larger one and furthermore, as pointed out earlier (p. 4), if these (or any) estimates can be related to other estimates based on different methods and assumptions then the reliability of the estimates is considerably strengthened. The mathematical background to this method is given by Bailey (1951) and it is described by Bailey (1952), Richards (1953) and MacLeod (1958). It was found particularly appropriate by Coulson (1962) in a study on craneflies, *Tipula*, with very short life expectancies, but series of triple-catch estimates may be used to estimate a population over a longer period (e.g. Iwao *et al.*, 1963). It seems that the results are most reliable when large numbers are marked and recaptured (Parr, 1965). For the purposes of illustration the sampling is said to take place on days 1, 2 and 3; the intervals between these sampling occasions may be of any length, provided they are long enough to allow for ample mixing of the marked individuals with the remainder of the population and not so long that a large proportion of the marked individuals have died.

With large samples the population on the second day is estimated:

$$\hat{N}_2 = \frac{a_2 n_2 r_{31}}{r_{21} r_{32}} \tag{3.18}$$

where a_2 = the number of newly marked animals released on the second day; n_2 = the total number of animals captured on the second day and r = recaptures with the first subscript representing the day of capture and the second the day of marking; thus r_{21} = the number of animals captured on the second day that had been marked on the first, r_{31} = the number of animals captured on the third day that had been marked on the first. It is clear that we are really concerned with the number of marks and hence the same animal could contribute to r_{31} and r_{32}. The logic of the above formula may be seen as follows:

$$\hat{N}_2 = \frac{\hat{a}_1 n_2}{r_{21}} \tag{3.19}$$

which is the simple Lincoln Index with $\hat{a}_1$ = the estimate of the number of individuals marked on day 1 that are available for recapture on day 2. Now if the death-rate is constant:

$$\frac{\hat{a}_1}{a_2} = \frac{r_{31}}{r_{32}}$$

$$\hat{a}_1 = \frac{a_2 r_{31}}{r_{32}} \tag{3.20}$$

Substituting for $\hat{a}_1$ in equation 3.19 above we arrive at the formula for the population given above. The large-sample variance of the estimate is:

$$\text{var } \hat{N}_2 = \hat{N}_2{}^2 \left(\frac{1}{r_{21}} + \frac{1}{r_{32}} + \frac{1}{r_{31}} - \frac{1}{n_2} \right) \tag{3.21}$$

Where the numbers recaptured are fairly small there is some advantage in using 'Bailey's correction factor', i.e. the addition of 1 so that:

$$\hat{N}_2 = \frac{a_2 (n_2 + 1) r_{31}}{(r_{21} + 1)(r_{32} + 1)} \tag{3.22}$$

with approximate variance:

$$\text{var } \hat{N}_2 = \hat{N}_2{}^2 - \frac{a_2{}^2 (n_2 + 1)(n_2 + 2) r_{31}(r_{31} - 1)}{(r_{21} + 1)(r_{21} + 2)(r_{32} + 1)(r_{32} + 2)} \tag{3.23}$$

The loss rate, which is compounded of the numbers actually dying and the numbers emigrating, is given by:

$$\gamma_{t=0 \to 1} = -\log_e \left(\frac{a_2 r_{31}}{a_1 r_{32}} \right)^{1/t_1} \tag{3.24}$$

where t_1 = the time interval between the first and second sampling occasions. The dilution rate, which is the result of births and immigration, is given by:

$$\beta_{t=1\to 2} = \log_e\left(\frac{r_{21}n_3}{n_2 r_{31}}\right)^{1/t_2}$$

(3.25)

where t_2 = the time interval between the second and third sampling occasions. Both these are measures of a rate per unit of time (the unit being the units of t); Bailey (1952) gives formulae for their variance.

Jolly-Seber stochastic method

The advantages of the stochastic model on which this method is based have already been discussed and its general utility has been indicated (p. 100).

The basic equation in Jolly's method is:

$$\hat{N}_i = \frac{\hat{M}_i n_i}{r_i}$$

(3.26)

where $\hat{N}_i$ = the estimate of population on day i, $\hat{M}_i$ = the estimate of the total number of marked animals in the population on day i (i.e. the counterpart of 'a' in the simple Lincoln Index), r_i = the total number of marked animals recaptured on day i and n_i = the total number captured on day i.

The procedure may be demonstrated by the following example taken from Jolly (1965). (The notation is slightly modified to conform with the rest of this section.)

(1) The field data are tabulated as in Table 3.3 according to the date of initial capture (or mark) and the date on which the animal was last captured. The columns are then summed to give the total number of (animals released on the ith occasion = s_i of Jolly), subsequently recaptured (R_i), e.g. for day 7, R = 108.

(2) Another table is drawn up (Table 3.4) giving the total number of animals recaptured on day i bearing marks of day j or earlier (Jolly's a_{ij}); this is done by adding each row in Table 3.3 from left to right and entering the accumulated totals. The number marked before time i which are not caught in the ith sample, but are caught subsequently (Z_i), is found by adding all but the top entry (printed in bold) in each column. Thus Z_7 is given by the figures enclosed in Table 3.3. The figures above the line, i.e. the top entry in each column, represent the number of recaptures (r_i) for the day on its right, e.g. $r_7 = 112$.

(3) Then the estimate of the total number of marked animals at risk in the population on the sampling day may be made:

$$\hat{M}_i = \frac{a_i Z_i}{R_i} + r_i$$

(3.27)

Table 3.3 The tabulation of recapture data according to the data on which the animal was last caught for analysis by Jolly's method (after Jolly, 1965)

Day of capture i	Total captured n_i	Total released a_i	\multicolumn{13}{c}{Day when last captured (j)}												
1	54	54	*1*												
2	146	143	10	*2*											
3	169	164	3	34	*3*										
4	209	202	5	18	33	*4*									
5	220	214	2	8	13	30	*5*								
6	209	207	2	4	8	20	43	*6*							
7	250	243	1	6	5	10	34	56	*7*						
8	176	175	0	4	0	3	14	19	46	*8*					
9	172	169	0	2	4	2	11	12	28	51	*9*				
10	127	126	0	0	1	2	3	5	17	22	34	*10*			
11	123	120	1	2	3	1	0	4	8	12	16	30	*11*		
12	120	120	0	1	3	1	1	2	7	4	11	16	26	*12*	
13	142		0	1	0	2	3	3	2	10	9	12	18	35	*13*
$R_i =$			80	70	71	109	101	108	99	70	58	44	35		

Thus
$$\hat{M}_7 = \frac{243 \times 110}{108} + 112 = 359.50$$

Similarly for other $\hat{M}_i$, the results being entered in Table 3.5 in which other population parameters are entered as calculated.

(4) The proportion of marked animals in the population at the moment of capture on day i is found and entered in the final table:

$$\alpha_i = \frac{r_i}{n_i} \tag{3.28}$$

Thus
$$\alpha_7 = \frac{112}{250} = 0.4480$$

(5) The total population is then estimated for each day (Table 3.5). (equations 3.26 and 3.28)

$$\hat{N}_i = \frac{\hat{M}_i}{\alpha_i}$$

(6) The probability that an animal alive at the moment of release of the ith sample will survive till the time of capture of the $i + 1$th sample is found:

$$\hat{\phi}_i = \frac{M_{i+1}}{\hat{M}_i - r_i + a_i} \tag{3.29}$$

Table 3.4 Calculated table of the total number of marked animals recaptured on a given day bearing marks of day or earlier (after Jolly, 1965)

Day i /						Day $i-1$						
1												
10	*2*											
3	**37**	*3*										
5	23	**56**	*4*									
2	10	23	**53**	*5*								
2	6	14	34	**77**	*6*							
1	7	12	22	56	**112**	*7*						
0	4	4	7	21	40	**86**	*8*					
0	2	6	8	19	31	59	**110**	*9*				
0	0	1	3	6	11	28	50	**84**	*10*			
1	3	6	7	7	11	19	31	47	**77**	*11*		
0	1	4	5	6	8	15	19	30	46	**72**	*12*	
0	1	1	3	6	9	11	21	30	42	60	**95**	*13*

$Z(i-1) + 1 = 14$ 57 71 89 121 110 132 121 107 88 60
Z_2 Z_3 Z_4 Z_5 Z_6 Z_7 Z_8 Z_9 Z_{10} Z_{11} Z_{12}

Survival rates estimates slightly over one may arise from sampling effects, but 'rates' greatly above this indicate a major error! Frequently it will be found that the marks of one occasion have been lost or were not recognized. This survival rate may be converted to a loss rate (the effect of death and emigration):

$$\hat{\gamma}_i = 1 - \hat{\phi}_i \tag{3.30}$$

(7) The number of new animals joining the population in the interval between the ith and $i + 1$th samples and alive at time $i + 1$ is given by:

$$\hat{B}_i = \hat{N}_{i+1} - \hat{\phi}_i(\hat{N}_i - n_i + a_i) \tag{3.31}$$

This may be converted to the dilution rate (β):

$$\frac{1}{\hat{\beta}} = 1 - \frac{\hat{B}_i}{\hat{N}_{i+1}} \tag{3.32}$$

(8) If desired the standard errors (the square roots of the variances) are obtained from:

$$\text{var}(\hat{N}_1) = \hat{N}_i(\hat{N}_i - n_i)\left[\frac{\hat{M}_i - r_i + a_i}{\hat{M}_i}\left(\frac{1}{R_i} - \frac{1}{a_i}\right) + \frac{1 - \alpha_i}{r_i}\right] + \hat{N}_i - \sum_{j=0}^{i=1}\frac{\hat{N}_i^2(j)}{\hat{B}_j} \tag{3.33}$$

The special problems involved in the computation of the summation term are discussed by Jolly (1965).

Table 3.5. The final table for a Jolly type mark and recapture analysis (after Jolly, 1965)

Day	Proportion of recaptures	No. marked animals at risk	Total population	Survival rate	No. of new animals	Standard errors			Standard errors due to errors in the estimation of parameter itself	
i	$\hat{\alpha}_i$	$\hat{M}_i$	$\hat{N}_i$	$\hat{\phi}_i$	$\hat{B}_i$	$\sqrt{\{V(\hat{N}_i)\}}$	$\sqrt{\{V(\hat{\phi}_i)\}}$	$\sqrt{\{V(\hat{B}_i)\}}$	$\sqrt{\{V(\hat{N}_i/N_i)\}}$	$\sqrt{\left\{V'(\hat{\phi}_i) - \dfrac{\hat{\phi}_i^2(1-\hat{\phi}_i)}{\hat{M}_{i+1}}\right\}}$
1	—	0	—	0.649	—	—	0.114	—	—	0.093
2	0.0685	35.02	511.2	1.015	263.2	151.2	.110	179.2	150.8	.110
3	.2189	170.54	779.1	0.867	291.8	129.3	.107	137.7	128.9	.105
4	.2679	258.00	963.0	.564	406.4	140.9	.064	120.2	140.3	.059
5	.2409	227.73	945.3	.836	96.9	125.5	.075	111.4	124.3	.073
6	.3684	324.99	882.2	.790	107.0	96.1	.070	74.8	94.4	.068
7	.4480	359.50	802.5	.651	135.7	74.8	.056	55.6	72.4	.052
8	.4886	319.33	653.6	.985	−13.8	61.7	.093	52.5	58.9	.093
9	.6395	402.13	628.8	.686	49.0	61.9	.080	34.2	59.1	.077
10	.6614	316.45	478.5	.884	84.1	51.8	.120	40.2	48.9	.118
11	.6260	317.00	506.4	.771	74.5	65.8	.128	41.1	63.7	.126
12	.6000	277.71	462.8	—	—	70.2	—	—	68.4	—
13	.6690	—	—	—	—	—	—	—	—	—

$$\text{var }(\hat{\phi}_i) = \phi_i{}^2 \left[\frac{(\hat{M}_{i+1} - r_{i+1})(\hat{M}_{i+1} - r_{i+1} + a_{i+1})}{\hat{M}_{i+1}{}^2} \left(\frac{1}{R_{i+1}} - \frac{1}{a_{i+1}} \right) \right.$$

$$\left. + \frac{\hat{M}_i - r_i}{\hat{M}_i - r_i + a_i} \left(\frac{1}{R_i} - \frac{1}{a_i} \right) + \frac{1 - \phi_i}{\hat{M}_{i+1}} \right] \tag{3.34}$$

$$\text{var }(\hat{B}_i) = \frac{\hat{B}_i{}^2 (\hat{M}_{i+1} - r_{i+1})(\hat{M}_{i+1} - r_{i+1} + a_{i+1})}{\hat{M}_{i+1}{}^2} \left(\frac{1}{R_{i+1}} - \frac{1}{a_{i+1}} \right) + \frac{\hat{M}_i - r_i}{\hat{M}_i - r_i - a_i}$$

$$\times \left[\frac{\phi_i a_i (1 - \alpha_i)}{\alpha_i} \right]^2 \left(\frac{1}{R_i} - \frac{1}{a_i} \right)$$

$$+ \frac{(\hat{N}_i - n_i)\hat{N}_{i+1} - \hat{B}_i)(1 - \alpha_i)(1 - \phi_i)}{\hat{M}_i - r_i + a_i}$$

$$+ \hat{N}_{i+1}(\hat{N}_{i+1} - n_{i+1}) \frac{1 - \alpha_{i+1}}{r_{i+1}} + \phi_i{}^2 \hat{N}_i (\hat{N}_i - n_i) \frac{1 - \alpha_i}{r_i} \tag{3.35}$$

Jolly (1965) also gives terms for the covariance. These equations for the variance are fairly complex and if standard errors are required for a long series of estimates, then they should if possible be programmed on a computer.

(9) Jolly shows that the variances of the population and survival rate estimates given above contain an error component due to the real variation in population numbers, apart from errors of estimation. If only errors of estimation are required the formulae are:

$$\text{var }(\hat{N}_i/N_i) = \hat{N}_i(\hat{N}_i - n_i) \left\{ \frac{\hat{M}_i - r_i + a_i}{\hat{M}_i} \left(\frac{1}{R_i} - \frac{1}{a_i} \right) + \frac{1 - \alpha_i}{r_i} \right\} \tag{3:36}$$

$$\text{var }(\hat{\phi}_i \,|\, \phi_i) = \text{var }(\hat{\phi}_i) - \frac{\phi_i{}^2 (1 - \phi_i)}{\hat{M}_{i+1}} \tag{3.37}$$

The errors that arise in this method may be thought of as having two causes: (a) 'small sample biases' inherent in the method and (b) biases due to the violation of the assumptions. The former will be minimal when the number of recaptures is large, and therefore Jolly suggests that when the cost of marking is high, it might be an advantage if the sampling for recaptures was separated from the sampling and marking of further animals to provide the next value of a_i. Sampling for recaptures could be carried out extensively by less experienced staff than are needed to mark.

Manly (1971b) and Roff (1973a) have shown that these errors often express themselves as underestimates of the variance, because of the correlation between the estimates and their variances. Some improvement is obtained if N and ϕ are transformed to logarithms before making the calculations (Manly, 1971).

The effect of unequal catchability on the estimates has been investigated by Carothers (1973) and Gilbert (1973). Their studies show that reasonable estimates may be obtained if certain additional conditions can be met. For example, if the population is closed and probabilities of capture remain constant, the mid-day of the sampling period will provide a less biased estimate; additional sampling days greatly reduce the bias; and provided all animals have a probability of capture > 0.5, few errors arise (Gilbert, 1973). Such a high probability of recapture is unusual but not impossible particularly in small populations [e.g. Southwood & Scudder's (1956) study of lacebugs] and it is in small populations that errors are particularly serious. Gilbert stresses that if this high probability of recapture can be achieved then unequal (heterogeneous) catchability is not in itself serious: for these special conditions, at least, a welcome counterblast to Roff's (1973a & b) pessimism! Carothers (1973) shows that violation of the equal catchability assumption is not so serious if the variations in the probability of capture are symmetrical both in regard to time and the individuals: however, for populations with, for example, a high degree of 'trapshyness' the problem is more severe. This emphasizes the importance of ensuring that the collecting method and/or the collectors skill does not vary during the experiment; if, for example, insects were being collected by sweeping (p. 240) a change in collector could significantly affect the probability of individuals living near the base of the vegetation being recaptured.

The Jolly-Seber method has been used with a number of different insects (e.g. Parr, 1965; Sheppard *et al.* 1969; Fletcher, 1973; Ito, Murai, Teruya, Hamada & Sugimoto, 1974). Computer programmes are provided by Davies (1971, programme 47) and White (1971).

Manly and Parr's Method

The assumption that survival probability is equal is often quite unjustified in entomological studies (Iwao *et al.* 1966). To overcome this Manly & Parr, (1968) devised a method based on the intensity of sampling, each animal being considered to have the same chance of capture on the i th occasion, so that

$$\hat{N}_i = \frac{n_i}{p_i} \tag{3.38}$$

where p_i = the sampling intensity. In practice:

$$\hat{p}_i = \frac{r_i}{\hat{M}_i} \tag{3.39}$$

where, as in the Jolly-Seber method, $\hat{M}_i$ is the estimate of the total number of marked animals present in the population on the ith occasion: a summation of the survivors from the various a's, so $\hat{M}_i \approx \hat{a}$ (as used in Bailey's Triple Catch). Thus Manly & Parr's equation becomes (equation 3.7):

$$\hat{N}_i = \frac{an_i}{r_i}$$

Absolute Population Estimates

the basic Lincoln-Petersen index. The animals must be individually marked or have date-specific marks. Manly & Parr prepared an individual animal table allocating the symbol x for the first or last occurrence of an individual mark, y for the intermediate occasion when individual mark was captured and z for occasions when it was there, but not recaptured, i.e. the blanks left between the two x's after the y's have been inserted. A greatly curtailed example is given in Table 3.6. The sampling intensity will clearly be the ratio of those animals known to be present before and after, captured out of the total number

Table 3.6 Manly & Parr's method of recording data for their method of estimation (obviously one would normally mark far more than three animals on each occasion).

		1	2	3	4	5	6
				Sampling Days			
1		x	y	z	x	–	–
2		x	z	y	z	x	–
3		x	–	–	–	–	–
4			x	z	x	–	–
5			x	y	z	y	x
6			x	z	y	x	–
7				x	y	z	x
8				x	z	x	–
9				x	y	z	x
$\Sigma y_i =$			1	2	3	1	
$\Sigma z_i =$			1	3	3	2	

available known to be present before and after. For any day these are given by Σy_i and $(\Sigma y_i + \Sigma z_i)$ respectively. Then:

$$\hat{p}_i = \Sigma y_i / (\Sigma y_i + \Sigma z_i) \tag{3.40}$$

or (from Table 3.6)

$$\hat{p}_3 = 2/(2+3) = 0.4$$

Thus the population estimate is (equations 3.38 and 3.40):

$$\hat{N}_i = \frac{n_i}{p_i} = \frac{n_i(\Sigma y_i + \Sigma z_i)}{\Sigma y_i} \tag{3.41}$$

survival may be estimated:

$$\phi_{i-(i+1)} = \frac{r_{i,\,i+1}}{n_i \times b_{i+1}} \tag{3.42}$$

where $r_{i,\,i+1}$ = the animals caught in both the ith and $(i+1)$ samples.

The births, number of new animals entering the population, are estimated by:

$$\hat{B}_{i-(i+1)} = \hat{N}_{i+1} - \hat{\phi}_i \hat{N}_i \qquad (3.43)$$

Manly (1969) has developed an expression for the variance of the population estimate:

$$\text{var } (\hat{N}_i) = \frac{\hat{N}_i(\hat{N}_i - n_i)(\hat{N}_i - C_i)}{n_i C_i} \qquad (3.44)$$

where $C_i = \Sigma y_i + \Sigma z_i$

As mentioned this method is not robust unless a fairly large sample is taken (Manly, 1970). Seber (1973) shows how the necessary data may be tabulated in a form similar to that of the Jolly-Seber method: this would undoubtedly be more convenient with large sets of data, but Manly & Parr's original arrangement is simpler for exposition.

3.2.2 Frequency of capture methods (Schnabel census)

In considering the assumption of equal probability of capture it was necessary to examine the frequency of capture, many models and tests have been developed of which a few are referred to above (p. 95). The information may be considered another way: if the frequency of capture follows a particular distribution (f_1 ... animals captured once, f_2 ... animals captured twice, f_3 ... f_x), then the term f_0 represents those animals that have not been marked or captured at all and the sum of all the terms will represent the total population. Theoretically problems do not arise from unequal catchability, a non-Poisson distribution may be fitted that describes the deviations from random in the probability of capture. Tanton (1965) fitted a zero truncated negative binomial to the frequency of recapture of wood mice, *Apodemus*. Further details of such models, particularly useful in small mammal work (e.g. Marten, 1970) are given by Seber (1973). (See also page 230 et seq.)

3.2.3 Craig's method: constant probability of capture

This method, devised for use with butterflies assumes that the population is closed (there are no births or deaths and the animal stays within its habitat) and there is a constant probability of capture. In practice the butterflies, or other highly mobile but colonial animals, are collected randomly, marked and immediately released, after recording the number of times, if any, the animal had previously been captured. The effective sample size is thus one. Craig (1953) assumed that the frequency of recapture could be described by two mathematical models: the truncated Poisson (p. 28) and the Stevens' distribution function. On the basis of these two models six different methods can be used to estimate the size of the population, three based on moments and three on maximum likelihood: Craig found, however, that they gave similar results. The two simplest are the moment estimates based on the Poisson; they are:

(1)* $$\hat{N} = (\Sigma x f_x{}^2/(\Sigma x^2 f_x - \Sigma x f_x)) \qquad (3.45)$$

where $\hat{N}$ = the estimate of population, x = the number of times an individual had been marked, f_x = the frequency with which individuals marked x times had been caught and thus $\Sigma x f_x$ = the total number of different times animals were captured (viz. $1 \times$ the number caught once $+ 2 \times$ the number caught twice $+ 3 \times$ the number caught thrice, etc). This method is simple, allowing a direct solution, but is subject to greater sampling error. It is useful for obtaining a trial value for use in solving the equation in method 2 (and in the other methods given by Craig).

(2) $$\log \hat{N} - \log(\hat{N} - \Sigma f_x) = \Sigma x f_x/N \qquad (3.46)$$

Where the symbols are as above, Σf_x being the total different individual animals caught (viz. the number caught once + the number caught twice + the number caught thrice, etc.). Craig (1953) gives formulae for the variances of these estimates:

for method 1 $$\text{var } \hat{N} = \frac{2N}{\lambda^2} \qquad (3.47)$$

where $\lambda = \Sigma x f_x$

for method 2 $$\text{var } \hat{N} = \frac{N}{(e^\lambda - 1 - \lambda)} \qquad (3.48)$$

where e is the base of natural (Napierian) logarithms and thus the value of e^λ is found by using a table of natural logs 'backwards'.

It is clear that this method is based on different assumptions from the Lincoln Index. It demands that the animals be very mobile so that their chances of recapture are virtually random almost immediately after release and yet they must not leave the habitat. It has been used with butterflies and might be applied to other large conspicuous flying or very mobile animals under certain circumstances. Phillips & Campbell (1970) showed that the Schnabel Census method is more robust when portions of the population are inaccessible.

3.2.4 Change in ratio methods (Kelker's selective removal)
This method uses the natural marks of a population, normally the difference between the sexes, but theoretically any other recognizable distinction could be used (e.g. the different morphs of a polymorphic species). The proportion of the different forms or components are determined, a known number of one form is removed ('selective removal') and the new ratio found; from the change in the ratio the total population can be estimated:

---

* The actual numbering of these methods is the reverse of that given by Craig, but it is most logical here to give the simplest method first.

$$\hat{N} = K_\alpha \div \left[D_{\alpha_1} - \frac{(D_{\beta_1} D_{\alpha_2})}{D_{\beta_2}} \right] \tag{3.49}$$

where α and β are the components (e.g. sexes) in a population, α is the component part of which is removed during the interval between times 1 and 2, and β is the component (or components) the member which are not removed, then K_α = number of component α that are killed; D = the proportion of the population represented by a component at time 1 or 2; thus D_{α_1} = the proportion of α in the population as a decimal before any were removed; D_{β_1} = the proportion of the population as a decimal represented by β after K individuals of α were removed.

Such an approach is clearly most applicable with game where individuals of certain sex or age class are killed (e.g. Kelker, 1940; Scattergood, 1954; Hanson, 1963); it is sometimes referred to as the dichotomy method (Chapman, 1954).

The mathematical background has been developed by Chapman (1955), who compared the results from this method with those given by the Lincoln type capture – recapture analysis; he showed that capture – recapture estimation procedure will yield more information for the same amount of effort. A further objection to the use of this method with insect populations is that the selective removal and the resulting atypical unbalance could seriously prejudice any further study of the population; in populations of game animals such removal is one of the 'normal' mortality factors.

This method does not depend on the assumption that initial capture does not alter the probability of subsequent recapture, and this is an advantage over those methods that utilize artifical marks and might indicate conditions for its use, perhaps only as a test of estimates derived by Lincoln type methods. Chapman (1955) gives a technique for combining the two approaches. Paulik & Robson (1969) and Seber (1973) provide comprehensive reviews of this approach.

REFERENCES

ABDEL-MALEK, A. A., 1961. The effect of radioactive phosphorus on the growth and development of *Culex pipiens molestus* Forsk. (Diptera, Culicidae). *Bull. ent. Res.* **52**, 701–8.

ABDEL-MALEK, A.A. and ABDEL-WAHAB, M.F., 1961*a*. Autoradiography as a technique for radioactive phosphorus, P-32, uptake in *Culex molestus* Forsk. (Diptera Culicidae). *Bull. Soc. ent. Egypt.* **45**, 409–18.

ABDEL-MALEK, A. A. and ABDEL-WAHAB, M. F., 1961*b*. Studies on the phosphorus-32 uptake in *Schistocerca gregaria* (Forsk.) and *Anacridium aegyptium* (L.) (Orthoptera: Acrididae). *Bull. Soc. ent. Egypt.* **45**, 419–25.

ANDREEV, S. V., *et al.*, 1963. Radioisotopes and radiation in animal and plant insect pest control. [In Russian with Eng., Fr. and Span. sum.] *Radiation and radioisotopes applied to insects of agricultural importance. Int. atomic Energy Ag.* STI/PUB/74, 115–32.

ANON., 1961, *Proceedings of the symposium on the use of biophysics in the field of plant protection.* [In Russian.] *Vsesoyuz. Inst. Zashch. Rast. Leningrad. (Rev. appl. Ent.* (A) **51**, 522–3; (B) **51**, 212).

ANON., 1963. Radioisotopes and ionizing radiations in entomology. *Bibl. Ser. Int. Atomic Energy Ag.* **9**, 414 pp.

AYRE, G. L., 1962. Problems in using the Lincoln Index for estimating the size of ant colonies (Hymenoptera: Formicidae). *J. N. Y. ent. Soc.* **70**, 159–66.

BABENKO, L. V., 1960. The use of radioactive isotopes for labelling ticks. [In Russian.] *Med. Parazitol.* **29**, 320–4.

BABERS, F. H., ROAN, C. C. and WALKER, R. L., 1954. Tagging boll weevils with radioactive cobalt. *J. econ, Ent.* **47**, 928–9.

BAILEY, N. T. J., 1951. On estimating the size of mobile populations from recapture data. *Biometrika* **38**, 293–306.

BAILEY, N. T. J., 1952. Improvements in the interpretation of recapture data. *J. Anim. Ecol.* **21**, 120–7.

BAILEY, S. F., ELIASON, D. A. and ILTIS, W. C., 1962. Some marking and recovery techniques in *Culex tarsalis* Coq. flight studies. *Mosquito News,* **22**, 1–10.

BALDWIN, W. F., ALLEN, J. R. and SLATER, N. S., 1966. A practical field method for the recovery of Blackflies labelled with Phosphorus-32. *Nature* **212**, 959–60.

BALDWIN, W. F., and COWPER, G., 1969. The use of radioactive platinum–iridium wire (Ir^{192}) as an internal tag for tracing insects. *Can. Ent.* **101**, 151–2.

BALDWIN, W. F., JAMES, H. G. and WELCH, H. E., 1955. A study of predators of mosquito larvae and pupae with a radioactive tracer. *Can. Ent.* **87**, 350–6.

BALDWIN, W. F., RIORDAN, D. F. and SMITH, R. W., 1958. Note on dispersal of radioactive grasshoppers. *Can. Ent.* **90**, 374–6.

BANKS, C. J., 1955. The use of radioactive tantalum in studies of the behaviour of small crawling insects on plants. *Br. J. Anim. Behav.* **3**, 158–9.

BANKS, C. J., BROWN, E. S. and DEZFULIAN, A., 1961. Field studies of the daily activity and feeding behaviour of Sunn Pest, *Eurygaster integriceps* Put. (Hemiptera: Scutelleridae) on wheat in North Iran. *Entomologia exp. appl.* **4**, 289–300.

BARNES, M. M., 1959. Radiotracer labelling of a natural tephritid population and flight range of the walnut husk fly. *Ann. ent. Soc. Am.* **52**, 90–2.

BARTHOLOMEW, R. M., 1967. A study of the winter activity of bobwhites through the use of radio telemetry. *Occ. Pap. C. C. Adams Cent. Ecol. Stud. Kalamazoo* **17**, 25 pp.

BATE, L. C., LYON, W. S. and WOLLERMAN, E. H., 1974. Gold tagging of elm bark beetles and identification by neutron activation analysis. *Radiochem. Radioanal. Letters* **17**(1), 77–85.

BEARD, R. L., 1965. Competition between DDT-resistant and susceptible house flies. *J. econ. Ent.* **58**, 584.

BENNETT, G. F., 1963. Use of P^{32} in the study of a population of *Simulium rugglesi* (Diptera: Simuliidae) in Algonquin Park, Ontario. *Can. J. Zool.* **41**, 831–40.

BERRY, W. L., STIMMANN, M. W. and WOLF, W. W., 1972. Marking of native phytophagous insects with Rubidium: a proposed technique. *Ann. ent. Soc. Am.* **65**, 236–8.

BEVERTON, R. J. H. and HOLT, S. J., 1957. On the dynamics of Exploited Fish populations. *Fishery investigations, ser. 2.* **19**, 533 pp. Min. Agric. Fish. Food Gt Britain, London. H.M.S.O.

BISHOP, J. A. and SHEPPARD, P. M., 1973. An evaluation of two capture–recapture models using the technique of computer simulation. In Bartlett, M. S. & Hiorns, R. W. (Ed.) *The Mathematical Theory of the Dynamics of Biological Populations,* pp. 235–52. Academic Press, London.

BLACK, J. B., 1963. Observations on the home range of stream-dwelling crawfishes. *Ecology* **44**, 592–5.

BLINN, W. C., 1963. Ecology of the land snails *Mesodon thyroidus* and *Allogona profunda. Ecology* **44**, 498–505.

BOHLIN, T. and SUNDSTROM, B., 1977. Influence of unequal catchability on population estimates using the Lincoln Index and the removal method applied to electrofishing. *Oikos* **28**, 123–9.

BORROR, D. J., 1934. Ecological studies of *Argia moesta* Hagen (Odonata: Coenagrionidae) by means of marking. *Ohio J. Sci.* **34**, 97–108.

BOWEN, H. J. M. and GIBBONS, D., 1963. *Radioactive Analysis.* Oxford University Press, Oxford.

BRUSSARD, P. F., 1971. Field techniques for investigations of population structure in a "ubiquitous" butterfly. *J. Lep. Soc.* **25**, 22–9.

BUCKNER, C. H., 1968. Reactions of small mammals to vital dyes. *Can. Ent.* **100**, 476–7.

BUGHER, J. and TAYLOR, M., 1949. Radiophosphorus and radiostrontium in mosquitoes. *Science* **110**, 146–7.

CALDWELL, A. H., 1956. Dry ice as an insect anaesthetic. *J. Econ. Ent.* **49**, 264–5.

CAROTHERS, A. D., 1971. An examination and extension of Leslie's test of equal catchability. *Biometrics* **27**, 615–30.

CAROTHERS, A. D., 1973. The effects of unequal catchability on Jolly-Seber estimates. *Biometrics* **29**, 79–100.

CHANG, H. T., 1946. Studies on the use of fluorescent dyes for marking *Anopheles quadrimaculatus* Say. *Mosquito News.* **6**, 122–5.

CHAPMAN, D. G., 1954. The estimation of biological populations. *Ann. math. Statist.* **25**, 1–15.

CHAPMAN, D. G., 1955. Population estimation based on change of composition caused by a selective removal. *Biometrika* **42**, 279–90.

CLARK, D. P., 1962. An analysis of dispersal and movement in *Phaulacridium vittatum* (Sjost.) (Acrididae). *Aust. J. Zool.* **10**, 382–99.

COLEMAN, D. C., 1968. A method for intensity labelling of fungi for ecological studies. *Mycologia* **60**, 960–1.

COLLINS, C. W. and POTTS, S. F., 1932. Attractants for the flying gypsy moths as an aid to locating new infestations. *U.S.D.A. Tech. Bull.* **336**, 43 pp.

COMAR, C. L., 1955. *Radioisotopes in biology and agriculture.* 481 pp., Academic Press, New York.

CONWAY, G. R., TRIPIS, M. and MCCLELLAND, G. A. H., 1974. Population parameters of the mosquito *Aedes aegypti* (L.) estimated by mark–release–recapture in a suburban habitat in Tanzania. *J. Anim. Ecol* **43**, 289–304.

COOK, L. M. and KETTLEWELL, H. B. D., 1960. Radioactive labelling of lepidopterous larvae: a method of estimating larval and pupal mortality in the wild. *Nature* **187**, 301–2.

CORBET, G. B., 1956. The life-history and host-relations of a Hippoboscid fly *Ornithomyia fringillina* Curtis. *J. Anim. Ecol.* **25**, 403–20.

CORBET, P. S., 1952. An adult population study of *Pyrrhosoma nymphula* (Sulzer); (Odonata: Coenagrionidae). *J. Anim. Ecol.* **21**, 206–22.

CORMACK, R. M., 1968. The statistics of capture–recapture methods. *Oceanogr. Mar. Bio. Ann. rev.* **6**, 455–506.

CORMACK, R. M., 1972. The logic of capture–recapture estimates. *Biometrics* **28**, 337–43.

CORMACK, R. M., 1973. Commonsense estimates from capture–recapture studies. In Bartlett, M. S. & Hiorns, R. W. (Ed.) *The Mathematical Theory of the Dynamics of Biological Populations,* pp. 225–34. Academic Press, London.

CORNWELL, P. B., 1955. Techniques for labelling trees with radioactive phosphorus. *Nature* **175**, 85–7.

CORNWELL, P. B., 1956. Some aspects of mealybug behaviour in relation to the efficiency of measures for the control of virus diseases of cacao in the Gold Coast. *Bull. ent. Res.* **47**, 137–66.

COTTAM, C., 1956. Uses of marking animals in ecological studies: marking birds for scientific purposes. *Ecology* **37**, 675–81.

COULSON, J. C., 1962. The biology of *Tipula subnodicornis* Zetterstedt, with comparative observations on *Tipula paludosa* Meigen. *J. Anim. Ecol.* **31**, 1–21.

CRAGG, J. B. and HOBART, J., 1955. A study of a field population of the blowflies *Lucilia ceasar* (L.) and *L. serricata* (MG.). *Ann. appl. Biol.* **43**, 645–63.

CRAIG, C. C., 1953. On the utilisation of marked specimens in estimating populations of flying insects. *Biometrika* **40**, 170–6.

CRUMPACKER, D. W., 1974. The use of micronized fluorescent dusts to mark adult *Drosophila pseudoobscura*. *Am. Midl. Nat.* **91**, 118–29.

CUNNINGHAM, M. P., HARLEY, J. M. and GRAINGE, E. B., 1963. The labelling of animals with specific agglutinins and the detection of these agglutinins in the blood meals of *Glossina*. *Rep. E. Afr. Tryp. Res. Org.* **1961**, 23–4.

DAHMS, P. A., 1957. Uses of radioisotopes in pesticide research. *Adv. pest cont. Res.* **1**, 81–146.

DALMAT, H. T., 1950. Studies on the flight range of certain Simuliidae, with the use of aniline dye marker. *Ann. ent. Soc. Am.* **43**, 537–45.

DARROCH, J. N., 1958. The multiple-recapture census. I. Estimation of a closed population. *Biometrika* **45**, 343–59.

DARROCH, J. N., 1959. The multiple-capture census II. Estimation when there is immigration or death. *Biometrika* **46**, 336–51.

DAUM, R. J., GAST, R. T. and DAVICH, T. B., 1969. Marking adult boll weevils with dyes fed in a cottonseed oil bait. *J. econ. Ent.* **62**, 943–4.

DAVEY, J. T., 1956. A method of marking isolated adult locusts in large numbers as an aid to the study of their seasonal migrations. *Bull. ent. Res.* **46**, 797–802.

DAVEY, J. T. and O'ROURKE, F. J., 1951. Observations on *Chrysops silacea* and *C. dimidiata* at Benin, Southern Nigeria. Part II. *Ann. trop. Med. Parasit., Liverpool.* **45**, 66–72.

DAVIES, R. G., 1971. *Computer Programming in Quantitative Biology*. Academic Press, London and New York.

DAVIS, J. M. and NAGEL, R. H., 1956. A technique for tagging large numbers of live adult insects with radioisotopes. *J. econ. Ent.* **49**, 210–1.

DE LURY, D. B., 1954. On the assumptions underlying estimates of mobile populations. *In* Kempthorne, O. *et al.* (eds.) *Statistics and Mathematics in Biology* 287–93. Hafner, New York.

DEMPSTER, J. P., 1971. The population ecology of the cinnabar moth, *Tyria jacobaeae* L. (Lepidoptera, Arctiidae). *Oecologia (Berl.)*, **7**, 26–67.

DINTHER, J. B. M. VAN., 1972. Evaluation of marking methods for studying adult activity of the cabbage root fly, *Hylemya brassicae* (Bche) in the field. *Meded. Landbouw. Weten. Gent.* **37**(2), 738–46.

DOBSON, R. M., 1962. Marking techniques and their application to the study of small terrestrial animals. *In* Murphy, P. W. (ed.) *Progress in Soil Zoology*, 228–39. Butterworths, London.

DOBSON, R. M. and MORRIS, M. G., 1961. Observations on emergence and life-span of wheat bulb fly, *Leptohylemyia coarctata* (Fall.) under field-cage conditions. *Bull. ent. Res.* **51**, 803–21.

DOBSON, R. M., STEPHENSON, J. W. and LOFTY, J. R., 1958. A quantitative study of a population of wheat bulb fly, *Leptohylemyia coarctata* (Fall.) in the field. *Bull. ent. Res.* **49**, 95–111.

DOW, R. P., 1959. Dispersal of adult *Hippelates pusio*, the eye gnat. *Ann. ent. Soc. Am.* **52**, 372–81.

DOW, R. P., 1971. The dispersal of *Culex nigripalpus* marked with high concentrations of radiophosphorus. *J. med. Ent.* **8**, 353–63.

DOWDESWELL, W. H., 1959. *Practical animal ecology*. 316 pp., Methuen, London.

DOWDESWELL, W. H., FISHER, R. A. and FORD, E. B., 1940. The quantitative study of populations in the Lepidoptera. 1. *Polyommatus icarus* Rott. *Ann. Eugen.* **10,** 123–36.

DUDLEY, J. E. and SEARLES, E. M., 1923. Color marking of the striped cucumber beetle (*Diabrotica vittata* Fab.) and preliminary experiments to determine its flight. *J. econ. Ent.* **16,** 363–8.

DUNCOMBE, W. G.,1959. An autoradiographic method for distinguishing samples labelled with phosphorus-32 and sulphur-35, *Nature* **183,** 319.

DUNN, P. H. and MECHALAS, B. J., 1963. An easily constructed vacuum duster. *J. econ. Ent.* **56,** 899.

EDDY, G. W., ROTH, A. R. and PLAPP, F. W., 1962. Studies on the flight habits of some marked insects. *J. econ. Ent.* **55,** 603–7.

EDGAR, W. D., 1971. The life-cycle, abundance and seasonal movement of the wolf spider, *Lycosa (Pardosa) lugubris,* in central Scotland. *J. Anim. Ecol.* **40,** 303–22.

EDWARDS, R. L., 1958. Movements of individual members in a population of the shore crab, *Carcinus maenas* L., in the littoral zone. *J. Anim. Ecol.* **27,** 37–45.

EDWARDS, R. L., 1961. Limited movement of individuals in a population of the migratory grasshopper, *Melanoplus bilituratus* (Walker) (Acrididae) at Kamloops, British Columbia. *Can. Ent.* **93,** 628–31.

EHRLICH, P. R. and DAVIDSON, S. E., 1960. Techniques for capture–recapture studies of Lepidoptera populations. *J. Lepidopterist's Soc.* **14,** 227–9.

ERICKSON, J. M., 1972. Mark–recapture techniques for population estimates of *Pogonomyrmex* and colonies: an evaluation of the ^{32}P technique. *Ann. ent. Soc. Am.* **65,** 57–61.

EVANS, W. G. and GYRISCO, G. G., 1960. The flight range of the European chafer. *J. econ. Ent.* **53,** 222–4.

FAIRES, R. A. and PARKS, B. H., 1958. *Radioisotope laboratory techniques.* 244 pp., Butterworths, London.

FALES, J. H., BODENSTEIN, P. F., MILLS, G. D. and WESSEL, L. H., 1964. Preliminary studies on face fly dispersion. *Ann. ent. Soc. Am.* **57,** 135–7.

FAY, R. W., KILPATRICK, J. W. and BAKER, J. T., 1963. Rearing and isotopic labelling of *Fannia canicularis. J. econ. Ent.* **56,** 69–71.

FISHER, R. A. and FORD, E. B., 1947. The spread of a gene in natural conditions in a colony of the moth *Panaxia dominula* L. *Heredity* **1,** 143–74.

FITZGERALD, P. J., 1958. Autoradiography in biology and medicine. *Second U.N. int. Conf. peaceful uses of atomic energy.*

FLETCHER, B. S., 1973. The ecology of a natural population of the Queensland fruit fly, *Dacus tryoni* IV. The immigration and emigration of adults. *Aust. J. Zool.* **21,** 541–65.

FLETCHER, T. B., 1936. Marked migrant butterflies, *Ent. Rec.* **48,** 105–6.

FOOTT, W. H., 1976. Use of fluorescent powders to monitor flight activities of adult *Glischrochilus quadrisignatus* [Coleoptera: Nitidulidae], *Can. Ent.* **108,** 1041–44.

FRANCIS, G. E., MULLIGAN, W. and WORMALL, A., 1959. *Isotopic tracers.* 2nd ed. London.

FRANKIE, G. W., 1973. A simple field technique for marking bees with fluorescent powders. *Ann. Ent. Soc. Am.* **66,** 690–1.

FRAZER, B. D. and RAWORTH, D. A., 1974. Marking aphids with rubidium. *Can. J. Zool.* **52,** 1135–6.

FREDEEN, F. J. H., SPINKS, J. W. T., ANDERSON, J. R., ARNASON, A. P. and REMPEL, J. G., 1953. Mass tagging of black flies (Diptera: Simuliidae) with radio-phosphorus. *Can. J. Zool.* **31,** 1–15.

FREDERICKSEN, C. F. and LILLY, J. H., 1955. Measuring wireworm reactions to soil insecticides by tagging with radioactive cobalt. *J. econ. Ent.* **48,** 438–42.

FREEMAN, B. E., 1964. A population study of *Tipula* species (Diptera, Tipulidae). *J. Anim. Ecol.* **33,** 129–40.

FRISCH, K. VON, 1950. *Bees, their vision, chemical senses, and language.* Ithaca, New York.

GANGWERE, S. K. CHAVIN, W. and EVANS, F. C., 1964. Methods of marking insects, with especial reference to Orthoptera (Sens. lat.). *Ann. ent. Soc. Am.* **57**, 662–9.

GARBY, L., YASUNO, M. and PHURIUETHAYA, Y., 1966. Labelling mosquitoes with radioactive iodine I. *Trans. R. Soc. trop. Med. Hyg.* **60**, 136.

GARDINER, J. E., 1963. A radioactive marking ink. *Nature* **197**, 414.

GARY, N. E., 1971. Magnetic retrieval of ferrous labels in a capture–recapture system for honey bees and other insects. *J. econ. Ent.* **64**(4), 961–5.

GASKELL, T. J. and GEORGE, B. J., 1972. A Bayesian modification of the Lincoln Index. *J. appl. Ecol.* **9**, 377–84.

GERARD, B. M., 1963. An earthworm labelled with radioactive tantalum. *Nature* **200**, 486–7.

GILBERT, R. O., 1973. Approximations of the bias in the Jolly-Seber capture–recapture model. *Biometrics* **29**, 501–26.

GILLIES, M. T., 1958. A simple autoradiographic method for distinguishing insects labelled with phosphorus-32 and sulphur-35. *Nature* **182**, 1683–4.

GILLIES, M. T., 1961. Studies on the dispersion and survival of *Anopheles gambiae* Giles in East Africa, by means of marking and release experiments. *Bull. ent. Res.* **52**, 99–127.

GODFREY, G. K., 1954. Tracing field voles (*Microtus agrestis*) with a Geiger-Müller counter. *Ecology* **35**, 5–10.

GODWIN, P. A., JAYNES, H. A. and DAVIS, J. M., 1957. The dispersion of radio-actively tagged white pine weevils in small plantations. *J. econ. Ent.* **50**, 264–6.

GOLLEY, F. B. and GENTRY, J. B., 1964. Bioenergetics of the southern harvester ant, *Pogonomyrmex badius, Ecology* **45**, 217–25.

GOODHART, C. B., 1962. Thrush predation on the snail *Cepaea hortensis. J. Anim. Ecol.* **27**, 47–57.

GRAHAM, B. F., 1957. Labelling pollen of woody plants with radioactive isotopes. *Ecology* **38**, 156–8.

GRAHAM, H. M. and MANGUM, C. L., 1971. Larval diets containing dyes for tagging pink bollworm moths internally. *J. econ. Ent.* **64**(2), 376–9.

GRAHAM, W. J. and AMBROSE, H. W., 1967. A technique for continuously locating small mammals in field enclosures. *J. Mammal.* **48**, 639–42.

GREEN, G. W., BALDWIN, W. F. and SULLIVAN, C. R., 1957. The use of radioactive cobalt in studies of the dispersal of adult females of the European pine shoot moth *Rhyacionia buoliana* (Schiff.). *Can. Ent.* **89**, 379–83.

GREENSLADE, P. J. M., 1964. The distribution, dispersal and size of a population of *Nebria brevicollis* (F.), with comparative studies on three other Carabidae. *J. Anim. Ecol.* **33**, 311–33.

GRIFFIN, D. R., 1952. Radioactive tagging of animals under natural conditions. *Ecology* **33**, 329–35.

GRÜM, L., 1959. Seasonal changes of activity of the Carabidae. *Ekol. Polska A* **7**, 255–68.

GULLAND, J. A., 1955. On the estimation of population parameters from marked members. *Biometrika* **42**, 269–70.

HAISCH, A., STARK, H. and FORSTER, S., 1975. Markierung von Fruchtfliegen und ihre erkennung durch indikatoraktivierung. *Entomologia exp. appl.* **18**, 31–43.

HANSON, W. R., 1963. Calculation of productivity, survival and abundance of selected vertebrates from sex and age ratios. *Wildl. Monogr.* **9**, 60 pp.

HARTLEY, P. H. T., 1954. Back garden ornithology. *Bird Study,* **1**, 18–27.

HAUSERMANN, W., FAY, R. W. and HACKER, C. S., 1971. Dispersal of genetically marked female *Aedes aegypti* in Mississippi. *Mosquito News* **31**, 37–51.

HAWKES, C., 1972. The estimation of the dispersal rate of the adult cabbage root fly (*Erioischia brassicae* Bouche) in the presence of a brassica crop. *J. appl. Ecol.* **9**, 617–32.

HAWKES, R. B., 1972. A fluorescent dye technique for marking insect eggs in predation studies. *J. econ. Ent.* **65**(5), 1477–8.

HAYES, J. T., 1969. A microcage for radioassay of live insects with automatic counting equipment. *Int. J. appl. Radiat. Isotopes* **20**, 603–4.

HENDRICKS, D. E., 1971. Oil-soluble blue dye in larval diet marks adults, eggs, and first stage F_1 larvae of the pink bollworm. *J. econ. Ent.* **64**(6), 1404–6.

HENDRICKS, D. E. and GRAHAM, H. M., 1970. Oil-soluble dye in larval diet for tagging moths, eggs, and spermatophores of tobacco budworms. *J. econ. Ent.* **63**, 1019–20.

HERON, R. J., 1968. Vital dyes as markers for behavioural and population studies of the larch sawfly, *Pristiphora erichsonii* (Hymenoptera: Tenthredinidae). *Can. Ent.* **100**, 470–5.

HEWLETT, P. S., 1954. A micro-drop applicator and its use for the treatment of certain small insects with liquid insecticide. *Ann. appl. Biol.* **41**, 45–64.

HINTON, H. E., 1954. Radioactive tracers in entomological research. *Sci. Progr., London* **42**, 292–305.

HOFFMAN, R. A., LINDQUIST, A. W. and BUTTS, J. S., 1951. Studies on treatment of flies with radioactive phosphorus. *J. econ. Ent.* **44**, 471–3.

HOLLANDER, J. ., PERLMAN, I. and SEABORG, G. T., 1953. Table of isotopes. *Rev. Mod. Physics* **25**, 469–651.

HOLT, S. J., 1955. On the foraging activity of the wood ant. *J. Anim. Ecol.* **24**, 1–34.

HORN, H. S., 1976. A clamp for marking butterflies in capture–recapture studies. *J. Lepidopterists' Soc.* **30**(2), 145–6.

HUBBELL, S. P., SIKORA, A. and PARIS, O. H., 1965. Radiotracer, gravimetric and calorimetric studies of ingestion and assimilation rates of an Isopod. *Health Physics* **11**(12), 1485–1501.

HUNTER, P. E., 1960. Plastic paint as a marker for mites. *Ann. ent. Soc. Am.* **53**, 698.

ITO, Y., 1970. A stable isotope, Europium–151 as a tracer for field studies of insects. *Appl. Ent. Zool.* **5**, 175–81.

ITO, Y., 1973. A method to estimate a minimum population density with a single recapture census. *Res. Pop. Ecol.* **14**, 159–68.

ITO, Y., MURAI, M., TERUYA, T., HAMADA, R. and SUGIMOTO, A., 1974. An estimation of population density of *Dacus cucurbitae* with mark-recapture methods. *Res. Pop. Ecol.* **15**, 213–22.

IWAO, S., KIRITANI, K. and HOKYO, N., 1966. Application of a marking and recapture method for the analysis of larval–adult populations of an insect, *Nezara viridula* (Hemiptera: Pentatomidae). *Res. Pop. Ecol.* **8**, 147–60.

IWAO, S., MIZUTA, K., NAKAMURA, H., ODA, T. and SATO, Y., 1963. Studies on a natural population of the large 28–spotted lady beetle, *Epilachna vigintioctomaculata* Motschulsky. 1. Preliminary analysis of the overwintered adult population by means of the marking and recapture method. *Jap. J. Ecol.* **13**, 109–17.

JACKSON, C. H. N., 1933*a*. On the true density of tsetse flies. *J. Anim. Ecol.* **2**, 204–9.

JACKSON, C. H. N., 1933*b*. On a method of marking tsetse flies. *J. Anim. Ecol* **2**, 289–90.

JACKSON, C. H. N., 1937. Some new methods in the study of *Glossina morsitans*. *Proc. zool. Soc. Lond.* **1936**, 811–96.

JACKSON, C. H. N., 1939. The analysis of an animal population. *J. Anim. Ecol.* **8**, 238–46.

JACKSON, C. H. N., 1940. The analysis of a tsetse fly population. *Ann. Eugen.* **10**, 332–369.

JACKSON, C. H. N., 1944. The analysis of a tsetse-fly population. II *Ann. Eugen.* **12**, 176–205.

JACKSON, C. H. N., 1948. The analysis of a tsetse-fly population. III. *Ann. Eugen.* **14**, 91–108.

JACKSON, C. H. N., 1953. A mixed population of *Glossina morsitans* and *G. swynnertoni*. *J. Anim. Ecol.* **22**, 78–86.

JAHN, E., LIPPAY, H., WEIDINGER, N. and SCHWACH, G., 1966. Untersuchungen uber die usbreitung von Nonnenfaltem durch Markierung mit Seltenen erden. *Anz. Schadlingskunst* **39**, 17–22.

JAMES, H. G., 1961. Some predators of *Aedes stimulans* (Walk.) and *Aedes trichurus* (Dyar.) (Dipt.: Culicidae) in woodland pools. *Can. J. Zool.* **39**, 533–40.

JENKINS, D. W., 1949. A field method of marking arctic mosquitoes with radiophosphorus. *J. econ. Ent.* **40**, 988–9.

JENKINS, D. W., 1963. Use of radionuclides in ecological studies of insects. *In* Schultz, V. and Klement, A. W. (eds) *Radioecology*, pp. 431–43. Rheinhold, New York.

JENKINS, D. W. and HASSETT, C. C., 1950. Radioisotopes in entomology. *Nucleonics* **6**, 5–14.

JENSEN, J. A. and FAY, R. W., 1951. Tagging of adult horse flies and flesh flies with radioactive phosphorus. *Am. J. trop. Med.* **31**, 523–30.

JEWELL, G. R., 1956. Marking of tsetse flies for their detection at night. *Nature* **178**, 750.

JEWELL, G. R., 1958. Detection of tsetse fly at night. *Nature* **181**, 1354.

JOHANSSON, T. S. K., 1959. Tracking honey bees in cotton fields with fluorescent pigments. *J. econ. Ent.* **52**, 572–7.

JOHNSON, C. G., 1940. The maintenance of high atmosphere humidities for entomological work with glycerol-water mixtures. *Ann. appl. Biol.* **27**, 295–9.

JOLLY, G. M., 1963. Estimates of population parameters from multiple recapture data with both death and dilution – a deterministic model. *Biometrika* **50**, 113–28.

JOLLY, G. M., 1965. Explicit estimates from capture-recapture data with both death and immigration – stochastic model. *Biometrika* **52**, 225–47.

KELKER, G. H., 1940. Estimating deer populations by a differential hunting loss in the sexes. *Proc. Utah Acad. Sci., Arts. Lett.* **17**, 65–9.

KETTLEWELL, H. B. D., 1952. Use of radioactive tracer in the study of insect populations (Lepidoptera). *Nature* **170**, 584.

KHUDADOV, G. D., 1959*a*. The method of marking insects by introducing radioactive isotopes incorporated into foodstuffs. [In Russian, with Eng. summary.] *Byull. mosk. Obshch. Ispÿt Prir.* (*N.S.*) *Otd. Biol* **64**(3), 35–45.

KHUDADOV, G. D., 1959*b*. Radioautographic method of finding insects and ticks marked with radioactive isotopes [In Russian.] *Med. Parazitol.* **28**, 60–4.

KLOCK, J. W., PIMENTEL, D. and STENBURG, R. L., 1953. A mechanical fly-tagging device. *Science* **118**, 48–9.

KLOFT, W., HOLLDOBLER, B. and HAISHCH, A., 1965. Traceruntersuchungen zur Abgrenzurg von Nestarealen Holzzerstorender Rossameisen (*Camponotus herculeanus* L. und *C. ligniperda* Satr.). *Entomologia exp. appl.* **8**, 20–6.

KNAPP, S. E., FARINACCI, C. J., HERBERT, C. M. and SAENGER, E. L., 1956. A method for labelling the Lone Star tick with a radioactive indicator (P^{32}). *J. econ. Ent.* **49**, 393–5.

KNIGHT, R. H., and SOUTHON, H. A. W., 1963. A simple method for marking haematophagous insects during the act of feeding. *Bull. ent. Res.* **54**, 379–82.

KRALL, J. H. and SIMMONS, G. A., 1977. Tree root injection of phosphorus-32 for labelling defoliating insects. *Environ. Ent.* **6**, 159–60.

KRUK-DE-BRUIN, M., ROST, L. C. M. and DRAISAMA, F. G. A. M., 1977. Estimates of the number of foraging ants with the Lincoln-index method in relation to the colony size of *Formica polyctena*. *J. Anim. Ecol.* **46**, 457–70.

KUENZLER, E. J., 1958. Niche relations of three species of Lycosid spiders. *Ecology* **39**, 494–500.

LABRECQUE, G. C., BAILEY, D. L., MEIFERT, D. W. and WEIDHAAS, D. E., 1975. Density estimates and daily mortality rate evaluations of stable fly (*Stomoxys calcitrans* (Diptera: Muscidae)) populations in field cages. *Can. Ent.* **107**, 597–600.

LAMB, R. J. and WELLINGTON, W. G., 1974. Techniques for studying the behavior and ecology of the European earwig, *Forficula auricularia* (Dermaptera: Forficulidae). *Can. Ent.* **106**, 881–8.

LE CREN, E. D., 1965. A note on the history of mark-recapture population estimates. *J. Anim. Ecol.* **34**, 453–4.

LEEUWEN, E. R. VAN, 1940. The activity of adult codling moths as indicated by captures of marked moths. *J. econ. Ent.* **33**, 162–6.

LESLIE, P. H., 1952. The estimation of population parameters from data obtained by means of the capture-recapture method II. The estimation of total numbers. *Biometrika* **39**, 363–88.

LESLIE, P. H. and CHITTY, D., 1951. The estimation of population parameters from data obtained by means of the capture-recapture method. I. The maximum likelihood equations for estimating the death-rate. *Biometrika* **38**, 269–92.

LEVIN, M. O., 1961. Distribution of foragers from honey bee colonies placed in the middle of a large field of alfalfa. *J. econ. Ent.* **54**, 431–4.

LEWIS, C. T. and WALOFF, N., 1964. The use of radioactive tracers in the study of dispersion of *Orthotylus virescens* (Douglas & Scott) (Miridae, Heteroptera). *Entomologia exp. appl.* **7**, 15–24.

LINCOLN, F. C., 1930. Calculating waterfowl abundance on the basis of banding returns. *U.S.D.A. Circ.* **118**, 1–4.

LINDQUIST, A. W., 1952. Radioactive materials in entomological research. *J. econ. Ent.* **45**, 264–70.

LINDROTH, A., 1953. Internal tagging of salmon smolt. *Inst. Freshw. Res. Drottningholm* **34**, 49–57.

LOMNICKI, A., 1969. Individual differences among adult members of a snail population. *Nature* **223**, 1073–4.

MCALLAN, J. W. and NEILSON, W. T. A., 1965. Labelling the apple maggot with strontium 89. *J. econ. Ent.* **58**, 168.

MCDONALD, W. A., 1960. Nocturnal detection of tsetse flies in Nigeria with ultra-violet light. *Nature* **185**, 867–8.

MACLEOD, J., 1958. The estimation of numbers of mobile insects from low-incidence recapture data. *Trans. R. ent. Soc. Lond.* **110**, 363–92.

MACLEOD, J. and DONNELLY, J., 1957. Individual and group marking methods for fly-population studies. *Bull. ent. Res.* **48**, 585–92.

MANLY, B. F. J., 1969. Some properties of a method of estimating the size of mobile animal populations. *Biometrika* **56**, 407–10.

MANLY, B. F. J., 1970. A simulation study of animal population estimation using the capture–recapture method. *J. appl. Ecol.* **7**, 13–39.

MANLY, B. F. J., 1971a Estimates of a marking effect with capture–recapture sampling. *J. appl. Ecol.* **8**, 181–9.

MANLY, B. F. J., 1971b A simulation study of Jolly's method for analysing capture–recapture data. *Biometrics* **27**(2), 415–24.

MANLY, B. F. J. and PARR, M. J., 1968. A new method of estimating population size survivorship, and birth rate from capture – recapture data. *Trans. Soc. Brit. Ent.* **18**, 81–9.

MARTEN, G. G., 1970. A regression method for mark–recapture estimates with unequal catchability. *Ecology* **51**, 291–5.

MEDER, O., 1926. Über die kennzeichnung von weisslingen zwecks Erfassung ihrer Wanderung. *Int. ent. Z.* **19**, 325–30.

MEDLEY, J. G. and AHRENS, E. H., 1968. Fluorescent dyes for marking and recovering fowl ticks in poultry houses treated with insecticides. *J. econ. Ent.* **61,** 81–4.

MEIJDEN, E. VAN DER., 1973. Experiments on dispersal, late-larval predation, and pupation in the Cinnabar moth (*Tyria jacobaeae* L.) with a radioactive label (^{192}Ir). *Netherl. J. Zool.* **23,** 430–45.

MELLANBY, K., 1939. The physiology and activity of the bed-bug (*Cimex lectularius* L.) in a natural infestation. *Parasitology* **31,** 200–11.

MICHENER, C. D., CROSS, E. A., DALY, H. V., RETTENMEYER, C. W. and WILLE, A., 1955. Additional techniques for studying the behaviour of wild bees. *Insectes Sociaux* **2,** 237–46.

MITCHELL, B., 1963. Ecology of two carabid beetles, *Bembidion lampros* (Herbst.) and *Trechus quadristriatus* (Schrank.) *J. Anim. Ecol.* **32,** 377–92.

MITCHELL, H. C., McGOVERN, W. L., CROSS, W. H. and MITLIN, N., 1973. Boll weevils: tagging for hibernation and field studies. *J. econ. Ent.* **66**(2), 563–4.

MONK, C. D., 1967. Radioisotope tagging through seed soaking. *Bull. Georgia Acad. Sci.* **25**(1), 13–7.

MONRO, J., 1968. Marking insects with dietary manganese for detection by neutron activation. *Ecology* **49,** 774–6.

MOTH, J. J. and BARKER, J. S. F., 1975. Micronized fluorescent dusts for marking *Drosophila* adults. *J. nat. Hist.* **9,** 393–6.

MUIR, R. C., 1958. On the application of the capture–recapture method to an orchard population of *Blepharidopterus angulatus* (Fall.) (Hemiptera-Heteroptera, Miridae). *Rep. E. Malling Res. Sta.* **1957,** 140–7.

MURDOCH, W. W., 1963. A method for marking Carabidae (Col.). *Ent. mon. Mag.* **99,** 22–4.

MUSGRAVE, A. J., 1949, The use of fluorescent material for marking and detecting insects. *Can. Ent.* **81,** 173.

MUSGRAVE, A. J., 1950. A note on the dusting of crops with flurorescein to mark visiting bees. *Can. Ent.* **82,** 195–6.

NAKAMURA, K., ITO, Y., MIYASHITA, K. and TAKAI, A., 1967. The estimation of population density of the green rice leafhopper, *Nephotettix cincticeps* Uhler, in spring field by the capture–recapture method. *Res. Pop. Ecol,* **9,** 113–29.

NAKAMURA, K., ITO, Y., NAKAMURA, M., MATSUMOTO, T. and HAYAKAWA, K., 1971. Estimation of population productivity of *Parapleurus alliaceus* Germar (Orthoptera: Acridiidae) on a *Miscanthus sinensis* Anders. Grassland. I Estimation of population parameters. *Oecologia (Berl.)* **7,** 1–15.

NIELSEN, E. T., 1961. On the habits of the migratory butterfly, *Ascia monuste* L. *Biol. Meddr. Dan. Vid. Selsk.* **23**(11), 81 pp.

NORRIS, K. R., 1957. A method of marking Calliphoridae (Diptera) during emergence from the puparium. *Nature* **180,**1002.

O'BRIEN, R. D. and WOLFE, L. S., 1964. *Radiation, radioactivity, and insects.* 211 pp., Academic Press, New York and London.

ODUM, E. P. and PONTIN, A. J., 1961. Population density of the underground ant, *Lasius flavus*, as determined by tagging with P^{32}, *Ecology* **42,** 186–8.

O'GRADY, J. J. and HOY, J. B., 1972. Rhodamine-B and other stains as markers for the mosquito fish, *Gambusia affinis. J. med. Ent.* **9,** 571–4.

ORIANS, G. H. and LESLIE, P. H., 1958. A capture–recapture analysis of a shearwater population. *J. Anim. Ecol.* **27,** 71–86.

ORPHANIDIS, P. S., SOULTANOPOULOS, C. D. and KARANEINOS, M. G., 1963. Essai preliminaire avec P^{32} sur la dispersion des adultes du *Dacus oleae* Gmel. *Radiation and radioisotopes applied to insects of agricultural importance. Int. atomic Energy Ag.* STI/PUB/74, 101–4.

OUGHTON, J., 1951. Tagging root maggot flies by means of radioactive phosphorus. *Ann. Rep. ent. Soc. Ontario* (81st) **1950,** 91–2.

PAJUNEN, V. I., 1962. Studies on the population ecology of *Leccorrhinia dubia* V. D. Lind. (Odon., Libellulidae). *Ann. Zool. Soc. 'vanamo'* **24**(4), 79 pp.

PAL, R., 1947. Marking mosquitoes with fluorescent compounds and watching them by ultra-violet light. *Nature* **160**, 298–9.

PARR, M. J., 1965. A population study of a colony of imaginal *Ischnura elegans* (van der Linden) (Odonata: Coenagriidae) at Dale, Pembrokeshire. *Fld. Stud.* **2**(2), 237–82.

PARR, M. J., GASKELL, T. J. and GEORGE, B. J., 1968. Capture–Recapture methods of estimating animal numbers. *J. Biol. Educ.* **2**, 95–117.

PAULIK, G. J. and ROBSON, D. S., 1969. Statistical calculations, for change-in-ratio estimators of population parameters. *J. Wildl. Manag.* **33**, 1–27.

PEER, D. F., 1957. Further studies on the mating range of the honey bee, *Apis mellifera* L. *Can. Ent.* **89**, 108–10.

PEFFLY, R. L. and LABRECQUE, G. C., 1956. Marking and trapping studies on dispersal and abundance of Egyptian house flies. *J. econ. Ent.*, **49**, 214–17.

PELEKASSIS, C. E. D., MOURIKIS, P. A. and BANTZIOS, D. N., 1962. Preliminary studies on the field movement of the olive fruit fly (*Dacus oleae* Gmel.)) by labelling a natural population with radioactive phosphorus(P^{32}). *Ann. Inst. Phytopath. Benaki* (N.S.) **4**, 170–9.

PENDLETON, R. C., 1956. The uses of marking animals in ecological studies: labelling animals with radioactive isotopes. *Ecology* **37**, 686–9.

PHILLIPS, B. F. and CAMPBELL, N. A., 1970. Comparison of methods of estimating population size using data on the whelk *Dicathais aegrota* (Reeve). *J. Anim. Ecol.* **39**, 753–9.

POLIVKA, J., 1949. The use of fluorescent pigments in a study of the flight of the Japanese bettle. *J. econ. Ent.* **42**, 818–21.

QUARTERMAN, K. D., MATHIS, W. and KILPATRICK, J. W., 1954. Urban fly dispersal in the area of Savannah, Georgia. *J. econ. Ent.* **47**, 405–412.

QUERCI, O., 1936. Aestivation of Lepidoptera. *Ent. Rec.* **48**, 122.

QURAISHI, M. S., 1963a. Water and food relationship of the eggs and first instar nymph of *Eurygaster integriceps* with the aid of P^{32}. *J. econ. Ent.* **56**, 666–8.

QURAISHI, M. S., 1963b. Use of isotopes for investigating the behaviour and ecology of insect pests in some recent studies. *Radiation and radioisotopes applied to insects of agricultural importance. Int. atomic Energy Ag.* STI/PUB/**74**, 93–9.

RADELEFF, R. D., BUSHLAND, R. C. and HOPKINS, D. E., 1952. Phosphorus-32 labelling of the screw-worm fly. *J. econ. Ent.* **45**, 509–14.

RADINOVSKY, S. and KRANTZ, G. W., 1962. The use of Fluon to prevent the escape of stored-product insects from glass containers. *J. econ. Ent.* **55**, 815–16.

RADIOCHEMICAL MANUAL., 1966. Radio Chemical Centre, Amersham, U.K. 327 pp.

RAHALKAR, G. W., MISTRY, K. B., HARNALKAR, M. R., BHARATHAN, K. G. and GOPAL-AYENGAR, A. R. 1971. Labelling adults of red palm weevil (*Rhynchophorus ferrugineus*) with Cerium for detection by neutron activation. *Ecology* **52**(1), 187–8.

RAKITIN, A. A., 1963. The use of radioisotopes in the marking of *Eurygaster integriceps* Put. [In Russian.] *Ent. Obozr.* **42**, 39–48. (Transl. *Ent. Rev.* **42**, 20–38).

REEVES, W. C., BROOKMAN, B. and HAMMON, W. M., 1948. Studies on the flight range of certain *Culex* mosquitoes, using a fluorescent-dye marker, with notes on *Culiseta* and *Anopheles*. *Mosquito News* **8**, 61–9.

RENNISON, B. D., LUMSDEN, W. H. R. and WEBB, C. J., 1958. Use of reflecting paints for locating tsetse fly at night. *Nature* **181**, 1354.

RIBBANDS, C. R., 1950. Changes in behaviour of honeybees, following their recovery from anaesthesia. *J. exp. Biol.* **27**, 302–10.

RICHARDS, O. W., 1953. The study of the numbers of the red locust, *Nomadacris septemfasciata* (Serville). *Anti-Locust Bull.* **15**, 30 pp.

RICHARDS, O. W. and WALOFF, N., 1954. Studies on the biology and population dynamics of British grasshoppers. *Anti-Locust Bull.* **17**, 182 pp.

RICHARDSON, R. H., WALLACE, R. J., GAGE, S. J., BOUCHEY, G. D. and DENNELL, M., 1969. XII. Neutron activation techniques for labelling *Drosophila* in natural populations. In Wheeler, M. R. (ed.) *Studies in genetics V*, 171–86. Univ. Texas Publ. 6918, Houston.

RICKER, W. E., 1956. Uses of marking animals in ecological studies: the marking of fish. *Ecology* **37**, 665–70.

RIEBARTSCH, K., 1963. Inaktive markierung von insekten mit dysproium. *Nachrbl. dt. Pflschutzdienst. Stuttg.* **15**, 154–7.

RINGS, R. W. and LAYNE, G. W., 1953. Radioisotopes as tracers in plum curculio behaviour studies. *J. econ. Ent.* **46**, 473–7.

RIORDAN, D. F. and PESCHKEN, D. P., 1970. A method for obtaining P^{32}–labelled eggs of the flea beetle *Altica carduorum* (Coleoptera: Chrysomelidae). *Can. Ent.* **102**, 1613–6.

ROBERTSON, G., 1973. The sensitivity of scale insects to $^{14}C\beta$-particles and to $^{60}C\gamma$- rays. *Int. J. Radiat. Biol.* **23**(4), 313–23.

ROBSON, D. S., 1969. Mark–recapture methods of population estimation. In Johnson, N. L. & Smith, H. (eds.) *New Developments in Survey Sampling* pp. 120–44. Wiley, New York.

ROER, H., 1959. Beitrag zur Erforschung der Migrationen des Distelfalters (*Vanessa cardui* L.) im paläarktischen Raum unter besonderer Berücksichtigung der Verhältnisse des Jahres 1958. *Decheniana* **111**, 141–8.

ROER, H., 1962. Zur Erforschung der Flug- und Wandergewohnheiten mitteleuropäischer Nymphaliden (Lepidoptera). *Bonn. zool. Beitr.* **10** (1959), 286–97.

ROFF, D. A., 1973a. On the accuracy of some mark–recapture estimators. *Oecologia (Berl.)* **12**, 15–34.

ROFF, D. A., 1973b. An examination of some statistical tests used in the analysis of mark–recapture data. *Oecologia (Berl.)* **12**, 35–54.

ROTH, A. R. and HOFFMAN, R. A., 1952. A new method of tagging insects with P^{32}. *J. econ. Ent.* **45**, 1091.

SANDERS, C. J. and BALDWIN, W. F., 1969. Iridium–192 as a tag for carpenter ants of the genus *Camponotus* (Hymenoptera: Formicidae). *Can. Ent.* **101**, 416–8.

SCATTERGOOD, L. W., 1954. In Kempthorne, O., *et al.* (eds.), *Statistics and mathematics in biology*, 273–85. Hafner, New York.

SCHJØTZ-CHRISTENSEN, B., 1961. Forplantngabiologien hos *Amara infirma* Dft. og *Harpalus neglectus* Serv. *Flora og Fauna* **67**, 8–12.

SCHOOF, H. F. and MAIL, G. A., 1953. The dispersal habits of *Phormia regina*, the black blowfly, in Charleston, West Virginia. *J. econ. Ent.* **46**, 258–62.

SCHOOF, H. F., SIVERLY, R. E. and JENSEN, J. A., 1952. House fly dispersion studies in metropolitan areas. *J. econ. Ent.* **45**, 675–83.

SCHROEDER, W. J., CUNNINGHAM, R. T., MIYABARA, R. Y. and FARIAS, G. J., 1972. A fluorescent compound for marking tephritidae. *J. econ. Ent.* **65**(4), 217–8.

SCOTT, J. D., 1931. A practical method of marking insects in quantitative samples taken at regular intervals. *S. Afr. J. Sci.* **28**, 371–5.

SEBER, G. A. F., 1965. A note on the multiple-recapture census. *Biometrika* **52**, 249.

SEBER, G. A. F., 1973. *The Estimation of Animal Abundance and Related Parameters.* 506 pp., Griffin, London.

SHEMANCHUK, J. A., SPINKS, J. W. T. and FREDEEN, F. J. H., 1953. A method of tagging prairie mosquitoes (Diptera: Culicidae) with radioœphosphorous. *Can. Ent.* **85**, 269–72.

SHEPPARD, P. M., 1951. Fluctuations in the selective value of certain phenotypes in the polymorphic land snail *Cepaea nemoralis* (L.). *Heredity* **5**, 125–54.

SHEPPARD, P. M., MACDONALD, W. W., TONN, R. J. and GRAB, B., 1969. The dynamics of an adult population of *Aedes aegypti* in relation to dengue haemorrhagic fever in Bankok. *J. Anim. Ecol.* **38**, 661–702.

SHEPPARD, D. C., WILSON, G. H. and HAWKINS, J. A., 1973. A device for self-marking of Tabanidae. *J. Environ. Ent.* **2**, 960–1.

SHURA-BURA, B. L. and GRAGEAU, U. L., 1956. Fluorescent analysis studies of insect migration. [in Russian.] *Ent. Obozr.* **35**, 760–3.

SHURA-BURA, B. L., SUCHOMLINOVA, O. I. and ISAROVA, B. I., 1962. Application of the radiomarking method for studying the ability of synanthropic flies to fly over water obstacles. [In Russian.] *Ent. Obozr.* **41**, 99–108.

SIMPSON, J., 1954. Effects of some anaesthetics on honeybees: nitrous oxide, carbon dioxide, ammonium nitrate smoker fumes. *Bee World* **35**, 149–55.

SKUHRAVY, V., 1957. Studium pohybu některých Střevlíkovitých značkováním jedincu. *Acta Soc. ent. Bohem.* **53**, 171–9.

SLACK, K. V., 1955. An injection method for marking crayfish. *Prog. Fish. Cult.* **17**(1), 36–8.

SMITH, M. V., 1958. The use of fluorescent markers as an aid in studying the forage behaviour of honeybees, *Proc. X int. Congr. Ent.* **4**, 1063.

SMITH, M. V. and TOWNSEND, G. F., 1951. A technique for mass-marking honeybees. *Can. Ent.* **83**, 346–8.

SMITTLE, B. J., HILL, S. O. and PHILIPS, F. M., 1967. Migration and dispersal patterns of ^{59}Fe–labelled Lone Star ticks. *J. Econ. Ent.* **60**, 1029–31.

SMITTLE, B. J., LOWE, R. E., FORD, H. R. and WEIDHAAS, D. E., 1973. Techniques for ^{32}P labelling and assay of egg rafts from field-collecting *Culex pipiens quinquefasciatus* Say. *Mosquito News* **33**, 215–20.

SONENSHINE, D. E., 1970. Fertility and fecundity of *Dermacentor variabilis* reared from radioisotope–tagged larvae. *J. econ. Ent.* **63**, 1675–6.

SONENSHINE, D. E. and YUNKER, C. E., 1968. Radiolabelling of tick progeny by inoculation of procreant females. *J. econ. Ent.* **61**, 1612–7.

SONLEITNER, F. J. and BATEMAN, M. A., 1963. Mark–recapture analysis of a population of Queensland fruit-fly, *Dacus tryoni* (Frogg.) in an orchard. *J. Anim. Ecol.* **32**, 259–69.

SOUTH, A., 1965. Biology and ecology of *Agriolimax reticulatus* (Müll.) and other slugs: spatial distribution. *J. Anim. Ecol.* **34**, 403–17.

SOUTHWOOD, T. R. E., 1965. Migration and population change in *Oscinella frit* L. (Diptera) on the oatcrop. *Proc. XII int. Congr. Ent. 420–1.*

SOUTHWOOD, T. R. E. and SCUDDER, G. G. E., 1956. The bionomics and immature stages of the thistle lace bugs (*Tings ampliata* H.–S. and *T. cardui* L.; Hem., Tingidae). *Trans. Soc. Brit. Ent.* **12**, 93–112.

STANILAND, L. N., 1959. Fluorescent tracer techniques for the study of spray and dust deposits. *J. agric. Eng. Res.* **4**, 110–25.

STEINER, L. F., 1965. A rapid method for identifying dye-marked fruit flies. *J. econ. Ent.* **58**, 374–5.

STERN, V. M. and MUELLER, A., 1968. Techniques of marking insects with micronized fluorescent dust with especial emphasis on marking millions of *Lygus hesperus* for dispersal studies. *J. econ. Ent.* **61**, 1232–7.

STRADLING, D. J., 1970. The estimation of worker ant populations by the mark–release–recapture method: an improved marking technique. *J. Anim. Ecol.* **39**, 575–91.

SUDIA, T. W. and LINCK, A. J., 1963. Methods for introducing radionuclides into plants. *In* Schultz. V. and Klement, A. W. (eds.) *Radioecology* 417–23. Rheinhold, New York.

SULLIVAN, C. R., 1953. Use of radioactive cobalt in tracing the movements of the whitepine weevil, *Pissodes strobi* Peck. (Coleoptera: Curculionidae). *Can. Ent.* **85**, 273–6.

SULLIVAN, C. R., 1961. The survival of adults of the white pine weevil, *Pissodes strobi* (Peck), labelled with radioactive cobalt. *Can. Ent.* **93**, 78–9.

SYLVEN, E. and LONSJO, H., 1970. Studies on a method of radioactive tagging of a gall midge, *Dasyneura brassicae* Winn. (Diptera, Cicedomyiidae). *Ent. scand.* **1**, 291–6.

TABER, R. D., 1956. Uses of marking animals in ecological studies: marking of mammals; standard methods and new developments. *Ecology* **37**, 681–5.

TAFT, H. M. and AGEE, H. R., 1962. A marking and recovery method for use in boll weevil movement studies. *J. econ. Ent.* **55**, 1018–19.

TANTON, M. T., 1965. Problems of live-trapping and population estimation for the wood mouse. *J. Anim. Ecol.* **38**, 511–29.

THORNBURN, C. C., 1972. *Isotopes and Radiation in Biology.* Butterworth, London.

TURNER, F. B., 1960. Tests of randomness in re-captures of *Rana pipretiosa*. *Ecology* **41**, 237–9.

URQUART, F. A., 1958. *The monarch butterfly.* 361 pp., University of Toronto Press, Toronto.

WALOFF, Z., 1963. Field studies on solitary and *transiens* desert locusts in the Red Sea area. *Anti-Locust Bull,* **40**, 93 pp.

WAVE, H. E., HENNEBERRY, T. J. and MASON, H. C., 1963. Fluorescent biological stains as markers for *Drosophila*. *J. econ. Ent.* **56**, 890–1.

WEATHERSBEE, A. A. and HASELL, P. G., 1938. On the recovery of stain in adults developing from anopheline larvae stained *in vitro*. *Am. J. trop. Med.* **18**, 531–43.

WELCH, H. E., 1960. Two applications of a method of determining the error of population estimates of mosquito larvae by the mark and recapture technique. *Ecology* **41**, 228–9.

WHITE, E. G., 1970. A self-checking coding technique for mark–recapture studies. *Bull. ent. Res.* **60**, 303–7.

WHITE, E. G., 1971. A versatile Fortran computer program for the capture–recapture stochastic model of G. M. Jolly. *J. Fish, Res. Bd. can.* **28**, 443–5.

WHITE, E. G., 1975. Identifying population units that comply with capture–recapture assumptions in an open community of alpine grasshoppers. *Res. Popul. Ecol.* **16**, 153–87.

WIEGERT, R. G. and LINDEBORG, R. G., 1964. A 'stem well' method of introducing radioisotopes into plants to study food chains. *Ecology* **45**, 406–10.

WILLIAMS, C. B., COCKBILL, G. F., GIBBS, M. E. and DOWNES, J. A., 1942. Studies in the migration of Lepidoptera. *Trans. R. ent. Soc. Lond.* **92**, 101–283.

WOJTUSIAK, R. J., 1958. (New method for marking insects.) *Folia biol. Krakow* **1958**, 6, 71–8.

WOLF, W. W. and STIMMANN, M. W., 1972. An automatic method of marking cabbage looper moths for release–recovery identification. *J. econ. Ent.* **65**(3), 719–22.

WOOD, G. W., 1963. The capture–recapture technique as a means of estimating populations of climbing cutworms. *Can. J. Zool.* **41**, 47–50.

WOODBURY, A. M., 1956. Uses of marking animals in ecological studies: marking amphibians and reptiles *Ecology* **37**, 670–4.

YATES, W. W., GJULLIN, C. M. and LINDQUIST, A. W., 1951. Treatment of mosquito larvae and adults with radioactive phosphorus. *J. econ. Ent.* **44**, 34–7.

YETTER, W. P. and STEINER, L. F., 1932. Efficiency of bait traps for the oriental fruit moth as indicated by the release and capture of marked adults. *J. econ. Ent.* **25**, 106–16.

ZACHARUK, R. Y., 1963. Vital dyes for marking living elaterid larvae. *Can. J. Zool.* **41**, 991–6.

ZAIDENOV, A. M., 1960. Study of house fly (Diptera, Muscidae) migrations in Chita by means of luminiscent tagging. [In Russian.] *Ent. Obozr.* **39,** 574–84. (Trans. *Ent. Rev.* **39,** 406–14.)

ZUKEL, J. W., 1945. Marking *Anopheles* mosquitoes with fluorescent compounds. *Science* **102,** 157.

4

Absolute Population Estimates by Sampling a Unit of Habitat – Air, Plants, Plant Products and Vertebrate Hosts

This is one of four approaches to the absolute population estimate; the other three being the spacing or nearest neighbour methods (p. 47), methods utilizing marked individuals (p. 70) and removal trapping (p. 230). In this approach the habitat is sampled and the contained animals along with it. Hence two separate measurements have to be made: the total number of animals in the unit of the habitat sampled and the total number of these units in the whole habitat of the population being studied. The second may involve the use of the techniques of the botanist, forester, surveyor or hydrologist and cannot be considered in detail here (see also Strickland, 1961). The first concerns the extraction of animals from the samples and sometimes the taking of samples; this and the next two chapters will be concerned mainly with these problems in two biotic (plants and vertebrate animals) and three physical habitats (air, soil and freshwater).

It is important to remember that if sampling is unbiased, the errors in the population estimates obtained by this method will nearly all lie on one side of the true value; all estimates will tend to be underestimates. This is because the efficiencies of the extraction processes will never be more than 100% – one cannot normally find more animals than are present! Usually one finds fewer. As the basis of the approach is to multiply the mean population figure per sampling unit by the number of such units in the habitat, usually a very large figure, the total final underestimation may, in terms of individuals, be large.

4.1 Sampling from the air

In some ways air is the simplest of the five habitats, it is homogeneous in all environments and permits a universal solution which has now been provided by the work of C. G. Johnson and L. R. Taylor. They have developed suction traps, standardized them and measured their efficiency so precisely that aerial populations can now be assessed with a greater level of accuracy than those in most other habitats. The only environmental variable for which allowance needs to be made is wind speed. There is no evidence that insects are in any way attracted or repelled by the traps (Taylor, 1962a), although precautions must be taken to avoid the development of a charge of static electricity on the traps (Maw, 1964). Only a very small proportion of the insects collected in suction traps are damaged by them.

The basic features of the suction trap are an electric fan that pulls or drives air through a fine gauze cone; this filters out the insects which are collected in a jar or cylinder. The trap may be fitted with a segregating device, which separates the catch according to a predetermined time interval (normally an hour), and this provides information, not only about the numbers flying, but also about the periodicity of flight. Few insects are active equally throughout 24 hours (Lewis & Taylor, 1965) and therefore the daily catch divided by the daily intake of air of the fan does not present as realistic a picture of aerial density as if each.hour is considered separately.

4.1.1 Sampling apparatus

Exposed cone type of suction trap
Here the air (and insects) pass through the fan first and subsequently into the copper gauze cone, mesh (26/in = 10.25/cm) (Fig. 4.1); the cone and the

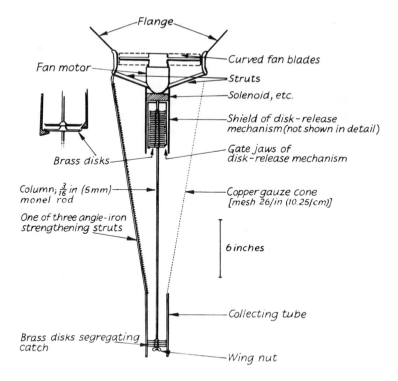

Fig. 4.1 The 9-in exposed cone type of suction trap.

mouth of the fan are both exposed to the effects of cross-winds. The most useful fan unit is the 9-in (23* cm) Vent-Axia ('Silent Nine'). With this fan, cross-winds of more than 14 m/h (22.4km/h) lead to significant reductions in the air intake of the mouth of the fan, the effects of cross-winds are less at speeds of under 10 m/h (16km/h) (Johnson, 1950; Taylor, 1955). If the trap is standing amongst a crop, so that the cone is sheltered from the effects of the cross-winds, these only impinging on the mouth of the fan, the delivery loss is only 0.8 % per m/h (0.5 % per km/h) cross-wind (Taylor, 1955).

The catch is collected in a brass tube at the base of the cone; the inside of this should be brushed with 1 % solution of pyrethrum in petrol ether (Johnson, 1950; Johnson & Taylor, 1955a). If the trap is fitted with the segregating device, a series of brass discs are released by the time switch operated solenoid, one say each hour, and these accumulate in the collecting tube; the catch for each hour being retained between the appropriate pair of discs. The undersurface of the disc has a circle of muslin slightly larger than itself attached to it; this serves to brush the insects down the collecting tube and retains them between the appropriate discs; it also helps to hold a sufficient quantity of pyrethrum within the collecting tube to kill the larger insects. Various improvements to the original suction trap (Johnson, 1950), more especially to the catch-segregating mechanism and collecting tube, were described in detail by Taylor (1951). The mechanism has also been described by Horsfall (1962). 9-in and 12-in Vent-Axia suction traps are now commercially produced by the Burkard Manufacturing Co., Rickmansworth, England, under the name 'Johnson-Taylor traps'.

A 6-in fan unit has also been used with an exposed cone type of trap (Johnson, 1950; Taylor, 1955, 1962a); this is a weaker fan and should only be operated in the shelter of vegetation. Bidlingmayer (1961) used a trap (referred to as 'powered aspirator') with a 5-in (12.7cm) fan for studying the flight of biting midges.

Enclosed cone types of suction trap

In these traps the fan unit is at the bottom of a metal cylinder and the metal gauze collecting cone opens to the mouth of the trap (Fig. 4.2) so that the insects are filtered out before the air passes through the fan. More important, for the performance in higher wind speeds, the cone and mouth of the fan are protected from cross-winds. Johnson & Taylor (1955a and b) developed three traps in this class:

(1)*The 18-in propeller trap*, powered by a Woods 230/50 V, 50 c/s, 1-phase, 400 W, 18in (46cm) diameter propeller fan mounted in a galvanized iron cylindrical duct; the air-filtering cone is of 32-mesh (= 12 mesh/cm) copper gauze and at its base the glass collecting jar screws into a socket; the jar

---

* Throughout this section 'rounded' metric equivalents are given: the traps were of course all designed in Imperial units and these names are retained.

contains 70% alcohol; to empty the trap the cone is removed and the jar unscrewed (Fig. 4.2).

(2) *The 12-in aerofoil trap*, powered by a Woods 230/50 V, 50 c/s, 440 W, 1-phase capacity start, capacitor run, 12 in (30.5 cm) diameter ventilator fan, mounted at the base of a sheet steel cylinder 14 in (35.5 cm) diameter, 4.5 ft (1.35 m) long; the air-filtering cone is of 28–mesh (11 mesh/cm) steel gauze. This trap is fitted with a catch-segregating device.

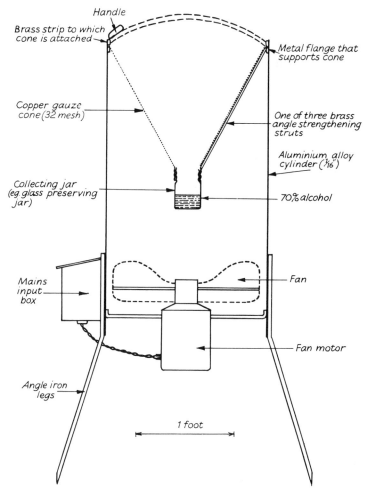

Fig. 4.2. The 18-in propeller enclosed cone type of section trap.

(3) *The 30-in air screw trap*, powered by a specially made light-weight, 30 in (75 cm) diameter fan at the base of a cylindrical cage 7.5 ft (2.3 m) high and 2.5 ft (76 cm) diameter. This trap, with its light construction and enormous air

delivery, was designed for operation on barrage balloons to sample the comparatively sparse aerial population at heights up to 1000 ft (304.8 m) above ground level.

(4) *Centrifugal flow traps* are valuable for continuous monitoring of insect abundance (Taylor & Palmer, 1972; Taylor 1974) because they operate uniformly over an even wider range of conditions. The cylinder enclosing the cone may be extended vertically to sample at heights of up to 12.2 m (called '40 feet traps') with a narrow inlet to accelerate sampling velocity and an expansion chamber to decelerate air speed before filtering. The collecting cone is at the base of the trap so that it may be emptied at ground level. The traps are powered by a 1.5 B.H.P., 1440 rev/min motor with direct drive to a 30.5 cm diameter horizontal discharge, forward-bladed centrifugal fan; the air filtering cone is of 15.74 mesh/cm gauze and the diameter of the cylinder is 25.4 cm.

Rotary and other traps

The rotary or whirligig trap consists of a gauze net that is rotated at a speed of about 10 m/h. on the end of an arm (Fig. 4.3); this type of trap has been developed a number of times, sometimes with a single net and sometimes with a net at each end of the arm (Williams & Milne, 1935; Chamberlin, 1940;

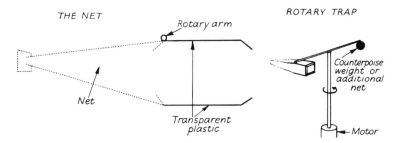

Fig. 4.3. A rotary trap, with an isokinetic net.

Stage, Gjullin & Yates, 1952; Nicholls, 1960; Vité & Gara, 1961; Gara & Vité, 1962; Prescott & Newton, 1963). The greatest efficiency will be obtained if the net is built so as to sample isokinetically, that is, the airflow lines in it are straight; Taylor (1962a) has described such a net (Fig. 4.3). Rotary traps sample more or less independently of wind speed, until the wind speed exceeds that of the trap; Taylor's (1962a) experiments indicate that they sample 85 % of the flying population, but Juillet (1963) has suggested that some insects may be able to crawl out and escape; this can be prevented by adequate baffles inside the net. It is also probable that strong flyers with good vision may avoid the area of the moving trap. It is not possible to segregate the catches of rotary traps, nor have their efficiencies on differently sized insects been studied; so for more precise work suction traps are to be preferred. However, the whirligig traps are of some value in comparative studies, especially in areas such as forests, where wind speeds may vary considerably from trap site to trap site.

Other traps for flying insects, sticky and water traps and suspended nets are greatly influenced by wind speed, their efficiencies being low at low wind speeds. Although Taylor (1962*b*) has shown how it is possible to convert sticky trap and suspended net catches into absolute densities if the wind speeds are known, these methods are used in general for the relative estimation of populations and are discussed in Chapter 7.

Comparison and efficiencies of the different types of suction traps

Three factors influence the choice of the type of suction trap: the density of the insects being studied, the wind speeds in the situation where sampling is proposed and the necessity or otherwise for information on periodicity. If information is required on periodicity, then a trap with a catch-segregating mechanism must be used, and this rules out the 18-in propeller trap. The other two factors generally give the same indications, for the sparser the insects the larger the desired air intake (so as to sample an adequate number), and the stronger the winds the stronger (larger) the trap should be. Now at ground level and up to 3 to 4 ft (91.5 – 142 cm) amongst vegetation insect populations are usually dense and wind speeds seldom exceed 6 m/h (10 km/h); under these conditions the 9-in exposed cone type is generally used. In more exposed situations (e.g. above short turf) or on towers or cables many feet above the ground the enclosed cone traps should be used; the greater the height the more powerful the winds and the sparser the insects and, therefore, the stronger the fan required. Where information on periodicity can be foregone the 18-in propeller trap provides a robust, simple and relatively inexpensive piece of equipment.

The average air deliveries of the traps* given by Taylor (1955 and pers. comm.) are: 9-in Vent-Axia, 7.84 m³/min; 12-in aerofoil trap, 36.79 m³/min; 18-in propeller trap, 77.37 m³/min; 30-in air screw trap 200.35 m³/min; centrifugal flow traps 97.6 m³/min: There is, however, a certain amount of individual variation, which over a long period could lead to significant differences in catches. Taylor (1962*a*) has suggested that unless the actual deliveries of the traps have been checked in the field, differences in catch should exceed 6 % for 9-in traps, 2 % for 12-in traps or 10 % for 18-in traps, before they be attributed to population differences.

The effects of wind speed on the absolute efficiencies of the traps vary according to the size of the insects as well as the performance of the fan. Taylor (1962*a*) has shown that the efficiency of a trap for a particular insect is given by the formula:

$$E = (W+3)(0.0082CE - 0.123) + (0.104 - 0.159 \log i) \qquad (4.1)$$

---

* In this section values have been converted to metric units and the Imperial equivalents (as given by Taylor) eliminated: these were not round numbers.

where E = the log efficiency of the trap, W = wind speed in m/h, CE = coefficient of efficiency of the trap and i = insect size in mm^2. The coefficient of efficiency of a trap = $(35.3 \times$ vol sampled in m^3/h$)^{\frac{1}{3}} \div 0.39$ inlet diameter of the fan in cm$)^{\frac{1}{2}}$; those for the above traps are: 9-in, 8.7; 12-in, 12.1; 18-in, 12.1; and 30-in, 13.4. Sizes (i.e. length $\times$ wing span) of various insects in mm^2 given by Taylor (1962a) are as follows:

Minute Phoridae and Sphaeroceridae	1–3 mm^2
Oscinella, Psychodidae and small Aphididae	3–10 mm^2
Drosophila, *Chlorops*, large Aphididae	10–30 mm^2
Fannia, *Musca*	30–100 mm^2
Calliphora, *Lucilia*	100–300 mm^2
Sarcophaga, *Tabanus*	300–1000 mm^2

4.1.2 Conversion of catch to aerial density

The numbers of insects caught is divided by the volume of air sampled; the actual crude delivery figures, given above, may be used for suction traps close to the ground if the wind speed seldom exceeds 5 m.p.h.; in other situations appropriate corrections should be made (Taylor, 1955). This figure is then corrected for the efficiency of extraction, of the particular insect, by use of the formula for efficiency given immediately above. This formula gives a negative value in logs and this is the proportion by which the actual catch is less than the real density.

Where catches have been segregated at hourly intervals the tables of conversion factors may be used (Table 4.1); Taylor (1962a) also gives conversion factors for 12-in and 30-in traps. The conversion factors are given in logs so that:

$$\text{log catch/hour} + \text{conversion factor} = \text{log density/}10^6 \text{ ft}^3 \text{ of air}$$

4.1.3 Conversion of density to total aerial population

Johnson (1957) has shown that the density of insects (f) at a particular height (z) is given by:

$$f_z = C(z + z_e)^{-\lambda} \tag{4.2}$$

where f_z = density at height z, C = a scale factor dependent on the general size of the population in the air, λ = an index of the profile and of the aerial diffusion process and z_e = a constant added to the actual height and possibly related to the height of the boundary layer: the height at which the insects' flight speed is exceeded by the wind speed (Taylor 1958).

These values should be determined as follows (Johnson, 1957). Firstly z_e must be found. The data are plotted on double logarithmic paper, when a

---

* Divide by 28 317 to get log density per m^3 of air.

Table 4.1 Conversion factors for suction trap catches (after Taylor, 1962)

FOR 9-IN VENT-AXIA TRAP

Insect size (mm²)	Wind speed (m/h)				
	0–2	2–4	4–6	6–8	8–10
			log *f*		
1– 3	1.89	2.00	2.12	2.23	2.34
3– 10	1.97	2.08	2.20	2.31	2.42
10– 30	2.05	2.16	2.28	2.39	2.50
30– 100	2.13	2.24	2.36	2.47	2.58
100– 300	2.21	2.32	2.44	2.55	2.66
300–1000	2.29	2.40	2.52	2.63	2.74

FOR 18-IN PROPELLER TRAP

Insect size (mm²)	Wind speed (m/h)						
	0–2	2–4	4–6	6–8	8–10	10–14	14–20
				log *f*			
1– 3	0.85	0.89	0.95	1.01	1.07	1.15	1.29
3– 10	0.93	0.97	1.03	1.09	1.15	1.23	1.37
10– 30	1.01	1.05	1.11	1.17	1.23	1.31	1.45
30– 100	1.09	1.13	1.19	1.25	1.31	1.39	1.53
199– 300	1.17	1.21	1.27	1.33	1.39	1.47	1.61
300–1000	1.25	1.29	1.35	1.41	1.47	1.55	1.69

curve will be obtained (Fig. 4.4). The observed heights are each increased by a constant until a straight line is obtained; the value that gives the greatest linearity is taken as z_e.

To find λ the densities at two heights near the ends of the graph are read off. Now:

$$\lambda = \log\left(\frac{f_{z_1}}{f_{z_2}}\right) \div \log\left(\frac{z_2 + z_e}{z_1 + z_e}\right) \tag{4.3}$$

where f_{z_1} = density of insects at height z_1 which is near or at the minimum height of observations and f_{z_2} = density at height z_2 which is near or at the maximum height of observations.

Then:

$$\log C = f_z + \lambda \log(z + z_e) \tag{4.4}$$

where f_z = density of insects at height z (which may be z_1 or z_2 or another).

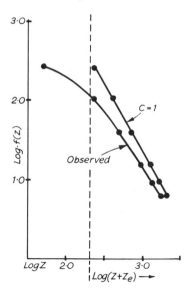

Fig. 4.4 The plotting of insect aerial density against height on a double logarithmic transformation. The observed curve is not linear until the constant Z_e has been added to the height (after Johnson, 1957).

The total number of insects between heights z_1 and z_2 (n) may then be integrated as follows

$$N = \frac{C}{1-\lambda}\left[(z_2 + z_e)^{1-\lambda} - (z_1 + z_e)^{1-\lambda}\right] \tag{4.5}$$

where the symbols are as above.

With individual species of insect, the profiles may be found to be rather irregular and approximate integrations may then be made graphically (Johnson, Taylor & Southwood, 1962). The density at each height sampled should be plotted on double logarithmic paper (i.e. plot log density against log height) and curves drawn in by eye. Then density estimates at various heights can be read off and these plotted against height on ordinary arithmetic graph paper, and the area under each curve, between say ground level and cloud base, may easily be found. Such an approach will, of course, give only approximate values and some measure of the variation can be obtained by drawing alternative curves, if they are possible, on the log/log plot and comparing the population figures derived from them.

4.2 Sampling from plants

In many ways this is the most difficult habitat to sample from; it differs from the soil and the air both in being much more heterogeneous and in continually changing. It is frequently convenient to take a part of the plant as the sampling unit; the resulting population estimate is not an absolute one, but a measure of *population intensity*. The distinction between these terms has already been discussed (p. 2), but the point is so vital that some further elaboration here is justified. If the amount of damage to the plant is the primary concern it may seem reasonable to make all estimates in terms of numbers of, say, mites per leaf. This is the intensity of mites that the tree has to withstand, and if mite populations were being related to some index of the health of the tree such an estimate of 'population intensity' is probably the most relevant. But if the study is concerned with the changes in the numbers of mites in a season a series of estimates of population intensity could easily be misleading. If the number of mites per leaf fell throughout the summer this could be due to an actual reduction in the mite population or to an increase in the numbers of leaves. In order to determine which of these explanations is correct the number of leaves per branch would have to be counted on each sampling date, and the mite population could then be expressed in terms of numbers per branch. This is sometimes called a 'basic population estimate', but it differs from the measure of population intensity only in degree and not fundamentally, for from year to year the number of branches or indeed the number of trees in the forest or orchard will alter. It is only when these units are also counted and the whole converted into numbers per unit surface area of soil, commonly per square metre, hectare or acre, that it is possible to make a valid comparison between populations differing in time (and space), which is the essence of life-table or budget construction.

The labour of obtaining the additional data on the density of the plant unit is often comparatively slight, and it cannot be stressed too strongly that, whenever possible, this should always be done as it makes the resulting data more meaningful and useful for testing a wider range of hypotheses.

4.2.1 Assessing the plant

The simplest condition is found in annual crops where the regular spacing and uniform age of the plants makes the estimation of the number of shoots or plants per length of row a relatively easy matter: the variance of even a few samples is often small and the determination of the number of row length units per unit of area (e.g. per m^2), is also straightforward. With natural herbaceous vegetation there will usually be far more variability both in the number of individual host plants per unit area and in their age and form. Although the number of plants per unit area is sometimes an adequate measure, the variation in size and age often means that weight of plant or the number of some part of the plant frequented by the insect (e.g. flower head) gives a more adequate measure. Occasionally botanical measures such as frequency, as measured by a point quadrat (Greig-Smith, 1964), are useful.

The most complex situation is found when one attempts to assess the habit of arboreal insects. The leaf is often, but not always the sampling unit (Harris, 1960; Richards & Waloff, 1961; Amman, 1969); its age, aspect and height above the ground need to be considered as well as tree age. With a citrus mealy bug Browning (1959) found the variance was minimal if each aspect of the tree was sampled in turn and individual leaves picked haphazardly rather than sampling the mealy bugs on all the leaves on a single branch.

The precise estimation of the number of leaves per branch is often a minor piece of research on its own. Actual numbers may be counted or dry weight used as the measure. In some trees the actual weight of the foliage may continue to increase after the leaves have ceased elongating (Ives, 1959). The diameter of the trunk may be taken as a cruder measure of foliage weight and of shoot number in young growing trees, especially conifers (Harris, 1960).

4.2.2 Determining the numbers of insects
Although a certain method may be suitable for a given insect, it is difficult to devise an 'all-species method' for synecological work, the more mobile insects frequently escaping (Chauvin, 1957).

DIRECT COUNTING
With large conspicuous insects it is sometimes possible to count all the individuals, for example Moore (1964) was able to make a census of the male dragonflies over a pond using binoculars. Smaller insects on distant foliage, e.g. the upper parts of trees, may be counted using a powerful tripod mounted telescope, capable of short range work and with a wide aperture (e.g. 'Questar'*); a transect may be made through the canopy of a tree by systematically turning the elevation adjustment of the telescope (L. E. Gilbert, pers. comm.). This should prove a very useful method for the survey of forest insects. Direct counts may also be used to follow a particular cohort (see p. 5) and these studies have the great advantage that the results are known to be true for that population; the difficulties of determining the true mean and other parameters, which may be severe in highly aggregated populations, are avoided.

More often a part of the habitat is delimited and the insects within the sample counted. Quadrats may be placed upon the ground and used to estimate fairly mobile, but relatively large, insects like grasshoppers (Richards & Waloff, 1954) or froghoppers (cercopids). These quadrats are made of wire and it is often useful to attach white marker-ribbons. They are placed in the field and are left undisturbed for some while before a count is made. Each quadrat is approached carefully and the numbers of insects within it counted as they fly or hop out. For smaller and less mobile animals the quadrat often needs wood or metal sides that may be driven into the soil to prevent the animals running out of the area while collecting or counting is in progress (Balogh & Loksa, 1956); in such cases and with other relatively immobile large

* Field model, Questar Inc., New Hope, Penn., U.S.A.

insects, e.g. the sunn pest, *Eurygaster integriceps* (Banks & Brown, 1962), or lepidopterous larvae and pupae (Arthur, 1962), a count is made as soon as the quadrat is in position.

Where the insects are restricted to particular plants or trees, leaves or other portions of these may be collected or examined and the numbers on each sample counted, but care must be exercised as many insects can be dislodged as the samples are taken (Satchell & Mountford, 1962) or small animals overlooked (Condrashoff, 1967). Forest and orchard pests are often assessed as numbers per shoot (e.g. Wildholtz *et al.*, 1956); crop insects as numbers per plant (Richards, 1940) or per leaf. It will commonly be found that the position of the shoot or leaf – whether it is in the upper, middle or lower part of the tree or plant – influences the numbers of insects upon it, and due allowance should be made for this stratification by a subdivision of the habitat (p. 17). When small abundant insects are actually being counted it is desirable that the sampling unit should be as small as possible; Shands, Simpson & Reed (1954) showed that various aphids on potatoes could be adequately estimated by counting, not every aphid on a selected leaf, but only those on the terminal and two basal leaflets; even a count based on a half of each of those leaflets could be satisfactory. Photographs (colour transparencies) have been found to be a practical and rapid method of recording aphids on rose shoots; they are taken from four aspects at a distance of about 40 cm and those aphids seen against the shoot (not silhouettes) counted by viewing under a dissecting microscope (Maelzer, 1976). The potential of photography for direct sampling (and recording) of insect populations has probably not been fully developed. Insects in stored products, especially cacao, are often estimated by 'snaking'. The bag is opened and the contents tipped on to a floor in a long wavy line; the insects may be counted directly and a large number will often be found in the last piece tipped out, known as the 'tail'. This method, although technically providing an absolute type of sample, is really more properly regarded as a relative method.

This type of work is often necessary to establish the basis of a presence or absence sampling programme (p. 55) for extensive work.

THE SEPARATION OF EXPOSED SMALL ANIMALS FROM THE FOLIAGE ON WHICH THEY ARE LIVING

Knockdown – by chemicals, jarring and heat
Knockdown by chemicals and jarring are discussed more fully below, as they are more generally used when the whole twig or tree is the sampling unit. Here we are concerned with their use with animals on herbaceous plants and with animals on trees when the leaf is the sampling unit. Some insects can be made to drop off foliage by exposing them to the vapours of certain chemicals. With the wide range of organic substances now available this approach could undoubtedly be extended; some recognized repellents (Jacobson, 1966) might

be screened. Aphids can be made to withdraw their stylets and will mostly drop off the plant if exposed to the vapour of methyl isobutyl ketone (Gray & Schuh, 1941; Helson, 1958; Alikhan, 1961) or γ-BHC (Way & Heathcote, 1966). Aphids that do not cover the host plant with a thick deposit of honey dew and are fairly active may be separated from the foliage by exposing them to this vapour and then shaking them in the type of sampling can described by Gray & Schuh (1941) (Fig. 4.5a). With the pea aphid *Acyrthosiphon pisum*, these authors shook the can 50 times following a 5-minute interval and found

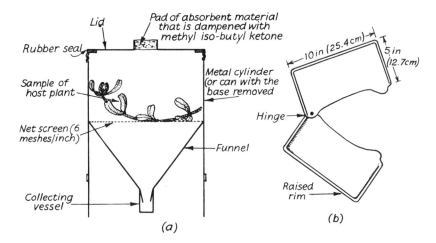

Fig. 4.5 *a*. Aphid sampling can (modified from Gray & Schuh, 1941). *b*. Sampling tray for flea-beetle, weevils and others on small crop plants (after la Croix, 1961).

that after one such treatment an average of 89.3 % of the aphids were extracted, after 10 minutes (and 100 shakes) 97 % and after 15 minutes 99.1 %. In the six samples tested the lowest extraction rate was 98.5 % after 15 minutes; the numbers extracted from each sample in the can were between 3000 and 7000. Walker, Cate, Pair & Bottrell (1972) found with *Schizaphis*, that because of entanglement with honeydew, extractions were in the region of 85 % after 15 minutes exposure, shaking more than 50 times did not increase the recovery efficiency (and the effect from 25 to 50 was only marginal). Laster & Furr (1962) found that the fauna of sorghum heads could be killed by insecticide and collected in a paper cone strapped round the stem and stabilized with a little sand in the base.

Thrips may be extracted from flowers and grassheads, as well as foliage, by exposing them to turpentine vapour (Evans, 1933; Lewis, 1960). A convenient apparatus has been described by Lewis who found the method was over 80 % efficient for adults, but confirmed Taylor & Smith's (1955) conclusion that it was unsatisfactory for larvae. Lewis & Navas (1962) obtained thrips and some other insects that were overwintering in bark crevices by breaking the bark sample into small pieces and sieving; this dislodges the majority, but to obtain 97–100 % extraction efficiencies the fragments were exposed to turpentine vapour.

Heat was used by Hughes (1963) to stimulate *Brevicoryne brassicae* to leave cabbage leaves, and Hoerner's (1947) onion thrips extractor worked on the same principle; these techniques can be considered modifications of the Berlese funnel (p. 185). Ultra-sonic radiation (of about 20 kHz generated by a Kerry Vibrason cell disruptor probe, 100 W output) was found to dislodge eriophyid mites from grass (Gibson, 1975). The foliage was placed in 50 or 75 % alcohol (ethanol) and exposed to the radiation for 45 s to 1 min (maximum). (See also p. 155).

Insects that fall from their host plant when disturbed, e.g. some beetles, can be sampled from young field crops by enclosing the base of the plant in a hinged metal tray (Fig. 4.5*b*). The plant is then tapped (jarred) and beetles from it may be counted as they fall on to the tray. This method was developed by la Croix (1961) for flea beetles, *Podagrica*, on cotton seedlings; they must be counted quickly before they hop off the tray. This is facilitated by painting the tray white.

Brushing

Mites, and apparently other small animals apart from insects, are removed from leaves if these are passed between two spiral brushes revolving at high speed in opposite directions (Henderson & McBurnie, 1943; Morgan *et al.*, 1955;Chant & Muir, 1955; Chant, 1962). The mites are collected on a glass plate, covered with adhesive, below the machine.* Clear varnish was the original adhesive, but various surface-active agents have been found more satisfactory if the mites are to be examined at once; e.g. Emcol 5100, an alkanolamine condensate (Morgan *et al.*, 1955), Tween 20 (polyoxyethylene sorbitan monolaurate) and Span (sorbitan monolaurate) (Cleveland, 1962). For permanent storage Muir in Chant (1962) recommends replacing the glass plate with a cardboard disc, repeatedly coated with varnish. Adult mites are usually removed slightly more efficiently than the eggs; in general the efficiency for fruit-tree mites on apple leaves has been found to be between 95 and 100 % (Morgan *et al.*, 1955; Chant & Muir, 1955), but it is much less on peach leaves because of their complex curvature (Putman, 1966).

Washing

Small animals, principally mites, aphids and thrips, can be washed off herbaceous plants or single leaves with various solutions. This approach can be more efficient, both in terms of accuracy and of cost, than direct counting. With aphids the extraction will be facilitated if they are first exposed to the vapour of methyl isobutyl ketone (=4-methyl 2-pentanone) (Gray & Shuh, 1941; Pielou, 1961). Dilute soap detergent or alcohol solutions are often

* Available from J. G. H. Edwards Co., R.R.I. Okanagan Falls, B.C., Canada.

adequate for washing (e.g. Newell, 1947; Taylor & Smith, 1955; Szalay-Marzsó, 1957), but with eggs a solvent must be used that will dissolve the cement. Benzene heated almost to boiling point, removes many of the eggs of mites on fruit-tree twigs (Morgan *et al.*, 1955), as does hot sodium hydroxide solution (Jones & Prendergast, 1937; Kobayashi & Murai, 1965). If conifer foliage with moth eggs is soaked for 30 hours in a 1.5 % solution of sodium hydroxide a 96 % recovery rate may be achieved (Condrashoff, 1967), and petrol has been found satisfactory for washing eggs of the corn earworm (*Heliothis zea*) from the silks of maize (Connell, 1959). Thrips may be washed off foliage with ethanol (Le Pelley, 1942), the solution filtered and the thrips separated from the other material by shaking in benzene and water; the thrips pass into the benzene where they may be counted (see also p. 179) (Bullock, 1963). Moth eggs may be freed from bark and moss by soaking for 45 minutes in 2 % bleach solution (sodium hypochlorite solution (5.25 % free chlorine) diluted with 98 parts water); the mixture is then agitated and passed through a sieve, which separates the larger debris (Otvos & Bryant, 1972).

The animals removed by washing are often too numerous to be counted directly by eye, and electronic counters have been devised by Lowe & Dromgoole (1958) and Hughes & Woolcock (1963). An alternative approach is to count a sample; this may be done by means of a counting grid or disc (Fig. 4.6) (Strickland, 1954; Morgan *et al.*, 1955). The animals are mixed with 70 % alcohol or some other solution, agitated in a Petri dish and allowed to settle. The dish is stood above a photographically produced counting disc (on celluloid) and illuminated from below. Strickland (1954) found that the percentage error of the count for aphids using the disc in Fig. 4.6 ranged from 5 (with about 3000 aphids/dish) to 15 (with less than 500 aphids/dish).

If the number of animals is too great for the counting grid to be used directly the 'suspension' of the animals in a solution may be aliquoted (Newell, 1947). The diluted samples may then be estimated with a counting disc or volumetrically or the two methods combined. In the volumetric method the animals are allowed to settle or are filtered out of the solution; they are then estimated as, for example, so many small specimen-tubefuls, the numbers in a few of the tubes are counted partially by the use of a counting disc or *in toto* (Banks, 1954; Walker *et al.*, 1972). If the population consists of very different sized individuals the volumetric method may lead to inaccuracies, unless special care is taken. This variation can, however, be used to aid estimation: Pielou (1961) showed that if apple aphids (*Aphis pomi*) were shaken with alcohol and the solution placed in Imhoff sedimentation cones (wide-mouthed graduates), the sediment stratifies into layers, firstly the larger living aphids, then the smaller living aphids, and finally dead aphids and cast skins; the volumes of these can easily be read off and this calibrated for the various categories. Another method of separating the instars based on the same principle has been developed by Kershaw (1964). A routine for subsampling when collections vary enormously in size is described by Corbet (1966).

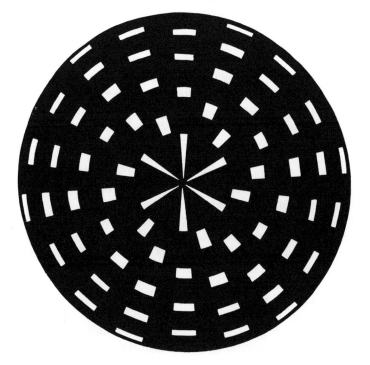

Fig. 4.6. A counting grid (after Strickland, 1954). All animals that are wholly within the clear areas or overlap the left-hand or outer margins should be counted and these will be one-sixth of the total.

Imprinting

This technique has been used only with mites and their eggs; the infested leaves are placed between sheets of glossy absorbent paper and passed between a pair of rubber rollers (e.g. a household wringer); where each mite or its egg has been squashed a stain is left (Venables & Dennys, 1941; Austin & Massee, 1947). Summers & Baker (1952) used the same method to record the mites beaten from almond twigs, and the advantage of this method is its speed and the provision of a permanent record; furthermore, with certain precautions, nearest neighbour type techniques (p. 47) might be applied to the resulting marks. However, Chant & Muir (1955) found with the fruit-tree red spider mite (*Panonychus ulmi*) that the number recorded by the imprinting method was significantly less than the number recorded by the brushing method (see above), which is therefore to be preferred in general.

THE EXPULSION OF ANIMALS FROM TALL VEGETATION

Jarring or beating

This is a collector's method and originally the tree was hit sharply with a stick and the insects collected in an umbrella held upside down under the tree! The umbrella is now replaced by a beating tray which is basically a cloth-covered frame, flat or slightly sloping towards the centre, that is large enough to collect all the insects that drop off the tree. The colour of the cloth should produce the maximum contrast with that of the insect being studied, and the insects are rapidly collected from the tray by an aspirator or pooter (Fig. 7.5). In general this is only a relative method, but with some insects such as leaf beetles (Chrysomelids), many weevils and lepidopterous larvae that fall from the host plant, if disturbed, a sufficiently high proportion may be collected for it to be regarded as an absolute method (e.g. Richards & Waloff, 1961; Legner & Oatman, 1962; Gibb & Betts, 1963; White, 1975). Weather affects the efficiency, differently for different species (Harris *et al.*, 1972). As very small animals, like mites, may be overlooked on a beating tray and active ones may escape, more accurate counts can sometimes be obtained by fastening a screen over a large funnel. The twigs are tapped on the screen and mites and insects funnelled into a container (Boudreaux, 1953; Steiner, 1962). The beating tray and funnel methods may be combined: Wilson (1962) attached a removal jar below a hole in the centre of the tray and Coineau (1962) placed a grid above a funnel in the centre of a net tray; this grid helps to separate twigs and other debris from the insects. Arthropods on row crops (e.g. soybeans) may be collected by quietly unrolling a length of oil-cloth ground sheet (or polythene) between two adjacent rows and briskly shaking the plants over it; the method is time consuming, but seems to provide absolute estimates of large easily dislodged species (e.g. pentatomids) (Rudd & Jensen, 1977).

Although the tree is generally hit sharply with a stick, Legner & Oatman (1962) suggest the use of a rubber mallet, and Richards & Waloff (1961) found that some host plants (e.g. *Sarothamnus*) could be damaged by such violence (so leading to a change in the habitat); they recommended vigorous shaking by hand.

A different approach was developed by Lord (1965) for sampling the mirid predators on apple trees: branches were cut from the tree over a cloth tray and then divided into smaller portions which were put in a 'shaker' sampler, a revolving rectangular cage. The animals are dislodged by this treatment; Lord found that 60–100 % of those present were dislodged after 500 revolutions in the sampler.

Chemical knockdown

The total number of insects on a tree have been estimated by shrouding it with polythene sheets or screens and applying an insecticide with a very rapid knockdown (e.g. pyrethrum) as a mist or aerosol; the dead insects are collected on a groundsheet below (Collyer, 1951; Gibbs, Pickett & Leston, 1968). The efficiency of the method is, of course, greatly impaired if the foliage is moist and unless shrouding can be quickly undertaken, almost simultaneously with the spraying, it is better omitted. The collection sheets should be left in position for up to an hour as certain types of insect do not fall immediately and efficiency can be further increased if, after about half an hour, the tree is jarred. Under some conditions with small insects the efficiency can be as low as 50 % (Muir & Gambrill, 1960). It also provides a useful method for removing insects from sample branches (Ives, 1967).

A method of estimating the population of leaf-feeding caterpillars on large forest trees was devised by Satchell & Mountford (1962). A systemic insecticide, phosdrin, was introduced into the tree and the falling caterpillars collected on a sheet. A watertight collar was formed round the trunk by a strip of rubber sheet attached at its lower end to a ring of smoothed bark by rubber solution, and supported by a wire ring held in position by skewers driven into the bark below the collar. A ring of horizontal chisel cuts at 2-cm intervals was made through the bark into the xylem near the bottom of this collar, and this was filled with the insecticide solution. Between 70 and 95 % of the larvae of the trees fell on to the sheets after this treatment, and therefore it is necessary to count the remaining larvae on a proportion of the shoots on the tree to obtain a complete estimate. A possible source of error in this type of method is the consumption by birds of moribund larvae.

Besides the objections to the above methods due to their only partial efficiency, it is seldom desirable in long-term ecological work to destroy all the insects from a sampling unit as large as a whole tree. The fauna of part of a tree or bush or from the tops of herbaceous plants may be sampled by the use of a hinged metal box (Fig. 4.7), whose leading edges are faced with sponge

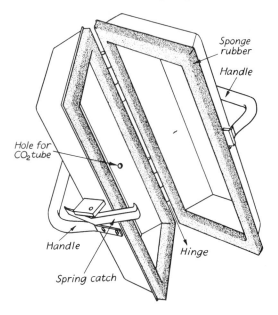

Fig. 4.7. A sampler for insects on woody vegetation (after Dempster, 1961).

rubber – this will accommodate branches up to $\frac{1}{2}$ in (1.3) cm) thick (Adam *in* Chauvin, 1957; Dempster, 1961). Once the branch is enclosed the insects are killed by a chemical: Adam used ethyl acetate, Dempster carbon dioxide from a small cylinder, released into the closed box through a small hole. The fauna becomes momentarily more active and then drops to the floor of the box, anaesthetized. The box may then be opened and the animals collected by a mouth-operated or an electric aspirator or pooter. This method is least efficient with those insects that tend to 'hang on' to the plant, e.g. mirid larvae, caterpillars. A net-covered sampler, based on the same principle, was used by Leigh, Gonzalez & van den Bosch (1970) to sample the fauna of a cotton bush. They termed it the 'clam-shell trap' and used it to standardize the efficiency of the 'D-vac sampler' (p. 150).

Collection of naturally descending animals
Many arboreal larvae pupate in the soil and these may be collected as they descend to provide a population estimate of the total numbers at that particular developmental stage (see p. 308). Cone (1963) found that the vine weevil (*Otiorrhynchus sulcatus*) dropped off its host plant at daybreak and the population could therefore be determined by placing funnel-type traps under the grape vines.

THE EXTRACTION OF ANIMALS FROM HERBAGE AND DEBRIS
Here one is concerned with animals living in the herbage layer just above the soil and the problem is not, as in the above section, to make them fall down, but to get them to 'come up', either artifically or naturally.

Suction apparatus
A number of types of machine have been devised to sample, by suction, the animals living on or amongst the herbage down to the soil surface. There are basically two types (Fig. 4.8); the narrow suction-hose type developed by Johnson *et al.* (1957) and others (Table 4.2), and the wide hose type of Dietrick Schlinger & van den Bosch (1959). In the narrow suction-hose model the area to be sampled is enclosed by forcing a small metal cylinder (e.g. a bottomless bucket or a special cylinder (Hower & Ferguson, 1972)) into the soil; the area delimited is then systematically worked over with the nozzle and a mixture of animals and plant debris are trapped in the collecting bag (Fig. 4.8). This may be sorted by hand, but many workers have found a Berlese type apparatus and/or flotation increased the ease and often the efficiency of recovery (Dietrick *et al.*, 1959; Turnbull & Nicholls, 1966) (pp. 141, 155 and Chapter 5). The suction is provided by a fan which can be driven by hand, by an electric motor or by a large or a small petrol motor (see Table 4.2); if high rates of extraction are to be obtained the wind speed at the nozzle must be about 60 m/h (96 km/h). In order to increase efficiency of the Dietrick sampler, Turnbull & Nicholls (1966) reduced nozzle and hose diameters getting a nozzle speed

Table 4.2 Basic features of various types of suction apparatus. (Note: Ordinary vacuum-cleaner motors are not usually satisfactory as they tend to race when the nozzle is impeded and models based on them are not claimed to give absolute estimates)

Authors	Fan unit and power source	Diameter of (in cm.)		Position of collecting chamber
		Nozzle	Suction hose	
Dietrick, Schlinger & van den Bosch (1959)	20B'Homelite, blower (single-cylinder, air-cooled petrol motor)		38	Separated from motor by 38 cm. diameter extensible (to about 1.5 m) air hose.
Dietrick (1961) ('D-vac' model)*	'Tecumsa' (single-cylinder, air-cooled petrol motor)	16.5–34	20	ditto
Weekman & Ball (1963)	Hoover 'Pixie' vacuum cleaner		3	ditto
Johnson, Southwood & Entwistle (1957)**	Wolf portable electric blower (Type NWBE) mains electricity	3	5	Fixed to motor casing
Southwood & Pleasance (1962)	Gear box, fan and casing of Procall Rex dusting machine (with blades of fan enlarged to increase air flow), hand-operated	2.5	5	ditto
Southwood & Cross (1969)	British vacuum cleaner Ltd, T172 with petrol engine	5	6.4	ditto
Whittaker (1965)	Smith's F350 centrifugal blower, operated from a 12-V battery	–	–	ditto
Arnold, Needham & Stevenson (1973)***	ILO35 two-stroke petrol engine	6.3–15	6.3	ditto
Remane (1958)	Blastor (Leer) 50-c.c petrol motor	–	–	–
Levi et al. (1959)	Pemz – 1 vacuum cleaner with 'Kiev' motor	3		Separated from fan by air hose

* Available from D-Vac Ltd., P.O. Box 2095, Riverside, California 92506, U.S.A.
** Available from Burkard Manufacturing Co. Rickmansworth, Herts. U.K.
*** Available from Burkard Manufacturing Co. ('Univac sampler').

approaching 90 m/h; likewise increased nozzle diameter was found by Richmond & Graham (1969) to reduce catch size, this reduction being especially marked if nozzle speed fell below 44 m/h. Extraction rates of 95–100 % were obtained by Johnson et al. (1957) for groups such as Hemiptera, adult Diptera, adult Hymenoptera and surface-dwelling Collembola, and the lowest rates (about 70–75 %) for larval Diptera and Coleoptera. An

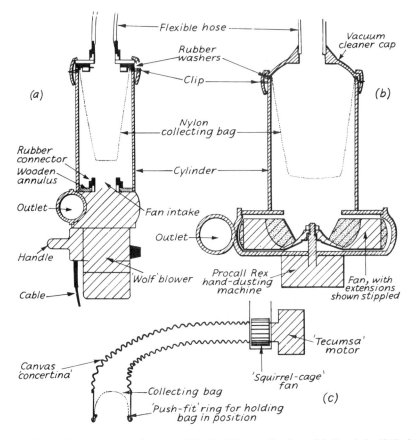

Fig. 4.8. Suction apparatus:*a*. electric model (after Johnson, Southwood & Entwistle, 1957); *b*. hand-operated model (after Southwood & Pleasance, 1962); *c*. the Dietrick model, 'D-Vac'.

extraction rate of 87 % for Homoptera was obtained by Whittaker (1965) with a battery-operated model. With the Arnold model ('Univac'), Henderson & Whitaker (1977) found extraction rates of 70 % to 98 % for most groups on short and medium (up 15 cm) length grass, but in long grass these were lower and the extraction of mites was always under 50 %. The efficiency of the Dietrick ('D-Vac') sampler is greatly affected by the mode of operation; the nozzle must be moved vertically into the foliage and down to the ground surface, although even when this is done, if the ground is subsequently searched carefully, a number of beetles, isopods and other 'heavy' arthropods will usually be found to have been missed. If the nozzle is swept horizontally across the foliage very small catches are obtained (Richmond & Graham, 1969). Several comparisons of the 'D-vac' with other methods have shown efficiencies around 50 % to 70 %; some of these studies have been concerned with coccinellid beetles and other predators on cotton [Turnbull & Nicholls

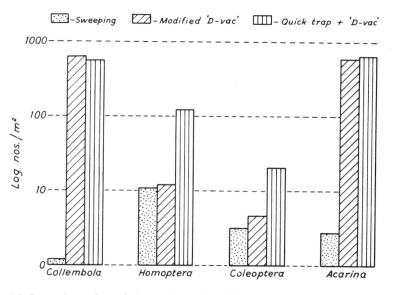

Fig. 4.9 Comparison of population estimates for different taxa, using three methods of sampling: sweep net, modified 'D-vac' sampler and 'Quick-trap' covering method + 'D-vac' (after Turnbull and Nicholls, 1968).

(1966) and Fig. 4.9; Leigh *et al.* 1970; Shepard *et al.* 1972; Pieters & Sterling, 1973; Smith *et al.* 1976.]

Owing to the formation of a water film, in which animals may become trapped, inside the suction hose, it is difficult to sample damp herbage with the narrow suction-hose types of apparatus, but the substitution of a wide polythene tube for the more normal wire and canvas hose at least enables a check to be kept on this potential source of error. Arnold *et al.* (1973) used 'Vacuflex' hose which although of clear plastic is reinforced with a spiral wire. When sampling arable habitats it may be an advantage to use two bags, one of nylon net which retains the insects and outside this a larger cloth bag to prevent the soil passing through the fan; whilst when sampling the surface fauna from woodland litter with a 'D-vac' it is useful to cover the ground with fine mesh wire netting: this prevents the collecting bag filling with leaf debris. The narrow suction-hose type of sampler, with its high nozzle wind speed and fan whose efficiency does not fall off when impeded, is particularly suited for work with the fauna of the lower parts of vegetation and the soil surface. The D-vac sampler (Dietrick, 1961) overcomes the problem of adhesion to the hose by having the collecting bag incorporated in the widened nozzle (Fig. 4.8) and a squirrel-cage type of fan produces a strong air current, so long as it is not impeded. This model is therefore particularly suited for use with animals on herbage.

In conclusion the efficiency of suction samplers is affected by several factors, but of particular importance are nozzle wind speed (for heavy and/or tenaceous species) and speed of enclosure (for active flying species). The 'D-vac' is excellent on the latter count, but the nozzle wind speed is, with several types of collecting head, on the borderline of efficient extraction. Leigh *et al.* (1970) consider that this sampler needs to be standardized for the particular conditions before it may be considered as an extraction method for absolute estimates. Other samplers seem more efficient under a wider range of conditions (Johnson *et al.*, 1957; Heikinheimo & Raatikainen, 1962). The method is more satisfactory than sweeping (p. 240), which can give biased results. Larger models seem efficient; Kirk & Bottrell (1969) devised a tractor-mounted model and the 'McCoy Insect Collector', that incorporated blowing [a principle first used by Santa (1961)] as well as sucking, is a 'lorry-sized' motor vehicle. Its efficiency is high and McCoy & Lloyd (1975) show it can detect 2 boll weevils in ten acres of cotton.

These vehicular machines that sample from such a large proportion of the habitat, overcome one of the main difficulties with smaller samplers: that is the high variance of the samples due to the patchy nature of many, especially natural, habitats. Theoretically the answer lies in taking more samples, but the worker easily becomes over-burdened with material to sort. With the large vehicular machines mechanized and computerized methods of sorting have been developed.

Small suction samplers have been used for rodents' nests (Levi *et al.*, 1959) and flowers (Kennard & Spencer, 1953).

Cylinder or covering method
The basis of this method is the enclosure of an area of herbage within a covered cylinder; the animals are collected by some method often after being knocked down by an insecticide. Kretzschmar (1948) used a modification of Romney's (1945) technique for sampling soya-bean insects; he placed plastic transparent base plates beneath the plants the day before sampling. These greatly facilitate the subsequent collection of the insects, but I would suggest that tests should be made to ensure that their presence does not influence the density of insects on the plants above. In order to prevent the escape of active insects, Balogh & Loksa (1956) attached the sampling cylinder to the end of a long pole and rapidly brought this down on to the crop. These workers killed the insects with powerful fumigants (carbon disulphide, hydrogen cyanide) and then removed them by hand. As it is difficult to reach and search the base of a tall cylinder Skuhravý *et al.* (1959, 1961) used two cylinders and removed the tall outer one after the insects had been knocked down. Cherry *et al.* (1977) combined this method with the 'tent' technique (see below).

The combination of the 'covering method' with a suction apparatus for collection probably provides the most efficient method for sampling the fauna of herbage. The most appropriate cylinder or tent (see below) and suction apparatus will depend on the animal, the vegetation and the availability of services (electricity, easy carriage of heavy equipment). Turnbull & Nicholls (1966) devised a 'quick trap' (Fig. 4.10) which they used in conjunction with a modified D-vac sampler. The quick trap is put into position several hours prior to sampling, with the net portion folded (the dotted outline in Fig. 4.10).

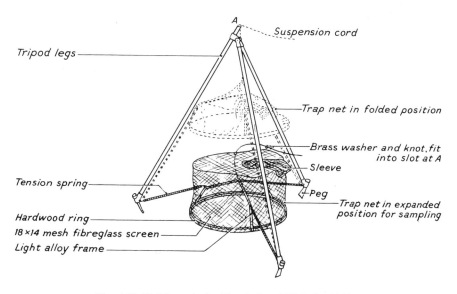

Fig. 4.10 'Quick-trap' after Turnbull and Nicholls (1966).

When the sample is to be taken the suspension cord, which can (and should) extend some distance from the trap, is pulled, which jerks it out of the slot at the top of tripod leg A. The tension springs expand the trap into position, but the operator should immediately check that the base ring is in close contact with the grund. The modified 'D-vac' sampler's 10 cm diameter hose or the 'Univac's' hose is inserted through the top of the trap and the area systematically worked over. The mixture of small arthropods and debris can be sorted by hand or by a combination of funnel extractors and flotation (Dondale *et al.* 1971) (see Chapter 5 and pp. 141 and 155).

Turnbull & Nicholls (1966) made most useful comparisons of this method with several others (Fig. 4.9). The extent of the advantage compared with the sweep net or normal 'D-vac' sampler varied from taxon to taxon. Low efficiencies for the 'D-vac' occur if the animals are very active (escaping as the sampler approaches) or if they can cling close to the soil surface, when the nozzle speeds of the normal 'D-vac' (*c.* 40 m/h) are inadequate to dislodge them.

Tents for sampling strongly phototactic animals

A large muslin covered cage or 'tent' (Fig. 4.11) is quickly put into position when the animals are least active, usually at dusk. When they are active again, the following morning, the animals (if large) may be directly collected from the sides of the tent; alternatively it may be covered with a black shroud leaving only the muslin-topped celluloid collecting cylinder exposed at the apex. After some time, about 15 minutes in bright sunshine, the majority of the animals will be in the cylinder, which can be quickly removed and is conveniently carried back to the laboratory by pushing its base into appropriately sized rings of 'Plasticine' on a metal tray. If possible someone should then get inside the 'tent' and remove the remaining animals from the insides of the walls; on a dull day quite a large proportion of the total population may fail to enter the collecting cylinder. This method has been used successfully with blowflies (*Lucilia*) (MacLeod & Donnelly, 1957), mosquitoes (Chinaev, 1959) and frit flies (*Oscinella frit*) (Southwood *et al.*, 1961). A miniature tent for use with mites is described by Jones (1950) and McGovern *et al.* (1973) have devised a model for separating phototactic (and resilient) animals (e.g. boll weevils) from the debris in suction apparatus samples.

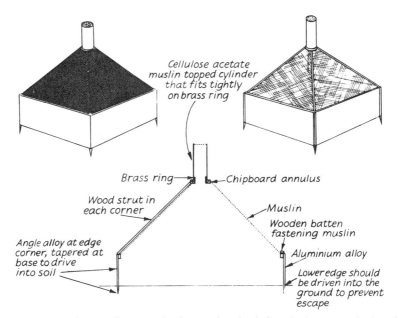

Fig. 4.11 A 'tent' for sampling strongly phototactic animals from herbage. A sagittal section through one coner and a side, with sketches of its appearance, shrouded and unshrouded.

An apparatus that is intermediate between a cylinder and a tent has been devised by Wiegert (1961) for sampling grasshoppers, cercopids and other insects that readily leave the vegetation if disturbed. It consists of a truncated nylon-covered strap-iron cone, topped with a metal funnel and a collecting jar. As with the tent or cylinder, this is placed rapidly over the vegetation which is then agitated. The sampling cone is then gradually moved on to a flat sheet, of wood or other substance, that is placed adjacent to it. Eventually most of the vegetation can be slid out of the cone whilst the active insects remain; it is then inverted and the insects jarred down into the collecting container. Wiegert considered the efficiency for the froghopper, *Philaenus*, greater than 90 %. The leafhopper, *Empoasca*, was efficiently collected by Cherry *et al.* (1977) in a rather similar device: a black plastic dust (garbage) bin, with a collecting cylinder (as used in tents, Fig. 4.11) in its base, was quickly lowered, upside down, suspended from the end of a long pole.

Extraction by heat drying and/or flotation
A further extension of the use of the animal's reactions to certain physical conditions leads logically to the removal of samples of herbage or debris to the laboratory, where they are subjected to drying by heat. This is the basis of the Berlese/Tullgren funnel. Although principally a method for soil and litter, it may be used with herbage and hence with its many modifications discussed in Chapter 5, it is being increasingly utilized to extract animals from herbage samples, especially the mixtures that result from suction sampling (see above) (e.g. Dietrick *et al.*, 1959; Dondale *et al.* 1971), and adult beetles and late instar larvae from grain samples (Smith, 1977). It is probable that it would be appropriate for many foliage insects that are moderately mobile: special attention would need to be given to the design of the funnel if they were strongly phototactic. Harris (1971) found the method fully efficient compared with hand-sorting, for the larvae of the pear psyllid, *Psylla pyricola*, provided the heat source was not too powerful. Flotation methods (Chapter 5) may also be useful for the material from a suction sampler (Dondale *et al.*, 1971). With animals that rest in litter, such as weevils and leaf-beetles, it may be possible (given a plentiful supply of water), to float the animals off the sample in the field. Wood & Small (1970) did this by enclosing the area in a solid metal frame and floating off the beetles and litter, through an overflow, on to a fine sieve.

METHODS FOR ANIMALS IN PLANT TISSUES
Although leaf-miners may be directly counted (p. 292), it is seldom possible to estimate eggs imbedded in plant tissue or larval or adult insects boring into the stems of herbaceous plants or trees so simply. The main impetus for the development of techniques for such animals has been provided by work on insect pests of stored grain and timber. Doubtless considerable advances will be made when the approaches developed in this field are applied and extended with insects in the tissues of herbaceous plants.

Dissection
This method is widely applicable, if tedious. Sometimes it is possible to limit

the dissection of stems, fruits or other parts to those that are obviously damaged (see Chapter 8).

Bleaching and/or selective staining

The plant tissue may be rendered transparent by treatment in 'Eau de Javelle', lactophenol (Koura, 1958; Carlson & Hibbs, 1962) or 10 % sodium hydroxide solution (Apt, 1950) so that the insects become visible; this method has been used with grains and lentils. Plant parasitic nematodes are assessed by simultaneously bleaching the plant tissues and staining the worms in a mixture of lactophenol and cotton blue (Goodey, 1957). The eggs of various Hemiptera in potato leaves can be detected by bleaching the leaves (boiled in water until limp and then in 95 % alcohol over a water bath) and then staining in saturated solution of methyl red and differentiating in a slightly alkaline solution (containing sodium hydroxide) of the same dye; the leaf tissues become orange or yellow, but the eggs remain bright red and may be counted by transmitted light (Curtis, 1942). However, Carlson & Hibbs (1962) found that the same eggs could be counted after merely boiling for one minute in lactophenol, when the leaf tissues become bleached and egg proteins coagulated. Insect egg plugs in grains have been found to be stained selectively by gentian violet (when moist) (Goosens, 1949) or by various alkaloids, derberine sulphate, chelidonium and primuline that fluoresce yellow, orange and light blue under ultra-violet light (Milner, Barney & Shellenberger, 1950). If the grain is crushed the insect protein may be recognized by the ninhydrin test (Dennis & Decker, 1962).

X-rays

In 1924 Yaghi suggested that these rays could be used in the detection of wood-boring insects, but it is only recently that the principle has been extensively applied, more particularly to insects in homogeneously textured materials: in grain (Milner, Lee & Katz, 1950, 1953; Pedersen & Brown, 1960) and other seeds (Simak, 1955); it is also of wide applicability to the assessment of insects in moist plant tissues: for wood-borers (Berryman & Stark, 1962; Bletchly & Baldwin, 1962; Amman & Rasmussen, 1969), stem-borers (Goodhue & van Emden, 1960) and bollworms (Graham *et al.*, 1964). The cavity made by the larva or adult is much more easily detected in the radiographs than the difference between animal and plant tissues. Where one has to rely on the latter distinction, as with the blackcurrant gall mite, *Cecidophytopsis ribis*, then the value of the method may be limited; for long exposures to maximize the distinction will lead to blurred images (Smith, 1960). It has generally been found that 'soft' X-rays (5–35 kV) are best for the work; the thicker the plant material the greater the kilovoltage required for the same exposure time or the greater the exposure time, which needs to be increased as the water content of the plant tissue rises (Amman & Rasmussen, 1969). Small portable X-ray units (e.g. 'Picker') with a thin (*c.* 2 mm) beryllium window are particularly suitable for ecological work. Further technical details are given in the papers cited above. It has been shown by De Mars (1963) that X-ray detection of a bark beetle, *Dendroctonus*, is virtually as efficient as dissection, eight times faster and less than a quarter the cost, even when allowance is made for materials. Wickman (1964) reached similar conclusions for its use with Siricidae. The interpretation of X-rays of bark insects is discussed by Berryman (1964).

Methods based on the different mass of the whole and the infested material
Where the plant material is of fairly uniform weight, e.g. grain, advantage may be taken of the fact that material in which insects have, or are, feeding will be lighter. The lighter grain can be made to float (e.g. in 2 % ferric nitrate solution – Apt, 1952) or to fall more rapidly when projected (Katz, Farrell & Milner, 1954). Such an approach naturally leads on to flotation techniques that are discussed in Chapter 5.

Aural detection
Insects living in plant material often make sounds, normally from their feeding. The aural detection of this sound, particularly in grain infesting species was first proposed by Adams *et al.* (1953). Modern electronic devices have facilitated this approach, which is non-destructive. Street (1971) describes a method of detection using a gramophone cartridge transducer, the stylus is replaced by about 5 cm of steel or copper wire with a small hook at the end. An adhesive coated cellulose acetate card (10×12 cm), to which 150 or more grains are attached, is suspended on this hook. Any sounds will be recorded from the stylus through appropriate amplification.

4.2.3 Special sampling problems with animals in plant material

The marking of turf samples
Van Emden (1963) has described a method whereby small areas of turf may be marked and subsequently found and recognized without impairing mowing or grazing or attracting the attention of vandals. Wire rings, marked with a colour code, were sunk into the soil around the samples and here they could remain for weeks or even years. Their approximate positions were recorded by reference to a grid and a diagonal tape, and their condition or other relevant ecological information noted. After the desired period, the rings may be rediscovered and the ecological changes in the precise area observed, by plotting the approximate position with reference to the grid and finding the actual ring with a mine-detector.

The sampling of bulk grain
When studying insect populations in stored grain, it is often desired to draw samples from known depths ensuring that they are not contaminated by

insects or grain from the outer layers. Burges (1960) has developed a method for doing this: a hollow spear with a lateral aperture is forced into the required position and then the desired sample sucked out through its shaft by a domestic vacuum cleaner. Other types of aspirator for sampling grain have been described by Chao & Peterson (1952) and Ristich & Lookard (1953).

The sampling of bark
A circular punch that cuts 0.1 ft^2 (92.9 cm^2) samples from bark has been devised by Furniss (1962). The punch is made from a segment of a 4-in (10 cm) diameter steel pipe fitted with a central handle so that it can be rapidly and symmetrically hammered into the bark. Very thick bark may need to have the outer layer shaved off first if Scolytidae are being sampled and the phloem layer required; but not, of course, if the crevice fauna is being studied (Lewis & Navas, 1962). There is also a problem of estimating the actual area sampled that will vary with the curvature of the trunk or bough (Dudley, 1971) (p. 21).

4.3 Sampling from vertebrate hosts
There are two important variables to be considered when attempting to obtain complete samples of vertebrate ectoparasites from their hosts. Firstly the readiness or otherwise with which they will leave their hosts, and secondly whether the host can be killed or not. Furthermore, if the host is killed, can the skin be destroyed whilst obtaining the parasites or must it be retained in good condition? Sampling from living animals will generally be more likely to provide absolute population counts of those parasites that are readily dislodged from the host (e.g. fleas, hippoboscids) than for those that are more firmly attached (e.g. lice, mites); to estimate these the host will often need to be killed and even the skin destroyed.

4.3.1 Sampling from living hosts
Care must be taken in handling the host, for Stark & Kinney (1962) found that many fleas would rapidly leave a struggling or agitated host.

Searching
This method is usually satisfactory only with relatively large parasites (e.g. ticks), active ones (e.g. hippoboscids – Ash, 1952) or those that sit on exposed and hairless parts of the body. In the rare cases where the host is almost completely hairless minute parasites can be counted by this method, e.g. on man (Johnson & Mellanby, 1942). Nelson, Slen & Banky (1957) compared various relative sampling techniques for the sheep ked, *Melophagus ovinus*, with the 'total live counts' (MacLeod, 1948) and with the 'picked-off' count; in the latter method every ked that can be found on the sheep is picked off. They showed that the 'total live count', in which the living keds are counted as the fleece is parted, underestimated the total population; the actual mean estimates for different breeds of sheep ranging from 61 to 81 % of the 'picked-

off' count. This experiment emphasizes the importance of confirming that apparently absolute methods are really giving estimates of the total population. Ash (1960) sampled lice from living birds by removing a proportion of the feathers from the area frequented; the lice on these were determined and the total feathers in the area counted. Buxton (1947) determined the head louse populations of men by shaving the scalp and dissolving the hair.

Combing

A fine-toothed comb may be used to remove the ectoparasites on living mammals; but with lice, only a small part of the population is removed (see below). It is often an advantage if the host is anaesthetized with ether (Mosolov, 1959), but this also affects the parasites and is strictly a combination of combing and fumigation (see below).

Fumigation

This technique is particularly successful for Hippoboscidae and fleas (Siphonaptera) and is used for obtaining these parasites from birds at ringing stations and elsewhere during ecological studies on vertebrates. The hosts are collected from a trap and immediately placed in a white cloth (linen) bag for transportation to the laboratory. Here they may be exposed to the vapour of ether (Mosolov, 1959; Janion, 1960) or chloroform (Williamson, 1954) that will dislodge the parasites. Janion merely sprinkled the bags containing the hosts (mice in his work) with ether and after a few seconds of gentle squeezing the mouse was released, the fleas remaining in the bag; Mosolov removed the rodents from the bag and shook and combed them with forceps and obtained a wide range of parasites. He points out that as larval ectoparasites are often pale and the adults dark they are most readily seen if they are shaken on to coloured paper. Malcomson (1960) dusted the plumage with pyrethrum powder and allowed the bird to flutter for about 5 min under an inverted paper cone. Chlordimeform causes ticks to detach from their hosts (Gladney, Ernst & Drummond, 1974).

In the 'Fair Isle apparatus' described by Williamson (1954) the bird is placed in an open-ended cylinder with its head projecting from one end and protected by a collar of oiled silk which also seals the top. The base of the cylinder is stood on a white tile and chloroform vapour is pumped into the cylinder. This is conveniently produced by bubbling air with a rubber bulb from inlet tubes (with ends of sintered glass) through chloroform (Fig. 4.12). The bird should be encouraged to flutter during this process, which lasts for about one minute. The parasites will be found on the tile and in the bag in which the bird was carried; in view of Mosolov's (1959) observations a pink or light blue tile might be preferable to the white tile of the original model.

Kalamarz (1963) found that lice were efficiently removed from the host after fumigation if it was vacuum cleaned with a special brush nozzle.

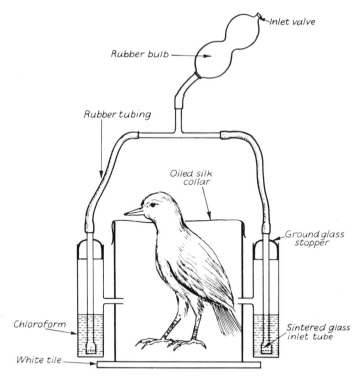

Fig. 4.12 The 'Fair Isle' apparatus for collecting bird ectoparasites.

4.3.2 Sampling from dead hosts
Mohr (1959) recommends the preservation of the host by formalin injection and its storage in a closed black cloth bag if the collection of ectoparasites cannot be carried out immediately.

Searching and combing
These methods combined with brushing in all directions with a stiff brush will provide many specimens from mammal pelts (Spencer, 1956), but often this is really only a relative method. With birds, however, Ash (1960) considered that a feather-by-feather search was the only efficient method for lice. Janzen (1963) used a novel principle, that may be of wider application, to obtain almost absolute samples of the beaver beetle (*Platypsyllus castoris*). The host's pelts were first frozen and then brought into a warm environment, the beetles would move away from the chilled pelts and the majority were found by searching; most of the remainder were extracted by combing.

Fumigation
If the mammals are trapped live, but can be sacrificed, fumigation with

hydrogen cyanide (produced from calcium cyanide) for about 15 minutes followed by combing will yield most parasites (Ellis, 1955; Murray, 1957). Bird lice on hens have been sampled by fumigation with methyl bromide followed by fluffing the feathers (Harshbarger & Raffensperger, 1959).

Dissolving
The pelt of the host is dissolved, most conveniently by incubating with the proteolytic enzyme trypsin for two days, followed by boiling in 10 % caustic potash for some minutes (Cook, 1954). After such treatment virtually only the ectoparasites remain and even minute Listrophorid mites can be recovered. This method is generally associated with Hopkins (1949), who records the following comparison with searching and beating; the latter method had yielded 31 lice (larvae and adults) from three pelts; when these were dissolved a further 1208 specimens were recovered. Although this method is very useful for mammal ectoparasites, Ash (1960) found it unsatisfactory for those on birds.

Clearing
The pelt is shaved and the hair placed in lysol or some other cleaning medium when the lice and their eggs may be counted under a microscope (Murray, 1961). If the exact distribution of the parasites is to be determined Murray recommends stunning the animals, soaking their coats with ether or chloroform to kill all lice *in situ* and then placing the animal in a closed jar until dead. The hair is then shaved from each area (these may be delimited by a grid) and mounted in Berlese's mounting medium.

Washing
Large numbers of ectoparasites, especially mites, can be removed by washing the pelt or the animal in a solution (< 5 %) of detergent (Lipovsky, 1951).

4.3.3 Sampling from vertebrate 'homes'
Birds' and rodents' nests, bat roosts and other vertebrate 'homes' may usually be treated as litter and their fauna extracted by modified Berlese funnels (Sealander & Hoffman, 1956) and the other methods described in the next chapter; however, Woodroffe (1953) concluded that there was no absolute method, he found a combination of warming and sieving most efficient. Drummond (1957) and Wasylik (1963) collected continuous samples of the mites in mammal and avian nests by placing funnels or gauze-covered tubes below the nests; such traps basically resemble pitfall traps (p. 247) and provide relative estimates.

REFERENCES

ADAMS, R. E. WOLFE, J. E., MILNER, M. and SHELLENBERGER, J. A., 1953. Aural detection of grain infested internally with insects. *Science* **118**, 163–4.

ALIKHAN, M. A., 1961. Population estimation techniques for studies on the black bean aphid, *Aphis fabae* Scop. *Annls. Univ. Mariae Curie-Sklodowska (C)* **14**, (1959), 83–92.

AMMAN, G. D. and RASMUSSEN, L. A., 1969. Techniques for radiographing and the accuracy of the X-ray method for identifying and estimating numbers of the Mountain Pine Beetle. *J. econ. Ent.* **62**, 631–4.

APT, A. C., 1950. A method for detecting hidden infestation in wheat. *Milling Production* **15** (5), 1.

APT, A. C., 1952. A rapid method of examining wheat samples for infestation. *Milling Production* **17** (5), 4.

ARNOLD, A. J., NEEDHAM, P. H. and STEVENSON, J. H., 1973. A self-powered portable insect suction sampler and its use to assess the effects of azinphos methyl and endosulfan on blossom beetle populations on oil seed rape. *Ann. app. Biol.* **75**, 229–33.

ARTHUR, A. P., 1962. A skipper, *Thymelicus lineola* (Ochs.) (Lepidoptera: Hesperiidae) and its parasites in Ontario. *Can. Ent.* **94**, 1082–9.

ASH, J. S., 1952. Records of Hippoboscidae (Dipt.) from Berkshire and Co. Durham in 1950, with notes on their bionomics. *Ent. mon. Mag.* **88**, 25–30.

ASH, J. S., 1960. A study of the Mallophaga of birds with particular reference to their ecology. *Ibis* **102**, 93–110.

AUSTIN, M. D. and MASSEE, A. M., 1947. Investigations on the control of the fruit tree red spider mite (*Metatetranychus ulmi* Koch) during the dormant season. *J. hort. Sci.* **23**, 227–53.

BALOGH, J. and LOKSA, I., 1956. Untersuchungen über die Zoozönose des Luzernenfeldes. Strukturzönologische Abhandlung. *Acta Zool. Hung.* **2**, 17–114.

BANKS, C. J., 1954. A method for estimating populations and counting large numbers of *Aphis fabae* Scop. *Bull. ent. Res.* **45**, 751–6.

BANKS, C. J. and BROWN, E. S., 1962. A comparison of methods of estimating population density of adult Sunn Pest, *Eurygaster integriceps* Put. (Hemiptera, Scutelleridae) in wheat fields. *Entomologia exp. appl.* **5**, 255–60.

BERRYMAN, A. A., 1964. Identification of insect inclusions in X-rays of Ponderosa pine bark infested by western pine beetle, *Dentroctonus brevicornis* Le Conte. *Can. Ent.* **96**, 883–8.

BERRYMAN, A. A. and STARK, R. W., 1962. Radiography in forest entomology. *Ann. ent. Soc. Am.* **55**, 456–66.

BIDLINGMAYER, W. L., 1961. Field activity studies of adult *Culicoides furens. Ann. ent. Soc. Am.* **54**, 149–56.

BLETCHLY, J. D. and BALDWIN, W. J., 1962. Use of X-rays in studies of wood boring insects. *Wood* **27**, 485–8.

BOUDREAUX, H. B., 1953. A simple method of collecting spider mites. *J. econ. Ent.* **46**, 1102–3.

BROWNING, T. O., 1959. The long-tailed mealybug *Pseudococcus adonidium* L. in South Australia. *Aust. J. agric. Res.* **10**, 322–39.

BULLOCK, J. A., 1963. Extraction of Thysanoptera from samples of foliage. *J. econ. Ent.* **56**, 612–14.

BURGES, H. D., 1960. A spear for sampling bulk grain by suction. *Bull. ent. Res.* **51**, 1–5.

BUXTON, P. A., 1947. *The Louse.* 2nd ed. 164 pp., Arnold, London.

CARLSON, O. V. and HIBBS, E. T., 1962. Direct counts of potato leafhopper, *Empoasca fabae*, eggs in *Solanum* leaves. *Ann. ent. Soc. Am.* **55**, 512–15.

CHAMBERLIN, J. C., 1940. A mechanical trap for the sampling of aerial insect populations. *U.S.D.A. Bur. Ent. Pl. Quar. E. T.* **163**, 12 pp.

CHANT, D. A., 1962. A brushing method for collecting mites and small insects from leaves. *Progress in Soil Zoology* **1**, 222–5.

CHANT, D. A. and MUIR, R. C., 1955. A comparison of the imprint and brushing machine methods for estimating the numbers of the fruit tree red spider mite, *Metatetranychus ulmi* (Koch), on apple leaves. *Rep. E. Malling Res. Sta.* (A) **1954**, 141–5.

CHAO, Y. and PETERSON, A., 1952. A new type of aspirator. *J. econ. Ent.* **45**, 751.

CHAUVIN, R., 1957. *Réflexions sur l'écologie entomologique.* Soc. Zool. Agricole Talence, 79 pp. [extract *Rev. Zool. Agric. appl.* 1956 (4–6, 7–9) and 1957 (1–3, 4–6)].

CHERRY, R. H., WOOD, K. A. and RUESINK, W. G., 1977. Emergence trap and sweep net sampling for adults of the potato leafhopper from Alfalfa. *J. econ. Ent.* **70**, 279.

CHINAEV, P. P., 1959. Methods in quantitative sampling of bloodsucking mosquitoes (Diptera, Culicidae). [In Russian.] *Ent. Obozr.* **38**, 757–65 (transl. *Ent. Rev.* **38**, 679–86).

CLEVELAND, M. L., 1962. Adhesives for holding mites to glass plates. *J. econ. Ent.* **55**, 570–1.

COINEAU, Y., 1962. Nouvelles méthodes de prospection de la faune entomologique des plantes herbacées et ligneuses. *Bull. Soc. ent. Fr.* **67**, 115–19.

COLLYER, E., 1951. A method for the estimation of insect populations of fruit trees. *Rep. E. Malling Res. Sta.* 1949–50, 148–51.

CONDRASHOFF, S. F., 1967. An extraction method for rapid counts of insect eggs and small organisms. *Can. Ent.* **99**, 300–3.

CONE, W. W., 1963. The black vine weevil, *Brachyrhinus sulcatus*, as a pest of grapes in South Central Washington. *J. econ. Ent.* **56**, 677–80.

CONNELL, W. A., 1959. Estimating the abundance of corn earworm eggs. *J. econ. Ent.* **52**, 747–9.

COOK, E. F., 1954. A modification of Hopkins' technique for collecting ectoparasites from mammal skins. *Ent. News* **15**, 35–7.

CORBET, P. S., 1966. A method for sub-sampling insect collections that vary widely in size. *Mosquito News* **26**, 420–4.

CROIX, E. A. S. LA, 1961. Observations on the ecology of the cotton-flea-beetles in the Sudan Gezira and the effect of sowing date on the level of population in cotton. *Bull. ent. Res.* **52**, 773–83.

CURTIS, W. E., 1942. A method of locating insect eggs in plant tissues. *J. econ. Ent.* **35**, 286.

DE MARS, C. J., 1963. A comparison of radiograph analysis and bark dissection in estimating numbers of western pine beetle. *Can. Ent.* **95**, 1112–16.

DEMPSTER, J. P., 1961. A sampler for estimating populations of active insects upon vegetation. *J. Anim. Ecol.* **30**, 425–7.

DENNIS, N. M. and DECKER, R. W., 1962. A method and machine for detecting living internal insect infestation in wheat. *J. econ. Ent.* **55**, 199–203.

DIETRICK, E. J., 1961. An improved back pack motor fan for suction sampling of insect populations. *J. econ. Ent.* **54**, 394–5.

DIETRICK, E. J., SCHLINGER, E. I. and BOSCH, R. VAN DEN, 1959. A new method for sampling arthropods using a suction collecting machine and modified Berlese funnel separator. *J. econ. Ent.* **52**, 1085–91.

DONDALE, C. D., NICHOLLS, C. F., REDNER, J. H., SEMPLE, R. B. and TURNBULL, A. L., 1971. An improved Berlese-Tullgren funnel and a flotation separator for extracting grassland arthropods. *Can. Ent.* **103**, 1549–52.

DRUMMOND, R. O., 1957. Observations on fluctuations of acarine populations from nests of *Peromyscus leucopus*. *Ecol. Monogr.* **27**, 137–52.

DUDLEY, C. O., 1971. A sampling design for the egg and first instar larval populations of the western pine beetle, *Dendroctonus brevicomis* (Coleoptera: Scolytidae). *Can. Ent.* **103**, 1291–313.

ELLIS, L. L., 1955. A survey of the ectoparasites of certain mammals in Oklahoma. *Ecology* **36,** 12–18.

EMDEN, H. F. VAN, 1963. A technique for the marking and recovery of turf samples in stem borer investigations. *Entomologia exp. app.* **6,** 194–8.

EVANS, J. W., 1933. A simple method of collecting thrips and other insects from blossom. *Bull. ent. Res.* **24,** 349–50.

FURNISS, M. M., 1962. A circular punch for cutting samples of bark infested with beetles. *Can. Ent.* **94,** 959–63.

GARA, R. I. and VITÉ, J. P., 1962. Studies on the flight patterns of bark beetles (Coleoptera: Scolytidae) in second growth Ponderosa pine forests. *Contrib. Boyce Thompson Inst.* **21,** 275–90.

GIBB, J. A. and BETTS, M. M., 1963. Food and food supply of nestling tits (Paridae) in Breckland pine. *J. Anim. Ecol.* **32,** 489–533.

GIBBS, D. G., PICKETT, A. D. and LESTON, D., 1968. Seasonal population changes in cocoa capsids (Hemiptera, Miridae) in Ghana. *Bull. ent. Res.* **58,** 279–93.

GIBSON, R. W., 1975. Measurement of eriophyid mite populations on ryegrass using ultrasonic radiation. *Trans. R. ent. Soc. Lond.* **127,** 31–2.

GLADNEY, W. J., ERNST, S. E. and DRUMMOND, R. O., 1974. Chlordimeform: a detachment-stimulating chemical for three-host ticks. *J. med. Ent.* **11,** 569–72.

GOODEY, J. B., 1957. Laboratory methods for work with plant and soil nematodes. *Tech. Bull. Min. Agric. Lond.* **2** (3rd ed.). London. H.M.S.O.

GOODHUE, R. D. and EMDEN, H. F. VAN, 1960. Detection of stem borers in Gramineae by X-rays. *Plant Path.* **9,** 194.

GOOSENS, H. J., 1949. A method for staining insect egg plugs in wheat. *Cereal Chem.* **26** (5), 419–20.

GRAHAM, H. M., ROBERTSON, O. T. and MARTIN, D. F., 1964. Radiographic detection of pink bollworm larvae in cottonseed. *J. econ. Ent.* **57,** 419–20.

GRAY, K. W. and SCHUH, J., 1941. A method and contrivance for sampling pea aphid populations. *J. econ. Ent.* **34,** 411–15.

GREIG-SMITH, P., 1964. *Quantitative Plant Ecology*. 2nd ed. 256pp. Butterworths, London.

HARRIS, J. W. E., COLLIS, D. G. and MAGAR, K. M., 1972. Evaluation of the tree-beating method for sampling defoliating forest insects. *Can. Ent.* **104,** 723–9.

HARRIS, M., 1971. Sampling pear foliage for nymphs of the pear Psylla, using the Berlese-Tullgren funnel. *J. econ. Ent.* **64,** 1317–8.

HARRIS, P., 1960. Number of *Rhyacionia buoliana* per pine shoot as a population index, with a rapid determination method of this index at low population levels. *Can. J. Zool.* **38,** 475–8.

HARSHBARGER, J. C. and RAFFENSPERGER, E. M., 1959. A method for collecting and counting populations of the shaft louse. *J. econ. Ent.* **52,** 1215–16.

HEIKINHEIMO, O. and RAATIKAINEN, M., 1962. Comparison of suction and netting methods in population investigations concerning the fauna of grass leys and cereal fields, particularly in those concerning the leafhopper, *Calligypona pellucida* (F.). *Valt. Maatalousk. Julk. Helsingfors* **191,** 31 pp.

HELSON, G. A. H., 1958. Aphid populations: ecology and methods of sampling aphids *Myzus persicae* (Sulz.) and *Aulacorthum solani* (Kltb). *N. Z. Entomologist* **2,** 20–3.

HENDERSON, C. F. and MCBURNIE, H. V., 1943. Sampling technique for determining populations of the citrus red mite and its predators. *U.S.D.A. Circ.* **671,** 11 pp.

HENDERSON, I. F. and WHITAKER, T. M., 1976. The efficiency of an insect sampler in grassland. *Ecol. Entomology* **2,** 57–60.

HOERNER, J. L., 1947. A separator for onion thrips. *J. econ. Ent.* **40,** 755.

HOPKINS, G., 1949. The host associations of the lice of mammals. *Proc. zool. Soc. Lond.* **119,** 387–604.

HORSFALL, W. R., 1962. Trap for separating collections of insects by interval. *J. econ. Ent.* **55**, 808–11.

HOWER, A. A. and FERGUSON, W., 1972. A square-foot device for use in vacuum sampling alfalfa insects. *J. econ. Ent.* **65**, 1742–3.

HUGHES, R. D., 1963. Population dynamics of the cabbage aphid *Brevicoryne brassicae* (L.). *J. Anim. Ecol.* **32**, 393–424.

HUGHES, R. D. and WOOLCOCK, L. T., 1963. The use of an electronic counter in population studies of the cabbage aphid (*Brevicoryne brassicae* (L.)). *N.Z.J. agric. Res.* **6**, 320–7.

IVES, W. G. H., 1959. A technique for estimating Tamarack foliage production, a basis for detailed population studies of the larch sawfly. *Can. Ent.* **91**, 513–19.

IVES, W. G. H., 1967. Relations between invertebrate predators and prey associated with Larch Sawfly eggs and larvae on Tamarack. *Can. Ent.* **99**, 607–22.

JACOBSON, M., 1966. Chemical insect attractants and repellants. *Ann. Rev. Ent.* **11**, 403–22.

JANION, S. M., 1960. Quantitative dynamics in fleas (Aphaniptera) infesting mice of Puszcza Kampinoska Forest. *Bull. acad. Pol. Sci.* II, **8** (5), 213–18.

JANZEN, D. H., 1963. Observations on populations of adult beaver-beetles, *Platypsyllus castoris* (Platypsyllidae: Coleoptera). *Pan. Pacif. Ent.* **32**, 215–28.

JOHNSON, C. G., 1950. A suction trap for small airborne insects which automatically segregates the catch into successive hourly samples. *Ann. appl. Biol.* **37**, 80–91.

JOHNSON, C. G., 1957. The distribution of insects in the air and the empirical relation of density to height. *J. Anim. Ecol.* **26**, 479–94.

JOHNSON, C. G. and MELLANBY, K., 1942. The parasitology of human scabies. *Parasitology* **34**, 285–90.

JOHNSON, C. G., SOUTHWOOD, T. R. E. and ENTWISTLE, H. M., 1957. A new method of extracting arthropods and Molluscs from grassland and herbage with a suction apparatus. *Bull. ent. Res.* **48**, 211–18.

JOHNSON, C. G. and TAYLOR, L. R., 1955a. The development of large suction traps for airborne insects. *Ann. appl. Biol.* **43**, 51–61.

JOHNSON, C. G. and TAYLOR, L. R., 1955b. The measurement of insect density in the air. *Lab. Pract.* **4**, 187–92, 235–9.

JOHNSON, C. G., TAYLOR, L. R. and SOUTHWOOD, T. R. E., 1962. High altitude migration of *Oscinella frit* L. (Diptera: Chloropidae). *J. Anim. Ecol.* **31**, 373–83.

JONES, B. M., 1950. A new method for studying the distribution and bionomics of trombiculid mites (Acarina: Trombidiidae) *Parasitology* **40**, 1–13.

JONES, L. S. and PRENDERGAST, D. T., 1937. Method of obtaining an index to density of field populations of citrus red mite. *J. econ. Ent.* **30**, 934–40.

JUILLET, J. A., 1963. A comparison of four types of traps used for capturing flying insects. *Can. J. Zool.* **41**, 219–23.

KALAMARZ, E., 1963. Badania nad biologia Mallophaga IV. Nowe methody zbierania ektopasozytów. *Ekol. Polska B* **9**, 321–5.

KATZ, R., FARRELL, E. P. and MILNER, M., 1954. The separation of grain by projection I. *Cereal Chem.* **31**, 316–25.

KENNARD, W. C. and SPENCER, J. L., 1955. A mechanical insect collector with high manœuverability. *J. econ. Ent.* **48**, 478–9.

KERSHAW, W. J. S., 1964. Aphid sampling in sugar beet. *Plant. Path.* **13**, 101–6.

KIRK, I. W. and BOTTRELL, D. G., 1969. A mechanical sampler for estimating Boll Weevil populations. *J. econ. Ent.* **62**, 1250–51.

KOBAYASHI, F. and MURAI, M., 1965. Methods for estimating the number of the Cryptomeria red mite, especially with the removal by solutions. *Res. Pop. Ecol.* **7**, 35–42.

KOURA, A., 1958. A new transparency method for detecting internal infestation in grains. *Agric. Res. Rev.* **36**, 110–13.

KRETZSCHMAR, G. P., 1948. Soy bean insects in Minnesota with special reference to sampling techniques. *J. econ. Ent.* **41**, 586–91.

LASTER, M. L. and FURR, R. E., 1962. A simple technique for recovering insects from sorghum heads in insecticide tests. *J. econ. Ent.* **55**, 798.

LEGNER, E. F. and OATMAN, E. R., 1962. Foliage-feeding Lepidoptera on young non-bearing apple trees in Wisconsin. *J. econ. Ent.* **55**, 552–4.

LEIGH, T. F., GONZALEZ, D. and VAN DEN BOSCH, R., 1970. A sampling device for estimating absolute insect populations on cotton. *J. econ. Ent.* **63**, 1704–6.

LE PELLEY, R. H., 1942. A new method of sampling thrips populations. *Bull. ent. Res.* **33**, 147–8.

LEVI, M. I., CHERNOV, S. G., LABUNETS, N. F. and KOSMINSKII, R. B., 1959. Aspiration method for the collection of fleas from rodents' nests. [In Russian.] *Med. Parazitol.* **28**, 64–9.

LEWIS, T., 1960. A method for collecting *Thysanoptera* from Gramineae. *Entomologist* **93**, 27–8.

LEWIS, T. and NAVAS, D. E., 1962. Thysanopteran populations overwintering in hedge bottoms, grass litter and bark. *Ann. appl. Biol.* **50**, 299–311.

LEWIS, T. and TAYLOR, L. R., 1965. Diurnal periodicity of flight by insects. *Trans. R. ent. Soc. Lond.* **116**, 393–469.

LIPOVSKY, L. J., 1951. A washing method of ectoparasite recovery with particular reference to chiggers. *J. Kansas ent. Soc.* **24**, 151–6.

LORD, F. T., 1965. Sampling predator populations on apple trees in Nova Scotia. *Can. Ent.* **97**, 287–98.

LOWE, A. D. and DROMGOOLE, W. V., 1958. The development of an electronic aphid counter. *N.Z.J. agric. Res.* **1**, 903–12.

McCOY, J. R. and LLOYD, E. P., 1975. Evaluation of airflow systems for the collection of Boll Weevils from cotton. *J. econ. Ent.* **68**, 49–52.

McGOVERN, W. L., LEGGETT, J. E., JOHNSON, W. C. and CROSS, W. H., 1973. Techniques for separating boll weevils and other small insects from samples taken with insect collecting machines. *J. econ. Ent.* **66**, 1332.

MACLEOD, J., 1948. The distribution and dynamics of ked populations, *Melophagus ovinus* Linn. *Parasitology* **39**, 61–8.

MACLEOD, J. and DONNELLY, J., 1957. Some ecological relationships of natural populations of Calliphorine blowflies. *J. Anim. Ecol.* **26**, 135–70.

MAELZER, D. A., 1976. A photographic method and a ranking procedure for estimating numbers of the rose aphid, *Macrosiphum rosae* (L.) on rose buds. *Aust. J. Ecol.* **1**, 89–96.

MALCOLMSON, R. O., 1960. Mallophaga from birds of North America. *Wilson Bull.* **72** (2), 182–97.

MAW, M. G., 1964. An effect of static electricity on captures in insect traps. *Can. Ent.* **96**, 1482.

MILNER, M., BARNEY, D. L. and SHELLENBERGER, J. A., 1950. Use of selective fluorescent stains to detect egg plugs on grain kernels. *Science* **112**, 791–2.

MILNER, M., KATZ, R., LEE, M. R. and PYLE, W. B., 1953. Application of the Polaroid-Land process to radiographic inspection of wheat. *Cereal Chem.* **30**, 169–70.

MILNER, M., LEE, M. R. and KATZ, R., 1950. Application of X-ray technique to the detection of insects infesting grain. *J. econ. Ent.* **43**, 933–5.

MOHR, C. O., 1959. A procedure for delayed collecting of ectoparasites from small captured hosts. *J. Parasit.* **45**(2), 154.

MOORE, N. W., 1964. Intra- and interspecific competition among dragonflies (Odonata). *J. Anim. Ecol.* **33**, 49–71.

MORGAN, C. V. G., CHANT, D. A., ANDERSON, N. H. and AYRE, G. L., 1955. Methods for estimating orchard mite populations, especially with the mite brushing machine. *Can. Ent.* **87**, 189–200.

MOSOLOV, L. P., 1959. A method of collecting the ectoparasites of rodents, without destroying the host population. [In Russian, Eng. Summary.] *Med. Parazitol.* **28**, 189–92.

MUIR, R. C. and GAMBRILL, R. G., 1960. A note on the knockdown method for estimating numbers of insect predators on fruit trees. *Ann. Rep. E. Malling Res. Sta.* **1959**, 109–11.

MURRAY, K. F., 1957. An ecological appraisal of host–ectoparasite relationships in a zone of epizootic plague in central California. *Am. J. trop. Med. Hyg.* **6**, 1068–86.

MURRAY, M. D., 1961. The ecology of the louse *Polyplax serrata* (Burm.) on the mouse *Mus musculus* L. *Aust. J. Zool.* **9**, 1–13.

NELSON, W. A., SLEN, S. B. and BANKY, E. C., 1957. Evaluation of methods of estimating populations of the sheep ked, *Melophagus ovinus* (L.) (Diptera: Hippoboscidae), on mature ewes and young lambs. *Can. J. Anim. Sci.* **37**, 8–13.

NEWELL, I. M., 1947. Quantitative methods in biological and control studies of orchard mites. *J. econ. Ent.* **40**, 683–9.

NICHOLLS, C. F., 1960. A portable mechanical insect trap. *Can. Ent.* **92**, 48–51.

OTVOS, I. S. and BRYANT, D. G., 1972. An extraction method for rapid sampling of eastern hemlock looper eggs, *Lambdina fiscellaria fiscellaria* (Lepidoptera: Geometridae). *Can. Ent.* **104**, 1511–4.

PEDERSEN, J. R. and BROWN, R. A., 1960. X-ray microscope to study behaviour of internal-infesting grain insects. *J. econ. Ent.* **53**, 678–9.

PIELOU, D. P., 1961. Note on a volumetric method for the determination of numbers of apple aphid, *Aphis pomi* DeG., on samples of apple foliage. *Can. J. Pl. Sci.* **41**, 442–3.

PIETERS, E. P. and STERLING, W. L., 1973. Sampling techniques for cotton arthropods in Texas. *Tex. Agr. Exp. Sta. MP* **1120**, 1–8.

PRESCOTT, H. W. and NEWTON, R. C., 1963. Flight study of the clover root Curculio. *J. econ. Ent.* **56**, 368–70.

PUTMAN, W. L., 1966. Sampling mites on peach leaves with the Henderson-McBurnie machine. *J. econ. Ent.* **59**, 224–5.

REMANE, R., 1958. Die Besiedlung von Grünlandflächen verschiedener Herkunft durch Wanzen und Zikaden im Weser – Ems – Gebiet. *Z. angew. Ent.* **42**, 353–400.

RICHARDS, O. W., 1940. The biology of the small white butterfly (*Pieris rapae*) with special reference to the factors controlling abundance. *J. Anim. Ecol.* **9**, 243–88.

RICHARDS, O. W. and WALOFF, N., 1954. Studies on the biology and population dynamics of British grasshoppers. *Anti-Locust Bull.* **17**, 184 pp.

RICHARDS, O. W. and WALOFF, N., 1961. A study of a natural population of *Phytodecta olivacea* (Forster) (Coleoptera, Chrysomeloidea). *Phil. Trans.* **244**, 204–57.

RICHMOND, C. A. and GRAHAM, H. M., 1969. Two methods of operating a vacuum sampler to sample populations of the Cotton Fleahopper on wild hosts. *J. econ. Ent.* **62**, 525–6.

RISTICH, S. and LOCKARD, D., 1953. An aspirator modified for sampling large populations. *J. econ. Ent.* **46**, 711–12.

ROMNEY, V. E., 1945. The effect of physical factors upon catch of the beet leafhopper (*Eutettix tenellus* (Bak.)) by a cylinder and two sweep methods. *Ecology* **26**, 135–47.

RUDD, W. G. and JENSEN, R. L., 1977. Sweep net and ground cloth sampling for insects in soybeans. *J. econ. Ent.* **70**, 301–4.

SANTA, H., 1961. [A method for sampling plant and leaf-hopper density in winter and early Spring.] [In Japanese.] *Plant Protection, Tokyo* **8**, 353–5.

SATCHELL, J. E. and MOUNTFORD, M. D., 1962. A method of assessing caterpillar populations on large forest trees, using a systemic insecticide. *Ann. appl. Biol.* **50**, 443–50.

SEALANDER, J. A. and HOFFMAN, C. E., 1956. A modified Berlese funnel for collecting mammalian and avian ectoparasites. *Southw. Nat., Dallas* **1**, 134–6.

SHANDS, W. A., SIMPSON, G. W. and REED, L. B., 1954. Subunits of sample for estimating aphid abundance on potatoes. *J. econ. Ent.* **47**, 1024–7.

SHEPARD, M., STERLING, W. and WALKER, J. K., 1972. Abundance of beneficial arthropods on cotton genotypes. *Environmental Ent.* **1**(1), 117–21.

SIMAK, M., 1955. Insect damage on seeds of Norway spruce determined by X-ray photography. *Medd. Stat. Skogsforsknings-Inst.*, Uppsala **41**, 299–310.

SKUHRAVÝ, V., NOVÁK, K. and STARÝ, P., 1959. Entomofauna jetele (*Trifolium pratense* L.) a jeji vývoj. *Rozpr. čsl. Akad. Věd.* **69** (7), 3–82.

SKUHRAVÝ, V. and NOVÁK, V., 1961. The study of field crop entomocenoses. [In Russian.] *Ent. Obozr.* **41**, 807–14 (transl. *Ent. Rev.* **41**, 454–8).

SMITH, B. D., 1960. Population studies of the black current gall mite (*Phytoptus ribis* Nal.). *Rep. agric. hort. Res. Sta. Bristol* **1960**, 120–4.

SMITH, J. W., STADELBACHER, E. A. and GANTT, C. W., 1976. A comparison of techniques for sampling beneficial arthropod populations associated with cotton. *Environmental Ent.* **5**, 435–44.

SMITH, L. B., 1977. Efficiency of Berlese-Tullgren funnels for removal of the Rusty grain beetle, *Cryptolestes ferrugineus*, from wheat samples. *Can. Ent.* **109**, 503–9.

SOUTHWOOD, T. R. E. and CROSS, D. J., 1969. The ecology of the partridge. III. Breeding success and the abundance of insects in natural habitats. *J. Anim. Ecol.* **38**, 497–509.

SOUTHWOOD, T. R. E., JEPSON, W. F. and VAN EMDEN, H. F., 1961. Studies on the behaviour of *Oscinella frit* L. (Diptera) adults of the panicle generation. *Entomologia exp. appl.* **4**, 196–210.

SOUTHWOOD, T. R. E. and PLEASANCE, H. J., 1962. A hand-operated suction apparatus for the extraction of Arthropods from grassland and similar habitats, with notes on other models. *Bull. ent. Res.* **53**, 125–8.

SPENCER, G. J., 1956. Some records of ectoparasites from flying squirrels. *Proc. Ent. Soc. B. C.* **52**, 32–4.

STAGE, H. H., GJULLIN, C. M. and YATES, W. W., 1952. Mosquitoes of the North-western states. *U.S.D.A., Agric. Handb.* no. **46**, 95 pp.

STARK, H. E. and KINNEY, A. R., 1962. Abandonment of disturbed hosts by their fleas. *Pan. Pacif. Ent.* **38**, 249–51.

STEINER, H., 1962. Methoden zur untersuchung der Populationsdynamik in Obstanlagen (inc. Musternahme und Sammeln). *Entomophaga* **7**, 207–14.

STREET, M. W., 1971. A method for aural monitoring of in-kernel insect activity. *J. Georgia ent. Soc.* **6**, 72–5.

STRICKLAND, A. H., 1954. An aphid counting grid. *Plant Path.* **3**, 73–5.

STRICKLAND, A. H., 1961 Sampling crop pests and their hosts. *A. Rev. Ent.* **6**, 201–20.

SUMMERS, F. M. and BAKER, G. A., 1952. A procedure for determining relative densities of brown almond mite populations on almond trees. *Hilgardia* **21**, 369–82.

SZALAY-MARZSÓ, L., 1957. Populációsdinamikai vizsgálatok egy répaöfld répalevéltetü (*Dorsalis fabae* Scop.) állományén. *Ann. Inst. Prot. Plant Hung.* **7** (1952–6), 91–101.

TAYLOR, E. A. and SMITH, F. F., 1955. Three methods for extracting thrips and other insects from rose flowers. *J. econ. Ent.* **48**, 767–8.

TAYLOR, L. R., 1951. An improved suction trap for insects. *Ann. appl. Biol.* **38**, 582–91.

TAYLOR, L. R., 1955. The standardization of air flow in insect suction traps. *Ann. appl. Biol.* **43**, 390–408.

TAYLOR, L. R., 1958. Aphid dispersal and diurnal periodicity. *Proc. Linn. Soc. Lond.* **169**, 67–73.

TAYLOR, L. R., 1962a. The absolute efficiency of insect suction traps. *Ann. appl. Biol.* **50**, 405–21.

TAYLOR, L. R., 1962b. The efficiency of cylindrical sticky insect traps and suspended nets. *Ann. appl. Biol.* **50**, 681–5.

TAYLOR, L. R., 1974. Monitoring change in the distribution and abundance of insects. *Rep. Rothamsted exp. Stn.* **1973**, 202–39.

TAYLOR, L. R. and PALMER, J. M. P., 1972. Aerial sampling. In van Emden, H. (Ed.) *Aphid Technology*, pp. 189–234. Academic Press, London and New York.

TURNBULL, A. L. and NICHOLLS, C. F., 1966. A "quick trap" for area sampling or arthropods in grassland communities. *J. econ. Ent.* **59**, 1100–4.

VENABLES, E. P. and DENNYS, A. A., 1941. A new method of counting orchard mites. *J. econ. Ent.* **34**, 324.

VITÉ, J. P. and GARA, R. I., 1961. A field method for observation on olfactory responses of bark beetles (Scolytidae) to volatile materials. *Contrib. Boyce Thompson Inst.* **21**, 175–82.

WALKER, A. L., CATE, J. R., PAIR, S. D. and BOTTRELL, D. G., 1972. A volumetric method for estimating populations of the Greenbug on grain sorghum. *J. econ. Ent.* **65**, 422–3.

WASYLIK, A., 1963. Metoda analizy ciaglej roztoczy gniazd ptasich. *Ekol. Polska B* **9**, 219–24.

WAY, M. J. and HEATHCOTE, G. D., 1966. Interactions of crop density of field beans, abundance of *Aphis fabae* Scop., virus incidence and aphid control by chemicals. *Ann. appl. Biol.* **57**, 409–23.

WEEKMAN, G. T. and BALL, H. J., 1963. A portable electrically operated collecting device. *J. econ. Ent.* **56**, 708–9.

WHITE, T. C. R., 1975. A quantitative method of beating for sampling larvae of *Selidosema suavis* (Lepidoptera: Geometridae) in plantations in New Zealand. *Can. Ent.* **107**, 403–12.

WHITTAKER, J. B., 1965. The distribution and population dynamics of *Neophilaenus lineatus* (L.) and *N. exclamationis* (Thun.) (Homoptera, Cercopidae) on Pennine moorland, *J. Anim. Ecol.* **34**, 277–97.

WICKMAN, B. E., 1964. A comparison of radiographic and dissection methods for measuring siricid populations in wood. *Can. Ent.* **96**, 508–10.

WIEGERT, R. G., 1961. A simple apparatus for measuring density of insect population. *Ann. ent. Soc. Am.* **54**, 926–7.

WILDHOLTZ, T., VOGEL, W., STRAUB, A. and GESLER, B., 1956. Befallskontrolle an apfelbäumen im Frühjahr 1955. *Schwiez. Z. Obst. u. Weinb.* **65**, 85–8.

WILLIAMS, C. B. and MILNE, P. S., 1935. A mechanical insect trap. *Bull. ent. Res.* **26**, 543–51.

WILLIAMSON, K., 1954. The Fair Isle apparatus for collecting bird ectoparasites. *Brit. Birds* **47**, 234–5.

WILSON, L. F., 1962. A portable device for mass-collecting or sampling foliage-inhabiting arthropods. *J. econ. Ent.* **55**, 807–8.

WOOD, G. W. and SMALL, D. N., 1970. A method of sampling for adults of *Chlamisus cribripennis*. *J. econ. Ent.* **63**, 1361–2.

WOODROFFE, G. E., 1953. An ecological study of the insects and mites in the nests of certain birds in Britain. *Bull. ent. Res.* **44**, 739–72.

YAGHI, N., 1924. Application of the Roentgen ray tube to detection of boring insects. *J. econ. Ent.* **17**, 662–3.

5

Absolute Population Estimates by Sampling a Unit of Habitat – Soil and Litter

The extraction methods described here can be used not only with soil and with plant and animal debris, litter and dung, but also with plant material collected by suction apparatus or other means, the nests of vertebrates and the mud from ponds and rivers. The actual methods for obtaining the samples from these other habitats are discussed in Chapters 4 and 6.

Much of the work in soil zoology was originally aimed at the extraction of a large segment of the fauna by a single method; however, most workers have now concluded that a method that will give an almost absolute estimate of one species or groups, will give at the most a rather poor relative estimate for another. Not only does the efficiency of extraction vary with the animal, but also with the soil, its nature, its water content and the amount of vegetable matter in it. Therefore, although with certain animals under certain conditions each of these methods will give absolute population estimates, none of them will provide such data under all conditions.

Further information on ecological methods in soil zoology is given in reviews by Balogh (1958), Murphy (1962*b* and *c*), Macfadyen (1955, 1962) and Kevan (1962) and in the collections of research papers edited by Kevan (1955) Murphy (1962*a*) and Phillipson (1970).

Bees, cicindelid and *Bledius* larvae, and other comparatively large insects that make holes or casts in bare ground, may be counted directly *in situ*. In most studies, however, it is necessary both to take a sample and to extract the animals.

5.1 Sampling

The number and size of the samples and the sampling pattern in relation to statistical considerations have been discussed in Chapter 2 and by Macfadyen (1962). The soil samples are usually taken with a corer; golf-hole borers or metal tubing sharpened at one end make simple corers, but it has been suggested that some animals may be killed by compression when the core is forced from such 'instruments', and furthermore it is highly desirable to keep the core undisturbed (especially for extraction by behavioural methods), so more elaborate corers have been developed. In general the larger the animal and the sparser its population the bigger the sample (e.g. Frick (1962) took 8-

in diameter cores for *Nomia* bees). The depth to which it is necessary to sample varies with the animal and the condition of the soil (e.g. Paris & Pitelka, 1962); it will be particularly deep in areas with marked dry seasons (Price, 1975); many soil animals have seasonal and diel vertical migrations (Erman, 1973). Termites and ants present special problems due to their patchy contribution and the difficulty of taking adequately deep samples before the disturbance has caused them to escape (Brian, 1970; Sands, 1972).

With the O'Connor (1957) split corer (Fig. 5.1) the risk of compressing the sample by forcing it out of the corer is avoided. After the core has been taken the clamping band can be loosened, the two aluminium halves of the cover separated and the sample exposed. Furthermore the sample can then be easily divided into the different soil layers: litter, humus, upper 2 cm soil, etc. Plastic or metal rings may be inserted inside the metal sheath of the corer, just behind

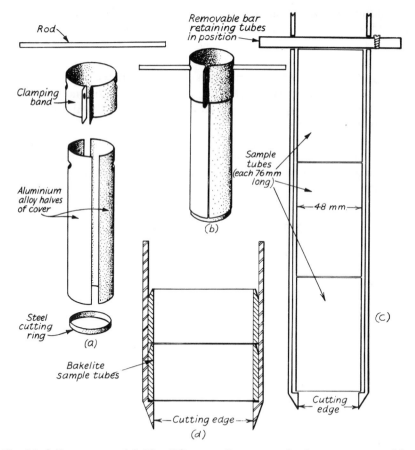

Fig. 5.1. Soil corers. *a* and *b*. The O'Connor split corer: *a*. showing compartments (after O'Connor, 1957); *b*. assembled. *c*. Soil corer with sample tubes (after Dhillon & Gibson, 1962). *d*. Soil corer for the canister extractor (after Macfadyen, 1961).

the cutting ring (Macfadyen, 1961; Dhillon & Gibson, 1962; Vannier & Vidal, 1965)(Fig. 5.1); these will enable the core to be extracted with its natural structure intact (see p. 184).

In order to penetrate hard tropical or frozen tundra soils it may be necessary to have equipment of the types described by Belfield (1956) and Potzger (1955). In contrast in soft humus rich situations, such as manure heaps, it is difficult to take an undisturbed sample; von Törne (1962*b*) has devised a sampler for these habitats consisting of two concentric tubes each with cutting teeth. The inner tube is pushed down firmly and held still, thereby protecting the sample, while the outer one is rotated and cuts through the compost. Tanton (1969) described the construction of a corer of this type, with a series of internal plastic rings, that can separate the sample into sub-samples. Another corer that allows sub-samples to be obtained from different depths is described by Vannier & Alpern (1968).

Semi-fluid substances, such as the contents of water-filled tree holes may be sampled with a device that includes an auger ('Worm bit') inside a normal corer; the turning of the auger takes the sample into the corer and retains it (Kitching, 1971).

Many pests of field crops lay their eggs on the bases of the plants and on the soil; the number of eggs usually falls off very rapidly with distance away from the plant (e.g. Lincoln & Palm, 1941; Abu Yaman, 1960). Suitable samples of young plants and the soil around them can be taken by the scissor type of sampler described by Webley (1957).

Fallen leaves and other debris are usually sampled with quadrats, such as a metal box with top and bottom missing and the lower edge sharpened (Gabbutt, 1959).

5.2 Mechanical methods of extraction

Mechanical methods have the advantages that theoretically they extract all stages, mobile and sedentary, and are in no way dependent on the behaviour of the animal or the condition of the substrate; samples for mechanical extraction may be stored frozen for long periods before use. Their disadvantages are that compared with behavioural methods the operator must expend a great deal of time and energy on each sample, that sometimes they damage the animals and that, as mobile and immobile animals are extracted, it may be difficult to distinguish animals that were dead at the time of sampling from those that were alive.

There are a number of distinct mechanical processes that can be used to separate animals from the soil and vegetable material: sieving, flotation, sedimentation, elutriation and differential wetting. The following account, however, is intended to be functional, rather than classificatory or historical; only the main types of processes will be outlined and it must be stressed that although there are already many different combination and variants (see

Murphy's, 1962c review) others will need to be developed for particular animals and substrates.

5.2.1 Dry sieving

This method may be of use for the separation of fairly large (and occasionally small) animals, from friable soil or fallen leaves. Its disadvantages are that small specimens are often lost and a considerable amount of time needs to be spent in hand-sorting the sieved material. The Reitter sifter (see Kevan, 1962) is a simple device, but this is really more a collector's tool than a means of estimating populations. Lane & Shirck's (1928) was the first mechanical sieve with a to-and-fro motion although it was worked by hand. The most complex apparatus of this type is the fully motorized self-propelled model of Lange, Akesson & Carlson (1955), which was used for surveying elaterid larvae in California.

Molluscs have been extracted by dry sieving methods (Økland, 1929; Jacot, 1935), but these have been shown to be inaccurate, some snails remaining in the leaves (Williamson, 1959).

Dry sieving has also been used to separate mosquito eggs from mud (Stage, Gjullin & Yates, 1952). A modified grain drier is the basis of this method; the samples are dried until almost dusty, passed through a mesh sieve and then into the top of four shaker sieves of the grain drier. Coarser particles are removed by the first three sieves, but the eggs and similar sized soil particles are held on the last one (80-mesh/in); they then fall through an opening on to a 60-mesh in roll screen; a carefully adjusted air current blows away the lighter particles during the fall. Because of their spindle shape the eggs, with only a very small quantity of soil, eventually pass 'end-on' through the roll screen into the 'catch pan'. Such a degree of mechanization is possible because only a single organism with a constant and regular shape and a uniform weight was required.

The active red-legged earth mite (*Halotydeus destructor*) lives amongst the grass and plant debris in Australian pastures; Wallace (1956) has shown that it can be very accurately sampled by taking shallow cores which are then inverted and rotated, still within the corer, over a sieve in a funnel. The sides of the metal corer are tapped a number of times and the mites fall through the sieve and may be collected in tubes below the funnel.

5.2.2 Soil washing (or wet sieving)

This technique is most often used in conjunction with other methods (see below); on its own it is of particular value where the organisms are much smaller than the particles of the substrate (e.g. small snails amongst freshly fallen leaves) or much larger (e.g.many mud-dwellers), and hence separation by size alone is sufficient. The sieves may be of basic designs: flat, revolving or 'three dimensional'.

Flat sieves may be built up into a tower (see Fig. 5.3) and a series of sieves was the basis of early uses of this method by N. A. Cobb and H. M. Morris. Sieving is often used in studies on the bottom animals of aquatic habitats for the separation of the mud from the organisms and stones that are retained by the sieve (Berg, 1938; Brundin, 1949; Jonasson, 1955; Hairston *et al.*, 1958); a single frame is used into which frames with different mesh phosphor-bronze screens may be fitted. Jonasson (1955) has shown that many benthic organisms pass through the 0.6-mm mesh gauge sieves frequently employed in such studies; with insect larvae the size of the head capsule determined whether or not it would pass through the mesh of a sieve and there was no evidence that a significant proportion would roll up and so be retained by coarse sieves. The smallest sieve used by Jonasson had a mesh gauze of 0.2 mm, but clearly in any particular study its size would be determined by the diameter of the organism being sampled.

Although most soil nematodes can be separated elutriation or the Baermann funnel (p. 189), for those in marine habitats these methods may not give good results (Capstick, 1959) and sieving is recommended to separate the largest worms, followed by agitation, to make a suspension of the remainder, which is subsampled volumetrically. The subsamples are examined directly.

Gastropod molluscs and possibly other small animals may be efficiently separated from fresh leaf litter by wet sieving, but here it is the organisms that pass through the sieve and the unwanted material that is retained (Williamson, 1959). The leaves are placed on a coarse sieve (mesh size 1 cm), above a fine cloth-covered grid, the whole immersed in a vessel of water (Fig. 5.2*a*) and the leaves are stirred from time to time; after about 15 minutes the vessel can be drained from below and the molluscs will be found on the cloth sheet, which is then inspected under the microscope. Clearly this method is not satisfactory for molluscs amongst soil or humus as much of these materials will pass through the upper sieve.

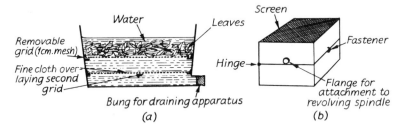

Fig. 5.2. *a*. Tank for the separation of molluscs from fallen leaves by wet sieving, based on Williamson's design. *b*. Simple sieve box for wet sieving sawfly cocoons, based on McLeod's design.

Simple revolving sieves were used by McLeod (1961) for the extraction of sawfly cocoons. They consisted of wooden boxes, hinged in the middle and with screen tops and bottoms (Fig. 5.2*b*); they were revolved once or twice while a jet of water from a hose was played on to them. More elaborate revolving models were designed by Horsfall (1956) and Read (1958); these are discussed below.

The so-called 'three-dimensional' or rod sieve was designed by Stewart (1974) to facilitate the separation, undamaged, of tipulid larvae or other large animals from grassy soil samples, prior to flotation. Stewart's apparatus consists of thirteen tiers of parallel galvanized steel rods set in an open-ended box. The rods are progressively closer in each tier (2.5 cm closing to 1.3 cm apart); the side of the box may be removed (for cleaning, a process assisted by a sliding panel). The sample is placed on the top row of rods and the box moved up and down in a tank of water; the inside of the tank is lined with a normal screen sieve. The vegetation will pass down between the rods, freeing insects and mineral material which remain in the tank. The three dimensional sieve is removed (and if necessary cleaned) and the sieve inside the tank removed and the normal flotation procedure followed (see below).

5.2.3 Soil washing and flotation

When the organisms to be extracted and the rest of the sample are of different particle-size, sieving alone is sufficient for extraction; however, in most situations after sieving the animals required remain mixed with a mass of similar sized mineral and vegetable matter. As the specific gravity of mineral and biotic material is frequently different the extraction may be taken a stage further by flotation of the animals; unfortunately plant material usually floats equally well. The combination of flotation, devised by A. Berlese many years before, with soil washing is usually associated with Ladell (1936). This approach, much modified and improved by Salt & Hollick (1944) and Raw (1955, 1962), is the basis of one of the most widely used, versatile and efficient techniques; it consists of four stages, three of which will be discussed under this heading.

(1) *Pretreatment.* In order to disperse the soil particles, particularly important if there is a high clay content, the cores may be soaked in water and deep frozen. Chemical dispersion may also be used: the core soaked in solutions of sodium citrate (d'Aguilar, Benard & Bessard, 1957) or sodium oxalate (Seinhorst, 1962). For the heavy clay soils it may be necessary to combine chemical and physical methods: the core is gently crumbled into a plastic container and covered with a solution of sodium hexametaphosphate (50 g) and sodium carbonate (20 g) in one litre of water (sold commercially as 'Calgon' or 'Sparkaleen'); the whole is then placed under reduced pressure in a vacuum desiccator for a time. After the restoration of atmospheric pressure the sample is frozen for at least 48 hours and may be stored in this condition (Raw, 1955).

(2) *Soil washing.* The sample is placed in the upper and coarsest, sieve of the washing apparatus (Fig.5.3*a*); it may be washed through by slow jets of water or single sieves 'dunked' in the settling can. The material retained in the sieves must be carefully teased apart, and thoroughly washed; a few large animals may be removed at this stage. The mesh of the lowest sieve* is such as to allow the animals required to pass through into

* If this sieve is finer its contents must also be tipped into the settling tank, its function being to aid dispersion.

the settling can, which is pivoted and is then tipped into the 'Ladell can'. A 'three-dimensional' sieve may be necessary for the above process if there is a lot of grass in the sample and large animals are being separated (see above). The Ladell, which resembles an inverted bottomless paraffin can, has a fine phosphor-bronze sieve and its lower opening immersed in the drainage tank; this tank should be arranged so that the level of water maintained in it, and hence in the Ladell, is slightly above the sieve of the latter – this minimizes blockage of the sieve. The Ladell is allowed to drain; this, often tedious, process may be aided by tapping the side of the can with the hand from time to time. The standard Ladell has a fine sieve (mesh size about 0.2 mm); for many animals this may be much finer than is necessary and to overcome this Stephenson (1962) has designed a Ladell, with a set of interchangeable sieve plates of various mesh sizes.

In conclusion it must be warned that soil washing is an invariably wet operation and is best carried out in a room with a concrete floor and the operator wearing rubber boots and suitable apparel.

(3) *Flotation*. The Ladell is removed from the tank to a stand (Fig. 5.3), allowed to drain completely and then the process of flotation is begun. The lower opening of the Ladell is closed with a bung and the flotation liquid introduced until the Ladell is about two-thirds full (Fig. 5.3b). Concentrated magnesium sulphate solution (specific gravity *c*. 1.2) is the most usual flotation liquid, but solutions of sodium chloride (Lincoln & Palm, 1941; Cockbill *et al.*, 1945; Golightly, 1952; Cohen, 1955), potassium bromide (d' Aguilar *et al.*, 1957) or zinc chloride (Sellmer, 1956) may be used. Air is then bubbled up from the bottom of the Ladell; this agitation is continued for $2 - 3$ minutes and serves to free any animal matter that may have been trapped on the sieve. More of the flotation solution is introduced from below until the liquid and 'float' passes over the lip and the latter is retained on the collecting tube, which usually consists of a glass tube with a piece of bolting silk held in place with a rubber band. A wash bottle may be used to help direct the float out of the Ladell. When the surface and sides of the Ladell are completely clean, the animal and plant material will all be in the collecting tube and the process is complete. The flotation liquid is then drained from the Ladell, for further use, by appropriate manipulations of the reservoir and pinchclips (see Fig. 5.3).

The animals may now be separated from the plant material by direct examination under the microscope or it may be necessary to carry out a further separation based on differential wetting and/or centrifuging (see below). However, before these are discussed other washing and flotation techniques must be reviewed.

The apparatus of d'Aguilar *et al.* (1957) first removes the smallest silt particles by agitating the sample with a water current in a cylinder with a fine gauze side; subsequently the material is passed through a series of sieves, followed by flotation. Edwards, Whiting & Heath (1970) have developed a fully mechanised soil washing process that in general is as efficient as the Salt-Hollick process described above.

The eggs and puparia of the cabbage root fly (*Erioischia brassicae*) may be separated from the soil by sieving and flotation in water (Abu Yaman, 1960; Hughes, 1960); with such insects therefore there is no need for the Ladell – the final sieve should be fine enough to retain them; this is immersed in water and the insects float to the surface. A certain amount of foam often develops in soil washing and hinders examination of the float; it may be dispersed with a small quantity of caprylic acid (Abu Yaman, 1960). To separate relatively large insects (root maggots) Read (1958) used a cylindrical aluminium screen sieve that was sprayed with water and rotated and half submerged in a tank; all the

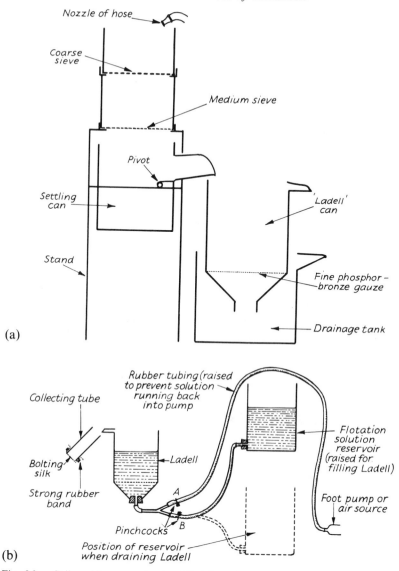

Fig. 5.3 *a*. Soil washing apparatus (modified from Salt & Hollick, 1944). *b*. Ladell can and associated equipment during the air agitation phase of flotation (diagrammatic).

fine soil passed out of the sieve, which could then be opened and the floating insects removed. Gerard (1967) found a combination of wet sieving and simple magnesium sulphate flotation efficient for earthworm sampling, although Nordstrom & Rundgren's (1972) studies suggest that the additional labour involved, compared with chemical extraction, is not always justified (see p. 178).

A revolving sieve was also used by Horsfall (1956) for mosquito eggs. Here the function of the drum, which consisted of three concentric sieves, was to disintegrate the sample and retain the larger materials, the eggs passing out and being collected in the finest of a series of sieves.

5.2.4 Flotation

If the sample is first dried flotation alone may be used for molluscs (Vágvölgyi, 1953) and nematode cysts (Goodey, 1957). Vágvölgyi's method has been modified by Mason (1970) who found it 84 % efficient. The sample is put into hot water and stirred, dead shells and litter float and are removed; badly broken shells, soil and the originally live snails, now killed by heat, sink. The sediment is then dried at 120°C for about 16 hours when the contents of the originally live snails contract; thus when the sample is carefully crumbled into a weak solution of detergent these now float and may be removed by careful searching. Badly broken shells, soil and other debris again sink. With nematode cysts drying, dry sieving and flotation in water is the usual sequence; various details and modifications are given by Goodey (1957) and in papers in Kevan (1955) and Murphy (1962a).

Often it may not be necessary to undertake the elaborate soil washing and flotation as described above. A very large range of benthic marine animals are efficiently extracted from grab-samples by merely stirring in a glass beaker with carbon tetrachloride (Birkett, 1957). The mineral matter remains as a sediment, but the animals float. This technique would seem to have many applications for the extraction of invertebrates from freshwater and terrestrial substrates, so long as there is not a great quantity of plant matter. Indeed there is a range of heavy organic solvents whose potentialities have been little explored for ecological work; but care should be taken over ventilation when working with them (Solomon, 1962). Sellmer (1956) used a 75 % solution of zinc chloride (sp. gravity 2.1) to float *Gemma*, a small bivalve mollusc, from marine sediment. Laurence (1954) separated insect larvae from dung by merely stirring the samples in 25 % solution of magnesium sulphate, and Iversen (1971) obtained $75 - 92 \%$ of mosquito eggs for temporary woodland pools by this method, though Service (1968) added centrifugal flotation (see below). The efficiencies of this method, for the extraction of all stages of the midge *Leptoconops* from sand, were determined by Davies & Linley (1966) as: eggs 18 %; 1st instar larvae 29 %; 2nd instar larvae 72 %; 3rd & 4th instar larvae 95 % and pupae 85 %. Slugs can be separated by wet sieving and flotation (Pinder, 1974) and sodium chloride (salt) solution or brine has been widely used (e.g. Dondale *et al.* 1971); calcium chloride for larval and pupal midges (Sugimoto, 1967) and sucrose solution (S.G.1.12) has been found effective for stream animals (Pask & Costa, 1971).

An interesting flotation technique using hydrogen peroxide is described by King (1975) for the separation of the eggs of the sugar-cane froghopper, *Aeneolamia varia*. After rotary sieving, the mixture of eggs, fine plant debris and sand was added to water containing one part per hundred of 6% hydrogen peroxide. Oxygen, produced from the hydrogen peroxide when it comes into contact with the plant debris, causes this to float to the surface. After its removal the eggs may be floated off in magnesium sulphate and glycerine solution (S.G.1.2).

5.2.5 The separation of plant and animal matter by differential wetting

A mixture of animal and plant material results from soil washing and flotation and also from various methods of sampling animals on vegetation (Chapter 4).

There are three possible approaches to their separation, one depending on the flotation effect of oxygen derived from decomposing hydrogen peroxide (see above) and two dependent on the lipoid and waterproof cuticle of arthropods. Either the arthropod cuticle can be wetted by a hydrocarbon 'oil' or the plant material can be waterlogged so that it will sink in aqueous solutions.

(1) *Wetting the arthropod cuticle.* If arthropods and plant material are shaken up in a mixture of petrol or other hydrocarbon and water and then allowed to settle, the arthropods, whose cuticles are wetted by the petrol, will lie in the pertrol layer above the water and the plant material in the water. As Murphy (1962c) has pointed out this separation will be imperfect if the plant material contains much air or if the specific gravities of the two phases are either too dissimilar or too close. The former may be overcome by boiling the suspension in water first (Cockbill *et al.*, 1945); the second by the choice of appropriate solutions. 60% ethyl alcohol may be used for the aqueous phase with a light oil. This concept was originally introduced into soil faunal studies by Salt & Hollick (1944). The float from the collection tube (see above) is placed in a wide-necked vessel with a little water, benzene is added, the whole shaken vigorously and allowed to settle. The arthropods will be in the benzene phase, which may be washed over into an outer vessel by adding further water below the surface from a pipette or wash bottle. Raw (1955) introduced the idea of freezing the benzene (m.p. $50°C$); the plug plus animals may then be removed and the benzene evaporated in a sintered glass crucible. (The fumes should be carried away by an exhaust fan as they are an accumulative poison.) A combination of decahydronophalene ('Dekalin') + carbon tetrachloride (S.G.1.2) and zinc sulphate solution (S.G.1.3) has been found particularly effective, the arthropods being floated off in the organic solvent mixture and the organic debris in the zinc sulphate solution run off in a separating funnel (Heath, 1965). Xylene and paraffin have also been used for the oil phase and gelatine, that solidifies on cooling, for the aqueous phase; in the latter case, it is the plant material that is removed in a plug. If the cuticle is not easily wetted by the benzene, extraction is improved if a solution with a specific gravity of about 1.3 is used instead of water and the initial mixture exposed to a negative pressure (Sugimoto, 1967).

(2) *Waterlogging the plant material.* This process, by boiling under reduced pressure (Hale, 1964) or vacuum extraction, may be used as part of the above technique. Repeated freezing and thawing also serves to impregnate the vegetable matter with water.

Danthanarayana (1966) has devised a method for the separation of the eggs of the weevil *Sitona* from the float by waterlogging the plant material by repeated freezing, filtering and transferring the whole float to a tube containing saturated sodium chloride solution; this is then centrifuged for five minutes (1500 r/m) and the eggs, and other animal matter, come to the surface whilst the plant material sinks. The eggs of some insects (e.g. the wireworm, *Ctenicera destructor*) sink rather than float, in sodium chloride (salt) solution because they are covered with small soil particles that have adhered to the coating from the female's glands. Doane (1969) found that this could be overcome by centrifuging in 20% alcohol solution, cleaning in 5% bleach (sodium hypochlorite) solution, followed by flotation and centrifuging in salt solution.

(3) *Grease-film separation*: Using the same principle as above, namely the resistance of arthropod cuticle to wetting by water and its lipophilic nature, Aucamp (1967) devised a 'grease-film machine', in which greased glass slides are revolved in an aqueous suspension of the litter or soil sample: the arthropods adhere to the grease. However, small stones remove the grease, so impairing the efficiency of the greased surface and providing an alternative surface. Shaw (1970) devised an apparatus to overcome this problem, in which the soil sample is enclosed in a cylinder of coarse mesh in the centre of a plastic box; the lid of the box is greased with anhydrous lanolin and the box filled with water containing a little sodium hexametaphosphate. Several such boxes may be revolved on a wheel, after which the grease and arthropods, can be washed from the lid into a dish with hot soapy water or other solvent. Fragments and many whole arthropods may be recovered on a large scale from litter by Speight's (1973) 'greased-belt machine' (Fig.5.4). The litter is poured in aqueous suspension onto a moving (9 – 12 m/min) fine nylon bolting cloth belt

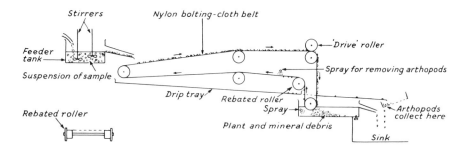

Fig. 5.4 Speight's greased belt machine for the separation of arthropods from litter (after Speight, 1973).

$(350 \times 15$ cm) (1.6 mm mesh), coated with petroleum jelly slightly thinned with liquid paraffin. At its lowest point the belt passes through a water trough when mineral and plant debris falls off. Subsequently the belt, with the arthropod material now on its undersurface passes over a special rebated roller, under a powerful jet of water which washes the arthropods off into a tray. Speight found that extraction efficiency for whole arthropods or fragments was 70–90%, except for whole organisms with diameters of less than 0.75 mm or greater than 10.0 mm for which it is unsatisfactory. About 1 litre of sample material can be extracted in one hour, the belt can run for 10 hours before 'regreasing' is necessary. The apparatus is particularly appropriate for dead organisms and fragments. To remove fine organic matter, Speight treated all samples with formalin, before wet sieving (1 mm mesh) and adding to the feeder tank.

5.2.6 Centrifugation
Although used widely in parasitological and pathological work, this technique has not been utilized to any extent in ecology, probably because only small samples can be treated at any time. However, it seems likely that it will be found to be of value for the final separation when the initial concentration has been by another method (see above). Müller (1962) found that Acarina and Collembola could be more efficiently extracted from soil by centrifugal flotation in saturated salt (sodium chloride) solution than by certain funnels and, after initially separating mosquito eggs from soil by washing and flotation in magnesium sulphate (see above), Service (1968) completed the extraction by this method. A similar technique has been used to separate nematode cysts from plant debris. Murphy (1962c) gives a useful summary, in tabular form, of previous work using centrifugal flotation to separate animals from foodstuffs, soil and excreta.

5.2.7 Sedimentation
The principle of this technique is to utilize the difference in settling rates between animal matter and the substrate. It is the basis of an unpublished method of Davies (*in* Macfadyen, 1955) for cleaning samples already separated from the soil; the sample is added to the top of a very long (4 m) glass tube, the base of which is in a rotating trough. However, the method is not very efficient as some mites settle at the same rate as soil particles. Seinhorst (1962) describes a sedimentation method ('Two Erlenmeyer method') for soil nematodes in which the flask containing the suspension was moved across a series of collecting vessels; the first of which was also inverted and the sediment in it further precipitated. The efficiency of the method was, however, only 60–75%.

Another sedimentation method for nematodes is described by Whitehead & Hemming (1965), but their 'tray method' (p. 190) was generally more efficient.

5.2.8 Elutriation

This is an extension of the sedimentation process in which the sedimentation takes place against a water current flowing in the opposite direction and the principle is incorporated in two widely used methods for the extraction of soil nematodes: the Oostenbrink and Seinhorst elutriators. In the Oostenbrink model (Fig. 5.5a) the nematodes are washed out of the sample and carried in suspension by the opposing water currents out of the overflow and through a series of sieves (Oostenbrink, 1954; Goffart, 1959, Murphy, 1962c).

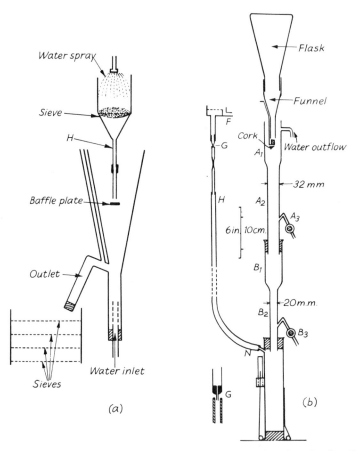

Fig. 5.5. Elutriators: *a*. Oostenbrink's model; *b*. Seinhorst's model (after Murphy, 1962*a*)

The Seinhorst elutriator is also part of a combined elutriation and sieving process; the soil is first passed through a coarse sieve and the suspension retained in a flask. This is inverted on top of the elutriator and the cork removed *in situ* (Fig. 5.5*b*); the upward flow of water (45 ml/min) coupled with the narrow sections A_2 and B_2 ensures that the nematodes are retained in sections A and B. After about half an hour the flask and the sections A and B may be drained through stopcocks A_3 and B_3. This material and that collected from the overflow is then wet sieved. A few large worms may pass into the soil-collecting container and this should be re-elutriated, but only for a brief period so that they will still be retained in vessel A (Goodey, 1957; Seinhorst, 1962). This method might well be adapted for single species studies on other organisms: those of a comparatively uniform size and mass, e.g. eggs, would be easily separated at a single level, whereas healthy, parasitized and dead individuals having different masses would separate at different levels.

Small arthropods such as Pauropoda and some Collembola that float in water may be extracted by von Törne's (1962a) elutriator and sieving process in which they are carried to the top of apparatus.

5.2.9 Sectioning
Small animals may be examined and counted in soil sections. Haarløv & Weis-Fogh (1953, 1955) have devised a method of fixing and freezing a soil core and then impregnating it with agar and sectioning. Gelatine is probably a more suitable substance for impregnation (Minderman, 1957; Anderson & Healey, 1970) and an automatic method for this is described by Vannier & Vidal (1964). Although the method is valuable for giving qualitative information on feeding sites and microdistribution, it is less suitable than other methods for population studies (Pande & Berthet, 1973). Vannier (1965) describes how microscopic observations on the biology of soil animals may conveniently be made in an apparatus, incorporating frigistors (= 'frigatrons') that allow the sample to be rapidly cooled.

5.3 Behavioural or dynamic methods

In these methods the animals are made to leave the substrate under some stimuli, e.g. heat, moisture (lack or excess) or a chemical. Their great advantage is that unlike the mechanical methods once the extraction has been set up it may usually be left, virtually unattended, and thus large quantities of material may be extracted simultaneously in batteries of extractors. Another important advantage is the ability to extract animals from substrates containing a large amount of vegetable material. The disadvantage is that, being based on the animal's behaviour, the extraction efficiency will vary with the condition of the animals and be influenced by changes in climate, water content, etc., experienced before and after sampling as well as by variations in these conditions in the apparatus itself. If samples have to be retained for several days, polythene bags seem the most suitable containers (Rapoport & Oros, 1969). Obviously eggs and other immobile stages cannot be extracted by this method.

The exact behavioural mechanism of extraction is not fully understood.

Temperature gradients are established from the start, but in some types of funnel the humidity gradient is not clearly marked in the lower part of the sample (Brady, 1969). The pattern of egress of the mites is irregular and there is often a marked 'flush' towards the end of the extraction (Brady, 1969), this has been associated with the moisture content reaching about 20 %, which triggers positive geotaxis (Nef, 1970, 1971). It seems that Oribatid mites are positively geotactic in dry conditions and negatively geotactic when the soil is moist; the adaptive significance of such responses is obvious and they are probably an important part of the mechanism of the vertical movements of soil fauna (p. 171). Whether all extractors act in this way, by the animal responding to a definite stimulus, rather than moving along a gradient, is uncertain. Many workers have sought to maximize the gradients (e.g. Macfadyen, 1961; Kempson *et al.* 1963) and these extractors have been found efficient (e.g. Edwards & Fletcher, 1970). There can be no doubt that for certain groups it is essential that the stimuli (usually heating and drying) are not applied too quickly (Lasebikan, 1971), but with the diversity of soil organisms it is not surprising that it is impossible to construct an extractor that is equally efficient for all groups. Therefore although dry funnels, in particular, have been used extensively for community studies, their efficiency varies from soil type to soil type and according to the animal groups (Nef, 1960; Macfadyen, 1961; Satchell & Nelson, 1962; Block, 1966; Edwards & Fletcher, 1970)

5.3.1 Dry extractors*

The basic apparatus is the Berlese – Tullgren funnel, a combination of the heated copper funnel designed at the turn of the century by the Italian entomologist, A. Berlese, and subsequently modified by the Swede, A. Tullgren, who used a light-bulb as a heat source. The funnel has been considerably modified and improved by many workers, as described in the reviews of Macfadyen (1955, 1962) and Murphy (1962*a*). Some of the most important innovations have been the demonstration by Hammer (1944) that when soil is being extracted the core should be retained intact and inverted, thereby enabling the animals to leave the sample by the natural passageways, and the discovery by Haarløv (1947) that serious losses could result from the animals becoming trapped in condensation from the core on the sides of the funnel; he recommended that the core should never touch the sides of the funnel and subsequent workers have sometimes referred to the space between the core and the sides of the funnel as the 'Haarløv passage'. When litter is being extracted and the 'passage' is difficult to maintain, it may be helpful to increase air circulation with wide plastic piping (Paris & Pitelka, 1962). Large numbers of small funnels were first grouped together by Ford (1937) and this approach has been much developed and improved by Macfadyen (1953, 1955,

---

* As Milne *et al.* (1958) correctly pointed out in connection with their hot water process, such methods really expel rather than extract the animals.

1961, 1962), who also introduced the concepts of steepening the heat gradient and arranging a humidity gradient. Murphy (1955), Newell (1955), Dietrick, Schlinger & van den Bosch (1959) and Kempson, Lloyd & Ghelardi (1963) have all introduced devices to reduce the fall of soil into the sample, thereby ensuring a cleaner extraction. A folding Berlese funnel for use on expeditions has been developed by Saunders (1959) and a simpler model, in which the funnels are constructed of oil cloth and the animals 'forced down' by the vapour from naphthalene held in cheese cloth bags above the sample, has been devised by Brown (1973).

There are many variants of the dry funnel in use, but most of them approximate to one of the following types.

Large Berlese funnel

This is used for extracting large arthropods. e.g. Isopoda, Coleoptera, from bulky soil or litter samples (Macfadyen, 1961) and also for the extraction of insects from suction apparatus and other samples containing much vegetation (Clark, Williamson & Richmond, 1959; Dietrick, *et al.*, 1959; Dondale *et al.*, 1971). Desirable features are an air circulation system, introduced by Macfadyen, which may be opened to ensure a rapid drying for the extraction of desiccation-resistant animals, such as beetles and ants, or partly closed for a slower 'wet regime' (but avoid condensation) for beetle larvae, *Campodea* and other animals that are susceptible to desiccation. Macfadyen recommends testing the humidity below the sample by cobalt thiocyanate papers; for the resistant animals humidities down to 70% are tolerable, for the others it should not fall below 90%. As phototactic animals, e.g. Halticine beetles, are often attracted upwards towards the light-bulb of the funnel and so fail to be extracted, this is replaced as a heat source by a Nichrome or similar wire-grid heating element – this also heats the sample more uniformly (Clark *et al.*, 1959). The extraction of positively phototactic animals has been further improved by Dietrick *et al.* (1959) who placed a 75 W spotlight below the collecting jar; this was switched on intermittently. An apparatus incorporating these and other features is illustrated in Fig. 5.6. As Macfadyen and others have frequently stressed, for most animals the heat should be applied slowly and therefore it may be useful to be able to control the voltage of the heating element through a rheostat (Dietrick *et al.*, 1959) or 'Simmerstat'. Extraction time will vary from a few days to over a month (e.g. Park & Auerbach, 1954) depending on the substrate and the animal.

Horizontal extractor

Designed by Duffey (1962) for extracting spiders from grass samples, this extractor (Fig. 5.7) would probably serve equally well for any rapidly moving litter animal. It has the advantages that steep gradients are built up and that debris cannot fall into the collecting trough; it is also relatively compact and built in paired units. Heat is provided by a 500-W fire heating element

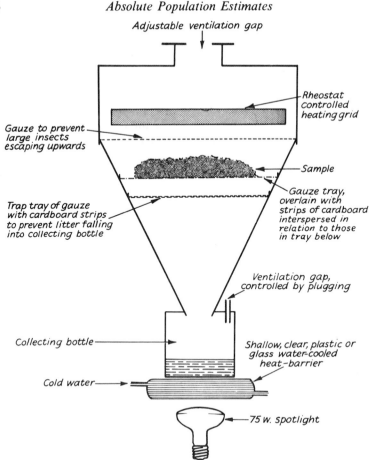

Fig. 5.6. A large Berlese funnel with modification.

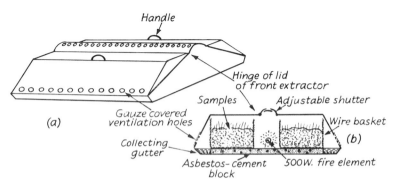

Fig. 5.7. A pair of horizontal extractors, based on Duffey's (1962) design: *a.* sketch; *b.* sectional view.

controlled by a 'Simmerstat'. Air passes in through the ventilation holes just above the aqueous solution of the collecting gutters; this helps to prevent too rapid desiccation. Duffey recommended a 0.001 % solution of phenylmercuric acetate, a fungicide and bactericide, with a few drops of a surface-active agent (detergent). The samples are held in wire trays 6 in (10 cm) wide. The extraction takes only a day or two.

High gradient (Multiple canister) extractor

Collembola, mites and other small animals may be obtained from small compact soil samples in this apparatus designed by Macfadyen (1961). The samples are taken with the special corer (see Figs. 5.1*d* and 5.8*a*), the samples removed, surrounded by the bakelite ring, and a sieve plate is fitted to each ring by pushing two springy wires into holes in the bakelite. The collecting canisters are cooled in a water bath and 95 % humidity is maintained below the samples during extraction (Fig. 5.8*b*). Heat is provided by two 160-W elements giving 10 W per sample, and the temperature of the upper surface of

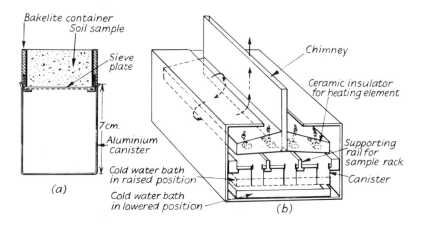

Fig. 5.8 High gradient extractor (after Macfayden, 1961): *a*. canister, core and sieve plate; *b*. whole apparatus.

the sample reaches $35°C$; the temperature of the lower surface is determined by the water bath (e.g. $15°C$), so that a very steep gradient is set up over the 3-cm deep sample. Extraction may be completed in 5 days, the maximum heat being applied after the first 24 hours. Macfadyen recommended a 0.3 % solution of the fungicide 'Nipagin' (which contains streptomycin and terramycin) be used in the collecting canisters, but the tests of Kempson *et al.* (1963) (see below) indicate that a saturated aqueous solution of picric acid with a little photographic wetting agent might be preferable. Constructional details are given by Macfadyen (1961), who has also (Macfadyen, 1962) described variants suitable for use under expedition conditions, the heating power being provided by kerosene or bottled gas, and the effect of various modifications on the physical conditions in the extractor (Macfadyen, 1968). Block (1966) found that this apparatus extracted 76 % of the acarina from mineral soils.

The Kempson bowl extractor (Fig.5.9)

Developed by Kempson, Lloyd & Ghelardi (1963), this method is suitable for the extraction of mites, Collembola, Isopoda and many other arthropods in woodland litter; extraction rates of 90–100 % being recorded for groups other than the larvae of holometabolous insects for which it is not efficient. The apparatus consists of a box or shrouded chamber, containing an ordinary light-bulb and an infra-red lamp (250 W) that is switched on in pulses, at first only for a few seconds at a time and gradually increased until it is on for about a third of the time. The pulsing of the lamp can be controlled by a 'Simmerstat', a time switch designed for electric cookers, and the ordinary bulb is connected across it so as to protect the lamp from surges. The extraction bowls are sunk into floor of the chamber. Each bowl

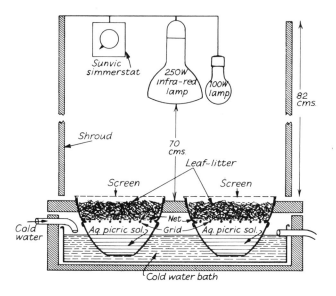

Fig. 5.9. Kempson bowl extractor (after Kempson, Lloyd & Ghelardi, 1963).

contains a preservative fluid and is immersed in a cold water bath. A saturated aqueous solution of picric acid, with some wetting agent, was used by Kempson *et al.* (1963), but Sunderland *et al.* (1976) have found a solution (at 80 g/l) of tri-sodium orthophosphate ($Na_3 PO_4.12H_2O$) a cheap, safe and suitable alternative. The litter is placed in a plastic tray above the bowl and is supported by two pieces of cotton fillet net laid on a coarse plastic grid. The top of the tray is covered by a fine black nylon screen. Kempson *et al.* (1963) give very full instructions for the construction of this apparatus. Its great advantages, besides its comparatively simple construction, are the high humidity maintained on the lower surface of the sample, the lack of debris among the extracted material (but the fillet net does not appear to present any barrier to the animals), the prevention by the black screen of animals escaping upwards and the simple mechanism of gradually increasing the amount of heat to which the sample is subjected. Extractions with this apparatus generally take about a week.

5.3.2 Wet extractors

The principle of this method is similar to that of the dry extractors, the animals being driven out of their natural substrate under the influence of a stimulus, possibly heat or, as the observations of Williams (1960*a*) would suggest for some cases, reduced oxygen tension. As the substrate is flooded with water in all variants of the method there is no risk of desiccation and the method is particularly successful with those groups that are but poorly extracted by the Berlese-type funnels, that is nematode and enchytraeid worms, insect larvae and various aquatic groups (Williams, 1960*a*). The method, which is faster than dry extraction, seems to have been discovered independently at least three times.

Baermann funnel

As originally designed this consisted of a glass funnel with a piece of metal gauze or screen resting in the funnel and a piece of rubber tubing with a pinchcock on the stem (Fig. 5.10*a*); the sample, which must not be very deep, is contained in a piece of muslin and partly flooded with warm water. Nematode worms leave the sample, and fall to the bottom of the funnel where they collect in the stem. They may be drawn off, in a little water, by opening the pinchcock (Peters, 1955). A number of refinements of this method have been introduced.

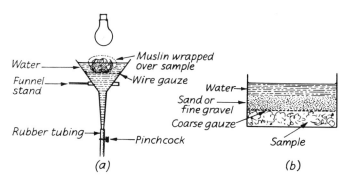

Fig. 5.10. *a*. Simple heated Baermann funnel. *b*. Sand extractor.

Nielsen (Overgaard, 1947 – 1948) ran a battery of Baermann funnels within a box with a lamp above; this heated the surface of the water in the funnels which was initially cold. Nematodes and rotifers could be extracted from soil and moss by this method, but it was not satisfactory for tardigrades. In order to extract Enchytraeidae, O'Connor (1955, 1962) used larger polythene funnels, spread the sample directly on to the gauze, without muslin, and submerged it completely; a powerful shrouded lamp heated the surface of the water to 45°C in three hours; dry samples must be moistened before being placed in the funnel and if the soil temperature is low samples must be gradually warmed to room temperature for a day. Extraction usually takes about three hours. In a series of tests O'Connor (1962) and Peachey (1962) have shown this method to be as efficient as the Nielsen inverted extractor (see below) for grassland soils and more efficient for peat and woodland soils. It is also very efficient for second and later instar tipulid larvae in moorland soils (Hadley, 1971), but Kiritani & Matuzaki (1969) found only 10 % of the larvae of the cotton root knot eelworm, *Meloidogyne incognita* were extracted after a week.

Whitehead & Hemming (1965) introduced a modification that has been termed the 'Whitehead tray'. The soil sample is placed in a small (23×33 cm) shallow tray of 8 mesh/cm phosphor-bronze wire cloth, the inside of the tray being lined with paper tissue ('Scotties') and the outside supported by a plastic coated wire basket. This stands in a plastic photographic tray or similar vessel and is flooded with water until the surface reaches the underside of the soil (i.e. it is not inundated as much as in the Baermann funnel (Fig. 5.10*a*)). After 24 hours at room temperature the wire basket is carefully lifted out and the water in the tray agitated and poured into vessels resembling separating funnels. After about 4 hours the nematodes collect at the tapered part of the funnel, which may be separated off by a rubber bung or diaphragm on a handle; the stop-cock is opened and the nematodes drawn off. The process may be repeated to further concentrate them. In comparative tests with clay, loam and sand and different nematodes, Whitehead & Hemming showed that this method usually extracted more efficiently than Seinhorst's 'Two Erlenmeyer method', elutriation or a sedimentation method they developed.

Hot water extractors
These have been developed by Nielsen (1952–53) for Enchytraeidae and by Milne, Coggins & Laughlin (1958) for tipulid larvae. Nielsen's apparatus involved the heating of the lower surface of the sample, contained in an earthenware vessel, by hot water; the animals moved up and entered a layer of sand, kept cool by a cold water coil, that was placed above the sample. The only advantage of this method over O'Connor's funnel (see above) is that it seems to extract more of the young worms (Peachey, 1962); under conditions with a large amount of humus it is less efficient (see above).

The larvae of many insects, notably of some Coleoptera and Diptera, earthworms, molluscs, and most motile animals seem to be efficiently and rapidly extracted from soil samples if these are heated from below. Schjøtz-Christensen (1957) recorded 85 % extraction of Elateridae using Nielsen's apparatus, modified by heating more strongly and omitting the cooling coil. Such an apparatus is not very different from that of Milne *et al.* (1958) (Fig. 5.11), which was found to be virtually 100 % efficient for active larvae and pupae. Essentially their device consits of two galvanized boxes, one within the other; the space between is a water bath heated by a thermostatically controlled immersion heater. The water temperature should not rise above 90°C. The inner box has a wire gauze base and holds the sample. The initial mode of operation is to fill the water bath until the level is just above the base of the sample, the heater is switched on until the temperature registered by the thermometer in the centre of the turf is 40°C; the water level is then raised to the level where it would just, eventually, flood the surface of the sample.

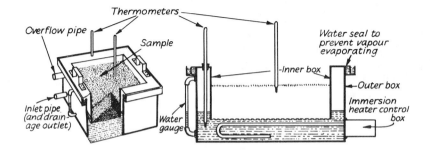

Fig. 5.11. Hot water extractor (after Milne *et al.*, 1958): *a.* sketch with segment 'removed'; *b.* sectional diagram.

Heating and observation is continued, the insects being picked up as they leave the turf (but not before or they will retreat). The light above the extraction apparatus should not be too bright; Milne and co-workers recommended as a maximum, a 4-ft 40-W natural fluorescent tube 3 ft (91.5cm) above the sample; however, other species of insects might be more or less sensitive. Extraction times varied from an hour in a light soil to nearly three hours in heavy peat-covered clay. The chief disadvantage of this method as at present designed is that the apparatus must be watched continuously and the animals removed as they appear; however, equivalent mechanical methods usually take longer.

Sand extractors
Extremely simple in design, consisting of a metal can, a piece of wire gauze and some sand (Fig. 5.10*b*), these devices seem to be remarkably efficient for extracting aquatic and semi-aquatic insect larvae and other animals from mud

and debris. The method was described by Bidlingmayer (1957) for the extraction of ceratopogonid midge larvae from salt marshes and has been improved and extended by Williams (1960*a* and *b*), who has found that the following freshwater invertebrates could be extracted from their soil substrate by this technique: Coelenterata, Turbellaria, Nematoda, Oligochaeta, Ostracoda, Acarina, Dipterous larvae, Gastropoda and Plecopoda. The sample, about 2 in thick, is cut so as to fill the container completely: it is then covered with 2 in of dry clean sand and the whole flooded with water. After 24 hours the sand may be scooped off and will contain large numbers of animals, but for 100 % extraction of *Culicoides* larvae it should be left for 40 hours (Williams, 1960*b*) and for other groups tests would need to be made. The animals may be separated from the sand as Bidlingmayer and Williams recommend by stirring in a black photographic tray when the pale moving objects easily show up against the dark background. If the promise of this method is justified it will clearly repay further development. The final separation might be facilitated by the adoption of a sedimentation, elutriation or flotation method, aided perhaps by the replacement of the sand by fine washed gravel as used by Nielsen (1952–53) in his extractor (see above – that really combines this method with the hot water process). The chief disadvantages of the method in its present form is that it is difficult to make a clean separation of the sand from the soil after extraction; the placing of a piece of wide-mesh wire gauze between the two substrates might aid this operation, or if the extraction was done in a square vessel a plate could be pushed across between the layers at the end of extraction and the top layer poured off.

Cold water extractor
South (1964) found that slugs could be efficiently extracted from 1-ft (30.5 cm) deep turves by slowly immersing these in water. Initially they are stood in 1 in (2.5 cm) of water; after about 17 h the water level is raised to half the depth of the turf (i.e. 6 in, 15.2 cm); after 2 days the water level is again raised, this time in several stages, until at the end of another 24 hours it is within $\frac{1}{2}$ in (1.2 cm) of the top of the turf. This approach has also been used to extract the collembolan, *Hypogastrura*, (Van der Kraan, 1973) and overwintering rice water weevils, *Lissorhoptrus* (Gifford & Trahan, 1969). The latter authors used a technique basically similar to the 'van Emden biscuit tin' (Fig. 8.8*a* p. 308), but constructed from plastic bleach bottles, in which the sample is flooded. Such a technique is, of course, only suitable for robust insects.

Mistifier
This is essentially a method of extracting active nematodes from plant tissue. The plant material is spread on a tray, broken up and continually sprayed with very fine jets of water ('mists') (Seinhorst, 1950; Webster, 1962). Combined with a cotton wool filter to collect the nematodes this method can be much

more efficient than other techniques for certain eelworms (de Guiran, 1966; Oostenbrink, 1970).

5.3.3 Chemical extraction

Chemical fumes may be used to drive animals from vegetation (p. 141); the extension of such methods to soil and litter animals has been successful only with aphids and thrips (Macfadyen, 1953–who used dimethyl phthalate and 2-cyclohexyl-4,6-dinitrophenol). Chemicals in a liquid form, notably potassium permanganate solution, orthodichlorobenzene and formalin, often with a 'wetter', have been use for the extraction of insect larvae and earthworms (Svendsen, 1955; Milne *et al.*, 1958; Raw, 1959). Formalin should be applied four times, with 15 minute intervals, at the rate of 75 ml of 4% formaldehyde/13.5 l water/1 m^2 ground (Raw, 1959; Nordstrom & Rundgren, 1972). Many workers have found these methods to be inefficient for earthworms (Satchell, 1955; Svendsen, 1955); for tipulids in the grasslands studied by Milne *et al.* (1958) the efficiency was 85%. However the method can be efficient, depending on the type of soil, the species of animal and the season (which will influence the activity level of the animal). The more porous the soil (Boyd, 1958) and the more active the earthworm the more satisfactory the method. Among earthworms it is as efficient as hand-sorting for *Lumbricus terrestris* during active periods; on sandy soils it seems suitable for *Lirabellus* and *Dendrobaena octaedra*, but is seldom reliable for *Allobophora* spp.; where it is efficient it is more reliable for small worms than hand sorting (Nordstrom & Rundgren, 1972).

5.3.4 Electrical extraction

By discharging a current from a water-cooled electrode driven into the soil, Satchell (1955) was able to expel large numbers of earthworms; but, as he has pointed out, this method suffers from the disadvantage that the exact limits of the volume of soil treated are unknown (Nelson & Satchell, 1962) and the efficiency of extraction seems directly related to the soil pH (Edwards & Lofty, 1975).

5.4 Summary of the applicability of the methods

It has already been pointed out that the efficiency of the different methods varies greatly with the animal group and the substrate, as is shown for example by the studies of Nef (1960), Raw (1959, 1960), Macfadyen (1961), O'Connor (1962) and Satchell & Nelson (1962). Some suggestions for different groups are given in a tabular form by Macfadyen (1962) and, on the basis of extensive comparative trials, by Edwards & Fletcher (1970). The comments made here should only be taken as general indications; more specific information is given under each method.

i. Substrate type. Because of the difficulties of separating plant and animal material the mechanical methods have generally been considered unsuitable for litter and for soils, pond mud and other substrates containing a large amount of organic matter. For these media behavioural methods are likely to be the most satisfactory, and the hot water and sand extraction techniques may well repay further development. There are, however, some exceptions; flotation is suitable for dung-dwelling insects (Laurence, 1954), Williamson's wet sieving (which may contain a behavioural element) works well for snails in fresh litter and Satchell & Nelson (1962) found that flotation was more efficient than dry funnels for the extraction of Scutacarid mites from a moder soil.

With friable and sandy soils or aquatic sediments mechanical methods are generally to be preferred and techniques as simple as wet sieving or flotation, perhaps using a heavy organic solvent, may be satisfactory for the study of a single species.

ii. Animal type. Obviously immobile stages can only be extracted by mechanical methods. Comparatively large and robust animals can often be estimated simply by sieving. When the specific gravity of the animal is very different from that of the substrate particles, possible techniques include elutriation and centrifugal separation, as well as the more widely used flotation. The two first named methods might also be useful for the separation of parasitized and unhealthy individuals from healthy ones and for the extraction of the eggs of a single species. Broadly speaking, Acarina and Collembola seem to be more efficiently extracted by an appropriate funnel technique and insects, except on peat soils, by a flotation method (Edwards & Fletcher, 1970).

Behavioural methods must naturally be adopted to fit the behaviour of the animal; fast-moving animals comparatively resistant to desiccation need to be extracted by the horizontal extractor or the large Berlese funnel; the last named, when modified as described here, is also the only method really suitable for positively phototactic species such as Lygaeid bugs and phytophagous beetles. Groups such as Collembola and most mites need a slower extraction in the Kempson or multiple canister extractors. Those animals that are very sensitive to desiccation (e.g. nematode and enchytraeid worms, dipterous and probably many coleopterous larvae) should be separated in wet extractors, which is also, of course, the behavioural method appropriate for animals in aquatic sediments. The degree of complexity for a single group is well illustrated by the earthworms, for while *Lumbricus terrestris*, but not other species, may be efficiently extracted from orchard soils by a surface application of fomalin and several species from shallow soils by wet sieving and flotation, in other habitats even hand-sorting fails (Raw, 1959, 1960; Nelson & Satchell, 1962).

iii. Cost. Mechanical methods do involve more work and time considerations may rule them out for an extensive survey. Hand-sorting, which is the 'last resort', is the most time-consuming method of all and suffers from human error.

REFERENCES

ABU YAMAN, I. K., 1960. Natural control in cabbage root fly populations and influence of chemicals. *Meded. LandbHoogesch. Wageningen* **60** (1), 1–57.

AGUILAR, J. D', BENARD, R. and BESSARD, A., 1957. Une méthode de lavage pour l'extraction des arthropodes terricoles. *Ann. Epiphyt. C* **8**, 91–9.

ANDERSON, J. M. and HEALEY, I. N., 1970. Improvements in the gelatine-embedding technique for woodlands soil and litter samples. *Pedobiologia* **8**, 108 – 20.

AUCAMP, J. L., 1967. Efficiency of the grease film extraction technique in soil micro-arthropod survey. *In* Graff, O. & Satchell, J. E. (eds.), *Progress in Soil Biology*, pp. 515–24. North-Holland, Amsterdam.

BALOGH, J., 1958. *Lebensgemeinschaften der Landtiere*. Budapest, 560 pp.

BELFIELD, W., 1956. The arthropoda of the soil in a West African pasture. *J. Anim. Ecol.* **25**, 275–87.

BERG, K., 1938. Studies on the bottom animals of Esrom Lake. *K.danske Vidensk. Selsk., Skr. Nat. Math. Afd. 9, Raekke* **8**, 1–225.

BIDLINGMAYER, W. L., 1957. Studies on *Culicoides furens* (Poey) at Vero Beach. *Mosq. News* **17**, 292–4.

BIRKETT, L., 1957. Flotation technique for sorting grab samples. *J. Comm. perm. int. Explor. Mer.* **22**, 289–92.

BLOCK, W., 1966. Some characteristics of the Macfadyen high gradient extraction for soil micro-arthropods. *Oikos* **17**, 1–9.

BOYD, J. M., 1958. Ecology of earthworms in cattle-grazed machair in Tiree, Argyll. *J. Anim. Ecol.* **27**, 147–57.

BRADY, J., 1969. Some physical gradients set up in Tullgren funnels during the extraction of mites from poultry litter. *J. appl. Ecol.* **6**, 391–402.

BRIAN, M. V., 1970. Measuring population and energy flow in ants and termites. In Phillipson, J. (ed.) *Methods of study in soil ecology*, 231–4. UNESCO, Paris.

BROWN, R. D., 1973. Funnel for extraction of leaf litter organisms. *Ann. ent. Soc. Am.* **66**, 485–6.

BRUNDIN, L., 1949. Chironomiden und andere Bodentiere der Südschwedischen Urgebirgsseen. *Rep. Inst. Freshw. Res. Drottningholm* **30**, 915 pp.

CAPSTICK, C. K., 1959. The distribution of free-living nematodes in relation to salinity and the middle and upper reaches of the river Blyth estuary. *J. Anim. Ecol.* **28**, 189–210.

CLARK, E. W., WILLIAMSON, A. L. and RICHMOND, C. A., 1959. C.A.,1959. A collecting technique for pink bollworms and other insects using a Berlese funnel with an improved heater. *J. econ. Ent.* **52**, 1010–12.

COCKBILL, G. F., HENDERSON, V. E., ROSS, D. M. and STAPLEY, J. H., 1945. Wire-worm populations in relation to crop production. 1.A large-scale flotation method for extracting wireworms from soil samples and results from a survey of 600 fields. *Ann. appl. Biol.* **32**, 148–63.

COHEN, M., 1955. Soil sampling in the national agricultural advisory service. *In* Kevan, D.K. McE. (ed.), *Soil Zoology*, 347-50. University of Nottingham School of Agriculture.

DANTHANARAYANA, W., 1966. Extraction of arthropod eggs from soil. *Ent. exp. appl.* **9**, 124–5.

DAVIES, J. B. and LINLEY, J. B., 1966. A standardised flotation method for separating *Leptoconops* (Diptera: Ceratopogonidae) and other larvae from sand samples. *Mosquito News* **26**, 440.

DHILLON, B. S. and GIBSON, N. H. E., 1962. A study of the Acarina and Collembola of agricultural soils. 1. Numbers and distribution in undisturbed grassland. *Pedobiologia* **1**, 189–209.

DIETRICK, E.J., SCHLINGER, E.I. and VAN DEN BOSCH, R., 1959. A new method for sampling arthropods using a suction collecting machine and modified Berlese funnel separator. *J. econ. Ent.* **52**, 1085–91.

DOANE, J.F., 1969. A method for separating the eggs of the prairie grain wireworm, *Ctenicera destructor*, from soil. *Can. Ent.* **101**, 1002–4.

DONDALE, C. D., NICHOLLS, C. F., REDNER, J. H., SEMPLE, R. B. and TURNBULL, A.L., An improved Berlese-Tullgren funnel and a flotation separator for extracting grassland arthropods. *Can. Ent.* **103**, 1549–52.

DUFFEY, E. (A.G.)., 1962. A population study of spiders in limestone grassland. Description of study area, sampling methods and population characteristics. *J. Anim. Ecol.* **31**, 571–99.

EDWARDS, C. A. and FLETCHER, K.E., 1970. Assessment of terrestrial invertebrate populations. *In* Phillipson, J. (ed.), *Methods of study in soil ecology*, UNESCO, Paris.

EDWARDS, C. A. and LOFTY, J. R., 1975. The invertebrate fauna of the Park Grass Plots. 1. Soil fauna. *Rep. Rothamsted exp. Sta.* 1974, Part 2, 133–54.

EDWARDS, C.A., WHITING, A.E. and HEATH, G.W., 1970. A mechanized washing method for separation of invertebrates from soil. *Pedobiologia* **10**, 141–8.

ERMAN, D.C., 1973. Invertebrate movements and some diel and seasonal changes in a Sierra Nevada peatland. *Oikos* **24**, 85–93.

FORD, J., 1937. Fluctuations in natural populations of Collembola and Acarina. *J. Anim. Ecol.* **6**, 98–111.

FRICK, K.E., 1962. Ecological studies on the alkali bee, *Nomia melanderi*, and its Bombyliid parasite, *Heterostylum robustum* in Washington. *Ann. ent. Soc. Am.* **55**, 5–15.

GABBUTT, P.D., 1959. The bionomics of the wood cricket, *Nemobius sylvestris* (Orthoptera: Gryllidae), *J. Anim. Ecol.* **28**, 15–42.

GERARD, B.M., 1967. Factors affecting earthworms in pastures. *J. Anim. Ecol.* **36**, 235–52.

GIFFORD, J.R. and TRAHAN, G.B., 1969. Apparatus for removing overwintering adult rice water weevils from bunch grass. *J. econ. Ent.* **62**, 752–4.

GOFFART, H., 1959. Methoden zur Bodenuntersuchung auf nichtzystenbildende Nematoden. *NachrBl. dtsch. PflSchDienst, Stuttgart* **11**, 49–54.

GOLIGHTLY, W.H., 1952. Soil sampling for wheat-blossom midges. *Ann. appl. Biol.* **39**, 379–84.

GOODEY, J. B., 1970. Laboratory methods for work with plant and soil nematodes. *Tech. Bull. Min. Agric. Lond.* **2** (5th Ed.), H.M.S.O., London.

GUIRAN, G. DE., 1966. Infestation actuelle et infestation potentielle du sol par nematodes phytoparasites du genre *Meloidogyne*. *C.r.Lebd. Seanc.Acad.Sci. (Paris)* **262**, 1754–6.

HAARLØV, N., 1947. A new modification of the Tullgren apparatus. *J. Anim. Ecol.* **16**, 115–21.

HAARLØV, N. and WEIS-FOGH, T., 1953. A microscopical technique for studying the undisturbed texture of soils. *Oikos* **4**, 44–7.

HAARLØV, N. and WEIS-FOGH, T., 1955. A microscopical technique for studying the undisturbed texture of soils. *In* Kevan, D. K. McE. (ed.), *Soil Zoology* 429–32. University of Nottingham School of Agriculture.

HADLEY, M. 1971. Aspects of the larval ecology and population dynamics of *Niolophilus ater* Meigen (Diptera:Tipulidae) on Pennine Moorland. *J. Anim. Ecol.* **40**, 445–66.

HAIRSTON, N. G., HUBENDICK, B., WATSON, J. M. and OLIVER, L. J., 1958. An evaluation of techniques used in estimating snail populations. *Bull. Wld Hlth Org.* **19,** 661–72.

HALE, W. G., 1964. A flotation method for extracting Collembola from organic soils. *J. Anim. Ecol.* **33,** 363–9.

HAMMER, M., 1944. Studies on the Oribatids and Collemboles of Greenland. *Medd. Grønland* **141,** 1–210.

HEATH, G. W., 1965. An improved method for separating arthropods from soil samples. *Lab. Pract.* April 1965.

HORSFALL, W.R., 1956. A method for making a survey of floodwater mosquitoes. *Mosquito News* **16,** 66–71.

HUGHES, R. D., 1960. A method of estimating the numbers of cabbage root fly pupae in the soil. *Plant Path.* **9,** 15–17.

IVERSEN, T.M., 1971. The ecology of the mosquito population (*Aedes communis*) in a temporary pool in a Danish beech wood. *Arch. Hydrobiol.* **69,** 309–32.

JACOT, A.P., 1935. Molluscan populations of old growth forests and rewooded fields in the Asheville basin of N. Carolina. *Ecology* **16,** 603–5.

JONASSON, P. M., 1955. The efficiency of sieving techniques for sampling freshwater bottom fauna. *Oikos* **6,** 183–207.

KEMPSON, D., LLOYD, M. and GHELARDI, R., 1963. A new extractor for woodland litter. *Pedobiologia* **3,** 1–21.

KEVAN, D. K. MCE. (ed.), 1955. *Soil Zoology, Proceedings of the University of Nottingham second Easter School in Agricultural Science, 1955.* London, 512 pp.

KEVAN, D. K. MCE., 1962. *Soil animals,* 237 pp., Witherby, London.

KING, A. B. S., 1975. The extraction, distribution and sampling of the eggs of the sugar-cane froghopper, *Aeneolamia varia saccharina* (Dist.) (Homoptera, Cercopidae). *Bull. ent. Res.* **65,** 157–64.

KIRITANI, K. and MATUZAKI, T., 1969. On the extraction rate of nematodes by the Baermann funnel. *Bull. Kochi Inst. Agr. For. Sci.* **2,** 25–30.

KITCHING, R. L., 1971. A core sampler for semi-fluid substances. *Hydrobiologia* **37,** 205–9.

KRAAN, C. VAN DER., 1973. Populationsokologische untersuchungen an *Hypogastrura vintica* Tullb. 1872 (Collembola) anf Schiermonnikoog. *Faun-okol. Mitt.* **4,** 197–206.

LADELL, W. R. S., 1936. A new apparatus for separating insects and other arthropods from the soil. *Ann. appl. Biol.* **23,** 862–79.

LANE, M. and SHIRCK, E., 1928. A soil sifter for subterranean insect investigations. *J. econ. Ent.* **21,** 934–6.

LANGE, W. H., AKESSON, N. B. and CARLSON, E. C., 1955. A power-driven self-propelled soil sifter for subterranean insects. *In* Kevan, D. K. McE. (ed.), *Soil Zoology* 351–5. University of Nottingham School of Agriculture.

LASEBIKAN, B. A., 1971. The relationship between temperature and humidity and the efficient extraction of Collembola by a dynamic-type method. *Rev. Ecol. Biol. Sol.* **8,** 287–93.

LAURENCE, B. R., 1954. The larval inhabitants of cow pats. *J. Anim. Ecol.* **23,** 234–60.

LINCOLN, C. and PALM, C. E., 1941. Biology and ecology of the Alfalfa Snout beetle. *Mem. Cornell Univ. agric. Exp. Sta.* **236,** 3–45.

MACFADYEN, A., 1953. Notes on methods for the extraction of small soil arthropods. *J. Anim. Ecol.* **22,** 65–78.

MACFADYEN, A., 1955. A comparison of methods for extracting soil arthropods. *In* Kevan, D. K. McE. (ed.). *Soil Zoology* 315–32. University of Nottingham School of Agriculture.

MACFADYEN, A., 1961. Improved funnel-type extractors for soil arthropods. *J. Anim. Ecol.* **30,** 171–84.

MACFADYEN, A., 1962. Soil arthropod sampling. *Adv. Ecol. Res.* **1,** 1–34.

MACFADYEN, A., 1968. Notes on methods for the extraction of small soil arthropods by the high gradient apparatus. *Pedobiologia* **8**, 401–6.

McLEOD, J. M., 1961. A technique for the extraction of cocoons from soil samples during population studies of the Swaine sawfly, *Neodiprion swainei* Midd. (Hymenoptera: Diprionidae). *Can. Ent.* **91**, 888–90.

MASON, C.F., 1970. Snail populations, beech litter production and the role of snails in litter decomposition. *Oecologia (Berl.)* **5**, 215–39.

MILNE, A., COGGINS, R. E. and LAUGHLIN, R., 1958. The determination of numbers of leatherjackets in sample turves. *J. Anim. Ecol.* **27**, 125–45.

MINDERMAN, G., 1957. The preparation of microtome sections of unaltered soil for the study of soil organisms in situ. *Plant Soil* **8**, 42–8.

MÜLLER, G., 1962. A centrifugal-flotation extraction technique and its comparison with two funnel extractors. *In* Murphy, P. W. (ed.), *Progress in Soil Zoology* 207–11. Butterworths, London.

MURPHY, P. W., 1955. Notes on processes used in sampling, extraction and assessment of the meiofauna of heathland. *In* Kevan, D. K. McE. (ed.), *Soil Zoology* 338–40. University of Nottingham School of Agriculture.

MURPHY, P. W. (ed.), 1962a. *Progress in Soil Zoology. Papers from a colloquium on Research methods organised by the Soil Zoology Committee of the International Society of Soil Science held at Rothamsted Experimental Station Hertfordshire 10–14th July, 1958.* 398 pp., Butterworths, London.

MURPHY, P. W., 1962b. Extraction methods for soil animals. I. Dynamic methods with particular reference to funnel processes. *In* Murphy, P. W. (ed.), *Progress in Soil Zoology* 75–114. Butterworths, London.

MURPHY, P. W., 1962c. Extraction methods for soil animals. II. Mechanical methods. *In* Murphy, P. W. (ed.), *Progress in Soil Zoology* 115–55. Butterworths, London.

NEF, L., 1960. Comparison de l'efficacité de différentes variantes de l'appareil de Berlese-Tullgren. *Z. angew. Ent.* **46**, 178–99.

NEF, L., 1970. Reactions des acariens a une dessication lente de la litiere. *Rev. Ecol. Biol. Sol.* **7**, 381–92.

NEF, L., 1971. Influence de l'humidite sur le geotactisme des Oribates (Acarina) dans l'entracteur de Berlese-Tullgren. *Pedobiologia* **11**, 433–45.

NELSON, J. M. and SATCHELL, J. E., 1962. The extraction of Lumbricidae from soil with special reference to the hand-sorting method. *In* Murphy, P. W. (ed.), *Progress in Soil Zoology* 294–9. Butterworths, London.

NEWELL, I., 1955. An autosegregator for use in collecting soil-inhabiting arthropods. *Trans. Amer. Microsc. Soc.* **74**, 389–92.

NIELSEN, C. OVERGAARD., 1952–53. Studies on Enchytraeidae 1. A technique for extracting Enchytraeidae from soil samples. *Oikos* **4**, 187–96.

NORDSTROM, S. and RUNDGREN, S., 1972. Methods of sampling lumbricids. *Oikos* **23**, 344–52.

O'CONNOR, F. B., 1955. Extraction of enchytraeid worms from a coniferous forest soil. *Nature* **175**, 815–17.

O'CONNOR, F. B., 1957. An ecological study of the Enchytraeid worm population of a coniferous forest soil. *Oikos* **8**, 162–99.

O'CONNOR, F. B., 1962. The extraction of Enchytraeidae from soil. *In* Murphy, P. W. (ed.), *Progress in Soil Zoology* 279–85. Butterworths, London.

ØKLAND, F., 1929. Methodik einer quantitativen Untersuchung der Landschnecken-fauna. *Arch. Molluskenk.* **61**, 121–36.

OOSTENBRINK, M., 1954. Een doelmatige methode voor het toetsen van aaltjesbestrijd-ingsmiddelen in grond met *Hoplolaimus uniformis* als proefdier. *Meded. Landb Hoogesch. Gent.* **19**, 377–408.

OOSTENBRINK, M., 1970. Comparison of techniques for population estimation of soil and plant nematodes. *In* Phillipson, J. (ed.), *Methods of study in soil ecology*, 249–55, UNESCO, Paris.

OVERGAARD, C. [NIELSEN, C. OVERGAARD],1947–48. An apparatus for quantitative extraction of nematodes and rotifers from soil and moss. *Natura jutl.* **1**, 271–8.

PANDE, Y. D. and BERTHET, P., 1973. Comparison of the Tullgren funnel and soil section methods for surveying Oribatid populations. *Oikos* **24**, 273–7.

PARIS, O. H. and PITELKA, F. A., 1962. Population characteristics of the terrestrial Isopod *Armadillidium vulgare* in California grassland. *Ecology* **43**, 229–48.

PARK, O. and AUERBACH, S., 1954. Further study of the tree-hole complex with emphasis on quantitative aspects of the fauna. *Ecology* **35**, 208–22.

PASK, W. M and COSTA, R. M., 1971. Efficiency of sucrose flotation in recovering insect larvae from benthic steam samples. *Can. Ent.* **103**, 1649–52.

PEACHEY, J. E., 1962. A comparison of two techniques for extracting Enchytraeidae from moorland soils. *In* Murphy, P. W. (ed), *Progress in Soil Zoology* 286–93. Butterworths, London.

PETERS, B. G., 1955. A note on simple methods of recovering nematodes from soil. *In* Kevan, D. K. McE. (ed.), *Soil Zoology* 373–4. University of Nottingham School of Agriculture.

PHILLIPSON, J., 1970. *Methods of study in soil ecology*. 303pp, UNESCO, Paris.

PINDER, L. C. V., 1974. The ecology of slugs in potato crops, with special reference to the differential susceptibility of potato cultivars to slug damage. *J. appl. Ecol.* **11**, 439–51.

POTZGER, J. E., 1955. A borer for sampling in permafrost. *Ecology* **36**, 161.

PRICE, D. W., 1975. Vertical distribution of small arthropods in a Californian Pine Forest soil. *Ann. ent. Soc. Am.* **68**, 174–80.

RAPOPORT, E. H. and OROS, E., 1969. Transporte y manipuleo de las muestras de suelo y su efecto sobre la micro y mesofauna. *Rev. Ecol. Biol. Soc.* **6**, 31–9.

RAW, F., 1955. A flotation extraction process for soil micro-arthropods. *In* Kevan, D. K. McE. (ed.), *Soil Zoology* 341–6. University of Nottingham School of agriculture.

RAW, F., 1959. Estimating earthworm populations by using formalin. *Nature* **184**, 1661–2.

RAW, F., 1960. Earthworm population studies: a comparison of sampling methods. *Nature* **187**, 257.

RAW, F., 1962. Flotation methods for extracting soil arthropods. *In* Murphy, P. W. (ed.), *Progress in Soil Zoology* 199–201. Butterworths, London.

READ, D. C., 1958. Note on a flotation apparatus for removing insects from soil. *Can. J. Pl. Sci.* **38**, 511–14.

SALT, G. and HOLLICK, F.S.J., 1944. Studies of wireworm populations. 1. A census of wireworms in pasture. *Ann. appl. Biol.* **31**, 53–64.

SANDS, W. A., 1972. Problems in attempting to sample tropical subterranean termite populations. *Ekologia Polska* **20**, 23–31.

SATCHELL, J. E., 1955. An electrical method of sampling earthworm populations. *In* Kevan, D. K. McE. (ed.), *Soil Zoology* 356–64. University of Nottingham School of Agriculture.

SATCHELL, J. E. and NELSON, J. M., 1962. A comparison of the Tullgren-funnel and flotation methods of extracting acarina from woodland soil. *In* Murphy, P. W. (ed.), *Progress in Soil Zoology* 212–16. Butterworths, London.

SAUNDERS, L. G., 1959. Methods for studying *Forcipomyia* midges, with special reference to cacao-pollinating species (Diptera, Ceratopogonidae). *Can. J. Zool.* **37**, 33–51.

SCHJØTZ-CHRISTENSEN, B., 1957. The beetle fauna of the Corynephoretum in the ground of the Mols Laboratory, with special reference to *Cardiophorus asellus* Er.

(Elateridae). *Natura Jutl.* **6–7**, 1–120.

SEINHORST, J. W., 1950. De betekenis van de toestand van de grond voor het optreden van aantasting door het stengelaaltje *Ditylenchus dipsaci* (Kuhn) Filipjev. *Tijdschr. Pl. Ziekt.* **56**, 269.

SEINHORST, J. W., 1962. Extraction methods for nematodes inhabiting soil. *In* Murphy, P. W. (ed.), *Progress in Soil Zoology* 243–56. Butterworths, London.

SELLMER, G. P., 1956. A method for the separation of small bivalve molluscs from sediments. *Ecology* **37**, 206.

SERVICE, M. W., 1968. A method for extracting mosquito eggs from soil samples taken from oviposition sites. *Ann. trop. Med. Parasit.* **62**, 478–80.

SHAW, G. G., 1970. Grease-film extraction of an arthropod: a modification for organic soils. *J. econ. Ent.* **63**, 1323–4.

SOLOMON, M. E., 1962. Notes on the extraction and quantitative estimation of Acaridiae (Acarina). *In* Murphy, P. W. (ed.), *Progress in Soil Zoology* 306–7. Butterworths, London.

SOUTH, A., 1964. Estimation of slug populations. *Ann. appl. Biol.* **53**, 251–8.

SPEIGHT, M. C. D., 1973. A greased-belt technique for the extraction of arthropods from organic debris. *Pedobiologia 13*, 99–106.

STAGE, H. H., GJULLIN, C. M. and YATES, W. W., 1952. Mosquitoes of the Northwestern States. U.S.D.A., *Agric. Handb.* **46**, 95 pp.

STEPHENSON, J. W., 1962. An improved final sieve for use with the Salt and Hollick soilwashing apparatus. *In* Murphy, P. W. (ed.), *Progress in Soil Zoology* 202–3. Butterworths, London.

STEWART, K. M., 1974. A three-dimensional net sieve for extracting Tipulidae (Diptera) larvae from pasture soil. *J. app. Ecol.* **11**, 427–30.

SUGIMOTO, T., 1967. Application of oil–water flotation method to the extraction of larvae and pupae of aquatic midges from a soil sample. *Jap. J. Ecol.* **17**, 179–82.

SUNDERLAND, K. D., HASSELL, M. and SUTTON, S. L., 1976. The population dynamics of *Philoscia muscorum* (Crustacea, Oniscoidea) in a dune grassland ecosystem. *J. Anim. Ecol.* **45**, 487–506.

SVENDESN, J. A., 1955. Earthworm population studies: a comparison of sampling methods. *Nature* **175**, 864.

TANTON, M. T., 1969. A corer for sampling soil and litter arthropods. *Ecology* **5**, 134–5.

TÖRNE, E. VON, 1962a. An elutriation and sieving apparatus for extracting micro-arthropods from soil. *In* Murphy, P. W. (ed.), *Progress in Soil Zoology* 204–6. Butterworths, London.

TÖRNE, E. VON, 1962b. A cylindrical tool for sampling manure and compost. *In* Murphy, P. W. (ed.), *Progress in Soil Zoology* 240–2. Butterworths, London.

VÁGVÖLGYI, J., 1953 (1952). A new sorting method for snails, applicable also for quantitative researches. *Ann. Hist. Nat. Mus. Nat. Hung.* **44** (N.S. 3), 101–4.

VANNIER, G., 1965. Enceinte réfrigérée par modules thermoélectriques à effet Pether (+ 30° C a − 40° C) permettant l'observation directe de la microfaune. *Rev. Ecol. Biol. Sol.* **2**, 489–506.

VANNIER, G. and ALPERN, I., 1968. Techniques de prélèvements pour l'étude des distribution horizontales et verticales des microarthropodes du sol. *Rev. Ecol. Biol. Sol.* **5**, 225–35.

VANNIER, G. and VIDAL, P., 1964. Construction d'un appareil automatique pour couper les inclusions de sol dans la gelatine. *Rev. Ecol. Biol. Sol.* **1**, 575–86.

VANNIER, G. and VIDAL, P., 1965. Sonde pédologique pour l'échantillonnage des microarthropodes. *Rev. Ecol. Biol. Sol.* **2**, 333–7.

WALLACE, M. M. H., 1956. A rapid method of sampling small free-living pasture insects and mites. *J. Aust. Inst. agric. Sci.* **22**, 283–4.

WEBLEY, D., 1957. A method of estimating the density of frit fly eggs in the field. *Plant Path.* **6**, 49–51.

WEBSTER, J. M., 1962. The quantitative extraction of *Ditylenchus dipsaci* (Kuhn) from plant tissues by a modified Seinhorst mistifier. *Nematologica* **8**, 245–51.

WHITEHEAD, A. G. and HEMMING, J. R., 1965. A comparison of some quantitative methods of extracting small vermiform nematodes from soil. *Ann. appl. Biol.* **55**, 25–38.

WILLIAMS, R. W., 1960a. A new and simple method for the isolation of fresh water invertebrates from soil samples. *Ecology* **41**, 573–4.

WILLIAMS, R. W., 1960b. Quantitative studies on populations of biting midge larvae in saturated soil from two types of Michigan bogs (Diptera: Ceratopogonidae). *J. Parasit.* **46**, 565–6.

WILLIAMSON, M. H., 1959. The separation of molluscs from woodland leaf-litter. *J. Anim. Ecol.* **28**, 153–5.

6

Absolute Population Estimates by Sampling a Unit of Habitat – Freshwater Habitats

Phyla other than the Arthropoda are well represented in this habitat, but it is not intended in this chapter to detail the methods for the study of the microfauna of inland waters; these are described in works such as Welch (1948) and Edmondson & Winberg (1971); whilst sampling aspects are reviewed by Elliott (1971) and Elliott & Décamps (1973). In contrast to terrestrial habitats, major difficulties in making absolute estimates lie in actually taking a sample of a known unit, as well as in the separation of the animals from the media. The problems of extraction are similar to those in terrestrial habitats and reference should therefore be made to Chapters 4 and 5.

Many of the methods described in this chapter have comparatively low efficiencies, so that the emphasis in the chapter heading must be on the use of a unit of habitat, rather than on the absoluteness of the estimate. When a method has been calibrated the data from it could be multiplied by an appropriate conversion factor to give an absolute estimate.

It is convenient to divide the freshwater habitat and its animals into:
  I. Open water – inhabited by surface dwellers and by plankton-type animals.
  II. Vegetation – animals living on or around submerged plants.
  III. Bottom fauna – animals living on or in the substrate.

6.1 Open water

No absolute quantitative method has been devised for estimating surface-dwelling insects. The larger forms may be counted directly *in situ*; their numbers and dispersion might also be studied by photography and possibly nearest neighbour techniques used in their estimation (p. 47).

Small free-swimming animals may be sampled by the many methods that have been developed in plankton studies. A full account of these is given by Welch (1948); there are four main types that are of significance in entomological studies.

Nets
The simplest type of plankton net, a bolting silk bag on a metal frame, is easy

202

to construct; towed behind a boat or hauled across a pond on a line, it collects and separates the animals in one operation. It is possible to calculate the volume of water that has to be filtered through the net, i.e. from which the insects have been extracted. Care must be taken to ensure that the net is not moved too fast so that it 'pushes aside' some water and the animals in it; this is particularly important with fine-mesh nets (Ricker, 1938; Fujita, 1956); however, if the net is moved too slowly the more agile animals with good vision may be able to avoid it. One should always use the coarsest net consistent with the size of the animal being studied, and the filtering surface should be large compared with the area of the mouth of the net, i.e. the net should be long.

The Birge cone net, as modified by Wolcott (1901) (Fig. 6.1*a*), is a convenient tow-net: the anterior wire-mesh cone ensures that water weed and other large debris do not clog the net, which is easily emptied by the removal of the bottom cap, conveniently made from the screw cap of a metal can.

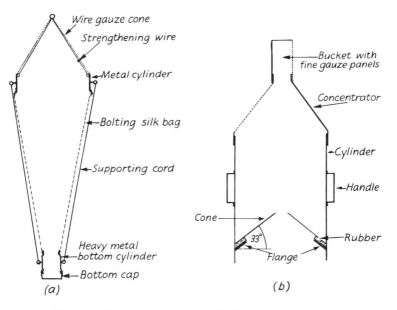

Fig. 6.1 *a*. Birge cone tow-net. *b*. Belleville mosquito sampler.

If nets are to be used to sample animals in the lower layers of water, then they must have a mechanism for closing them as they pass through the surface layer (Ackefors, 1971). Wickstead (1953) has described one for sampling the animals in the immediate vicinity of the sea or lake bottom.

The 'dipping' method using a long-handled ladle is usually regarded as a relative method (p. 240), but Croset *et al.* (1976) found that provided the ladle was large enough (so that 95 % of the samples contain larvae), this method could be regarded as sampling from a unit of the habitat (the volume of the ladle), and provided population estimates of culicid larvae that were comparable to those from mark and recapture or removal trapping.

Pumps and baling

If a pump is used, a hose is inserted in the water to a known depth, the water pumped out and the organisms extracted by sieves or screens. The disadvantages of this technique are that the precise depth from which the water is drawn is unknown and some animals may react to the current and so avoid capture (Welch, 1948). A portable unit, fitting exactly into a small boat, has been described by Griffith (1957).

With small bodies of water, e.g. mosquito breeding pools, it may be possible to empty the complete pond, carefully separating the animals by a series of sieves, the contents of each sieve being sorted in a large pan. As developed by Christie (1954), the method allows the mosquito larvae to be returned to the pond after enumeration; although the water level is lowered with a pump, the residue containing the insects is baled out and the sieves are kept full of water so that the larvae are not damaged by violent contact with them.

The Clarke–Bumpus plankton sampler

This apparatus consists of a normal cone-shaped plankton net; in its mouth is a metal cylinder with a propeller blade (that records the throughput of water), two stabilizing vanes and a shutter mechanism. It is described in detail by Clarke & Bumpus (1940), Welch (1948) and Tonolli (1971).

The Belleville mosquito sampler

Because mosquito larvae dive to the bottom of a pond when disturbed and then gradually make their way up to the surface again, sampling by dipping with a net or strainer will give variable results depending on the skill of the collector. Welch & James (1960) have, however, used this habit in the Belleville mosquito sampler (Fig. 6.1*b*). The sampler consists of a cylinder, a cone, a concentrator and a bucket. The cylinder is placed in the pond, its base firmly pushed on to the substrate, the cone slipped inside and the apparatus left for 20 minutes. The concentrator and bucket are then fixed to the top; the whole rapidly reversed. Most of the larvae are collected in the bucket, as the water drains out through it and the concentrator; a few may remain stuck on the sides, so the apparatus should be rinsed.

The angle of the cone may have to be adjusted for different species; for Canadian *Aedes* Welch & James found the minimum was 33°. The greater the angle of the cone, the greater its height; as the water must always be deeper than the height of the cone this value will limit the depth of water in which the sampler can be used. Since the samplers need to be left for some time, a practical set of equipment for field work would consist of ten cylinders and cones with a single concentrator and bucket. Laboratory tests gave disappointingly low efficiencies, in the region of 30–40 %; and increasing the 'rising time' from 5 to 20 minutes did not markedly increase the number of larvae extracted. Welch & James suggest that the sampler be standardized by comparing the catch for a 24-hour period with that for a shorter time. The series of catches may then be corrected to give an absolute estimate.

6.2 Vegetation

Animals among floating vegetation are frequently sampled by dipping with a pond net or a strainer; such methods usually give only relative estimates of population density. For absolute estimates it is necessary to enclose a unit volume of the vegetation and associated water; when this has been strained the organisms are usually separated by hand-sorting; however, some of the methods described in Chapter 4 might be found useful (e.g. clearing and staining for eggs in water plants). Extraction by spraying the material with a fine mist, as used in plant nematode work, might also be found applicable (p. 192).

Sampling cylinder for floating vegetation
This was described by Hess (1941); it consists of a stout galvanized cylinder with copper mesh screening (Fig. 6.2b) and is lowered under the vegetation,

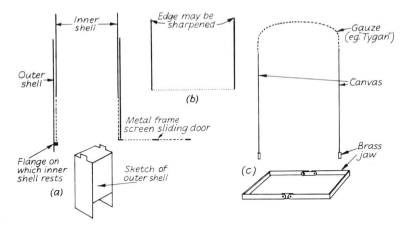

Fig. 6.2 *a*. Gerking sampler, sketch of outer shell and section. *b*. Hess's sampling cylinder for floating vegetation. *c*. Wisconsin trap, sketch of jaws and section.

moved into position, raised and any plant stems crossing its edge cut by striking the edges of the cylinder with a wooden paddle. If necessary this could be aided by sharpening the upper edges of the cylinder. The cylinder is then raised from the water and the plants and animals retained within it.

Wisconsin trap

As described by Welch (1948) this is simply a canvas and gauze net with a closable mouth (Fig. 6.2c). The trap is lowered over the vegetation; the jaws of the trap are closed just above the substrate, the plants being uprooted or cut off. The whole is then raised from the water and drained.

The Gerking sampler

Designed by Gerking (1957) for the sampling of littoral macrofauna, this equipment would seem to be satisfactory, apart from the labour involved, for obtaining absolute samples of all but the most active animals. It consists of two galvanized iron shells, each with a square cross-section; they may be nearly 1 m high if necessary (Fig. 6.2a). The two shells of the sampler are placed in position, the lower edges of the outer shell being forced into the mud. One side of the base of the outer shell is open (see sketch, Fig. 6.2) and through this opening the stems of the water plants are quickly cut with a pair of grassclippers. The metal and screen sliding door is then inserted and this effectively closes the inner shell, which may be slowly raised. The outer cylinder is left in position and Gerking used an Ekman dredge to remove the substrate from within the same area; alternatively it could be scooped out (see below). If it is not desired to sample the substrate simultaneously then only a single shell need be used; it should correspond in shape to the outer shell, but have a number of gauze panels. In some ways this apparatus is a modification of the Wilding (1940) square foot sampler.

McCauley's samplers

McCauley (1975) described two samplers for the macro-invertebrates on aquatic vegetation. The general model (Fig. 6.3) for submerged and floating vegetation can be conveniently operated from the back of a boat. The lower plate of the sampler has a cutting edge, it is moved into position as close to the bottom as possible. When this has been done the sampling cylinder is released (triggered) and the pair of powerful springs force it down on to the lower plate. The plants are removed and the contents pumped out, using a hand operated plastic bilge pump, and filtered through a fine mesh sieve. The sampler may then be removed and the inside washed down to remove any remaining animals.

McCauley's sampler for rushes consists of a length of acrylic tubing that encloses the sample and a sharpened brass plate that is released by a trigger and cuts the bases of the rushes, rather like the sliding plate of the Gerking sampler.

Sampling cages

A simpler approach along the same lines as the Gerking and McCauley's samplers, that is enclosing a column of water of a known volume, is the sampling cage of James & Nicholls (1961). It is a screen cage that is pushed

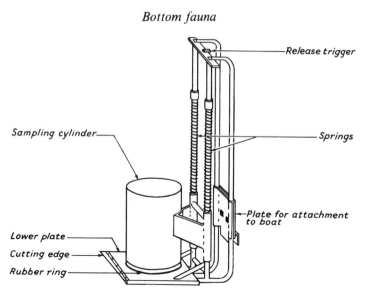

Fig. 6.3 McCauley's sampler for submerged and floating vegetation (after McCauley, 1975).

down into the mud, the enclosed water is hand-sieved and then the substrate dredged up and sorted on sorting trays attached to the sides of the cage. There is always the danger with this method that some animals will be missed in the hand-sieving. Earlier versions of this sampling device are described by Bates (1941) and Goodwin & Eyles (1942).

6.3 Bottom fauna

A large number of methods have been developed for sampling the bottom fauna of freshwater habitats; one reviewer has commented that the number of samplers is nearly proportional to the number of investigators (Cummins, 1962)! This fact arises because five variables affect the choice of sampler:

(1) The animal
(2) The nature of the bottom substrate – soft or hard
(3) The current
(4) The depth of the water
(5) The object of the study, e.g. a survey for pollution, an assessment of the food potentially available for fish, or an intensive ecological investigation for the development of a life-table for a single species.

A few generalizations are possible. Fast-flowing water has the advantage that the current may be used to carry animals, disturbed from the substrate, into a sampler (as with the Surber sampler, p. 211); however, it limits the use of devices that enclose a unit area, for as these are lowered the increased current immediately beneath them may scour the organisms from the very area that is to be sampled.

After the sample has been taken the problem remains of separating the animals from the substrate. If the sample is too large or contains too many organisms it may be divided and subsampled (see also p. 144). The material needs to be thoroughly agitated for satisfactory subsampling. Huckley (1975) describes a 'sample-splitter' that uses compressed air for mixing. Hand sorting is the most widely used method (Frost, 1971); its accuracy may often be checked by staining any remaining animals with rhodamine B and viewing under UV light (Eckblad, 1973) (see also p. 77). Aquatic biologists could undoubtedly make use of many of the methods described in Chapter 5 (Hynes, 1971): i.e. wet sieving, flotation in saturated solutions of various salts (e.g. $MgSO_4$, NaCl, $CaCl_2$, $ZnCl_2$), heavy chlorinated hydrocarbons or sucrose (sugar) solution. As Kajak, Dusage & Prejs (1968) point out, the latter has the advantage of maintaining the animals alive; the inorganic salts cause dehydration and death; however, sugar solutions do deteriorate, and the recovery of living animals is only occasionally required. Elutriation (p. 182) and the sand extractor (p. 191) have been used to a limited extent in freshwater studies but the potential is probably considerable.

Sampling in stony streams has recently been reviewed by Macan (1958) and the general problem of benthic sampling by Albrecht (1959), Longhurst (1959) and Cummins (1962). Cummins discusses the type of sampling programme and pattern; he stresses the importance of simultaneous sampling and characterizing the substrate in any study of the distribution of benthic animals. Kajak (1963) discusses the problems of sample unit size and the number of samples (see Chapter 2); with the tremendous variability of microhabitats within a small area of stream bottom the variance of a series of samples is often extremely large (Usinger & Needham, 1954).

The 'planting' of removable portions of the substrate
One of the most accurate methods of sampling the bottom fauna is to place a bag, tray or box in the stream or pond bed and either replace the substrate or allow the sediment to accumulate naturally (Moon, 1935; Wene & Wickliff, 1940; Usinger & Needham 1954; Ford, 1962). An elaborate apparatus of this type is the box of Ford. This has two fixed wooden sides and a bottom and is placed in a hole in the bed of the stream, the two sides being parallel to the course of the water. After a suitable time (Ford left it for 6 weeks) the other two sides, which are made of 'Perspex',* are slid into position, the box is then made watertight and may be lifted from the stream bed with the sample undisturbed. In Ford's model it was possible to divide the sample horizontally so that each stratum could be separately analysed: one of the wooden sides had a series of slots in it; the egress of mud and water through these was prevented by rubber flaps on the inside, the other sides were grooved at the corresponding level; 'Perspex' sheets could then be pushed through these slots

---

* 'Plexiglass' is a similar transparent plastic material.

dividing the sample. Simpler 'basket samplers' have been widely used and are generally found to give satisfactory assessments of the macro-invertebrate fauna (Mason *et al.*, 1973; Crossman & Cairns, 1974). A larger framed net, that may be rapidly pulled to the surface, has been designed for use in the shallow waters of the Florida Everglades (Higer & Kolipinski, 1967).

The fauna of rocks or concrete substrates may be studied by placing easily removable blocks or plates of similar composition. Britt (1955) placed blocks of concrete, heavily scored on the undersurface, on a rubble and gravel bottom in deep water. The blocks can be easily located and raised with the help of the buoy and cord, but some animals may be lost as the block is raised. This loss is overcome in Mundie's (1956) method, devised for the study of Chironomidae on the sloping slides of artificial reservoirs. The artificial substrate is a plate of asbestos–cement composition (Fig. 6.4); attached to the centre of the plate is a guide wire which must be attached to a line above the water surface. The plates are left for the period of time necessary for them to be indistinguishable from the surrounding substrate (1–3 months) and then retrieved. Mundie's ingenious method of retrieval was as follows: a retrieving cone (Fig 6.4*b*) is slid down the guide line; this cone has flap-valves on the upper surface and these open as it descends ensuring that there is no surge of water to disturb the sample as the cone settles; rubber flanges around these valves and the base of the cone protect the sample from disturbance as the plate is gradually hauled up by the guide wire. Before the sample breaks the surface a gauze-bottomed bucket (Fig 6.4*c*) is placed below it; this retains any organisms washed out as the water in the sampler drains through the flap-valves of the cone.

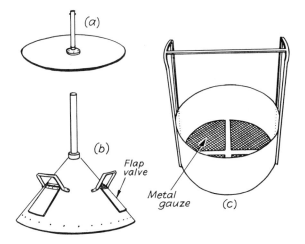

Fig. 6.4 Artificial substrate and sampling apparatus (after Mundie, 1956): *a*. artificial substrate-plate with central pipe to which wire is attached; *b*. retrieving cone; *c*. gauzebottomed landing bucket.

Completely artificial substrates may also be used, but these give only relative estimates as settlement on them will be different to that on the natural substrate. These standardized artificial substrate units (SASU's) have been found particularly useful for estimating relative changes in the numbers of blackflies (Simulidae) (Wolfe & Peterson, 1958; Williams & Obeng, 1962) and Disney (1972) points out that the numbers obtained will depend on (i) the absolute population of the blackflies (ii) the relative area of other suitable substrata, (iii) the length of time the artificial substrate units are exposed (iv) the intensity and nature of the factors that cause the larval blackflies to move from the substrate where they are already established (v) the nature of the substrate and its acceptability. The latter effect will vary with the animal and the substrate (Glime & Clemons, 1972; Mason *et al.* 1973; Benfield, Hendricks & Cairns, 1974).

Lifting stones

The majority of the animals in fast-flowing streams will be underneath the stones on the bottom. Scott & Rushforth (1959) have investigated from the mathematical angle the influence of stone size and spacing on the area of the stream bed covered by stones. They propose the symbol C_v for this parameter and discuss its value. An estimate of the absolute population of animals may be obtained by picking up the stones from an area of the bottom; a net should be held on the downstream side to catch those animals washed off (Macan, 1958; Ulfstrand, 1968). If an estimate of population intensity (p. 2) is to be made, it will be necessary to make an estimate of the effective surface area of the stones. Calow (1972) showed that this could be calculated if longest perimeter of the stone (x) was found, then

surface area $= 2.22\ (\pm 0.26)x$

Some animals may be especially difficult to remove from the stones and Britt (1955) found that these would become active if placed in a very weak acid alcohol solution (2–5 % alchol, 0.03–0.06 % hydrochloric acid). A flotation solution (p. 178) might serve the same purpose and separate the animals as well.

Alternatively a net may be fixed in position; the stones and remainder of the substrate are disturbed from a unit area (one square foot) upstream and the animals caught in the net. The simplest approach is the 'kicking technique' when the stones immediately upstream are 'kicked'. Frost, Huni & Kershaw (1971) found that the first kick disturbed about 60 % of the fauna yielded by 10 kicks, this is essentially a catch per unit effort (p. 236). The same principle, but with a more precise delimitation of the area, is used in the Surber sampler (Surber, 1936) (Fig. 6.5). The variability of this technique, with operator and stream conditions, and its efficiency have been tested by Usinger & Needham (1954; and Needham & Usinger, 1956). As some stones will lie partly inside

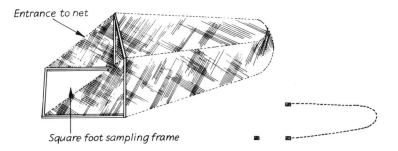

Fig. 6.5 Surber sampler; sketch and section.

and partly outside the square-foot frame the selection and rejection of these is a matter of personal judgement, and Usinger & Needham found that one operator continually sampled half as much area again as the other four operators, but the latter were consistent. Comparison of the results from the Surber sampler with absolute counts of animals from buried trays (see above) showed that it caught only about a quarter of the population. The exact proportion will vary from site to site, but it is obvious that the Surber sampler cannot be used for absolute population estimates from a shallow stream without a careful test of its efficiency to determine what correction factor should be applied (Kroger; 1972, 1974).

Cylinders and boxes for delimiting an area

As pointed out above, when a box or cylinder is lowered into flowing water, as it approaches the bottom it is likely to cause this to be scoured. This problem is least serious with the Hess circular square-foot sampler (Hess, 1941) (Fig. 6.6*a*), which is made of a fairly coarse mesh and may be rapidly turned so as to sink a little way into the bottom. The smaller organisms may be lost through

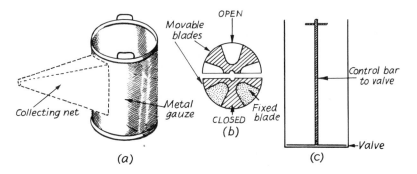

Fig. 6.6 *a*. Hess circular square-foot sampler. *b*. and *c*. Inner cylinder of the Wilding sampler (simplified): *b*. diagram of rotary valve: upper half, open position; lower half, closed position; *c*. section of cylinder.

the coarse grid. If a finer grid is used scouring becomes progressively more serious, but on the other hand drifting animals may pass through a coarse net and add cumulatively to the sample. Therefore Waters & Knapp (1961) designed a circular sampler with fine-mesh ('Nitex') screening; the emptying of the collecting bag was facilitated by attaching it to the sleeve with a zip-fastener. The use of these samplers is restricted to streams, as a current is necessary to carry the organisms into the collecting bag. The Wilding sampler (Wilding, 1940; see also Welch, 1948) may be used in still or moving water. It consists of two cylinders; the outer cylinder has finely perforated sides and a band of saw teeth along the lower edge to aid penetration into the bed of the stream or lake (these are not essential). This cylinder delimits the area and the larger organisms, rocks and other debris are removed; then the whole is stirred and the inner cylinder (of slightly smaller diameter) lowered inside. The inner cylinder has a rotary valve at the bottom through which virtually all the water in the outer cylinder enters it (Fig. 6.6b and c). When it is resting on the bottom the rotary valve is closed and the inner cylinder, containing all the water and smaller animals, removed. Its contents can be passed through a nest of sieves by opening the valve slowly.

The Gerking sampler and cages described above (p. 206) may also be used for delimiting an area of bed for sampling, and simple metal frames and cylinders were used for this purpose by Scott (1958) and Dunn (1961). Neill's (1938) sampler is intermediate in design between the Hess and Wilding models.

Movable nets – drags

The use of the Surber sampler and the various sampling cylinders is limited to shallow waters (i.e. of depth of an arm's length or less); in deeper waters either a moving net (drag, scoop, shovel) or a metal sampling box (dredge, grab, etc. – see below) must be used. Several movable nets have been described (see Welch, 1948; Macan, 1958; Albrecht, 1959); they may be pushed or pulled – because they are themselves moved they are not dependent on a current in the water. One of the most robust models is that devised by Usinger & Needham (1954, 1956) (Fig. 6.7) for sampling stone or gravel bottoms (none of these movable nets are really suitable for soft mud). The tines on the mouth disturb the substrate, but prevent large stones and debris from entering the bag, and the weights and the heavy steel frame ensure that the tines really scour the bottom. The bag is attached by means of a brass zip-fastener and protected from tearing by a canvas sleeve. It is probable that a bag of one of the recent synthetic materials, e.g. 'Tygan' or 'Nitex', would be more resistant to tearing than the nylon or silk originally used. Usinger & Needham (1954) found that, under the conditions they were working, this drag caught about one-quarter of the animals from the area it traversed. Therefore, unless standardized this is only a relative method. A modification of this drag, particularly suitable for sampling unionid mussels, has the tines removed, an adjustable blade attached

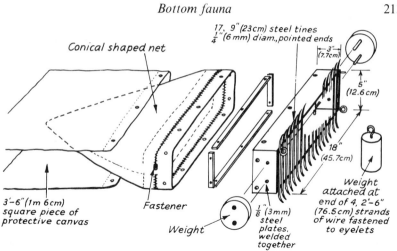

Fig. 6.7 Usinger & Needham's drag (after Usinger & Needham, 1956).

to the lower edge and a vertical handle (in addition to the tow rope) fixed to the top leading edge (Negus, 1966).

Dredges and other devices for removing portions of the substrate
A wide variety of devices have been described for sampling a portion of the bed of a river or lake and bringing this to the surface (Table 6.1). The variety

Table 6.1 Summary of some of the types of equipment for sampling portions of the benthic substrate – dredges, grabs, piston samplers, etc. Those in italics give undisturbed samples for the study of stratification

	Substrate texture	May be used in water up to 1–2 m deep	May be used in water of 'any' depth
Very soft ↓ Hard	Finely divided ↓		*Shapiro core-freezer* *Elgmork's sampler* *Kajak–Brinkhurst sampler* *Brown's piston sampler* Ekman dredge (rope)
		Dendy inverting sampler	
			Hayward orange-peel bucket
		Ekman dredge (pole) Allan grab	Petersen grab
			Smith & McIntyre sampler
	Contains large particles – rocks, thick sticks	*O'Connor corer* (Ch. 5)	

has been due to the different emphasis placed by those primarily interested in physico-chemical investigations of the substrate from those working on benthic organisms; Cummins (1962) pleads for the combination of the two approaches. Further variation is due to the nature of the bottom and the depth of water in which the sampler must work (see Table 6.1). A number of comparative tests have been made (Flaunagan, 1970, Pearson, Litterick & Jones, 1973). The Ekman dredge has been compared with various corers for sampling chironomid larvae in mud: Milbrink & Wiederholm (1973) found little difference, but Karlsson *et al.* (1976) who used flotation (p. 178) with material taken by the corer, recovered more larvae by this method, than from hand-sorted dredge samples.

(1) *Dredges and grabs*. These all take a comparatively shallow sample, which is more or less disturbed by the time it has reached the surface. Therefore they are not suitable for studies on stratification. One of the earliest pieces of equipment in this category is the Ekman dredge (Ekman, 1911), which is still widely used for sampling soft bottoms (Fig. 6.8). It should be lowered gently

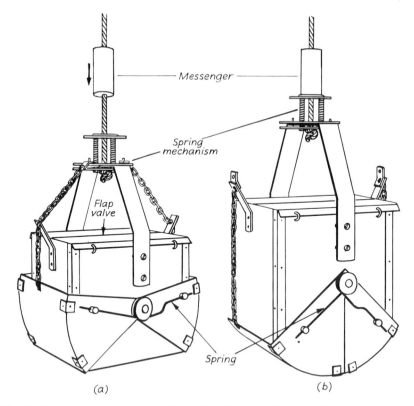

Fig. 6.8 Ekman dredge: *a.* open position while being lowered; *b.* closed after messenger has released jaws.

into the water and, as it is lowered, the lid flaps will rise allowing it to settle into the mud without too much scouring. When the dredge has reached the bottom and settled, the messenger is dropped down the rope, and when this impacts on the spring mechanism above the box the two chains holding the jaws are released; the jaws are shut by the spring on their side. The dredge is then hauled to the surface. It is clear from the mode of operation that it relies on its own weight to sink and on the comparatively weak spring to close the jaws, therefore this dredge is limited to finely divided muddy or peaty bottoms; large bivalves or sticks will interfere with the closing of the jaws. In general better results are obtained if the Ekman dredge is mounted on a pole, instead of a rope; this is essential if it is to be used on hard bottoms (Hynes, 1971). Wigley (1967) used motion pictures to determine the effectiveness of closing of various designs of Petersen's grab.

Slightly less homogeneous substrates, i.e. those containing fine gravel or small sticks, may be sampled by the Dendy (1944) inverting sampler (Fig. 6.9).

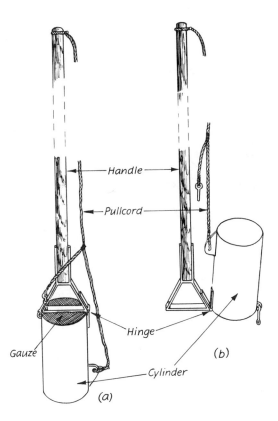

Fig. 6.9 Dendy inverting sampler (after Dendy, 1944): *a.* position when entering mud; *b.* inverted, ready to raise to surface.

This apparatus consists of a brass cylinder of about 8 cm diameter, its top is covered with a piece of brass gauze. The cylinder is on the end of a long handle (this limits the depth of operation), which is used to drive it into the substrate; the pull cord is then tugged, the cylinder inverted and lifted to the surface. Only material that can pass out through the gauze is lost in the ascent (apart from a small amount from the very bottom of the sample). The small size of this sampler relative to the Ekman dredge may indicate its use in preference to the latter from cost and sampling pattern considerations (see p. 19), but it does have the disadvantage of disturbing the light layer on the surface of the bottom and for this reason, Kajak (1971) recommends against its use.

The commercially manufactured 'Hayward orange-peel bucket' may be used for bottom sampling of harder substrates in deep waters. Reish (1959) describes a simple modification whereby the use of a dropping messenger or second cable for closure is obviated. Its main disadvantages are that both very soft and rocky bottoms interfere with the closing mechanism.

The Petersen grab (Petersen, 1911) is a veteran piece of equipment that works in sand, gravel or marl (Fig. 6.10a). It is lowered to the bottom on a

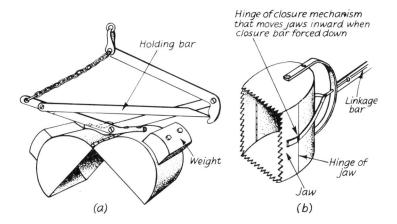

Fig. 6.10 *a*. Petersen grab (after Welch, 1948). *b*. Allan grab.

cable and when it reaches the bottom the release of tension frees the holding bar so that the jaws start to 'bite' into the substrate under its own weight (about 35 lb); when the dredge is hoisted, leverage forcing the closure of the jaws aids the bite. The main disadvantage of this dredge is its weight and the hoisting apparatus necessary for its use. A modification has been described by Lisitsyn & Udintsev (1955) in which the messenger is replaced by a counterpoised weight (Fig. 6.11).

In relatively shallow water samples may be taken from stony or vegetation-clothed bottoms by the Allan (1952) grab (Fig. 6.10b), which, like the Dendy sampler, is attached to the end of a long handle (tubular steel); this has a

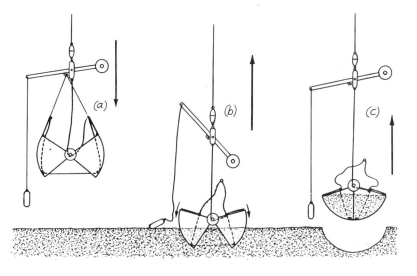

Fig. 6.11 A method of closing the Petersen grab using a counterpoised weight (after Lisitsyn & Udintsev, 1955).

double advantage. The operator may exert direct force, through the linkage bar, on the closure mechanism, and will be aware when a large rock or piece of wood has jammed the jaws; secondly the dredge is held rigidly and so will enter the substrate vertically whatever the water current (rope-suspended dredges may be deflected).

No dredge is suitable for sampling really rocky beds; one of the most powerful is that designed by Smith & McIntyre (1954); its jaws are closed by strong springs.

(2) Piston-type samplers. These enable a vertical core to be taken from the bottom and are ideal for the study of stratification of micro-organisms and for the assessment of the physical and chemical properties of the various layers. One of the earliest was the Jenkin surface mud sampler (Jenkin & Mortimer, 1938; Mortimer, 1942; Goulder, 1971; Milbrink, 1971).

Elgmork (1962) modified the Freidinger water sampler to take samples of soft mud; after sampling a piston was inserted and the sample slowly pushed out. The stratification of *Chaoborus punctipennis* was studied with this apparatus. The Kajak-Brinkhurst sampler* is very suitable for soft sediments, the lid may be closed by a messenger after the sample is taken; it is described by Kajak (1971) and has found to be nearly as efficient as using a diver (Holopainen & Sarvala, 1975). A multiple corer has been developed by Hamilton *et al.* (1970).

Livingstone (1955) described a piston sampler particularly well adapted for obtaining the lower layers of sediment from a muddy bottom; Vallentyne

* The 'K–B sediment sampler', available from Wildlife Supply Co., Saginaw, Michigan, USA.

(1955) and Rowley & Dahl (1956) introduced modifications that made the construction simpler and less costly, and Brown (1956) has modified it to sample the mud – water interface as well and, by the use of a clear plastic tube, to allow the inspection of the intact core.

A different approach is that of Shapiro (1958), who surrounded the sampling tube (corer) with a tapered jacket. Just before lowering the jacket is filled with crushed solid carbon dioxide and *n*-butyl alcohol. It is lowered quickly through the water, allowed to settle, left for about five minutes and retrieved. The freezing mixture is replaced with water; the frozen core may soon be slid out and divided and handled in the solid state As pointed out by Kajak (1971) soil corers, especially O'Connor's (p. 171), are very useful for hard bottoms in shallow water.

Air-lift and suction devices

When compressed air is discharged under water, an air-water mixture is formed that rises to the surface and beyond it. This technique has been used in marine archaeology, mining, etc. and the fundamental principles are well known (e.g. Dumas, 1962). It is clearly very suitable for the collection of benthic samples. Mackey (1972) describes a simple apparatus, essentially for taking core samples: compressed air lead opens at the centre of the base of a length of plastic drain pipe, the top of the pipe is angled and opens in a collecting net or sieve. A more elaborate device, in which the sampling area is delimited by a rectangular frame and the compressed air is liberated through a series of openings, was developed by Pearson, Litterick & Jones (1973). They compared its efficiency with the Suber sampler and the Allen grab; overall there were not great differences, but particular animals were sampled in greater numbers by one or other method. The air lift sampled significantly larger numbers of *Polycelis*, *Ephemerella*, *Sphaerium* and *Caenis*.

Another approach is to suck the animals off the bottom (paralleling the suction sampler for terrestrial habitats (p. 148). Remarkably little has been done in this respect, although Gulliksen & Deras (1975) describe a diver-operated model for marine situations.

REFERENCES

ACKEFORS, H., 1971. A quantitative plankton sampler. *Oikos* **22,** 114–8.

ALBRECHT, M.-L., 1959. Die quantitative Untersuchung der Bodenfauna fliessender Gewässer (Untersuchungsmethoden und Arbeitsergebnisse). *Z. Fisch.* (N.F.) **8,** 481–550.

ALLAN, I. R. H., 1952. A hand-operated quantitative grab for sampling river beds. *J. Anim. Ecol.* **21,** 159–60.

BATES, M., 1941. Field tudies of the anopheline mosquitoes of Albania. *Proc. ent. Soc. Wash.* **43,** 37–58.

BENFIELD, E. F., HENDRICKS, A. C. and CAIRNS, J., 1974. Proficiencies of two artificial substrates in collecting stream macroinvertebrates. *Hydrobiologia* **45,** 431–40.

BRITT, N. W., 1955. New methods of collecting bottom fauna from shoals or rubble bottoms of lakes and streams. *Ecology* **36**, 524–5.

BROWN, S. R., 1956. A piston sampler for surface sediments of lake deposits. *Ecology* **37**, 611–13.

CALOW, P., 1972. A method for determining the surface area of stones to enable quantitative density estimates of littoral stone dwelling organisms to be made. *Hydrobiologia* **40**, 37–50.

CHRISTIE, M., 1954. A method for the numerical study of larval populations of *Anopheles gambiae* and other pool-breeding mosquitoes. *Ann. trop. Med. Parasit., Liverpool* **48**, 271–6.

CLARKE, G. L. and BUMPUS, D. F., 1940. The Plankton sampler–an instrument for quantitative plankton investigations. *Limnol. Soc. Amer. Spec. Publ.* **5**, 1–8.

CROSET, H., PAPIEROK, B., RIOUX, J. A., GABINAUD, A., COUSSERANS, J. and ARNAUD, D., 1976. Absolute estimates of larval populations of culicid mosquitoes: comparison of capture–recapture removal and dipping methods. *Ecological Entomology* **1**, 251–6.

CROSSMAN, J. S. and CAIRNS, J., 1974. A comparative study between two different artificial substrate samplers and regular sampling techniques. *Hydrobiologia* **44**, 517–22.

CUMMINS, K. W., 1962. An evaluation of some techniques for the collection and analysis of benthic samples with special emphasis on lotic waters. *Am. Midl. Nat.* **67**, 477–503.

DENDY, J. S., 1944. The fate of animals in streamdrift when carried into lakes. *Ecol. Monogr.* **14**, 333–57.

DISNEY, R. H. L., 1972. Observations on sampling pre-imaginal populations of blackflies (Dipt., Simuliidae) in West Cameroon. *Bull. ent. Res.* **61**, 485–503.

DUMAS, F., 1962. *Deep water archaeology*, London.

DUNN, D. R., 1961. The bottom fauna of Llyn Tegid (Lake Bala), Merionethshire. *J. anim. Ecol.* **31**, 267–81.

ECKBLAD, J. W., 1973. Population studies of three aquatic gastropods in an intermittent backwater. *Hydrobiologia* **41**, 199–219.

EDMONDSON, W. T., and WINBERG, G. G. (eds:) 1971. *A manual on methods for the assessment of secondary productivity in freshwaters*. I.B.P. Handbook 17. 358 pp., Blackwells, Oxford and London.

EKMAN, S., 1911. Die Bodenfauna des Vättern, qualitativ und quantitativ untersucht. *Int. Revue ges. Hydrobiol. Hydrogr.* **7**, 146–204.

ELGMORK, K., 1962. A bottom sampler for soft mud. *Hydrobiologia* **20**, 167–72.

ELLIOTT, J. M., 1971. Some methods for the statistical analysis of samples of benthic invertebrates. *Sci. Publ. Freshw. Biol. Assoc.* **25**, 148 pp.

ELLIOTT, J. M. and DÉCAMPS, H., 1973. Guide pour l'analyse statistique des échantillons d'invertébrés benthiques. *Annls. limnol.* **9**, 79–120.

FLANNAGAN, J. F., 1970. Efficiencies of various grabs and corers in sampling freshwater benthos. *J. Fish. Res. Bd. Canada* **27**, 1691–1700.

FORD, J. B., 1962. The vertical distribution of larvae Chironomidae (Dipt.) in the mud of a stream. *Hydrobiologia* **19**, 262–72.

FROST, S., 1971. Evaluation of a technique for sorting and counting stream invertebrates. *Can. J. Zool.* **49**, 878–83.

FROST, S., HUNI, A. and KERSHAW, W. E., 1971. Evaluation of a kicking technique for sampling stream bottom fauna. *Can. J. Zool.* **49**, 167–73.

FUJITA, H., 1956. The collection efficiency of a plankton net. *Res. Popul. Ecol.* **3**, 8–15.

GERKING, S. D., 1957. A method of sampling the littoral macrofauna and its application. *Ecology,* **38**, 219–26.

GLIME, J. M. and CLEMONS, R. M., 1972. Species diversity of stream insects on *Fontinalis* spp. compared to diversity on artificial substrates. *Ecology* **53**, 458–64.

GOODWIN, M. H. and EYLES, D. E., 1942. Measurements of larval populations of *Anopheles quadrimaculatus*, Say *Ecology* **23**, 376.

GOULDER, R., 1971. Vertical distribution of some ciliated protozoa in two freshwater sediments. *Oikos* **22**, 199–203.

GRIFFITH, R. E., 1957. A portable apparatus for collecting horizontal plankton samples. *Ecology* **38**, 538–40.

GULLIKSEN, B., and DERAS, K. M., 1975. A diveroperated suction sampler for fauna on rocky bottoms. *Oikos* **26**, 246–9.

HAMILTON, A. L., BURTON, W. and FLANNAGAN, J. F., 1970. A multiple corer for sampling profundal benthos. *J. Fish. Res. Bd. Canada* **27**, 1867–9.

HESS, A. D., 1941. New limnological sampling equipment. *Limnol. Soc. Amer. Spec. Publ.* **6**, 1–15.

HIGER, A. L. and KOLIPINSKI, M. C., 1967. Pull-up trap: a quantitative device for sampling shallow-water animals. *Ecology* **48**, 1008–9.

HOLOPAINEN, I. J. and SARVALA, J., 1975. Efficiencies of two corers in sampling soft-bottom invertebrates. *Ann. Zoo. Fenn* **12**, 280–4.

HUCKLEY, P., 1975. An apparatus for subdividing benthos samples. *Oikos* **26**, 92–6.

HYNES, H. B. N., 1971. Benthos of flowing water. *In* Edmondson, W. I. & Winberg, G. G. (eds.). *I.B.P. Handbook* **17**, 66–80. Blackwells, Oxford and London.

JAMES, H. G. and NICHOLLS, C. F., 1961. A sampling cage for aquatic insects. *Can. Ent.* **93**, 1053–5.

JENKIN, B. M. and MORTIMER, C. H., 1938. Sampling lake deposits. *Nature* **142**, 834.

KAJAK, Z., 1963. Analysis of quantitative benthic methods. *Ekologia Polska A* **11**, 1–56.

KAJAK, Z., 1971. Benthos of standing water. *In* Edmondson, W. T. & Winberg, G. G. (ed.) *I.B.P. Handbook* **17**, 25–65. Blackwells, Oxford and London.

KAJAK, Z., DUSOGE, K. and PREJS, A., 1968. Application of the flotation technique to assessment of absolute numbers of benthos. *Ekologia Polska* **16**, 607–20.

KARLSSON, M., BOHLIN, T. and STENSON, J., 1976. Core sampling and flotation: two methods to reduce costs of a chironomid population study. *Oikos* **27**, 336–8.

KROGER, R. L., 1972. Underestimation of standing crop by the Surber sampler. *Limnol. Oceanogr.* **17**, 475–8.

KROGER, R. L., 1974. Invertebrate drift in the Snake River, Wyoming. *Hydrobiologia* **44**, 369–80.

LISITSYN, A. P. and UDINTSEV, G. B., 1955. New model dredges. [In Russian.] *Trudy vses. gidrobiol. Obsch.* **6**, 217–22.

LIVINGSTON, D. A., 1955. A lightweight piston sampler for lake deposits. *Ecology* **36**, 137–9.

LONGHURST, A. R., 1959. The sampling problem in benthic ecology. *Proc. N. Z. ecol. Soc.* **6**, 8–12.

MACAN, T. T., 1958. Methods of sampling the bottom fauna in stony streams. *Mitt. int. Verein. theor. angew. Limnol.* **8**, 1–21.

MACKEY, A. P., 1972. An air-lift for sampling freshwater benthos. *Oikos* **23**, 413–5.

MCCAULEY, V. J. E., 1975. Two new quantitative samplers for aquatic phytomacrofauna. *Hydrobiologia* **47**, 81–9.

MASON, W. T., WEBER, C. I., LEWIS, P. A. and JULIAN, E. C., 1973. Factors affecting the performance of basket and multiplate macro-invertebrate samplers. *Freshwater Biol.* **3**, 409–36.

MILBRINK, G., 1971. A simplified tube bottom sampler. *Oikos* **22**, 260–3.

MILBRINK, G. and WIEDERHOLM, T., 1973. Sampling efficiency of four types of mud bottom sampler. *Oikos* **24**, 479–82.

MOON, H. P., 1935. Methods and apparatus suitable for an investigation of the littoral region of Oligotrophic Lakes, *Int. Revue ges.Hydrobiol. Hydrogr.* **32**, 319–3.

MORTIMER, C. H., 1942. The exchange of dissolved substances between mud and water in lakes. III and IV. *J. Ecol.* **30**, 147–201.

MUNDIE, J. H., 1956. A bottom sampler for inclined rock surfaces in lakes. *J. Anim. Ecol.* **25**, 429–32.

NEEDHAM, P. R. and USINGER, R. L., 1956. Variability in macrofauna of a single riffle in Prosser creek, California, as indicated by the Surber sampler. *Hilgardia* **24**(14), 383–409.

NEGUS, C. L., 1966. A quantitative study of growth and production of unionid mussels in the River Thames at Reading. *J. Anim. Ecol.* **35**, 513–32.

NEILL, R. M., 1938. The food and feeding of the brown trout (*Salmo trutta* L.) in relation to the organic environment. *Trans. R. Soc. Edinb.* **59**, 481–520.

PEARSON, R. G., LITTERICK, M. R. and JONES, N. V., 1973. An air-lift for quantitative sampling of the benthos. *Freshwater Biol.* **3**, 309–15.

PETERSEN, C. G. J., 1911. Valuation of the sea. I *Rep. Dan. biol. Stn* **20**, 1–76.

REISH, D. J., 1959. Modification of the Hayward orange peel bucket for bottom sampling. *Ecology* **40**, 502–3.

RICKER, W. E., 1938. On adequate quantitative sampling of the pelagic net plankton of a lake. *J. Fish. Res. Bd Can.* **4**, 19–32.

ROWLEY, J. R. and DAHL, A., 1956. Modifications in design and use of the Livingstone piston sampler. *Ecology* **37**, 849–51.

SCOTT, D., 1958. Ecological studies on the Trichoptera of the River Dean, Cheshire. *Arch. Hydrobiol.* **54**, 340–92.

SCOTT, D. and RUSHFORTH, J. M., 1959. Cover on river bottoms. *Nature* **183**, 836–7.

SHAPIRO, J., 1958. The core-freezer—a new sampler for lake sediments. *Ecology* **39**, 758.

SMITH, W. and MCINTYRE, A. D., 1954. A spring-loaded bottom-sampler. *J. mar. biol. Ass. U.K.* **33**, 257–64.

SURBER, E. W., 1936. Rainbow trout and bottom fauna production in one mile of stream. *Trans. Am. Fish. Soc.* **66**, 193–202.

TONOLLI, V., 1971. Methods of collection. Zooplankton. *In* Edmondson. W.T. & Winberg, G. G. (eds.) *I.B.P. Handbook* **17**, 1–20. Blackwells, Oxford and London.

ULFSTRAND, S., 1968. Benthic animal communities in Lapland streams. *Oikos (suppl.)* **10**, 1–120.

USINGER, R. L. and NEEDHAM, P. R., 1954. *A plan for the biological phases of the periodic stream sampling program* (Mimeographed) *Final Rep. to Calif. St. Wat. Pollution Cont. Bd.* 59 pp.

USINGER, R. L. and NEEDHAM, P. R., 1956. A drag-type riffle-bottom sampler. *Progve Fish. Cult.* **18**, 42–44.

VALLENTYNE, J. R., 1955. A modification of the Livingstone piston sampler for lake deposits. *Ecology* **36**, 139–41.

WATERS, T. F. and KNAPP, R. J., 1961. An improved stream bottom fauna sampler. *Trans. Am. Fish. Soe.* **90**, 225–6.

WELCH, H. E. and JAMES, H. G., 1960. The Belleville trap for quantitative samples of mosquito larvae. *Mosquito News* **20**, 23–6.

WELCH, P. S., 1948. *Limnological methods.* 381 pp., McGraw-Hill, New York.

WENE, G. and WICKLIFF, E. L., 1940. 'Basket' method of bottom sampling. *Can. Ent.* **72**, 131–5.

WICKSTEAD, J., 1953. A new apparatus for the collection of bottom plankton. *J. mar. biol. Ass. U.K.* **32**, 347–55.

WIGLEY, R. L., 1967. Comparative efficiencies of van Veen & Smith-McIntyre grab samplers as revealed by motion pictures. *Ecology* **48**, 168–9.

WILDING, J. L., 1940. A new square-foot aquatic sampler. *Limnol. Soc. Am. Spec. Publ.* **4,** 1–4.

WILLIAMS, T. R. and OBENG, L., 1962. A comparison of two methods of estimating changes in *Simulium* larval populations, with a description of a new method. *Ann. trop. Med. Parasit., Liverpool* **56,** 359–61.

WOLCOTT, R. H., 1901. A modification of the Birge collecting net. *Joppl. Microsc. Lab. Meth.* **4,**(8), 1407–9.

WOLFE, L. S. and PETERSON, D. G., 1958. A new method to estimate levels of infestations of black-fly larvae (Diptera: Simulidae) *Can. J. Zool.* **36,** 863–7.

7

Relative Methods of Population Measurement and the Derivation of Absolute Estimates

Most of these relative methods require only comparatively simple equipment and, as they often serve to concentrate the animals, they provide impressive collections of data from situations where few animals will be found by absolute methods. From entirely statistical considerations the plentiful data from relative methods is preferable to the hard-won, often scanty, information from unit area sampling. Most traps will collect specimens continuously, providing a relatively large return for the amount of time spent working with them; i.e. the cost (see p. 20) of the data is low. With all these apparent advantages it is hardly surprising that these methods have been extensively used and developed; there are probably more accounts of their design and use in the literature than references to all the other topics in this book (therefore the list at the end of this chapter is highly selective).

7.1 Factors affecting the size of relative estimates

The biological interpretation of relative population estimates (p. 3) is extremely difficult. Their size is influenced by the majority or all of the following factors:

(1) Changes in actual numbers – population changes.
(2) Changes in the numbers of animals in a particular 'phase'.
(3) Changes in activity following some change in the environment.
(4) The responsiveness of that particular sex and species to the trap stimulus.
(5) Changes in efficiency of the traps or the searching method themselves.

It is clear, therefore, that the estimation of absolute population by relative methods is difficult; what one is really estimating is the proportion of those members of the population that were in the 'phase' to respond to the trap and that did so under the prevailing climatic conditions and the current level of efficiency of the trap. The influence of factors 2–5 on these relative methods must be considered further.

Theoretically it is possible for a trap to catch such a large proportion of the population that it actually reduces the numbers being assessed. The failure of many attempts to control insect populations by light traps (e.g. Stahl, 1954; Stanley & Dominick, 1957) confirm the conclusions of Williams *et al.* (1955) that in general the previous nights' trapping has little effect. However, in isolated communities, e.g. small oceanic islands, or when the trap contains a powerful attractant (Petruska, 1968), effects may sometimes be detected.

7.1.1 The 'phase' of the animal

The susceptibility of an animal to being caught or observed may alter with age, because the behavioural attributes and responses of an animal vary from age to age. Many relative methods rely to some extent on the movement of the insect; insect movements are basically of two types, migratory and trivial (Southwood, 1962), and there is evidence from many species that migratory movements occur mainly early in adult life or between reproductive periods (Johnson, 1960, 1963; Kennedy, 1961). Trivial movements, during which the insect will be especially responsive to bait, may occur mostly in later life. The effects of these phenomena on trap catches had in fact been recorded before they themselves were recognized: Geier (1960) showed with the codling moth that the majority of the females taken in light-traps were in the pre-reproductive phase, whilst bait-trap catches were predominantly mature (i.e. egg-laying) or post-reproductive females. Another example is the fall-off in numbers of *Culicoides* midges caught on sticky traps which occurs before the actual population starts to decline (Nielsen, 1963).The reaction of animals to stimuli, important in many trapping techniques, also varies greatly with phase: this is particularly true of blood-sucking insects (Gillies, 1974).

7.1.2 The activity of the animal

The level of activity of an insect will be governed by its diurnal cycle, some insects flying by day, others at night (Lewis & Taylor, 1965), and the expression of this activity will be conditioned by the prevailing climatic conditions. The separation of changes in trap catch due to climate from those reflecting population change has long exercised entomologists. Williams (1940) approached the problem by taking running-means of the catches; the variation in these running-means (i.e. the long-term variation) reflected population changes and the departures of the actual catch from them reflected the influence of climate. Working with groups of species, such as the larger Lepidoptera, Williams was able to demonstrate both the long-term effects of climate through population change and the short-term effects on activity. When studying the populations of airborne aphids, Johnson (1952) found that the running-means reflected current climatic conditions and the deviations population trends; this was due to the relatively short period of time any given aphid spends in the air.

Thus the running-mean technique is unsafe on biological grounds, being influenced by the relative frequency of population and climatic change; furthermore its use places severe restrictions on the number of degrees of freedom available for the calculation of significance levels. Even if no attempt is made to separate population trend and activity and the regression of the actual catch of a group of animals on temperature is calculated, highly significant results are obtained, thereby emphasizing the role of temperature in the determination of catch size (e.g. Williams, 1940: Southwood, 1960).

Although it is occasionally possible to obtain a significant regression of catch size on temperature for a single species (e.g. Southwood, 1960), in general the relationship does not hold at the species level, and Taylor (1963) has propounded and demonstrated the idea that there are upper and lower thresholds for flight. When the temperature is above the lower threshold (and below the upper threshold) the insect will be flying, when below (or above the upper threshold) it will be inactive. The thresholds may be determined by classifying each trapping period as either 1, one or more insects of the particular species were caught or 0, none were caught. Then the trapping periods are grouped according to the prevailing temperature and the percentage of occasions with flight plotted against temperature (Fig. 7.1). For example if there were twenty-five trapping periods when the temperature was 16°C and one or more specimens were collected on ten occasions, a point would be entered at the 40 % flight occurrence (Fig. 7.1). Thresholds for light and other physical conditions may be determined in the same way. The

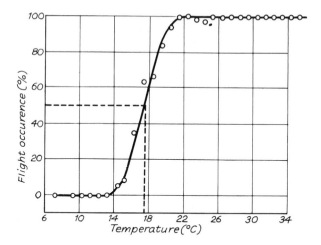

Fig. 7.1 The graphical determination of the flight threshold of a species (after Taylor, 1963).

transition from 0 % flight occurrence to 100 % is not sharp, presumably because of the variation in the individual animals and the microclimate of the sites from which they have flown. Taylor (1963) found that the mouse moth (*Amphipyra tragopoginis*) appeared to have two lower thresholds for temperature; but his data could also be interpreted as showing a steady increase in % flight occurrence to a maximum. Both interpretations are so contrary to the findings with other insects that this and any similar cases should be viewed with suspicion until further data has been obtained.

The influence of activity may therefore be expressed in terms of the various thresholds; once these have been determined fluctuations in numbers between the lower and upper thresholds for flight may be considered as due to other causes.

Taylor (1963) demonstrates elegantly that, when mixed populations of several species are considered, the series of thresholds will lead to an apparent regression of activity on temperature of the type demonstrated by Williams and others, therefore he concludes that regression analysis as a means of interpreting the effect of temperature on insect flight should be limited to multispecific problems. Flight thresholds for temperature and wind velocity have been demonstrated in water beetles, *Helophorus* (Landin, 1968; Landin & Stark, 1973), for temperature and light in aphids (Halgren & Taylor, 1968; Dry & Taylor, 1970).

7.1.3 Variation in the reponsiveness of different sexes and species to trap stimuli

In many groups significantly more of one sex or the other are caught in light-traps (Williams, 1939; Masaki, 1959). Take for example the Miridae; males make up the majority of light-trap catches and there is some evidence that male mirids do engage in significantly more trivial movement, 'flits', than females (Southwood, 1960; Waloff & Bakker, 1963), but the excess of male Miridae is greater in ultra-violet than in tungsten filament light-traps (Southwood, 1960), so that at least a proprtion of this predominance must be due to a selective effect of the traps between the sexes, rather than a real difference in flight activity.

An interesting case of a difference between sexes is the large number of male relative to female tsetse flies usually taken on 'fly rounds' (Glasgow & Duffy, 1961); the biological interpretation of this seems to be that newly emerged female flies usually feed on moving prey and the early pairing, desirable in this species, is achieved by numbers of males following moving bait (Bursell, 1961).

With such marked difference between the sexes it is hardly surprising that different species respond very differently. Eastop (1955) and Heathcote (1957a) have compared the ratios of the numbers of different species of aphids caught in yellow sticky traps to the numbers caught in suction traps, presumably the true population. Eastop, for example, found ratios varying from 31 to 0.5 and even within a genus (*Macrosiphum*) they ranged from 14.7 to 0.8. Analogous observations on the relative numbers of different species of mosquitoes caught in light-traps compared with the results from a rotary trap have been made by Love & Smith (1957), and they found ratios (that they called the 'index of attraction') from over 7 down to 0.24.

7.1.4 The efficiency of the trap or searching method

The efficiency of a method of population estimation is the percentage of the animals actually present that are recorded. The efficiency of a searching method will clearly depend on the skill of the observer and also on the habitat; for example, any observer is likely to see tiger beetles (Cicindellidae) far more easily on lacustrine mud flats than on grass-covered downlands.

Several types of traps – sticky, water and flight – catch insects that are carried into or on to them by the wind and Johnson (1950) and Taylor (1962) have shown that the efficiency of such traps varies with wind speed and the size of the insect (Table 7.1). Thus such traps may give widely different counts for the same aerial population if the wind speeds on the sampling occasions differ, and the results of Juillet (1963) are almost certainly an example of this. Taylor (1962) has, however, shown that it is possible to correct the catches from these traps if the wind speed is known (see p. 229). Light traps catch fewer specimens of most insects on nights when the moon is full. In general this seems to be due to a fall in efficiency of the traps (Williams, Singh & el-Ziady, 1956; Miller, Stryker, Wilkinson & Esah, 1970; Bowden & Church, 1973). The efficiency of baits varies from many causes: the ageing and fermentation of artificial or non-living baits (Kawai & Suenaga, 1960) affecting their 'attractiveness', whilst living baits may differ in unexplained ways – Saunders (1964) found that two apparently identical black zebu oxen trapped different numbers of tsetse flies. The effectiveness of a given bait may vary from habitat to habitat (Starr & Shaw, 1944).

The efficiency of the various traps will be discussed further for each type, but the examples already given are sufficient to indicate that variation in efficiency provides a very real limitation on the value of relative estimates even for comparative purposes.

7.2 The uses of relative methods

It is apparent from the above section that the actual data from relative methods should be used and interpreted with far more caution than has often been shown. Comparisons of different species and different habitats are particularly fraught with dangers.

7.2.1 Measures of the availability

This is the most direct approach; the availability of the population of an animal is the result of the response to the stimuli, the activity and the abundance; that is the product of factors 1–3 and 5 in the list above (p. 223); it may be defined as the ratio of total catch to total effort. Thus, assuming the effciency of the trap or search does not change, the raw data of catch per unit time or effort will provide a measure of availability, or what Heydemann (1961) terms 'Aktivitatsdicht'.

True availability is meaningful in many contexts: it is most easily interpreted with natural 'baits', e.g. the availability of bloodsucking insects to their normal prey (in a bait trap) is a measure of their 'biting level', and the availability of flying insects to colonize or oviposit on a trap host plant is a measure of these parameters. Extrapolations from more artificial situations can only be made in the light of additional biological knowledge. The availability of codling moth females to a light-trap gives a convenient indication of the magnitude and phenology of emergence and oviposition, and this may be used to time the application of control measures. But the peak of availability of the females to the bait traps indicates that the main wave of oviposition has passed (Geier, 1960), and therefore in warm climates, where the eggs will hatch quickly, this information may be too late for effective control measures.

In general, measures of availability may be used for the immediate assessment of the 'attacking' or colonization potential of a population and its phenology. Over a long period of time the changes in the species composition of the catches of the same trap in the same position may be used to indicate changes in the diversity of the fauna (see p. 420). It is only under exceptional circumstances that the trapping seems to affect population size, i.e. when the traps are powerful and the area is localized.

7.2.2 Indices of absolute population

When the efficiency of the trap and the responsiveness of the animal to it can be regarded as constant and if the effects of activity can be corrected for, then the resulting value is an index of the size of the population in that particular phase. The effects of activity due to temperature or other physical factors on the catch size of sticky, flight, pitfall and similar traps may be eliminated by the determination of the thresholds using Taylor's (1963) method (see above). With net catches the regression technique of Hughes (1955) serves the same purpose (p. 242). The animals' diel periodicity cycle may need to be known before these corrections can be made and the index derived. Such an index may be used in place of actual absolute population in damage assessment and in studies on the efficiency of pest control measures. The value of independent estimates of population size has been stressed (p. 4) and a series of such indices may be compared with actual population estimates: if the ratio of one to the other is more or less constant the reliability of the estimates has been confirmed. Indeed these indices may be used in place of absolute population estimates for any comparative purpose, but are of course of no value in life-table construction.

7.2.3 Estimates of absolute population

It is possible to derive estimates of absolute population from what are essentially relative methods by two approaches, each of which subdivides: the correction of the data, by 'calibration' with absolute estimates or by measurement of the efficiency, and its extension to determine density from the frequency of encounters (line transect theory) or from the rate by which trapping reduces the sizes of successive samples (removal trapping).

'CALIBRATION' BY COMPARISON WITH ABSOLUTE ESTIMATES

When a series of indices of absolute population have been obtained simultaneously with estimates of absolute population by another method, the regression of the index on absolute population may be calculated. This can then be used to give estimates of population directly from the indices, but such 'corrections' should only be made under the same conditions that held during the initial series of comparisons.

CORRECTING THE CATCH TO ALLOW FOR VARIATIONS IN TRAP EFFICIENCY

This approach is really a refinement of that above; trap catches and absolute estimates are made simultaneously under a variety of conditions that are known to affect trap efficiency. A table of the correction terms for each condition can then be drawn up and will give much greater precision than a regression coefficient for all conditions (as in 1 above). A valuable approach along these lines has been made by Taylor (1962), who used the data from Johnson's (1950) comparison of sticky trap and tow-net with a suction trap to determine the efficiency of the two former. The efficiency of the suction trap had already been determined by him (see p. 135) and this gave a direct measure of the absolute density of the aerial population. Taylor was therefore able to construct a table of correction terms (Table 7.1); they are of course only applicable to the particular traps Johnson used and it is important to note that the sticky trap was white, not yellow. Taylor also computed the expected fall in efficiency of the traps allowing for the different volumes of air passing at different winds and the effect of this on the impaction of the insects on the sticky trap, assuming that they behave as inert particles; he found that a large part of the fall in efficiency at low wind speeds could be accounted for by these factors.

Table 7.1 The efficiency of a cylindrical sticky trap for different insects at different wind speeds (after Taylor, 1962)

| | Wind speed (m/h) | | | | |
	1	2	6	10	20
Small aphid (*Jacksonia*)	20	45	64	66	68
Small fly (*Drosophila*)	35	58	72	73	74
House fly (*Musca*)	45	64	76	76	76
Bumble bee (*Bombus*)	52	68	78	78	78

LINE TRANSECTS

The basic method of the line transect is that an observer walks at a constant speed through a habitat and records the number of animals he sees. This number will be a reflection of the density of the animals, their speed of movement, the distance over which the observer perceives them and the observer's speed of movement. If the other factors are known the density can be arrived at. The method is developed in detail later (p. 237); it may perhaps be regarded as the mobile analogue of the nearest neighbour and other spacing methods (p. 47).

REMOVAL TRAPPING OR COLLECTING

The principle of removal trapping or collecting is that a known number of animals are removed from habitat on each trapping occasion, thus affecting subsequent catches. The rate at which trap catches fall off will be directly related to the size of the total population (unknown) and the number removed (known). Although Le Pelley (1935) first demonstrated that this fall of catch is geometric in observations on the hand-picking of the coffee bug, *Antestiopsis*, the method has been developed by mammalologists, starting in 1914 (Seber, 1973). They demonstrated the theory underlying this approach, which, like that of the line transect, is borrowed from physical chemistry and concerns the chances of collision of gas molecules.

For the application of the method the following conditions must be satisfied (Moran, 1951):

(1) The catching or trapping procedure must not lower (or increase) the probability of an animal being caught. For example, the method will not be applicable if the insects are being caught by the sweep net and after the first collection the insects drop from the tops of the vegetation and remain around the bases of the plants, or if the animals are being searched for and, as is likely, the most conspicuous ones are removed first (Kono, 1953).

(2) The population must remain stable during the trapping or catching period; there must not be any significant natality, mortality (other than by the trapping) or migration. The experimental procedure must not disturb the animals so that they flee from the area. As Glasgow (1953) has shown, if the trapping is extended over a period of time, immigration is likely to become progressively more significant as the population falls.

(3) The population must not be so large that the catching of one member interferes with the catching of another. This is seldom likely to be a problem with insects where each trap can take many individuals, but may be a significance in vertebrate populations where one trap can only catch one animal.

(4) The chance of being caught must be equal for all animals. This is the most serious limitation in practice. Some individuals of a population, perhaps those of a certain age, may never visit the tops of the vegetation and so will not

be exposed to collection by a sweep net. In vertebrates 'trap-shyness' may be exhibited by part of the population.

Zippin (1956, 1958) has considered some of the specific effects of failures in the above assumptions. If the probability of capture falls off with time the population will be underestimated, but if the animals become progressively more susceptible to capture the population will be overestimated. Changes in susceptibility to capture will arise not only from the effect of the experiment on the animal, but also from changes in behaviour associated with weather conditions or a diel periodicity cycle.

The practice of the method is not dissimilar to that of Kelker's selective removal method when both sexes are hunted (p. 114); the approaches differ mainly in the method of analysis: in Kelker's method the total population is estimated from changes in the ratios of the different constituents of the population.

There are three different approaches to the analysis of removal trapping data. The simplest is the regression method (Fig. 7.2). The number caught on the ith occasion are plotted against the previous total catch and the line may be fitted by eye (Hayne, 1949; Menhinick 1963) or the regression line calculated (De Lury, 1947; Zippin, 1956; Kemp-Turnbull, 1960; Wada, 1962).

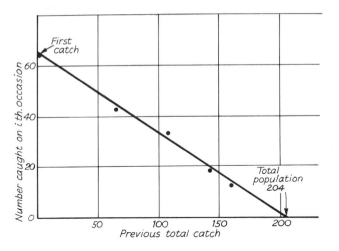

Fig. 7.2 The estimation of population by removal trapping – the fitting of the regression line by eye.

Fitting by eye is acceptable only when the points lie fairly close to a straight line. Fitting by the regression equation is not really acceptable as the two values are not independent.

Working with the same basic approach, but apparently unaware of earlier work, Emsley (1957) estimated the total population by plotting accumulated catch against the trapping occasion; this describes a curve the asymptote of which indicates the level of the total population. As Menhinick (1963) points out, this will be something of an underestimate. Emsley, however, found that for a certain species in a given habitat the percentage of the total population obtained in the first catch was more or less constant, so that this single catch (actually made with a sweep net) could be taken as a fixed percentage of the total population. This is the calibration of a relative method (see above).

The second method, also simple but approximate, is that of Kono (1953) and was referred to by him as 'time–unity' collecting. He postulated that the exponential relationship between the number collected and time, which is the basis of all methods, may be discovered by the consideration of the catches at just three time points (t_1, t_2 and t_3), such that $\frac{1}{2}(t_1 + t_2) = t_3$. Under these conditions:

$$\hat{N} = \frac{n_3{}^2 - n_1 n_2}{2n_3 - (n_1 + n_2)} \tag{7.1}$$

where n_1, n_2 and n_3 = the accumulated catches at times t_1, t_2 and t_3 as defined above, Kono showed that it was important that the collectors are familiar with the animal before the start, so that the collecting efficiency does not improve during the estimation and that the greater the value of t_2, i.e. the closer n_2 approximates towards N, the more accurate the estimator $\hat{N}$.

Taking our previous example (Fig. 7.2) with successive catches of 65, 43, 34, 18 and 12 at equal time intervals, then the accumulated catches are 65, 108, 142, 160 and 172. If $t_1 = 1$ and $t_2 = 5$, then $t_3 = (1 + 5)/2 = 3$, hence $n_1 = 65$, $n_2 = 172$ and $n_3 = 142$ and therefore:

$$\hat{N} = \frac{142^2 - 65 \times 172}{2 \times 142 - (65 + 172)} = 191$$

The third and most accurate method, which also provides an estimate of the standard error, is that based on maximum likelihood. First developed by Moran (1951), its application has been considerably simplified by Zippin (1956, 1958). Modifications for procedures involving at the most three samples, but containing a large proportion of the population – as occurs in electric fishing – have been considered by Seber & Le Cren (1967) and Seber & Whale (1970). These methods might be appropriate for use in connection with, for example, electrical or chemical methods for extraction of soil invertebrates (p. 193).

Zippin's procedure will be illustrated with the previous example. The total catch T is calculated

$$T = 65 + 43 + 34 + 18 + 12 = 172$$

Then the value of $\sum_{i=1}^{k}(i-1)y$ is found, where k = the number of occasions (unrelated to 'k' elsewhere in this book) and y_i = the catch on the ith occasion.

Therefore

$$\sum_{i=1}^{k} (i-1)y_i = (1-1)65 + (2-1)43 + (3-1)34 + (4-1)18 + (5-1)12$$

$$= 0 + 43 + 68 + 54 + 48$$
$$= 213$$

Next the ratio R is determined:

$$R = \frac{\sum_{i=1}^{k} (i-1)y_i}{T} \tag{7.2}$$

Therefore

$$R = \frac{213}{172} = 1.238$$

Now

$$R = \frac{q}{p} - \frac{kq^k}{(1-q^k)} \tag{7.3}$$

where p = the probability of capture on a single occasion and

$$q = 1 - p$$

and the estimate of the total population is given by the equation:

$$\hat{N} = \frac{T}{(1-q^k)} \tag{7.4}$$

The mathematics of these last steps may be circumvented for $k = 3, 4, 5$ or 7 by the use of Zippin's charts (Figs. 7.3 and 7.4).

Therefore in the present example for $k = 5$ and $R = 1.24$ the value of $(1 - q^k)$ is read off Fig. 7.3 as 0.85, so that:

$$\hat{N} = 172 \div 0.85 = 202$$

The standard error of $\hat{N}$ is given by

$$\text{S.E. of } \hat{N} = \sqrt{\frac{\hat{N}(\hat{N}-T)T}{T^2 - \hat{N}(\hat{N}-T)[(kp)^2/(1-p)]}} \tag{7.5}$$

where the notation is as above and p is read from Fig. 7.4
In our example

$$\text{S.E. of } \hat{N} = \sqrt{\frac{202(202-172)172}{172^2 - 202(202-172)[(5 \times 0.33)^2/(1-0.33)]}}$$

$$= 14.46$$

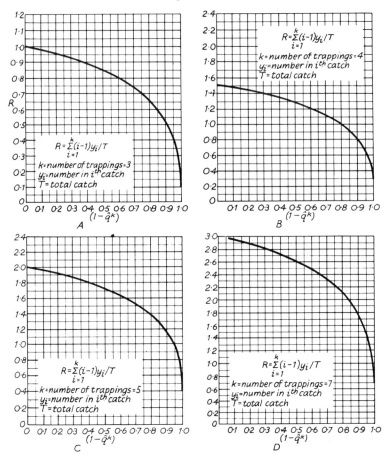

Fig. 7.3. Graphs for the estimation of $(1 - q^k)$ from ratio R in removal trapping (after Zippin, 1956).

Therefore the 95% confidence limits of the estimate are:

$$202 \pm 2 \times 14.46$$
$$= 202 \pm 28.9$$

It will be noted that both the $\hat{N}$ obtained in Fig. 7.2 by the visual fitting of the line and that obtained by Kono's method lie within these limits.

It has been shown by Zippin (1956, 1958) that a comparatively large proportion of the population must be caught to obtain reasonably precise estimates. His conclusions are presented in Table 7.2, from which it may be seen that, to obtain a coefficient of variation (C.V. = Estimate/Standard error × 100) of 30%, more than half the animals would have to be removed from a population of less than 200. For this reason Turner (1962) found it

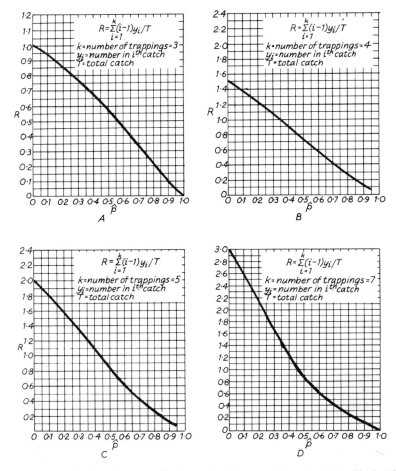

Fig. 7.4. Graphs for the estimation of p from ratio R in removal trapping (after Zippin, 1956).

Table 7.2 Proportion of total population required to be trapped for specified coefficient of variation of $\hat{N}$ (after Zippin, 1966)

	Coefficient of variation			
	30%	20%	10%	5%
	Proportion (to nearest .05) of population to be captured (in 100 or fewer trappings)			
200	.55	.60	.75	.90
300	.50	.60	.75	.85
500	.45	.55	.70	.80
1,000	.40	.45	.60	.75
10,000	.20	.25	.35	.50
100,00	.10	.15	.20	.30

impractical for estimating populations of insects caught in pitfall traps; the proportion of the population caught was too low. It is clearly desirable that, where this approach is used in life-table studies, the method of catching does not involve killing the animals, so that they may be kept captive and then released at the end of the estimation.

7.2.4 Collecting

Relative methods may be used as collecting methods, e.g. for animals for mark and recapture and for age determination in the construction of time-specific life-tables, provided the chances of all age and sex groups being captured are equal (e.g. Davies *et al.*, 1971, who found that parous rates in mosquitoes collected in light traps and at baits were only slightly different). They may also be used to determine the proportion of forms in a population (e.g. Bishop, 1972).

7.3 Relative methods – catch per unit effort

Methods grouped here are those in which the movement or action that results in the capture or observation of the insect is made by the observer.

7.3.1 Visual observation

This is the simplest approach; the observer collects or counts *in situ* all the animals he can see in a fixed time or area. Because the efficiency of search is bound to differ in different habitats, contrary to the views of Mann (1955), fixed-time collecting is not very satisfactory for the comparison of faunas of different habitats, although it may be the only method available (see also van der Drift, 1951). Within a given habitat it may give estimates that approach the measurement of absolute population and has been used widely, for example by Macan (1958) in stony streams, by Barnes & Barnes (1954) and Duffey (1968) in studies on spiders and by Murray (1963) for mosquitoes; the last named author found this method more reliable than light-traps. Hughes (1977) used it to give an 'abundance index' for the bushfly, *Musca*, but corrected for the effects of temperature on activity: the index being catch/min $X°C$ above the threshold ($12.5°C$ for this species).

The aspirator or 'pooter' (Fig. 7.5) is convenient for the rapid collection of small insects in fixed-time estimations (e.g. Jepson & Southwood, 1958), although now that so much attention is given to possible health risks, the slight danger to the operator from inhaling small particles has to be recognised. To overcome these, which will only arise in exceptional circumstances, Evans (1975) has devised two models that work on a 'Venturi' principle, the operator blowing. The one figured (Fig. 7.5*b*) is emptied by carefully removing the top cork and the gauze filter; this operation and the re-alignment of the tubes before re-use must be carried out carefully.

Searching a fixed area, if completely efficient, provides an absolute estimate

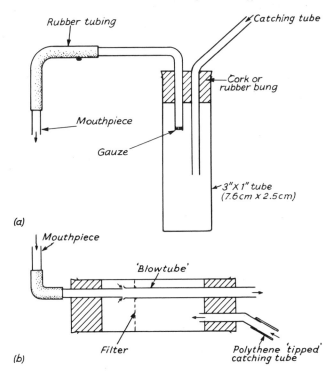

*Fig. 7.5 The aspirator or 'pooter': *a.* simple model *b.* a blow-model (after Evans, 1975). (Mouthpiece 9 mm external dia. tube; 'blow tube' 11 mm external dia. tube; filter (in 3 cm dia. tube) of fine copper gauze held in position with 'Araldite' or similar adhesive; catching tube 9 mm external dia.).

by a direct count (p. 140). However, methods that seem promising in this respect, for example Yeo & Foster's (1958) for the estimation of the coreid, *Pseudotheraptus wayi*, on coconut palms, may prove to be unreliable owing to variations in availability under different weather conditions and at different times of the day (Vanderplank, 1960). But the search is seldom as biased as the trap and sampling by searching the resting sites has provided valuable information in tsetse fly studies (Isherwood, 1957; Glasgow & Duffy, 1961).

7.3.2 Population estimates from line transects
If the observer moves through the habitat the number of animals he sees will clearly be related to their density. Basically there are two different models. The static model that assumes that the animals are not moving: the observer may 'sight' them or actually disturb them causing them to move – 'flushing'. The dynamic model assumes that both the observer and the animal are moving.

The simplest approach to estimation from flushing is to record the number disturbed from a known width of habitat. If the proportion disturbed is constant then the numbers themselves give an index of absolute population and, incorporating the proportion and the area covered, an estimate of absolute population density. If the efficiency of flushing varies only a relative measure of availability is given.

This method has been used for the assessment of populations of the red locust (*Nomadacris septemfasciata*) in its outbreak areas. The observer moved across the area in a motor vehicle (Land Rover) and counted the insects disturbed ahead (Scheepers & Gunn, 1958). Although the response of a locust to flushing depends on its condition (Nickerson, 1963), this method appears to have a fairly constant efficiency, about 75 % of the locusts within the path of the vehicle rising (Symmons, Dean & Stortenbeker, 1963). In an attempt to cover a larger area more readily, flushing by a low-flying aircraft was tried, but the efficiency of flushing, even when increased by spraying a noxious chemical, was low in sparse populations; it was suggested that a siren might be used to disturb a higher proportion of locusts (Symmons *et al.*, 1963).

A more sophisticated approach to the static model has been developed in studies on bird populations. More than six different estimators have been developed (Gates, Marshall & Olson, 1968; Gates, 1969), but an unbiased estimator with minimal variance is considered to be that of Gates (1969) (Kovner & Patil, 1974), although some others may be more robust. This estimator is:

$$\hat{D} = [2n - 1]2L\bar{r} \tag{7.6}$$

where n = number of animals sighted, L = length of the line transect and $\bar{r}$ = average radial distance over which the observer encountered animals. The assumptions are made that encounter distances (r) follow a negative exponential distribution and that the flushing of one animal will not effect another. The effect of varying the two latter assumptions has been explored by Sen, Tourigny & Smith (1974) with several sets of data for birds, and they concluded that, although r often followed a negative exponential function, a Pearson type III distribution was more general. They also concluded that the flushing of one animal was not independent of the flushing of another, and the best fit was obtained if cluster size was considered to follow a Poisson distribution (i.e. the distribution of 'flushings' would fit the negative binomial). The estimation of each r can, of course, be a major source of error.

The dynamic model is the basis of Yapp's (1956) estimator:

$$\hat{D} = Z/2rV \tag{7.7}$$

where Z = number of encounters/unit time, i.e. n/t, V = average velocity of the organism relative to the observer, which is given by:

$$V^2 = \bar{u}^2 + \bar{w}^2 \tag{7.8}$$

where $\bar{u}$ = average velocity of the observer and $\bar{w}$ = average velocity of the animal.

This estimator is based on the kinetic theory of gases and assumes that the observer walks in a straight line at a constant speed, the animals move at random at a constant speed and there is a fixed and known radius r within which the observer can recognize the animals.

The applicability of this approach has been investigated by Skellam (1958), who concluded that equation 7.7 was valid, but that the derivation of V contained an approximation the effect of which would be most serious when the speed of the observer and the animal were equal and almost negligible when the two were very dissimilar; Skellam also showed that equation 7.8 could be regarded as a Poisson variable provided the animals did not move back on their tracks or move in groups; under these conditions the variance would equal the mean. Aggregation will increase the variance. The major practical difficulty in the application of this method – for the estimation of populations of butterflies, dragonflies, birds and other animals for which it would seem to be appropriate – is the determination of the average speed of the animal. Care must of course be taken with the units: if Z is expressed as the number of encounters/hour and r in metres, the speed, must be expressed in metres/hour.

7.3.3 Observation by radar

Microwave radar techniques have been found to provide powerful tools for the study of insect movement in the upper air, both in daylight and after dark (Riley 1974, 1975; Schaefer, 1976). This is still a relatively complex and expensive technique for entomological work, but it would appear to have great potential. It is basically an absolute method, but specific identification for the radar cross section (the radar 'signature') is not possible; when, as with locusts, aerial flights consist of virtually a single species, 'echoes' of the appropriate size may reasonably be assumed to represent that species. Types of insects may be distinguished (Riley, 1975; Schaefer, 1976) and occasionally the swarms may be sampled by aeroplane and actual specimens obtained.

7.3.4 Aural detection

Strictly speaking, sounds are 'products of animals' and this paragraph should come in the next chapter; however, this method has mostly been used with birds and the approach is similar to line transect estimation. A simple estimator, due to W. H. Petrabough, is:

$$\hat{D} = \bar{h}/L\pi r_a^2 \tag{7.9}$$

where $\bar{h}$ = average number of songs heard per stop, L = the total number of listening stops made and r_a = the radius of audibility. The estimation of r_a, like the corresponding terms in line transects, presents a major source of error. A further problem is that birds and insects have a definite periodicity of calling; Gates & Smith (1972) developed a time-dependent model to allow for this.

However, with many insects one 'caller' stimulates others, so that the probability of hearing a song might be better described by a skewed rather than a Poisson distribution with time. This method might be used with some insects (male cicadas, acrids) as second or third method to compare with capture-recapture or other estimates.

7.3.5 Exposure by plough

The surveying of large but aggregated soil animals by conventional methods (Chapter 5) is often extremely laborious. Roberts & Smith (1972) found that a rapid survey of scarabaeid larvae could be made by ploughing a transect across grassland and counting those insects exposed in the furrow or on the overturned turf. Such transects would also detect patterns of aggregation; in general there was a linear relationship between numbers per unit plough transect and absolute population as determined from core samples. Relatively more are, however, exposed in dense populations, because they cause the soil to break up more.

7.3.6 Collecting with a net or similar device

A number of approaches are included under this heading: in aquatic habitats dipping with a net or strainer is a widely practised relative method (Shemanchuk, 1959; Zimmerman, 1960); animals on terrestrial vegetation may be collected with a sweep net or, for those on trees, a beating tray. The latter approaches an absolute method and is discussed in Chapter 4 (p. 145). Anderson & Poorbaugh (1964) have obtained indices of the populations of dung-frequenting Diptera by collecting over a known area with a Dietrick suction sampler (p. 148).

Aerial insects may be collected by random strokes through the air with a light net (Parker, 1949; Linsley *et al.*, 1952; Nielsen, 1963) or with net or gauze cones on a car (Stage *et al.*, 1952; Almand, Stirling and Green, 1974), ship (Yoshimoto & Gressitt, 1959; Yoshimoto *et al.*, 1962), or an aeroplane (Glick, 1939; Odintsov, 1960; Gressitt *et al.*, 1961); because of the impedance to airflow due to the gauze, which becomes more severe at higher speeds, it is difficult – but not impossible – to obtain direct measure of aerial density from such sampling cones.

The sweep net is perhaps the most widely used piece of equipment for sampling insects from vegetation; its advantages are its simplicity and speed – a high return for a small cost – and it will collect comparatively sparsely dispersed species. However, only those individuals on the top of the vegetation that do not fall off or fly away on the approach of the collector are caught. The influence of these behavioural patterns on the efficiency of the method has been investigated by the comparison of sweep net samples with those from cylinders or suction apparatus (p. 148) (Beall, 1935; Romney, 1945; Johnson, Southwood & Entwistle, 1957; Race, 1960; Heikinheimo & Raatikainen, 1962; Rudd & Jensen, 1977), sweep net samples with capture–recapture estimates (Nakamura *et al.* 1967), or by a long series of sweeps in the same habitat (Hughes, 1955; Fewkes, 1961; Saugstad, Bram & Nyguist, 1967). Changes in efficiency may be due to:

(1) different habitat or changes in the habitat
(2) different species
(3) changes in the vertical distribution of the species being studied
(4) variation in the weather conditions
(5) the influence of the diel cycle of vertical movements

A sweep net cannot be used on very short vegetation. Once plants become more than about 30 cm tall further increases in height mean that the net will be sampling progressively smaller proportions of any insect whose vertical distribution is more or less random.

Even related species may differ in their availability for sweeping (Johnson *et al.*, 1957; Heikinheimo & Raatikainen, 1962) and therefore the method is unsuitable for synecological work; these same workers also showed with a nabid bug and a leafhopper that the vertical distribution of the various larval instars and the adults differed (Fig. 7.6). Within the adult stage the vertical distribution in vegetation may alter with age; ovipositing females of the grass mirid, *Leptopterna dolabrata*, are relatively unavailable to the sweep net; the

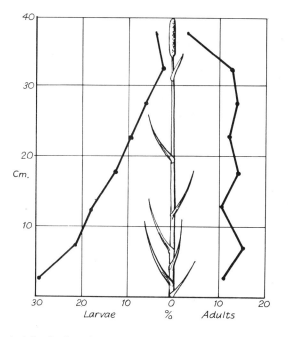

Fig. 7.6 The vertical distribution of larval instars IV and V and adults of *Javesella pellucida* on Timothy grass (after Heikinheimo & Raatikainen, 1962).

same appears to be true of mature healthy males and females of the leafhopper, *Javesella* (= *Calligypona*) *pellucida*, in cereals, although parasitized individuals do not change their behaviour with age and so the percentage of parasitism would be overestimated by sweeping (Heikinheimo & Raatikainen, 1962). (see also p. 311)

Weather factors profoundly affect the vertical distribution and hence the availability of insects to the sweep net. These have been studied by Romney (1945), Hughes (1955) and Saugsted *et al.* (1967); Hughes found with the chloropid fly, *Meromyza*, that the regressions of numbers swept against the various factors gave the following coefficients: wind speed − 0.1774, time since saturation − 0.0815, air temperature + 0.0048, radiation intensity + 0.0010 and radiation penetration − 0.0367. Thus the major influences on the catches of this fly are the first two factors listed, maximum efficiency being achieved at low wind speeds and immediately after a shower. Many arthropods have a diel periodicity of movement up and down the grass, the maximum numbers being on the upper parts a few hours after sunset (Romney, 1945; Fewkes, 1961; Benedek, Erdelyi & Fesus, 1972).

The bags of sweep nets are usually made of linen, thick cotton or some synthetic fibre; the mouth is most often round; a square-mouthed net does not give more consistent results (Beall, 1935), but a D-shaped mouth is useful for collecting from short vegetation, especially young crops. There may be considerable variation in the efficiency of different collectors; usually the more rapidly the net is moved through the vegetation the larger the catch (Balogh & Loksa, 1956).

The number of sweeps necessary to obtain a mean that is within 25 % of the true value has been investigated by Gray & Treloar (1933) for collections from lucerne (alfalfa), and found to vary from taxon to taxon, with an average of 26 units each of 25 sweeps. These authors suggested that this high level of variability was due to heterogeneity in the insect's spatial distribution and much of this may have been attributable to the diurnal periodicity demonstrated by Fewkes (1961). In contrast Luczak & Wierzbowska (1959) considered 10 units of 25 sweeps adequate for grassland spiders and Banks & Brown (1962) found that sweep net catches of the shield bug, *Eurygaster integriceps*, on wheat had sampling errors of only 10 % and reflected absolute population differences determined by other methods.

7.4 Relative methods – trapping

The methods described in this section, in contrast to the last, are those in which it is the animal rather than the observer that makes the action that leads to its enumeration. Basically traps may be divided into those that catch animals randomly and those that attract them in some way. (The word 'attract' is used in its widest sense without any connotation of desire: the studies of Verheijen (1960), for example, have shown that animals are trapped by artificial light through interference with the normal photic orientation and not strictly because of attraction.) It is important to disinguish between these two types as those that are based on attraction allow the possibility of a further source of error. But a strict division is impossible as some traps, more particularly water and sticky traps, are intermediate in position.

7.4.1 Interception traps

These are traps that intercept the animals, more or less randomly as they move through the habitat: air, water or land. Indices of absolute population may theoretically be obtained more easily from this type of trap than from others, as there is no variation due to attraction.

AIR – FLIGHT TRAPS

Here we are concerned with stationary flight traps that are not believed to attract the insects; those that have some measure of attraction are discussed below (p. 249); others described elsewhere are moving (rotary) nets which may give absolute samples (p. 134) or, if the quantity of air they filter is uncertain, they are regarded as aerial sweeps (p. 240).

One of the simplest type is the suspended cone net as used by Johnson (1950); this may have a wind vane attachment to ensure that it swivels around to face into the wind; Taylor (1962) has shown that its efficiency at different wind speeds may be calculated (p. 290). Nets are particularly useful for weak flyers such as aphids (Davis & Landis, 1949) or at heights or in situations where the wind speed is always fairly high (Gressitt *et al.*, 1960) so that the insects cannot crawl out after capture. Similar traps have been used on moving ships (p. 240).

The Malaise trap is more elaborate; it consists basically of an open-fronted tent of cotton or nylon net, black or green in colour; the 'roof' slopes upwards to the innermost corner at which there is an aperture leading to a trap. It was developed by Malaise (1937) as a collector's tool; modified designs have been described by Gressitt & Gressitt (1962), Townes (1962) and Butler (1965), and a basically similar trap by Leech (1955). Gressitt's (Fig. 7.7) and Butler's models are much simpler to construct and transport than Townes'. Gressitt's is large (7 m long and 3.6 m high). Butler's is smaller, being made from a bed-mosquito-net by cutting out part of one side and a hole in the roof into which the collecting trap, a metal cylinder and a polythene bag, is placed. Townes (1962) gives very full instructions for the construction of his model, which is more durable and traps insects from all directions. When maximum catches are desired (for collecting) Malaise traps should be placed across 'flight paths' such as woodland paths, but in windy situations they cannot be used. The studies of Juillet (1963), who compared a Malaise trap with others including the rotary which is believed to give unbiased catches, suggest that for the larger Hymenoptera and some Diptera this trap is unbiased, but it is unsatisfactory for Coleoptera and Hemiptera. However, Roberts (1970, 1972)

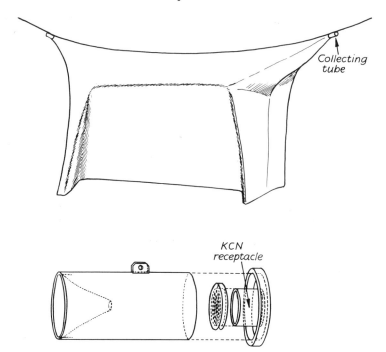

Fig. 7.7 Malaise trap: *a*. sketch of the Gressitt type; *b*. plastic collecting tube (after Gressitt & Gressitt, 1962).

found that even for Diptera the form and colour of the trap would greatly influence the catch.

A very large 'net-tent' (Filet-tente) extending in an arc for 37 m was designed by Aubert (1969) for trapping migratory insects in mountainous areas. An 'electrified net' has been used, sometimes with bait, for tsetse flies (Rogers & Smith, 1977).

Flying Coleoptera and other insects that fall on hitting an obstacle during flight may be sampled with a window trap, which is basically a large sheet of glass held vertically with a trough, containing water with a wetting agent and a little preservative below it (Chapman & Kinghorn, 1955; Van Huizen, 1977)(Fig. 7.8). Living insects may be collected by replacing the trough with opaque cylinders whose ends have transparent collecting tubes: insects move towards the light and accumulate in the tubes (Nijholt & Chapman, 1968.)

One of the most important uses of stationary flight traps is to determine the direction of flight (see p. 342). Nielsen (1960) has observed that migratory butterflies will enter stationary nets and remain trapped, whereas those engaging in trivial movement will fly out again. The direction of migration can be determined by a number of stationary flight traps arranged so as to sample from four fixed directions. Stationary nets may be used (Nielsen, 1960) or a

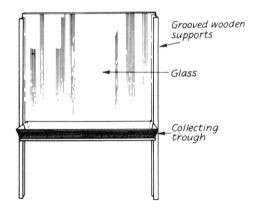

Fig. 7.8 Window trap.

Malaise trap modified to collect the insects from each side separately (Roos, 1957). A more extensive trap is the 'robot observer' which was used in work on blowflies and consisted of a zigzag wall of net, with 10 V-shaped bays each with a 2m square aperture and ending in a fly trap; the catches from the 10 opposing bays on each side are directly comparable (J. Macleod, pers. comm.).

Special traps have been designed to collect and distinguish between the insects entering and leaving the 'homes' of vertebrates, principally burrows and houses (Myers, 1956; Service, 1963; Muirhead-Thomson, 1963). Exit-traps are usually net cones that are inverted above the exit hole; entrance-traps have to be inconspicuous from the outside and a metal gauze cylinder with a dark canvas 'skirt' that could be spread out to prevent passage round it was used in rabbit burrows by Myers, who concluded, however, that the entrance-trap did not sample as randomly as the exit-trap. In houses, ingress and egress may be measured by identical traps, but Service's (1963) results show that in this habitat the former is measured less efficiently. The traps consist of a net cone leading into a net or net and 'Perspex' box; Saliternik (1960) trapped the mosquitoes leaving cesspits with a similar trap. This could be a measure of emergence (see Chapter 9).

WATER – AQUATIC TRAPS

Methods of studying organisms drifting in streams have been reviewed by Waters (1969) and Elliott (1970). The simplest type is based on Waters' (1962) design (Fig. 7.9a), a 1m long 'Nitex' net with a square mouth and a board sunk flush with the stream bed in front of the mouth. Nets of this type may become clogged, but if the mouth is narrowed (Fig. 7.9b) this is alleviated and hence backflow is prevented and filtration increased (Cushing, 1964: Mundie 1964; Anderson, 1967). A collecting vessel may be placed at the end of the net.

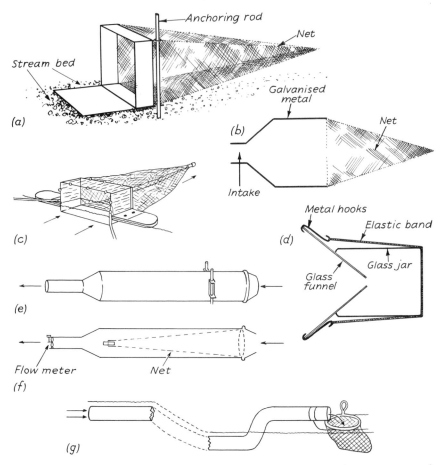

Fig. 7.9 Aquatic interception traps: *a*. Waters type for stream drift animals; *b*. Cushing type trap for stream drift animals; *c*. Elliott floating trap (after Elliott, 1970); *d*. Pieczynski trap for water mites; *e*. and *f*. Müller type tubular trap with flow-meter (after Elliott, 1970) (e-external view; f-section view); *g*. 'Pipe trap' (after Elliott, 1970) (the dotted portion represents a length of pipe (and stream) sufficient for drop in level).

Elliott (1967) devised a floating trap (Fig. 7.9*c*), similar to Waters' but having the advantage for 'drift sampling' that crawling forms were definitely excluded.

If the volume of water passing through a drift sampler is measured it then provides an absolute population estimate, analogous to a suction trap. Elliott (1970) was able to measure the water flow through his floating net using an off-current meter, C1. The flow meter may be incorporated into the sampler (Fig. 7.9*e* and *f*); originally designed by K. Müller, they have been modified by Elliott (1967, 1970) and Cloud and Stewart (1974). A third approach to the measurement of drift is to pipe a portion over a sufficient distance so that it can be brought above the surface and filtered (Fig. 7.9*g*) (Elliott, 1970). Such methods are also particularly appropriate by weirs and falls. Anderson & Lehmkuhl (1968) funnelled the entire stream-flow through their trap. Sampling period is important because drift often shows a diel periodicity (Elliott, 1969), which is influenced by moonlight and other variations in light intensity (Anderson, 1966).

Pieczynski (1961) has found that what is in effect an unbaited glass lobster-pot (Fig. 7.9*d*) will trap numbers of water mites and other free swimming animals. Usinger & Kellen (1955) used a similar trap constructed of plastic screen rather than glass and caught large numbers of aquatic insects; their trap is figured in Usinger (1963; p. 55); a modification with a collecting jar reached through a narrower entrance on the roof has been found particularly useful for water beetles (James & Redner, 1965). These traps are the aquatic equivalents of the pitfall trap.

LAND–PITFALL AND OTHER TRAPS

Like the lobster-pot, the pitfall trap was an adaptation by the ecologist of the technique of the hunter; basically it consists of a glass, plastic or metal container, sunk into the soil so that the mouth is level with the soil surface (e.g. Gist & Crossley, 1973; Luff, 1975). Many cursorial animals fall into the trap and are unable to escape. Pitfall traps may be emptied with a hand-operated or mechanical suction apparatus (p. 148), thereby avoiding the disturbance to the surroundings that would result from continued removal and re-sinking.

Pitfall traps have been used extensively for studies on surface dwellers such as spiders, Collembola, centipedes, ants and beetles, especially Carabidae (e.g. Barber, 1931; Tretzel, 1955; Boyd, 1957; Grüm, 1959; Doane, 1961; Duffey, 1962; Kaczmarek, 1963; Vlijm & Kessler-Geschiere, 1967; Petruska, 1969; Hayes, 1970; Ahearn, 1971; Luff, 1973; Kowalski 1976; Uetz, 1977). More elaborate traps have been designed to facilitate emptying (Rivard, 1962); for use under snow (Steigen, 1973), with rain-guards (Fichter, 1941; Steiner *et al.*, 1963) or timing devices that allow the catch from each time period to be segregated (Williams, 1958). Artificial (Stammer, 1949) or natural (Barber, 1931; Walsh, 1933) baits can be used in pitfall traps and baited traps have been found useful for collecting beetles from the burrows of mammals (Welch, 1964). However, the effects of baits are variable and in many cases, as their attractiveness will change with age, they may well introduce a further source of error, and simple traps are generally to be preferred in population studies (Greenslade & Greenslade, 1971). Preservatives such as formalin or alcohol have sometimes been used in traps, but these may affect the catches of species differentially (Luff, 1968; Greenslade & Greenslade, 1971): aqueous solutions of trisodium phosphate or picric acid might be more suitable (p. 188).

Pitfall traps have many advantages: they are cheap (empty food or drink containers may be used), they are easy and quick to operate and a grid of traps can provide an impressive set of data (e.g. Gist & Crossley, 1973). However, many studies have shown that catch size is influenced by a wide range of factors, apart from population size (Grüm, 1959; Briggs, 1961; Mitchell, 1963; Greenslade, 1964; 1973; Petruska, 1969; Hayes, 1970; Luff, 1975). Pitfall traps must therefore be used with extreme caution, but as Gist & Crossley (1973), Uetz & Unzicker (1976) and others have found they may provide valuable information provided proper attention is paid to potential sources of variation.

An interesting model of the action of pitfall traps has been developed by Jansen & Metz (1977) who consider that the number of animals trapped depends on: (a) their population density, (b) their movement, assumed to be Brownian, (c) the boundary of the pitfall and its 'absorbtiveness', (d) the outer boundary of the area and the extent to which the animals penetrate it (the probability of absorption).

Movement of cursorial arthropods is affected by temperature, moisture and other weather conditions (Grüm, 1959; Mitchell, 1963), food supply (Briggs, 1961), the characters of the habitat (e.g. the amount of impedence by the vegetation) and the age, sex and condition of the individuals (Petruska, 1969; Hayes, 1970). If due account is taken of the thresholds, corrections can be made for weather conditions (p. 225): the other variables need to be kept in view, they may or may not be significant in the particular study.

The catches immediately after a pitfall trap is placed in position are commonly found to be higher than those subsequently achieved; Joosse (1965) and Greenslade (1973) termed these 'digging-in effects'. Jansen & Metz (1977) obtained similar effects from their theoretical model due to depletion of the individuals, moving in a Brownian manner, around the trap. If further studies confirm this, the magnitude and duration of the 'digging-in effect' could, taken with knowledge of daily movement, provide a local estimate of population density.

The boundary of the pitfall trap influences the catch both in respect of its magnitude and its 'capture efficiency' (the probability of absorption). Traps are generally circular and catch size will theoretically be related to the length of the perimeter; if approach, in the region of the trap, is effectively linear then rectangular or any non-reflexed polygonal trap aperture will also catch in proportion to its diameter (Luff, 1975). However Luff's (1975) studies showed that the capture efficiency of trap boundaries varied both with respect to the nature of the trap (e.g. glass jar or tin) and the species of ground beetle. The only generalization that was possible was that smaller aperture traps seemed to have higher efficiencies for the smaller beetles, whilst the larger traps caught a higher proportion of the larger beetles that encountered the trap boundary (aperture).

Gist & Crossley (1973) used exclusion barriers (p. 319) to provide a 'non-absorptive' (in Jansen & Metz' sense) outer boundary, the grid of pitfalls 'trapped out' the area and removal trapping was used to estimate population size (p. 230). Such studies and the 'digging-in effect' suggest that within a discrete habitat, a grid of pitfall traps could influence the population density of a highly mobile animal.

Two other factors may influence the efficiency of pitfall traps: the level of the trap lip and the retaining efficiency. Greenslade (1964) found that trap catches were qualitatively or quantitatively affected by the precise placement of the level of lip of the trap container, whether at the surface of the soil, or the surface of the litter. Retaining efficiency depends on escape; the smoother the side the less likely this is. Glass and most plastic containers are therefore to be preferred to metal traps that quickly corrode, providing footholds. Small winged species may escape from large aperture traps by flight or they may be eaten by larger species: it is, of course, for these reasons that investigators have often used preservative solutions.

Walking animals may also be caught on sticky traps, bands placed round tree trunks (see also p. 308) and glass plates lain on the ground. Mellanby (1962) found that the latter caught numbers of springtails and other animals; he suggested a gauze roof over the trap to prevent flying insects being caught. Details of the design of sticky traps are given below.

7.4.2 Flight traps combining interception and attraction

The distinction between water and sticky traps and those flight traps described above (p. 243) should not be regarded as rigorous for, as Taylor (1962) has shown, a white sticky trap catches insects almost as if they are inert particles, whereas a net may sample selectively. The two approaches may be combined in the sticky net described by Provost (1960). However, many aphids are particularly attracted to yellow (Broadbent, 1948; Moericke, 1950; Hottes, 1951), ceratopogonid midges to black (Hill, 1947); flower-dwelling thrips to white (Lewis, 1959), frit flies to blue (Mayer, 1961) and bark beetles to red (Entwistle, 1963); therefore the quality and magnitude of water and sticky trap catches will be greatly influenced by the colour of the trap and this will be an additional source of variation in the efficiency of these traps. It is probable that the attraction of different colours to a given species may vary with both age and sex.

STICKY TRAPS

These are an extension of the 'fly-paper'; the animal settles or impacts on the adhesive surface and is retained. A variety of adhesives may be utilized; those resins and greases developed for trapping moths ascending fruit trees have proved particularly useful*; castor oil may be satisfactory for minute insects

---

* Some proprietary brands of sticky material that may be used are 'Bird Tanglefoot' (Tanglefoot Co., Grand Rapids, Michigan, US), 'Sticken Special' (Michel & Pelton Co., California, US), 'Boltac' (ICI, Farnham, Surrey, U.K.), 'Corry's Tree Grease' (Corry & Co. Ltd., New bury, Berks, U.K.), 'Tack trap' (Animal Repellents Inc., P. O. Box 999, Griffin, Georgia, US), 'Vaseline' (Chesebrough Ponds Ltd., UK & US) and High vacuum grease (Dow Corning Co., US).

(Dipeolu, Durojaiye & Sellers, 1974). The insects are separated from most fruit-tree banding resins by warming and then scraping the resin plus insects into an organic solvent like trichlorethylene or hot paraffin from which they may be filtered. The separation from the greases is easier; a mixture of benzene and isopropyl alcohol rapidly dissolves the adhesive. Thus, when possible, a grease will be used in preference to a resin, but only weak insects will be trapped by a grease – mosquitoes (Provost, 1960), mites (Staples & Allington, 1959) and aphids (Close, 1959). For these a grease may be more efficient, as the effective area of the trap is not reduced by the numbers of large insects that are also trapped if a powerful adhesive is used (Close, 1959). Greases may become too fluid at high temperatures.

Sticky traps are of several basic designs: Até strands, an African birdlime, were used by Golding (1941, 1946) in warehouses, and Provost (1960) used sticky nets, but most traps have been either large screens or small cylinders, boxes or plates. Large screens consisting of a wooden lattice or series of boards have been used to measure movement at a range of heights of aphids and beetles (Dudley, Searles & Weed, 1928; Moreland, 1954; Prescott & Newton, 1963; Taft & Jernigan, 1964). Individual boards, white and yellow in colour, were used by Roesler (1953) and Wilde (1962) for studies of various flies and psyllids.

Ibbotson (1958) introduced the idea of glass plates for sticky traps. These were 8 in. square, the upper surface coated with the adhesive and the under painted yellow (or as desired). Glass has the advantage that the adhesive is easily scraped off with a knife or the whole plate immersed in a solvent. Ibbotson used his trap for the frit fly; Staples & Allington (1959) used grease-coated microscope slides for trapping mites; Chiang (1973) an oil-coated strip over a white disc for midges; Wakerley (1963) found Ibbotson's trap suitable for the carrot fly, *Psila rosae,* and Maxwell (1965) a similar trap for the apple maggot fly, *Rhagoletis.* These plates may be exposed either vertically or horizontally; the exposure will affect the catch (Table 7.3), (Heathcote 1957a); the influence of colour is less with vertical plates where most insects are caught by wind impaction, i.e. the catches are less biased, but horizontal plates catch more aphids in the landing phase (A'Brook, 1973) and gave high catches of *Adelges piceae* (Lambert & Franklin, 1967). However, fixed vertical plate traps will sample a different proportion of the passing air depending upon the wind direction, and eddies will develop at the sides of the plates and around sticky boxes (Fröhlich, 1956). Small glass plates may be mounted at right angles to a windvane and so present a constant exposure to the wind.

A revolving trap was found most efficient for mosquitoes (Dow & Morris, 1972), but most *Stomoxys* were caught on the leeward side of white sticky traps (Williams, 1973). A good design, sampling at random from the passing air, is the cylinder sticky trap of Broadbent *et al.* (1948) (Fig. 7.10), which consists basically of a piece of plastic material covering a length of stove-pipe. Taylor (1962) showed that if the wind speed was known the catches of small

Table 7.3 The catches of some species of aphid on three types of trap expressed as a ratio of the number caught by a suction trap (after Heathcote, 1957a)

	Water	Cylindrical sticky	Flat sticky
Aphis fabae group	2.28	1.92	0.84
Tuberculoides			
annulatus	3.91	0.45	0.18
Cavariella aegopodii	0.36	2.00	0.27
Myzus persicae	1.34	0.34	0.78
Hyalopterus pruni	0.33	1.50	0.40
Drepanosiphum			
platanoidis	1.29	0.25	0.01
Brevicoryne brassicae	0.47	0.06	0.07
Sitobion spp.	0.08	0.17	0.06

insects on a white sticky trap of this type could be coverted to a measure of aerial density (see Table 7.1, p. 180); however, yellow traps catch more aphids than white ones, and these catch more than black (Broadbent, 1948; Taylor & Palmer, 1972; A'Brook, 1973). Care should therefore be exercised in applying these corrections to a species whose reaction to the colour of the trap is unknown. Colour will presumbly have the greatest effect at low wind speeds.

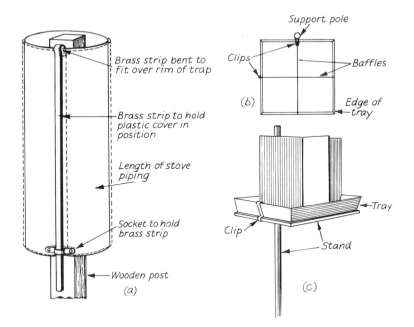

Fig. 7.10 a. Cylindrical sticky trap (after Broadbent *et al.*, 1948). *b.* and *c.* Water trap with baffle (Coon & Rinicks' design): *b* plan from above; *c.* general view.

Cylindrical sticky traps have been utilized in studies on the cacao mealy bug (Cornwell, 1960) and small ones of 1-in (2.5 cm) diameter were used by Lewis (1959) for trapping thrips.

The relation of the size of the sticky trap to the catch has been investigated by Heathcote (1957b) and Staples & Allington (1959), who found that although catch usually increases with size, it is not proportional to size so that the smallest trap catches the largest number per unit area (Table 7.4).

Table 7.4 The effect of trap size on the catch of black cylindrical sticky traps (after Heathcote, 1957b)

| | Trap diameter | | |
	3 cm	6 cm	12 cm
Total catch	837	1473	2017
Catch/unit area (3-cm trap = 1)	837	736.5	504.2

Simpson & Berry (1973) exposed successive portions of a 'Tanglefoot'* coated strip of plastic on the leading edge of a trap, with a vane, so that it always faced into the wind: in this way they were able to get a 24-hour record of wind-borne aphids.

Mention was made above to the attraction of aphids and other insects in the landing phase to particular coloured horizontal plates. Prokopy (1968, 1977) and Kring (1970) have shown with fruit flies (Trypetidae = Tephritidae) that yellow panels and red spheres attract flies in different behavioural conditions: the yellow plate is a 'super leaf', the red sphere a fruit. The use of relative methods sensitive to behavioural phase to partition the population in this way is a powerful approach.

WATER TRAPS

These are simply glass, plastic or metal bowls or trays filled with water to which a small quantity of detergent and a preservative (usually a little formalin) have been added. Omission of the detergent will more than halve the catch (Harper & Story, 1962). The traps may be transparent or painted various colours and placed at any height. Yellow bowls were used by Moericke (1951) and others for trapping aphids and by Fröhlich (1956) and Fritzsche (1956) for a weevil; white trays were used by Southwood *et al.* (1961) for the frit fly; fluorescent yellow trays caught twice as many cabbage-root flies, *Erioischia*, as yellow trays (Finch & Skinner, 1974). A variety of colours were tested by Harper and Story (1962) for the sugar beet fly, *Tetanops myopaeformis*: the total numbers taken in the different coloured traps were as follows: yellow 330, white 264, black 202, red 107, blue 64, green 53.

* see footnote p. 249.

Sticky traps would be chosen in preference to water traps because the relationship between wind speed and the catch of the water trap is likely to prove less simple than that for the sticky trap investigated by Taylor (1962) and because water traps must be frequently attended, or they overflow in heavy rain or dry out in the sun; although a model has been designed with a reservoir for automatically maintaining a constant level (Adlerz, 1971). On the other hand water traps have certain advantages: the insects that are caught are in good condition for identification, as the catch is easily separated by straining or individual insects picked out with a pipette or forceps. Futhermore, when the population is sparse a water trap will, with aphids at least, make catches when a sticky trap of a manageable size would not (Heathcote, 1957a). The 'catching power' of a water trap can be further increased by standing two upright baffles of aluminium sheeting at right angles to each other to form a cross in the tray (Coon & Rinicks, 1962), Fig. 7.10b and c). These divisions enable one to separate the insect according to the quadrant in which it has been captured; this may be related to the direction of flight at the time of capture (see also p. 244). The results of Coon & Rinicks show that the effect of the baffle varies from species to species; in the four cereal aphids that they studied it increased the catch of *Rhopalosiphum maidis* tenfold, nearly doubled the very low catch of *Schizaphis agrostis* (= *Toxoptera graminum*) but had no effect on those of the two others, *Macrosiphum granarium* and *Rhopalosiphum fitchii*.

A floating water trap that collected the insects settling on the surface of ponds was devised by Grigarick (1959a), the animals being drowned through the action of the detergent. The tray of his trap was surrounded by a wooden frame, therefore it would seem that the contrast presented by this surround might bias the catch and a more transparent float (e.g. an air-filled polythene tube) would be preferable.

7.4.3 Light and other visual traps

MODE OF ACTION AND LIMITATIONS

Light-traps are probably the most widely used insect traps and there are several hundred references to them. Originally, paraffin and acetylene lamps were used by collectors (Frost, 1952); later the tungsten filament electric light and, after its development by Robinson & Robinson (1950), the ultraviolet trap became widely used.

The uses of light-traps in ecology are subject to all the general limitations of relative methods outlined above (p. 223); but the variation in efficiency of the trap from insect to insect, from night to night and from site to site is more serious than in almost every other type of trap because light-traps are entirely artificial, relying on the disturbance of normal behaviour for their functioning. It is not justified to claim on *a priori* grounds, as Mulhern (1953) has done, that, being mechanical, light-traps are more reliable than hand collection per unit of habitat. Furthermore, in the discussion of the pros and cons of the many models of light-trap it is often implied that the bigger the catch, the 'better' the trap: although it is true that the larger figures are often more acceptable for statistical analysis it is unwise to assume that they are biologically more valuable.

The exact mechanisms that lead to an insects' capture by a light trap are far from clear. Mikkola (1972) has shown that electroretinograms evoked by different types of light were at variance to the actual responses of the moths to traps. It seems reasonable to suppose that there is an area, sometimes called the radius of the trap or 'catchment area', within which the insect comes into the influence of the light. The brighter the trap the greater this area and hence when there is bright moonlight the amount of contrast between the trap and its surroundings will be lessened, and the catchment area reduced (Williams *et al.* 1956; Miller *et al.* 1970). Bowden & Church (1973) showed that the extent of the reduction was consistent with the random, rather than fixed, capture of the insect within the catchment area. The behavioural studies of Verheijen (1960) and Mikkola (1972) suggest that the result may appear random because even if there are zones of attraction, corresponding perhaps to the insect seeking open flight routes through the vegetation, there are probably 'annuli of repulsion' and 'zones of dazzle' when the normal photic orientation is disturbed.

In a series of studies, Bowden has been concerned with the changes in the catchment area ('trap radius') with time – the lunar cycle; he has shown how the precise distribution of moonlight throughout the night needs to be considered in relation to the insects periodicity of flight (Bowden, 1973; Bowden & Church, 1973). Using this information, standardized catches may be calculated, these eliminate the variations due to the changes in the catchment area brought about by the effect of moonlight (Bowden & Morris, 1975). When this is done it is found that there are some insects that are more active on nights with full moon (Bidlingmayer, 1964; Anderson, 1966; Bowden & Gibbs, 1973).

Variation in the catchment area in space, with regard to the height of flight of the moths, the position and design of the trap and the role of wind in moving a column of air across the trap have been studied by Taylor & Brown (1972) and Taylor & French (1974). They postulate a volume equivalent to the amount of air originally occupied by the catch i.e. 100 % efficient extraction. This will, of course, be less than the catchment area, but one can visualize that the two approaches might be combined to make light traps a measure of absolute population.

Light traps have been found useful in survey work, both for particular species or groups (Bogush, 1958; Geier, 1960; Otake & Kono, 1970) or on an extensive scale (Taylor & Taylor, 1977).

THE EFFECT OF TRAP DESIGN ON CATCH

From the above we can conclude that catches in a light trap will be influenced by the following design features:

(1) The amount of contrast between the light source and surroundings: the greater the contrast, the greater the 'catchment area'.

(2) Most animals will have a tendency to withdraw from the high light intensity immediately adjacent to the lamp.

(3) The extent to which an animal may be able to change from approach to avoidance will depend on its flight speed; the faster, heavier flyers are unable to stop and change course quickly. Indeed Stanley & Dominick (1970) found with a Pennsylvania-type trap with nested funnels that the large moths were trapped in the central smallest funnel, whilst coccinellids were most frequently caught in the widest outer funnel.

The effect of the illumination of the environment on the catch was utilized by Common (1959), who designed a transparent light-trap for Lepidoptera. The main advantage of this over the Robinson trap, of which it is a modification, is that the number of Coleoptera (especially Scarabaeidae) caught is greatly reduced; large numbers of beetles will damage many of the moths in a catch.

An increase in the intensity of the lamp, frequently brought about by the substitution of an ultra-violet lamp for an incandescent one, usually leads to an increased catch (e.g. Williams, 1951; Williams *et al.*, 1955; Barr *et al.*, 1963; Belton & Kempster, 1963; Breyev, 1963). Because of the role of intensity in the mechanism of trapping, the actual size of the light source is of less importance (Belton & Kempster, 1963) and a small trap may often be substituted for a larger one without the catch being reduced in proportion to the change in light output (Smith *et al.*, 1959).

Many workers (e.g. Gui *et al.*, 1942) have considered that the differences in quality of the light, from incandescent lamps, from mercury-quartz lamps emitting ultra-violet and visible light, and from black (or 'blue') lamps emitting only or mainly ultra-violet, affect the catch; but Mikkola's (1972) study shows this to be very complex. Whatever the mechanism, there are considerable differences; for example, many Diptera and Miridae are taken in the largest numbers in traps with incandescent (tungsten filament) lamps; Corixidae, many Lepidoptera (especially noctuids) are more abundant in 'UV traps' and are even better represented in 'black-light traps', as are Trichoptera (Frost, 1953; Williams *et al.*, 1955; Breyev, 1958; Southwood, 1960; Tshernyshev, 1961).

Some of these differences are explained by the increased intensity of the ultra-violet traps leading to the repulsion of the insect when close to the lamp; slow flyers such as many Diptera Nematocera and geometrid and pyralid moths stop their approach before they enter the trap, whilst the heavier, faster flyers, like noctuid and sphingid moths, pass straight into the trap. Suction fans have therefore been incorporated in light-traps to catch those insects that avoid capture at the 'last moment' (e.g. Glick *et al.*, 1956; Downey, 1962). The same phenomenon was demonstrated in the opposite way by Hollingsworth *et al.* (1961), who found that a suitably placed windbreak could increase the catch of trap: in the exposed situation many species would avoid trapping by being blown off course, but in the lee of the windbreak this would not occur and the larger moths would be caught particularly efficiently. The addition of baffles to the trap also serves to catch those insects whose repulsion by the high intensity close to the lamp would otherwise cause them to escape; Frost (1958a) almost doubled the numbers caught by a trap by adding four intersecting baffles to it.

When differences due to flight speed and momentum have been eliminated, other specific differences in susceptibility to trapping remain: e.g. amongst the mosquitoes, *Aedes* seem especially susceptible to trapping (Love & Smith, 1957; Loomis, 1959b). Males are often taken in much greater numbers than females at light-traps (e.g. Southwood, 1960).

The height above the vegetation at which the trap is exposed will influence the catch. As the aerial density of most insects decreases with height, in general the higher the trap the smaller the catch (Frost, 1958c; Taylor & Brown, 1972). There are a few exceptions; Taylor & Carter (1961) have shown that some Lepidoptera may occur at a maximum density at as much as 30 ft above the ground and many ornithophilic mosquitoes and blackflies also have maxima some distance above the ground.

TECHNIQUES AND TYPES OF TRAP

Many light-traps retain the insects alive; Merzheevskaya & Gerastevich (1962) did this by covering the collecting funnel of a Pennsylvania type trap with a cloth bag and inserting another shallow cone in the neck. The Robinson trap retains the insects in the drum of the trap; but Belton & Kempster (1963) showed that unless the opening was under $\frac{1}{2}$ in in diameter, many small moths, like geometrids, would leave the trap in the morning, more especially if the sun shone on it; however, such a small cone opening will restrict the entry of other species.

A large number of killing agents have been used in light-traps; some of the chlorinated hydrocarbons such as tetrachlorethylene, trichlorethylene and tetrachlorethane are particularly useful. They may be poured on to a block of plaster of Paris at the bottom of the killing bottle, where their heavy vapours will be slowly evolved and remain relatively concentrated (Williams, 1948; Haddow & Corbet, 1960). Excessive inhalation of their vapours over a long period may be dangerous to man. Delicate insects such as Ephemeroptera or Psocoptera may be collected straight into alcohol and chafer beetles into kerosene (Frost, 1964). Potassium and sodium cyanides are traditional, but dangerous poisons; the last named is more suitable in moist climates (Frost, 1964).

White (1964) has designed a killing tin to preserve the light-trap catch in good condition (Fig. 7.11). It has the further advantage that only specimens too large to pass through the screen across the top of the funnel are retained; the mesh of this screen may be so fine as to retain the whole catch or, for example, it may be of such a size that the unwanted Diptera and small Trichoptera escape (White, pers. comm.). When only the smaller species are required, the larger ones may be excluded and so prevented from damaging the catch by covering the entrance to the trap with a coarse screen (e.g. Downey, 1962). If all specimens are to be captured a series of graded screens in the killing bottle will tend to separate the insects by size and aid in the subsequent sorting (Frost, 1964). In connection with the latter operation the sorting tray described by Gray (1955) may be useful.

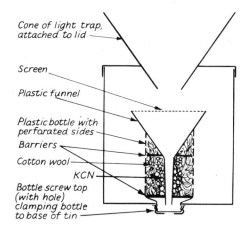

Cone of light trap, attached to lid

Screen

Plastic funnel

Plastic bottle with perforated sides

Barriers

Cotton wool

KCN

Bottle screw top (with hole) clamping bottle to base of tin

Fig. 7.11 White's light-trap killing tin.

The Rothamsted trap (Fig. 7.12a)

Originally used by C. B. Williams as a research tool for studying the influence of climate on the numbers of flying insects, this trap takes moderate numbers of most groups, including nematocerous flies and the larger moths; it would seem to be less selective than most other light-traps and is useful in studies on the diversity of restricted groups. It is described in detail by Williams (1948): it has a roof, which allows its operation in all weather, but reduces the catch. The collections from different periods of the night may be separated by an automatic device (Williams, 1935). A simple transportable non-ultra-violet trap has been developed by Jalas (1960).

The Robinson trap (Fig. 7.12b)

This was the first trap using ultra-violet light and was designed by Robinson & Robinson (1950) to make maximum catches of the larger Lepidoptera. The

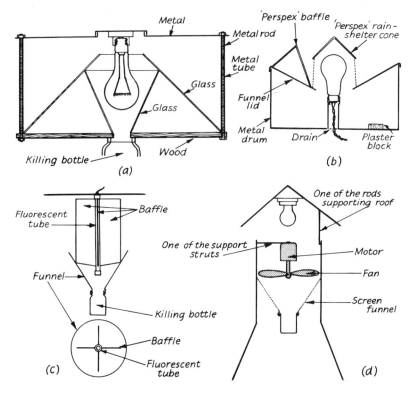

Fig. 7.12 Light-traps: *a*. Rothamsted trap (after Williams, 1948); *b*. Robinson type trap; *c* Pennsylvanian trap (sectional view and plan); *d*. New Jersey trap.

trap is without a roof and although the bulb is protected by a celluloid cone and there is a drainage hole in the bottom, it cannot be used in all weathers. The drum is partly filled with egg boxes or similar pieces of cardboard in which the insects can shelter. The catch may be retained alive or killed by placing blocks of plaster of Paris saturated with a killing agent on the floor. Large numbers of fast-flying nocturnal insects (e.g. Noctuidae, Sphingidae, Corixidae, Scarabaeidae) are trapped (Williams, 1951; Williams *et al.*, 1955); but Common's (1959) transparent trap catches fewer beetles. The 'Muguga trap', widely used in eastern Africa, is a modification of this type of trap (Brown *et al.*, 1969; Siddorn & Brown, 1971).

The Pennsylvanian and Texas traps (Fig. 7.12c)
These traps are basically similar, consisting of a central fluorescent tube surrounded by four baffles; below the trap is a metal funnel and a collecting jar (Frost, 1957; Hollingsworth *et al.*, 1963). The Pennsylvanian trap has a circular roof to prevent the entry of rain into the killing bottle and it is

presumably a reflection of the climates of the two States that the Texas trap is roofless. A wire mesh grid directly below the funnel in the Texas trap will allow rain and small insects to escape, whilst the larger specimens are deflected into a collecting container (Dickerson, Gentry & Mitchell, 1970). Frost's (1958b) experiments suggest that the largest catches would be obtained with a 15-W black-light fluorescent tube; Graham *et al.* (1961) used three argon-glow lamps in the Texas trap in survey work on the pink bollworm, *Platyedra gossypiella*, and found moths were trapped over a radius of 200 ft from the lamp. Fluorescent tubes may be run for many hours or even years from batteries (Clark & Curtis, 1973) and King *et al.* (1965) have devised a mechanism for separating the catches from different time periods, by changing the killing bottle at fixed intervals.

The Minnesota light-trap is basically similar to the Pennsylvanian model, but the roof is cone-shaped rather than flat and this appears to reduce the catch (Frost, 1952, 1958b). The effects of many different modifications of this type of trap on the catches of small moths have been reported by Tedders & Edwards (1972).

The New Jersey trap (Fig. 7.12d)

Developed by T. J. Headlee this is primarily a trap for sampling mosquitoes; it combines light and suction – the lamp causes the insects to come into the vicinity of the trap and they are drawn in by the suction of the fan (Mulhern, 1942). The trap is therefore particularly useful for weak flyers that may fail to be caught by the conventional light-trap, as when close to the lamp they are repelled by its high intensity. Fitted with an ultra-violet lamp the New Jersey trap has been used to collect other groups of Nematocera, as well as mosquitoes (Zhogolev, 1959). Kovrov & Monchadskii (1963) found that if the trap was modified to emit polarized light many insects, but not mosquitoes, were caught in larger numbers. Like the Rothamsted and Pennsylvanian traps, this trap may be fitted with an automatic interval collector; one is described by Bast (1960).

A tiny battery-operated model, the CDC (Centre for Disease Control) trap has been developed (Sudia & Chamberlain, 1962; Buckley & Stewart, 1970; Vavra *et al.* 1974). For comparative studies using different traps it is important to standardize air flow and direction (Loomis, 1959a; Wilton & Fay, 1972), although, as has been pointed out several times, relative trapping methods, and especially light-traps, may sample different proportions of the population in different areas even when the traps are identical. Acuff (1976) found that light traps collected a wider range of mosquito species than bait and Malaise traps. For bloodmeal and arbovirus studies the insects should be collected into saline (7.5g/1) with some antiseptic (Walker & Boreham, 1976).

The Haufe–Burgess visual trap (Fig. 7.13a)

As Verheijen's (1960) studies have shown that the trapping effect of light depends on the degree of contrast between the lamp and the surroundings, it is not surprising that light-traps are ineffective in the twilight night of the arctic, and Haufe & Burgess (1960) designed the present trap for mosquitoes; it has been found to catch other groups as well. Basically the trap is an exposed cone type of suction trap on the outside of which is a revolving cylinder painted with $1\frac{5}{8}$-in wide black and white stripes. A horizontal disc extends out at the top of the cylinder and the insects are drawn through slits at the junction between the cylinder and the disc. Harwood (1961) retained the light-bulb above the trap (as in the New Jersey light-trap) and exposed various animals bait below the trap, but the combination of so many attractants makes the interpretation of the results very difficult.

The Manitoba or canopy trap (Fig. 7.13b)

Tabanidae are attracted to the highlights of a black or red sphere (Bracken, Hanec & Thorsteinson, 1962) and large numbers may be collected by

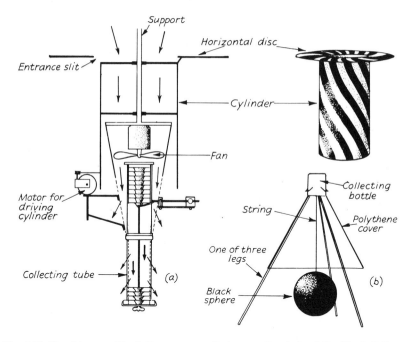

Fig. 7.13 Visual traps: *a.* Haufe-Burgess mosquito trap, sectional view (after Haufe & Burgess, 1960) and external view of the attraction cylinder; *b.* Manitoba or canopy trap.

suspending a sphere of these colours (e.g. a balloon) beneath a polythene collecting cone (Thorsteinson, Bracken & Hanec, 1965 Brachen & Thorsteinson, 1965). The collecting container may be of the type used on 'tents' (Fig. 4.11) (Adkins *et al.*, 1972) and with modifications flies can be marked (see p. 75).

Aquatic light-traps

There are two types: floating and submerged. Baylor & Smith (1953) have described an ingenious floating trap, principally for Cladocera, which utilizes their 'attraction' to yellow to get them into the trap and their 'repulsion' to blue to drive them into the collecting net (Fig. 7.14*b*). It is important that the blue light is not visible laterally; thus the level of the blue 'Cellophane' must not extend above the lip of the funnel. Baylor & Smith suggest that the trap could be improved with refinements, such as the polarization of the yellow light that would increase the 'fishing area' without increasing the intensity (see also Kovrov & Monchadskii, 1963).

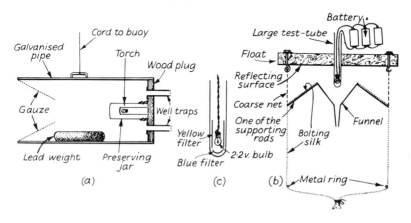

Fig. 7.14 Aquatic light-traps: *a.* Hungerford's subaquatic light-trap; *b.* and *c.* Baylor & Smith's floating trap: section (after Baylor & Smith, 1953); *c.* detail of test tube and light source.

A floating light trap, especially suitable for emerging insects, may be made out of plastic bucket with a black light inside the lid and a funnel replacing the base (Carlson, 1971). A model for use in shallow water is described by Apperson & Yows (1976).

A subaquatic light-trap was designed by Hungerford, Spangler & Walker (1955). The body of the trap is a length of wide galvanized piping and the light source a torch, held inside a glass fruit-preserving jar (Fig. 7.14*a*). When the trap is hauled to the surface, the water is drained out through the 'well traps' and the animals retained in these. The well traps are pieces of metal pipe with gauze at one end and flange at the other.

7.4.4 Traps that attract the animals by some natural stimulus or a substitute
When these traps are entirely natural they have the advantage over artificial ones in that the variations in efficiency reflect real changes in the properties of the population: the changes in the numbers of a phytophagous insect colonizing a plant or of a bloodsucking species biting the host may not reflect

population changes, but they do reflect accurately changes in the colonizing or feeding rates of the species. That is, with the possible exception of artificial bait traps, these traps give measures of availability that are biologically meaningful.

SHELTER TRAPS

Cubical boxes (1 ft^3 (283 cm^3) in volume) with one open side and painted red have been used in mosquito surveys; they are often termed 'artificial resting units' (Goodwin, 1942; Goodwin & Love, 1957; Burbuits & Jobbins, 1958). As might be expected, only certain mosquitoes settle in these shelters, particularly species of *Culex, Culiseta* and *Anopheles*. Loomis & Sherman (1959) found that visual counts of the resting mosquitoes with a torch were accurate to within 10 % of the true value. Dales (1953) found that large numbers of *Tipula* would shelter in and be unable to escape from small metal truncated cones (15 cm aperture at top, 25 cm high and 28 cm at base), and Vale (1971) tested a range of artificial refuges for tsetse flies.

Ground dwellers (Cryptozoa) such as woodlice, centipedes and carabids that shelter beneath logs and stones, may be trapped by placing flat boards in the habitat. The 'cryptozoa boards' can be placed systematically in the habitat and used to study the dispersion of the animals (Cole, 1946); however, the proportion of the population that shelters beneath them will vary with soil moisture aspect, and other factors (Paris, 1965; Jenson, 1968).

Many insects shelter in cracks or under bark scales on tree trunks, for overwintering or as a mode of life. Artificial shelters consisting of several grooved boards bolted together have been placed in trees to trap earwigs (Chant & McLeod, 1952), in the field to trap overwintering *Bryobia* mites (Morgan & Anderson, 1958) and screwed on to the tree bark beside a fungus to trap the beetle, *Tetratoma fungorum* (Paviour-Smith, 1964). Fager (1968) used oak boxes filled with sawdust, with holes drilled in them, as 'synthetic logs' in studying the fauna of decaying wood.

Trap nests, made by drilling holes in woody stems (e.g. sumac) or in pieces of dowelling which are split and then bound together, will be colonized by solitary Hymenoptera if the open ends are exposed in a board (Fig. 7.15); the binding may be removed and the nest contents easily exposed (Medler & Fye, 1956; Levin, 1957; Medler, 1964; Fye, 1965; Freeman & Jayasingh, 1975).

TRAP HOST PLANTS

Insect-free potted host plants exposed in a habitat may be colonized by insects; these can be removed and counted and their numbers will be directly proportional to the colonization potential of the population in that habitat. Trap host plants are therefore most useful for measuring emergence and colonization (Fritzsche, 1956; Grigarick, 1959b; Smith, 1962; Waloff & Bakker, 1963; Bucher & Cheng, 1970) and susceptible logs will record the same phenomena in bark beetles (Chapman, 1962). The actual adult insects

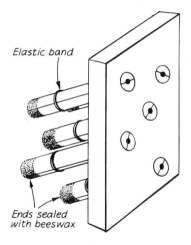

Elastic band

Ends sealed
with beeswax

Fig. 7.15 Levin's trap nests for solitary Hymenoptera.

or the resulting eggs or larvae may be counted. Michelbacher & Middlekauff (1954) studied population changes in *Drosophila* in tomato fields by placing as 'traps' ripe tomatoes that had been slit vertically on the sides and squeezed slightly; the number of eggs laid in the flesh on either side of the cuts could be easily ascertained in the laboratory. It is of course important to ensure that the plants, fruit or logs remain in the condition where they are attractive to the insects.

THE USE OF VERTEBRATE HOSTS OR SUBSTITUTES AS BAIT
The methods described in this section are used principally to determine the biting rate of intermittent ectoparasites under various conditions, the host range and the relative importance of different hosts and vectors and other measurements of significance in epidemiological studies. Some give unbiased indices of these behavioural characteristics, but in others the presence of the trap may repel some insects or attract others independently of the presence of the host. Dyce & Lee (1962) found that the presence of the drop-cone trap about 1 ft above the rabbit would deter certain species of mosquito, but not others, from biting it. In contrast, Colless (1959) showed that from one-third to half the catch of a Malayan trap is independent of the presence of the host. Comparisons of the responsiveness of tsetse flies to stationary and mobile baits showed that whereas visual and olfactory stimuli are important with the former, visual cues predominate with moving bait (Vale, 1974). As there is often a marked diurnal periodicity in biting rate (Haddow, 1954) it is important that assessments should be made throughout the 24 hours; indeed the 'biting cycle' is often longer (McClelland & Conway, 1971) so that if there is an unusual age distribution estimates need to be made over at least one full cycle.

These methods therefore give measures of availability and when 'attractiveness' is constant indices of absolute population (i.e. relative abundance). Under certain circumstances they might be used to measure absolute population density by the removal technique. One must note that the attractiveness of individual baits differs (Saunders, 1964; Khan *et al.*, 1970). Many special traps are reviewed by Service (1976).

Reference should also be made to the Manitoba or canopy trap (p. 260), the carbon dioxide trap (p. 270) and perhaps even the Haufe–Burgess mosquito trap (p. 259) which could have been included here on the grounds that they are substitutes for vertebrate hosts.

Moving baits

As ectoparasite incidence is often low and non-random, and as landing rate may vary in relation to the time of exposure in one spot (Fig. 7.16), a moving bait may give more extensive data than one stationary over a long period. The

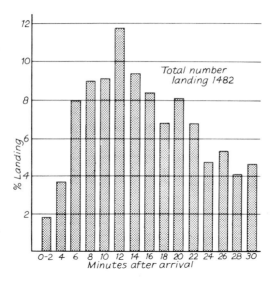

Fig. 7.16 The variation in landing rate of *Culicoides* with time after arrival of human bait (after Jamnback & Watthews, 1963).

'fly-round' first developed by W. H. Potts has been used extensively in work on the ecology of the tsetse fly. In this, the catching party consisting of the human bait and the collectors would walk through the habitat, making a number of stops at which flies were caught. Ford *et al.* (1959) proposed a transect fly-round that followed arbitrary straight lines with equidistantly placed halting sites. These transect rounds have advantages in the statistical analysis necessary for the determination of dispersion, but the catches are still variable and Glasgow (1961) concluded that a 7500-yd fly-round done once a week could not detect less than a fivefold change in the mean catch (i.e. in the index of absolute population). More males than females are usually (Glasgow & Duffy, 1961), but not always (Morris, 1960), caught on fly-rounds. This appears to be due to the fact that only teneral females feed on moving hosts, mature females are therefore seldom taken; furthermore the males follow the moving objects (even vehicles) not so much to feed (except those that are teneral) as to be in the vicinity to pair with the young females (Bursell, 1961). Oxen and other animals may be used instead of man as the bait on fly-rounds; the species of *Glossina* taken will be influenced by the bait animal (Jordan, 1962). Studying the biting habits of *Chrysops,* Duke (1955) found that biting rate per individual of a group of human baits was greater than that for a single individual working on his own. Single human baits were used by Murray (1963) in studies on *Aedes* and found to give a better picture of seasonal variation than light-traps.

Ticks and fleas sit on grass and other vegetation and attach themselves to their hosts as they pass by. Rothschild (1961) and Bates (1962) sampled fleas with what they termed artificial rabbits and birds: a skin with an internal heating device (a hot-water bottle) that was pulled across the ground. Ticks have been sampled by dragging blankets and pieces of cloth over the vegetation; Wilkinson (1961) has compared various methods and concludes that the 'hinged-flag sampler' a modification of Blagoveschenskii's (1957) technique, was most efficient and gave a good correlation with counts on cattle in the same enclosure. This sampler consists of a piece of plywood (50 cm square), covered with cloth and with a handle strongly hinged to the centre of one side. The sampler is held as rigidly as possible with the lower edge making contact with the grass. In very rough country this cannot be used and cloth-covered leggings are the only available method. Ticks may be removed from these samplers, retained and stored on transparent adhesive tape (e.g. 'Sellotape').

Stationary baits

The most natural situation is achieved when the host is freely exposed in the normal habit; when man is the host this is of course possible, and it may be feasible for one individual to act as both bait and collector. In order to obtain a complete collection the bait may need to enclose himself periodically in a cage. Klock & Bidlingmayer (1953) devised one that was essentially an umbrella with blinds that could be released and would rapidly drop to the ground. Blagoveschenskii *et al.* (1943) dropped a bell-shaped cover over their human bait at intervals and Myers (1956) extended this principle to the rabbit with the 'drop-cone trap': a muslin cone is suspended by a rope over a pulley about 1 ft above a tethered bait (Fig. 7.17a); it may be quickly lowered by an observer some distance away. But as Dyce & Lee (1962) showed, the presence of the drop-cone trap just above the bait may affect the behaviour of some species. These authors also suggest that the effect of man handling the traps in

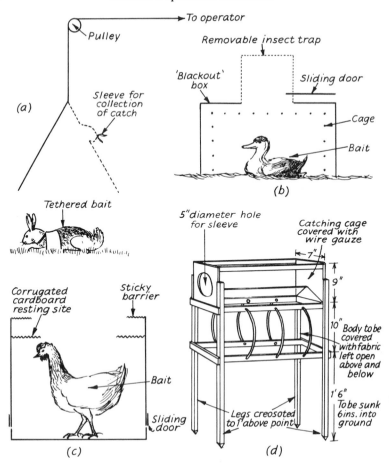

Fig. 7.17 *a* The Myers drop-cone trap. *b* Anderson & Defoliart's method for trapping blackflies. *c* Harrison's trap for poultry mites. *d* The Morris trap.

which other species are exposed may invalidate comparisons between the biting rate on man and these other animals.

Blackflies (Simuliidae) are less easily disturbed when feeding on their hosts, and Anderson & Defoliart (1961) investigated their host preferences by exposing various birds and mammals in net or wire cages for 15 minutes, and then covering these with a blackout box with removable insect trap on top (Fig. 7.17*b*). The flies would soon be attracted to the light and move into the muslin trap. The blackout box was conveniently made of a strong cardboard box strengthened with wood. Disney (1972) lowered the box on a cantilever: it then becomes a type of 'drop-cone trap' (see above).

Apart from any influence due to the trap, the catches at a bait may be influenced by the length of time it has been present (Fig. 7.16) (Jamnback &

Watthews, 1963) and there are considerable local variations due to site (e.g. Saunders, 1964). Selection due to the influence of the trap itself will be greater with those methods where the animal is more or less enclosed; four types of trap for flying ectoparasite do this. They are:

(1) The stable trap originally designed by E. H. Magoon and W. C. Earle
(2) Cage traps for small animals
(3) The Malayan trap originally designed by B. A. R. Gater
(4) The Lumsden trap developed by W. H. R. Lumsden

The stable trap design has been modified by Bates (1944) to retain a large proportion of the mosquitoes that had fed on the donkey, mule or other bait, and in a modified form has been used with night-herons as bait and hoisted into trees by Flemings (1959). Basically the stable trap consists of a portable sectional wood and screen shed into which mosquitoes may enter by a horizontal slit in the lower part of the walls; most are then trapped, as only a few fly downwards after feeding and so manage to escape. Small animals may be exposed in cage traps that are covered with fine netting above and open below (Worth & Jonkers, 1962; Turner 1972) or in traps resembling the carbon dioxide trap (Fig. 7.18).

The Malayan human bait trap consists of a mosquito-net cage, the flaps of which can be dropped by pulling a string (this is usually done every hour). Wharton *et al.* (1963) used a similar cage for exposing monkeys to mosquitoes. It has some resemblance both to Klock & Bidlingmayer's (1953) umbrella with blinds and to a Malaise trap (p. 192); in view of its similarity to the latter it is not surprising that an unbaited trap catches up to half as many mosquitoes as a baited one (Colless, 1959).

Lumsden's trap (Lumsden, 1958) is constructed of a 'Perspex'* hood leading to a suction fan, a gauze funnel and a collecting tube ('an inverted suction trap'). A bait animal is tethered, or placed in a cage, below the hood. The fan is switched on periodically and the resulting air current carries the insects feeding on the bait into the collecting tube. Portable, battery- or generator-operated versions have been developed by Snow *et al.* (1960) and Minter (1961); these catch large numbers of *Culicoides* and *Phlebotomus*, but the airflow seems insufficient to retain many of the mosquitoes.

A special trap has been constructed by Harrison (1963) for studying the poultry red mite, *Dermanyssus*. It cosists of a box containing the bait chicken with a number of holes near the base (Fig. 7.17c); these may be closed with a time switch or hand-operated barrier. The trap could be used for other ectoparasites that leave the host after feeding (e.g. Cimicidae).

R. H. T. P. Harris devised a trap that apparently attracted and retained tsetse flies; his design was considerably modified and improved by Morris & Morris (1949) and this trap is now generally known as the 'Morris trap' (Fig.

* 'Plexiglass' is a similar material.

7.17*d*). A portable folding version has been described by Morris (1961). The flies enter the traps from below and only a few seem able to escape; however, if the traps were emptied less than once every 24 hours losses of trapped flies occurred due to predation by ants (Smith & Rennison, 1961, part IV). A high proportion of the flies caught in the Morris trap are females and this is their great advantage compared with the fly-round; however, the dispersion pattern suggested by trap catches does not coincide with that found by fly-rounds or searching the resting sites (Glasgow & Duffy, 1961). Thus their value as an index of absolute population must be doubted and the time of day of maximum catch in the Morris trap seems to differ from that at oxen (Smith & Rennison, 1961, part II). Several species of *Glossina* and many tabanids are caught by Morris traps, but others are not taken even when they attack tethered oxen in the vicinity (Morris, 1960, 1961; Jordan, 1962; Vale 1971). Black-coloured cloth on the trap seemed to attract more tabanids (Morris, 1961) and, in the wet season more tsetse flies, than did brown hessian (Morris, 1960). The variations in catch in these traps remain something of a mystery in spite of extensive tests (Hargrove, 1977). Fredeen (1961) used basically similar traps (termed silhouette traps) to collect Simuliidae, which, as Wenk & Schlörer (1963) have shown, are attracted to the silhouette of their hosts.

BAIT TRAPS

The study of the behavioural responses of insects to scents from either their own species (pheromones) or their food sources (kairomones) is one of the most recent and most active fields of entomological research (Shorey, 1973; Birch, 1974; Roelofs, 1975; Kennedy, 1976). Such scents are used by man for a variety of purposes: for confusing the pest in its approach to the prey (plant) or in traps, either to measure or monitor the population or to reduce it by killing large numbers, directly or by the incorporation of a sterilant into the bait. An enormous variety of traps have been described; the present account must be regarded as merely an outline of some of the types, with an emphasis on those most useful in ecological work. (Traps utilizing vertebrates as bait are described in the previous section).

Traps

The designs have generally been empirical, although current work is taking account of the flight behaviour of the insect (Kennedy & Marsh, 1974) and the shape of the scent plumes (Lewis & Macaulay, 1976). Basically traps can be divided into the two categories:

(1) 'Lobster-pot' traps (Fig. 7.18*a*, *c* & *d*); these lure in insects through an inverted funnel when they are generally captured alive. Traps of this type have been used widely, with carbon dioxide as bait, for blood sucking insects (Bellamy & Reeves, 1952; Wilson & Richardson, 1970); with carrion as bait for flies (Dodge & Seago, 1954; MacLeod & Donnelly, 1956; Gillies, 1974) or with various extracts for bark beetles, when they are termed 'field olfacto-

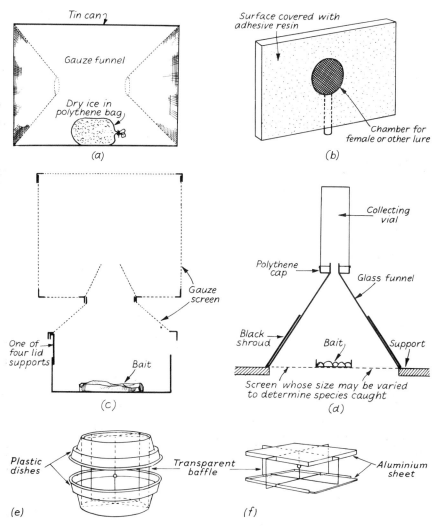

Fig. 7.18 Bait traps: *a.* carbon dioxide trap; *b.* sticky trap (Coppel *et al.*, 1960); *c.* MacLeod's carrion trap; *d.* Citrus Experimental Station general bait trap; *e.* water trap (after Lewis and Macaulay, 1976). *f.* 'lantern trap', (after Lewis and Macaulay, 1976). (Chemical bait is exposed in the centre of the baffles that are covered with adhesive resin in the water and lantern traps).

meters' (Vité & Gara, 1962; Chararas, 1977).

(2) Impaction-traps (Fig. 7.18, *b*, *e* & *f*): in these the insect is held generally in an adhesive resin or grease* or water trap in proximity to the lure. Designs of this type have been widely used in studies on chemical lures (Coppel *et al.*

* See footnote p. 249

1960; Campion, 1972; Howell, 1972; Otake, Takigawa & Oyama, 1974; Howell *et al.*, 1975; Lewis & Macaulay, 1976).

Baits or lures
The baits used have ranged from the actual biological material (e.g. virgin female insects or rotting fruit) to the extracted chemicals or their 'mimics'. Broadly, a distinction may be made between kairomones and pheromones, although some behavioural processes, such as the attraction of scolytids to a host tree, may involve both. Furthermore some of the lures have been discovered empirically and their classification is a matter of conjecture.

(1) Kairomones: carrion is used in trapping flies, especially blowflies (Dodge & Seago, 1954); the freshness of the bait, the precise position of the trap and the presence of other flies have been found to affect the results (MacLeod & Donnelly, 1956; Fukuda, 1960; Kawai & Suenaga, 1960). Rotting fruit will attract certain butterflies (Sevastopulo, 1963), bananas or vinegar with yeast, *Drosophila* (da Cunha *et al.*, 1951; Mason, 1963), hydrolysed proteins and similar compounds, fruit flies (McPhail, 1939; Neilson, 1960; Ruffinelli *et al.*, 1962; Bateman, 1972) and proteinaceous materials, eye gnats, *Hippelates* (Mulla *et al.*, 1960). Various chemicals will act as lures; these are probably either the actual kairomone (as with carbon dioxide) or are related to it. Various fruit flies have been attracted by a variety of chemicals (Bateman, 1972), but complex synthetic lures e.g. 'Medlure' and 'Cuelure' are particularly effective (Beroza *et al.* 1961; Fletcher, 1974). Many scolytids are attracted by turpinoids (Chararas, 1977), the cupesid beetle, *Priacma serrata* by chlorine (Atkins, 1957), the cabbage rootfly, *Erioschia brassicae*, by allylisothiocyanata (Finch & Skinner, 1974) and the pine weevil, *Hylobius abietis* by the methyl ester of linoleic acid (Hesse *et al.*, 1955). Carbon dioxide, conveniently supplied as 'dry-ice', has been used as a lure for many blood sucking insects: mosquitoes (Bellamy & Reeves, 1952; Gillies, 1974; Service, 1976), ticks (Garcia, 1962), tsetse flies (Rennison & Robertson, 1959), sandflies (Rioux & Golvan, 1969) and blackflies (Fallis & Smith, 1964). Carbon dioxide has been combined with a Malaise trap for tabanids (Blume *et al.*, 1972; Roberts, 1976); although Wilson & Richardson (1970) found *Chrysops* were less attracted to it than were *Tabanus*.

(2) Pheromones: These may mostly be categorized as sex attractants and the simplest approach is to expose a virgin female as the lure (Coppel *et al.*, 1960; Lewis, Snow & Jones, 1971; Goodenough & Snow, 1973; Otake *et al.* 1974). Several insect sex pheromonoes have now been extracted and sometimes synthesized; in other instances synthetic analogues are available. Particular progress has been made with the gypsy moth (*Lymantria dispar*) (Jacobson *et al.*, 1961; Schneider *et al.*, 1974), with various tortricoid moths (Bartell & Shorey, 1969; Roelofs, 1975; Wall, Greenway & Burt, 1976) and with the noctuid moth, *Spodoptera* (Campion, Bettany & Steedman, 1974). In

bark beetles (Scolytidae) there is a complex series of signals: kairomones, pheromones and sounds (Wood *et al.*, 1967; Chararas, 1977; Rudinsky, 1977).

SOUND TRAPS

Many insects respond to sounds of their own and other species: recordings could be used as the lure in traps (Belton, 1962; Cade, 1975).

REFERENCES

A'BROOK, J., 1973. Observations on different methods of aphid trapping. *Ann. appl. Biol.* **74**, 263–77.

ACUFF, V. R., 1976. Trap biases influencing mosquito collecting. *Mosquito News* **36**, 173–6.

ADKINS, T. R., EZELL, W. G., SHEPPARD, D. C. and ASKEY, M. M., 1972. A modified canopy trap for collecting Tabanidae (Diptera). *J. med. Ent.* **9**, 183–5.

ADLERZ, W. C., 1971. A reservoir-equipped Moericke Trap for collecting aphids. *J. econ. Ent.* **64** (4), 966–67.

AHEARN, G. A., 1971. Ecological factors affecting population sampling of desert tenebrionid beetles. *Am. Midl. Nat.* **86**, 385–406.

ALMAND, L. K., STERLING, W. L. and GREEN, C. L., 1974. A collapsible truck-mounted aerial net for insect sampling. *Tex. Agr. Exp. Sta. Misc. Pub.* **1189**, 1–4.

ANDERSON, J. R. and DEFOLIART, G. R., 1961. Feeding behaviour and host preferences, of some black flies (Diptera: Simulidae). *Ann. ent. Soc. Am.* **54**, 716–29.

ANDERSON, J. R. and POORBAUGH, J. H., 1964. Observation on the ethology and ecology of various Diptera associated with northern California poultry ranches. *J. med. Ent.* **1**, 131–47.

ANDERSON, N. H., 1966. Depressant effect of moonlight on activity of aquatic insects. *Nature* **209**, 319–20.

ANDERSON, N. H., 1967. Biology and downstream drift of some Oregon Trichoptera. *Can. Ent.* **99**, 507–21.

ANDERSON, N. H. and LEHMKUHL, D. M., 1968. Catastrophic drift of insects in a woodland stream. *Ecology* **49**, 198–206.

APPERSON, C. S. and YOWS, D. G., 1976. A light trap for collecting aquatic organisms. *Mosquito News* **36**, 205–6.

ATKINS, M. D., 1957. An interesting attractant for *Priacma serrata* (Lec.) (Cupesidae: Coleoptera). *Can. Ent.* **89**, 214–19.

AUBERT, J., 1969. Un appareil de capture de grandes dimensions destiné au marquage d'insectes migrateurs. *Mitt. Schweiz. entom. Gesell.* **42**, 135–9.

BALOGH, J. and LOKSA, I., 1956. Untersuchungen über die Zoozönose des Luzernenfeldes. *Acta Zool. Acad. Sci. Hung.* **2** (1–3), 17–114.

BANKS, C. J. and BROWN, E. S., 1962. A comparison of methods of estimating population density of adult Sunn Pest *Eurygaster integriceps* Put. (Hemiptera: Scutelleridae) in wheat fields. *Entomologia exp. appl.* **5**, 255–60.

BARBER, H., 1931. Traps for cave-inhabiting insects. *J. Elisha Mitchell sci. Soc.* **46**, 259–66.

BARNES, B. M. and BARNES, R. D., 1954. The ecology of the spiders of maritime drift lines. *Ecology* **35**, 25–35.

BARR, A. R., SMITH, T. A., BOREHAM, M. M. and WHITE, K. E., 1963. Evaluation of some factors affecting the efficiency of light traps in collecting mosquitoes. *J. econ. Ent.* **56**, 123–7.

BARTELL, R. J. and SHOREY, H. H., 1969. A quantitative bioassay for the sex pheromone of *Epiphyas postvittana* (Lep.) and factors limiting male responsiveness. *J. Insect Physiol.* **15**, 33–40.

BAST, T. F., 1960. An automatic interval collector for the New Jersey light trap. *Proc. N.J. Mosq. Exter. Ass.* **47**, 95–104.

BATEMAN, M. A., 1972. The ecology of fruit flies *Ann. Rev. Ent.* **17**, 493–518.

BATES, J. K., 1962. Field studies on the behaviour of bird fleas. 1. Behaviour of the adults of three species of bird flea in the field. *Parasitology* **52**, 113–32.

BATES, M., 1944. Notes on the construction and use of stable traps for mosquito studies. *J. Natn. Malar. Soc.* **3**, 135–45.

BAYLOR, E. R. and SMITH, F. E., 1953. A physiological light trap. *Ecology* **34**, 223–4.

BEALL, G., 1935. Study of arthropod populations by the method of sweeping. *Ecology* **16**, 216–25.

BELLAMY, R. W. and REEVES, W. C., 1952. A portable mosquito bait trap. *Mosquito News* **12**, 256–8.

BELTON, P., 1962. Effects of sound on insect behaviour. *Proc. ent. Soc. Manitoba* **18**, 1–9.

BELTON, P. and KEMPSTER, R. H., 1963. Some factors affecting the catches of Lepidoptera in light traps. *Can. Ent.* **95**, 832–7.

BENEDEK, P., ERDELYI, C. and FESUS, I., 1972. General aspects of diel vertical movements of some arthropods in flowering lucerne stands and conclusions for the integrated pest control of lucerne. *Acta Phytopath. hung.* **7**, 235–49.

BEROZA, M., GREEN, N., GERTLER, S. I., STEINER, L. R. and MIYASHITA, D. H., 1961. New attractants for the Mediterranean fruit fly. *J. agric. Fd. Chem.* **9**, 361–5.

BIDLINGMAYER, W. L., 1964. The effect of moonlight on the flight activity of mosquitoes. *Ecology* **45**, 87–94.

BIRCH, M. C. (ed.), 1974. *Pheromones.* Academic Press, New York.

BISHOP, J. A., 1972. An experimental study of the cline of industrial melanism in *Biston betularia* (L.) (Lepidoptera) between urban Liverpool and rural North Wales. *J. Anim. Ecol.* **41**, 209–43.

BLAGOVESCHENSKII, D. I., 1957. Biological principles of the control of ixodid ticks. [In Russian.] *Ent. Obozr.* **36**, 125–33.

BLAGOVESCHENSKII, D. I., SREGETOVA, N. G. and MONCHADSKII, A. S., 1943. Activity in mosquito attacks under natural conditions and its diurnal periodicity. [In Russian.] *Zool. Zhur.* **22**, 138–53.

BLUME, R. R., MILLER, J. A., ESCHLE, J. L., MATTER, J. J. and PICKENS, M. O., 1972. Trapping tabanids with modified Malaise traps baited with CO_2. *Mosquito News* **32**, 90–5.

BOGUSH, P. P., 1958. Some results of collecting click-beetles (Coleoptera, Elateridae) with light-traps in Central Asia. [In Russian.] *Ent. Obozr.* **31**, 347–57 (transl. *Ent. Rev.* **37**, 291–9).

BOWDEN, J., 1973. The influence of moonlight on catches of insects in light-traps in Africa. Part 1. The moon and moonlight. *Bull. ent. Res.* **63**, 113–28.

BOWDEN, J. and CHURCH, B. M., 1973. The influence of moonlight on catches of insects in light traps in Africa. Part II. The effect of moon phase on light trap catches. *Bull. ent. Res.* **63**, 129–42.

BOWDEN, J. and GIBBS, D. G., 1973. Light-trap and suction-trap catches of insects in the northern Gezira, Sudan, in the season of southward movement of the Inter-Tropical Front *Bull. ent. Res.* **62**, 571–96.

BOWDEN, J. and MORRIS, M., 1975. The influence of moonlight on catches of insects in light-traps in Africa. III. The effective radius of a mercury-vapour light trap nd the analysis of catches using effective radius. *Bull. ent. Res.* **65**, 303–48.

BOYD, J. M., 1957. Comparative aspects of the ecology of Lumbricidae on grazed and ungrazed natural maritime grassland. *Oikos* **8**, 107–21.

BRACKEN, G. K., HANEC, W. and THORSTEINSON, A. J., 1962. The orientation of horseflies and deer flies (Tabanidae: Diptera). *Can. J. Zool.* **40** (5), 685–95.

BRACKEN, G. K. and THORSTEINSON, A. J., 1965. Orientation behaviour of horse flies; influence of some physical modifications of visual decoys on orientation of horse flies. *Entomologia exp. appl.* **8**, 314–8.

BREYEV, K. A., 1958. On the use of ultra violet light-traps for determining the specific composition and numbers of mosquito populations. [In Russian.] *Parazit. Sborn.* **18**, 219–38 (abs. *Rev. appl. Ent. B* **50**, 88).

BREYEV, K. A., 1963. The effect of various light sources on the numbers and species of blood-sucking mosquitoes (Diptera: Culicidae) collected in light traps. [In Russian.] *Ent. Obozr.* **42**, 280–303 (transl. *Ent. Rev.* **42**, 155–68).

BRIGGS, J. B., 1961. A comparison of pitfall trapping and soil sampling in assessing populations of two species of ground beetles (Col.: Carabidae). *Rep. E. Malling Res. Sta.* **1960**, 108–12.

BROADBENT, L., 1948. Aphis migration and the efficiency of the trapping method. *Ann. appl. Biol.* **35**, 379–94.

BROADBENT, L., DONCASTER, J., HULL, R. and WATSON, M., 1948. Equipment used for trapping and identifying alate aphids. *Proc. R. ent. Soc. Lond. A* **23**, 57–8.

BROWN, E. S., BETTS, E. and RAINEY, R. C., 1969. Seasonal changes in the distribution of the African armyworm, *Spodoptera exempta* (Wlk.) (Lep., Noctuidae) with special reference to eastern Africa. *Bull. ent. Res.* **58**, 661–728.

BUCHER, G. E. and CHENG, H. H., 1970. Use of trap plants for attracting cutworm larvae. *Can. Ent.* **102**, 797–8.

BUCKLEY, D. J. and STEWART, W. W. A., 1970. A light-activated switch for controlling battery operated light traps. *Can. Ent.* **102**, 911–2.

BURBUTIS, P. P. and JOBBINS, D. M., 1958. Studies on the use of a diurnal resting box for the collection of *Culiseta melanura* (Coquillet). *Bull. Brooklyn ent. Soc. (N.S.).* **53**, 53–8.

BURSELL, E., 1961. The behaviour of tsetse flies (*Glossina swynnestoni* Austen) in relation to problems of sampling. *Proc. R. ent. Soc. Lond. A* **36**, 9–20.

BUTLER, G. D., 1965. A modified Malaise insect trap. *Pan. Pacif. Ent.* **41** (1), 51–3.

CADE, W., 1975. Acoustically orienting parasitoids: fly phototaxis to cricket song. *Science* **190**, 1312–13.

CAMPION, D. G. 1972. Some observations in the use of pheromone traps as a survey tool for *Spodoptera littoralis*. *COPR Misc. Rep.* **4**.

CAMPION, D. G., BETTANY, B. W. and STEEDMAN, R. A. 1974. The arrival of male moths of the cotton leaf-worm *Spodoptera littoralis* (Boisd.) (Lepidoptera, Noctuidae) at a new continuously-recording pheromone trap. *Bull. ent. Res.* **64**, 379–86.

CARLSON, D., 1971. A method for sampling larval and emerging insects using an aquatic black light trap. *Can. Ent.* **103**, 1365–9.

CHANT, D. A. and MCLEOD, J. H., 1952. Effects of certain climatic factors on the daily abundance of the European earwig, *Forficula auricularia* L. (Dermaptera: Forficulidae), in Vancouver, British Columbia. *Can. Ent.* **84**, 174–80.

CHAPMAN, J. A., 1962. Field studies on attack flight and log selection by the Ambrosia beetle, *Trypodendron lineatum* (Oliv.) (Coleoptera: Scolytidae). *Can. Ent.* **94**, 74–92.

CHAPMAN, J. A. and KINGHORN, J. M., 1955. Window-trap for flying insects *Can. Ent.* **82**, 46–7.

CHARARAS, C., 1977. Attraction chimique exercée sur certains Scolytidae par les pinacees et les cupressacées. *Comp. Insect. Milieu Trop., Coll. Int. C.N.R.S.* **265**, 165–86.

CHIANG, H. C., 1973. A simple trap for certain minute flying insects. *Ann. ent. Soc. Am.* **66**, 704–5.

CLARK, J. D. and CURTIS, C. E., 1973. A battery-powered light trap giving two years' continuous operation. *J. econ. Ent.* **66** (2), 393–6.

CLOSE, R., 1959. Sticky traps for winged Aphids. *N.Z.J. agric. Res.* **2**, 375–9.

CLOUD, T. J. and STEWART, K. W., 1974. Seasonal fluctuations and periodicity in the drift of Caddis fly larvae (Trichoptera) in the Brazos River, Texas. *Ann. ent. Soc. Am.* **67**, 805–11.

COLE, L. C., 1946. A study of the Cryptozoa of an Illinois woodland. *Ecol. Monogr.* **16**, 49–86.

COLLESS, D. H., 1959. Notes on the culicine mosquitoes of Singapore. VI. Observations on catches made with baited and unbaited trap-nets. *Ann. Trop. Med. Parasit., Liverpool* **53**, 251–8.

COMMON, I. F. B., 1959. A transparent light trap for the field collection of Lepidoptera. *J. Lepid. Soc.* **13**, 57–61.

COON, B. F. and RINICKS, H. B., 1962. Cereal aphid capture in yellow baffle trays. *J. econ. Ent.* **55**, 407–8.

COPPEL, H. C., CASIDA, J. E. and DAUTERMAN, W. C., 1960. Evidence for a potent sex attractant in the introduced pine sawfly, *Diprion similis* (Hymenoptera: Diprionidae). *Ann. ent. Soc. Am.* **53**, 510–12.

CORNWELL, P. B., 1960. Movements of the vectors of virus diseases of cacao in Ghana. *Bull. ent. Res.* **51**, 175–201.

CUSHING, C. E., 1964. An apparatus for sampling drifting organisms in streams. *J. Wildl. Mgmt.* **28**, 592–4.

DA CUNHA, A. B., DOBZHANSKY, T. and SOKOLOFF, A., 1951. On food preferences of sympatric species of *Drosophila*. *Evolution* **5**, 97–101.

DALES, R. P., 1953. A simple trap for Tipulids (Dipt.) *Ent. mon. Mag.* **89**, 304.

DAVIES, J. B., CORBET, P. S., GILLIES, M. T. and McCRAE, A. W. R., 1971. Parous rates in some Amazonian mosquitoes collected by three different methods. *Bull. ent. Res.* **61**, 125–32.

DAVIS, E. W. and LANDIS, B. J., 1949. An improved trap for collecting aphids. *U.S.D.A. Bur. Ent. Pl. Quar. E.T.* **278**, 3. pp.

DE LURY, B., 1947. On the estimation of biological populations. *Biometrics* **3** (4), 145–67.

DICKERSON, W. A., GENTRY, C. R. and MITCHELL, W. G. 1970. A rainfree collecting container that separates desired Lepidoptera from small undesired insects in light traps. *J. econ. Ent.* **63**, 1371.

DIPEOLU, O. O., DUROJAIYE, O. A. and SELLERS, K. C., 1974. Preliminary observations on the distribution of *Culicoides* species (Diptera: Cerntopogonidae) on the teaching and research farm of the University of Ibadan. *Nigerian J. Ent.* **1** (1), 35–42.

DISNEY, R. H. L., 1972. Observations on chicken-biting blackflies in Cameroon with a discussion of parous rates of *Simulium damnosum*. *Ann. trop. Med. Parasit.* **66**, 149–58.

DOANE, J. F., 1961. Movement on the soil surface, of adult *Ctenicera aeripennis destructor* (Brown) and *Hypolithus bicolor* Esch. (Coleoptera: Elateridae), as indicated by funnel pitfall traps, with notes on captures of other arthropods. *Can. Ent.* **93**, 636–44.

DODGE, H. R. and SEAGO, J. M., 1954. Sarcophagidae and other Diptera taken by trap and net on Georgia mountain summits in 1952. *Ecology* **35**, 50–9.

DOW, R. P. and MORRIS, C. D., 1972. Wind factors in the operation of a cylindrical bait trap for mosquitoes. *J. med. Ent.* **9**, 60–6.

DOWNEY, J. E., 1962. Mosquito catches in New Jersey Mosquito traps and ultra-violet light traps. *Bull. Brooklyn ent. Soc.* **57**, 61–3.

DRIFT, J. VAN DER, 1951. Analysis of the animal community in a beech forest floor. *Tijdschr. Ent.* **94**, 1–168.

DRY, W. W. and TAYLOR, L. R., 1970. Light and temperature thresholds for take-off by aphids. *J. Anim. Ecol.* **39**, 493–504.

DUDLEY, J. E., SEARLES, E. M. and WEED, A., 1928. Pea aphid investigations. *Trans. IV int. Congr. Ent.* **2**, 608–21.

DUFFEY, E. (A. G.), 1962. A population study of spiders in limestone grassland. Description of study area, sampling methods and population characteristics. *J. Anim. Ecol.* **31**, 571–599.

DUFFEY, E., 1968. An ecological analysis of the spider fauna of sand dunes. *J. Anim. Ecol.* **37**, 641–74.

DUKE, B. O. L., 1955. Studies on the biting habits of *Chrysops*. III. *Ann. trop. Med. Parasit. Liverpool* **49** (4), 362–7.

DYCE, A. L. and LEE, D. J., 1962. Blood-sucking flies (Diptera) and myxomatosis transmission in a mountain environment in New South Wales. II. Comparison of the use of man and rabbit as bait animals in evaluating vectors of myxomatosis. *Aust. J. Zool.* **10**, 84–94.

EASTOP, V., 1955. Selection of aphid species by different kinds of insect traps. *Nature* **176**, 936.

ELLIOTT, J. M., 1967. Invertebrate drift in a Dartmoor stream. *Arch. Hydrobiol.* **63**, 202–37.

ELLIOTT, J. M., 1969. Diel periodicity in invertebrate drift and the effect of different sampling periods. *Oikos* **20**, 524–8.

ELLIOTT, J. M., 1970. Methods of sampling invertebrate drift in running water. *Ann. Limnol.* **6**, 133–59.

EMSLEY, M. G., 1957. A coarse method of estimating mirid populations in the field. *Emp. Cotton Grow. Rev.* **34**, 191–5.

ENTWISTLE, P. F., 1963. Some evidence for a colour sensitive phase in the flight period of Scolytidae and Platypodidae. *Entomologia exp. appl.* **6**, 143–8.

EVANS, L. J., 1975. An improved aspirator (pooter) for collecting small insects. *Proc. Br. ent. nat. Hist. Soc.* **8**, 8–11.

FAGER, E. W., 1968. The community of invertebrates in decaying oak wood. *J. Anim. Ecol.* **37**, 121–42.

FALLIS, A. M. and SMITH, S. M., 1964. Ether extracts from birds and CO_2 as attractants for some ornithophilic simuliids. *Can. J. Zool.* **42**, 723–30.

FEWKES, D. W., 1961. Diel vertical movements in some grassland Nabidae (Heteroptera). *Ent. mon. Mag.* **97**, 128–30.

FICHTER, E., 1941. Apparatus for the comparison of soil surface arthropod populations. *Ecology* **22**, 338–9.

FINCH, S. and SKINNER, G., 1974. Some factors affecting the efficiency of water-traps for capturing cabbage root flies. *Ann. Appl. Biol.* **77**, 213–26.

FLEMINGS, M. B., 1959. An altitude biting study of *Culex tritaeniorhynchus* (Giles) and other associated mosquitoes in Japan. *J. econ. Ent.* **52**, 490–2.

FLETCHER, B. S., 1974. The ecology of a natural population of the Queensland fruit fly, *Dacus tryoni* IV. *Aust. J. Zool.* **21**, 541–65.

FORD, J., GLASGOW, J. P., JOHNS, D. L. and WELCH, J. R., 1959. Transect fly-rounds in field studies of *Glossina*. *Bull. ent. Res.*, **50**, 275–85.

FREDEEN, F. J. H., 1961. A trap for studying the attacking behaviour of black flies *Simulium arcticum* Mall. *Can. Ent.* **93**, 73–8.

FREEMAN, B. E. and JAYASINGH, D. B., 1975. Population dynamics of *Pachodynerus nasidens* (Hymenoptera) in Jamaica. *Oikos* **26**, 86–91.

FRITZSCHE, R., 1956. Untersuchungen zur Bekämpfung der Rapsschädlinge. IV. Beiträge zur Ökologie und Bekämpfung des Grossen Rapsstengelrüsslers (*Ceuthorrhynchus napi* Gyll.). *NachrBl. dt. PflSchntzdienst, Berl.* **10**, 97–105.

FRÖHLICH, G., 1956. Methoden Zur Bestimmung der Befalls-bzw. Bekämpfungstermine verschiedener Rapsschädlinge, insebesondere des Rapsstengelrüsslers (*Ceuthorrhynchus napi* Gyll.). *NachrBl. dt. PflSchutzdienst, Berl.* **10**, 48–53.

FROST, S. W., 1952. Light traps for insect collection, survey and control. *Bull. Pa agric. Exp. Sta.* **550**, 32 pp.

FROST, S. W., 1953. Response of insects to black and white light. *J. econ. Ent.* **46** (2), 376–7.

FROST, S. W., 1957. The Pennsylvania insect light trap. *J. econ. Ent.* **50**, 287–92.

FROST, S. W., 1958a. Insects captured in light traps with and without baffles. *Can. Ent.* **90**, 566–7.

FROST, S. W., 1958b. Traps and lights to catch night-flying insects. *Proc. X int. Congr. Ent.* **2**, 583–7.

FROST, S. W., 1958c. Insects attracted to light traps placed at different heights. *J. econ. Ent.* **51**, 550–1.

FROST, S. W., 1964. Killing agents and containers for use with insect light traps. *Ent. News* **75**, 163–6.

FUKUDA, M., 1960. On the effect of physical condition of setting place upon the number of flies collected by fish baited traps. *Endemic Dis. Bull. Nagasaki Univ.* **2** (3), 222–8.

FYE, R. E., 1965. The biology of the Vespidae, Pompilidae and Sphecidae (Hymenoptera) from trap nets in Northwestern Ontario. *Can. Ent.* **97**, 716–44.

GARCIA, R., 1962. Carbon dioxide as an attractant for certain ticks (Acarina: Argasidae and Ixodidae). *Ann. ent. Soc. Am.* **55**, 605–6.

GATES, C. E., 1969. Simulation study of estimators for the line transect sampling method. *Biometrics* **25** (2), 317–28.

GATES, C. E., MARSHALL, W. H. and OLSON, D. P., 1968. Line transect method of estimating grouse population densities. *Biometrics* **24** (1), 135–45.

GATES, C. E. and SMITH, W. B., 1972. Estimation of density of mourning doves from aural information. *Biometrics* **28** (2), 345–59.

GEIER, P. W., 1960. Physiological age of codling moth females (*Cydia pomonella* L.) caught in bait and light traps. *Nature* **185**, 709.

GILLIES, M. T., 1974. Methods for assessing the density and survival of blood sucking Diptera. *Ann. Rev. Ent.* **19**, 345–62.

GIST, C. S. and CROSSLEY, D. A., 1973. A method for quantifying pitfall trapping. *Environ. Ent.* **2**, 951–2.

GLASGOW, J. P., 1953. The extermination of animal populations by artificial predation and the estimation of populations. *J. Anim. Ecol.* **22**, 32–46.

GLASGOW, J. P., 1961. The variability of fly-round catches in field studies of *Glossina*. *Bull. ent. Res.* **51**, 781–8.

GLASGOW, J. P. and DUFFY, B. J., 1961. Traps in field studies of *Glossina pallidipes* Austen. *Bull. ent. Res.* **52**, 795–814.

GLICK, P. A., 1939. The distribution of insects, spiders and mites in the air. *U.S.D.A. Tech. Bull.* **673**, 150 pp.

GLICK, P. A., HOLLINGSWORTH, J. P. and EITEL, W. J., 1956. Further studies on the attraction of pink bollworm moths to ultra-violet and visible radiation. *J. econ. Ent.* **49**, 158–61.

GOLDING, F. D., 1941. Two new methods of trapping the cacao moth (*Ephestia cautella*). *Bull. ent. Res.* **32**, 123–32.

GOLDING, F. D., 1946. A new method of trapping flies. *Bull. ent. Res.* **37**, 143–54.

GOODENOUGH, J. L. and SNOW, J. W., 1973. Tobacco budworms: nocturnal activity of adult males as indexed by attraction to live virgin females in electric grid traps. *J. econ. Ent.* **66** (2), 543–4.

GOODWIN, M. H., 1942. Studies on artificial resting places of *Anopheles quadrimaculatus* Say. *J. natn. Malar. Soc.* **1**, 93–9.

GOODWIN, M. H. and LOVE, G. J., 1957. Factors influencing variations in populations of *Anopheles quadrimaculatus* in southwestern Georgia. *Ecology* **38**, 561–70.

GRAHAM, H. M., GLICK, P. A. and HOLLINGSWORTH, J. P., 1961. Effective range of argon glow lamp survey traps for pink bollworm adults. *J. econ. Ent.* **54**, 788–9.

GRAY, H. and TRELOAR, A., 1933. On the enumeration of insect populations by the method of net collection. *Ecology* **14**, 356–67.

GRAY, P. H. H., 1955. An apparatus for the rapid sorting of small insects. *Entomologist* **88**, 92–3.

GREENSLADE, P. J. M., 1964. Pitfall trapping as a method for studying populations of Carabidae (Coleoptera). *J. Anim. Ecol.* **33**, 301–10.

GREENSLADE, P., 1973. Sampling ants with pitfall traps: digging-in effects. *Insectes Soc.* **20** (4), 343–53.

GREENSLADE, P. and GREENSLADE, P. J. M., 1971. The use of baits and preservatives in pitfall traps. *J. Aust. Ent. Soc.* **10**, 253–60.

GRESSITT, J. L. and GRESSITT, M. K., 1962. An improved Malaise trap. *Pacific Insects* **4**, 87–90.

GRESSITT, J. L., LEECH, R. E. and O'BRIEN, C. W., 1960. Trapping of air-borne insects in the antarctic area. *Pacific Insects* **2**, 245–50.

GRESSITT, J. L., SEDLACEK, J., WISE, K. A. J. and YOSHIMOTO, C. M., 1961. A high speed airplane trap for air-borne organisms. *Pacific Insects* **3**, 549–55.

GRIGARICK, A. A., 1959*a*. A floating pan trap for insects associated with the water surface. *J. econ. Ent.* **52**, 348–9.

GRIGARICK, A. A., 1959*b*. Bionomics of the rice leaf miner, *Hydrellia griseola* (Fallen.), in California (Diptera: Ephydridae). *Hilgardia* **29** (1), 80 pp.

GRÜM, L., 1959. Sezonowe zmiany aktywności biegaczowatych (Carabidae). *Ekol. Polska* A **7** (9), 255–68.

GUI, H. L., PORTER, L. C. and PRIDEAUX, G. F., 1942. Response of insects to colour intensity and distribution of light. *Agric. Engng. St. Joseph, Mich.* **23**, 51–8.

HADDOW, A. J., 1954. Studies on the biting habits of African mosquitoes. An appraisal of methods employed, with special reference to the twenty-four-hour catch. *Bull. ent. Res.* **45**, 199–242.

HADDOW, A. J. and CORBET, P. S., 1960. Observations on nocturnal activity in some African Tabanidae (Diptera). *Proc. R. ent. Soc. Lond.* A **35**, 1–5.

HALGREN, L. A. and TAYLOR, L. R., 1968. Factors affecting flight responses of alienicolne of *Aphis fabae* Scop. and *Schizaphis graminum* Rondeni (Homoptera: Aphididae). *J. Anim. Ecol.* **37**, 583–93.

HARGROVE, J. W., 1977. Some advances in the trapping of tsetse (*Glossina* spp.) and other flies. *Ecol. Ent.* **2**, 123–37.

HARPER, A. M. and STORY, T. P., 1962. Reliability of trapping in determining the emergence period and sex ratio of the sugar-beet root maggot *Tetanops myopaeformis* (Röder) (Diptera: Otitidae). *Can. Ent.* **94**, 268–71.

HARRISON, I. R., 1963. Population studies on the poultry red mite *Dermanyssus gallinae* (Deg.). *Bull. ent. Res.* **53**, 657–64.

HARWOOD, R. F., 1961. A mobile trap for studying the behaviour of flying bloodsucking insects. *Mosquito News* **21**, 35–9.

HAUFE, W. O. and BURGESS, L., 1960. Design and efficiency of mosquito traps based on visual response to patterns. *Can. Ent.* **92**, 124–40.

HAYES, W. B., 1970. The accuracy of pitfall trapping for the sand-beach isopod *Tylos punctatus. Ecology* **51**, 514–6.

HAYNE, D. W., 1949. Two methods of estimating populations from trapping records. *J. Mammal.* **30** (4), 399–411.

HEATHCOTE, G. D., 1957*a*. The comparison of yellow cylindrical, flat and water traps and of Johnson suction traps, for sampling aphids. *Ann. appl. Biol.* **45**, 133–9.

HEATHCOTE, G. D., 1957b. The optimum size of sticky aphid traps. *Plant Path.* **6**, 104–7.

HEIKINHEIMO, O. and RAATIKAINEN, M., 1962. Comparison of suction and netting methods in population investigations concerning the fauna of grass leys and cereal fields, particularly in those concerning the leafhopper, *Calligypona pellucida* (F.) *Valt. Maatalourk. Julk., Helsingfors*, **191**, 31 pp..

HESSE, G., KAUTH, H. and WÄCHTER, R., 1955. Frasslockstoffe beim Fichtenrüsselkäfer *Hylobius abietis. Z. angew. Ent.* **37**, 239–44.

HEYDEMANN, B., 1961. Untersuchungen über die Aktivatäts und Besiedlungsdichte bei epigaischen spinnen. *Verh. dt. zool. Ges.*, 538–56.

HILL, M. A., 1947. The life-cycle and habits of *Culicoides impuctatus* Goet. and *C. obsoletus* Mg., together with some observations on the life-cycle of *Culicoides odibilis* Aust., *Culicoides pallidicornis* Kief., *Culicoides cubitalis* Edw. and *Culicoides chiopterus* Mg. *Ann. trop. Med. Parasit.* **41**, 55–115.

HOLLINGSWORTH, J. P., BRIGGS, C. P., GLICK, P. A. and GRAHAM, H. M., 1961. Some factors influencing light trap collections. *J. econ. Ent.* **54**, 305–8.

HOLLINGSWORTH, J. P., HARTSOCK, J. G. and STANLEY, J. M., 1963. Electrical insect traps for survey purposes. *U.S.D.A. Agric. Res. Serv.* (*ARS*) **42**–3–1, 10 pp.

HOTTES, F. C., 1951. A method for taking aphids in flight. *Pan-Pacif. Ent.* **27**, 190.

HOWELL, J. F., 1972. An improved sex attractant trap for Codlin moth. *J. econ. Ent.* **65**, 609–11.

HOWELL, J. F., CHEIKH, M. and HARRIS, E. J., 1975. Comparison of the efficiency of three traps for the Mediterranean Fruit Fly baited with minimum amounts of Trimedlure. *J. econ. Ent.* **68** (2), 277–79.

HUGHES, R. D., 1955. The influence of the prevailing weather conditions on the numbers of *Meromyza variegata Meigèn* (Diptera: Chloropidae) caught with a sweep net. *J. Anim. Ecol.* **24**, 324–35.

HUGHES, R. D., 1977. The population dynamics of the bush fly: the elucidation of population events in the field. *Aust. J. Ecol.* **2**, 43–54.

HUIZEN, T. H. P., VAN 1977. The significance of flight activity in the life cycle of *Amara plebeja* Gyll. (Coleoptera, Carabidae). *Oecologia* **29**, 27–41.

HUNGERFORD, H. B., SPANGLER, P. J. and WALKER, N. A., 1955. Sub-aquatic light traps for insects and other animal organisms. *Trans. Kansas Acad. Sci.* **58**, 387–407.

IBBOTSON, A., 1958. The behaviour of frit fly in Northumberland. *Ann. appl. Biol.* **46**, 474–9.

ISHERWOOD, F., 1957. The resting sites of *Glossina swynnertoni Aust.* in the wet season. *Bull. ent. Res.* **48**, 601–6.

JAMES, H. G. and REDNER, R. L., 1965. An aquatic trap for sampling mosquito predators. *Mosquito News* **25**, 35–7.

JAMNBACK, H. and WATTHEWS, T., 1963. Studies of populations of adult and immature *Culicoides sanguisuga* (Diptera: Ceratopogonidae). *Ann. ent. Soc. Am.* **56**, 728–32.

JACOBSON, M., BEROZA, M. and JONES, W. A., 1961. Insect sex attractants. 1. The isolation, identification, and synthesis of the sex attractant of the gypsy moth. *J. Am. chem. Soc.* **83**, 4819–24.

JALAS, I., 1960. Eine leichtgebaute, leichttransportable Lichtreuse zum Fangen von Schmetterlingen. *Suom. hyönt. Aikak.* (*Ann. ent. Fenn.*) **26**, 44–50.

JANSEN, M. J. W. and METZ, J. A. J., 1979. How many victims will be pitfall make? *Acta Biotheoretica* **28**, 2, 98–122.

JENSON, P., 1968. Changes in Cryptozoan numbers due to systematic variation of covering boards. *Ecology* **49** (3), 409–18.

JEPSON, W. F. and SOUTHWOOD, T. R. E., 1958. Population studies on *Oscinella frit* L. *Ann. appl. Biol.* **46**, 465–74.

JOHNSON, C. G., 1950. The comparison of suction trap, sticky trap and tow-net for the quantitative sampling of small airborne insects. *Ann. appl. Biol.* **37**, 268–85.

JOHNSON, C. G., 1952. The role of population level, flight periodicity and climate in the dispersal of aphids. *Trans. IX int. Congr. Ent.* **1**, 429–31.

JOHNSON, C. G., 1960. A basis for a general system of insect migration and dispersal flight. *Nature* **186**, 348–50.

JOHNSON, C. G., 1963. Physiological factors in insect migration by flight. *Nature* **198**, 423–7.

JOHNSON, C. G., SOUTHWOOD, T. R. E. and ENTWISTLE, H. M., 1957. A new method of extracting arthropods and molluscs from grassland and herbage with a suction apparatus. *Bull. ent. Res* **48**, 211–18.

JOOSSE, E. N. G., 1975. Pitfall-trapping as a method for studying surface dwelling Collembola. *Z. Morph. Oekol. Tiere* **55**, 587–96.

JORDAN, A. M., 1962. The ecology of the *fusca* group of tsetse flies (*Glossina*) in Southern Nigeria. *Bull. ent. Res.* **53**, 355–85.

JUILLET, J. A., 1963. A comparison of four types of traps used for capturing flying insects. *Can. J. Zool.* **41**, 219–23.

KACZMAREK, W., 1963. An analysis of interspecific competition in communities of the soil macrofauna of some habitats in the Kampinos National Park. *Ekol. Polska A* **11** (17), 421–83.

KAWAI, S. and SUENAGA, O., 1960. Studies of the methods of collecting flies. III. On the effect of putrefaction of baits (fish). [In Japanese, Eng. summary.] *Endemic Dis. Bull. Nagasaki Univ.* **2**, 61–6.

KEMP-TURNBULL, P. ST J., 1960. Quantitative estimations of populations of the river crab *Potamon (Potamonautes) perlatus* (M. Edw.) in Rhodesian trout streams. *Nature* **185**, 481.

KENNEDY, J. S., 1961. A turning point in the study of insect migration. *Nature* **189**, 785–91.

KENNEDY, J. S., 1976. Olfactory responses to distant plants and other odour sources. *In* McKelvey, J. J. & Shorey, H. H. (eds.) *Chemical Control of Insect Behaviour: Theory & Application* (in press)

KENNEDY, J. S. and MARSH, D., 1974. Pheromone-regulated anemotaxis in flying moths. *Science* **184**, 999–1001.

KHAN, A. A., MAIBACH, H. I., STRAUSS, W. G. and FISHER J. L., 1970. Differential attraction of the yellow fever mosquito to vertebrate hosts. *Mosquito News* **30**, 43–7.

KING, W. V., BRADLEY, G. M., SMITH, C. N., and MCDUFFIE, W. C., 1960. *A handbook of the mosquito of the southeastern United States.* Agric. Handb. U. S. Dep. Agric. **173**, 1–188.

KLOCK, J. W. and BIDLINGMAYER, W. L., 1953. An adult mosquito sampler. *Mosquito News* **13**, 157–9.

KONO, T., 1953. On the estimation of insect population by time unit collecting. [In Japanese.] *Res. Popul. Ecol.* **2**, 85–94.

KOVNER, J. L. and PATIL, S. A., 1974. Properties of estimators of wildlife population density for the line transect method. *Biometrics* **30** (2), 225–30.

KOVROV, B. G. and MONCHADSKII, A. S., 1963. The possibility of using polarized light to attract insects. [In Russian.] *Ent. Obozr.* **42**, 49–55 (transl. *Ent. Rev.* **42**, 25–8).

KOWALSKI, R., 1976. Obtaining valid population indices from pitfall traps. *Bull. Acad. Pol. Sci., Ser. Sci. Biol. II*, **23**, 799–803.

KRING, J. B., 1970. Red spheres and yellow panels combined to attract apple maggot flies. *J. econ. Ent.* **63**, 466–69.

LAMBERT, H. L. and FRANKLIN, R. T., 1967. Tanglefoot traps for detection of the balsam woolly aphid. *J. Econ. Ent.* **60**, 1525–29.

LANDIN, J., 1968. Weather and diurnal periodicity of flight by *Helophorus brevipalpis* Bedel (Col.: Hydrophilidae). *Opusc. ent.* **33**, 28–36.

LANDIN, J. and STARK, E. 1973. On flight thresholds for temperature and wind velocity, 24-hour flight periodicity and migration of the beetle, *Helophorus brevipalpis* Bedel (Col. Hydrophilidae). *Zoon. Suppl.* **1**, 105–14.

LEECH, H. B., 1955. Cheesecloth flight trap for insects. *Can. Ent.* **85**, 200.

LE PELLEY, R. H., 1935. Observations on the control of insects by hand-collection. *Bull. ent. Res.* **26**, 533–41.

LEVIN, M. D., 1957. Artificial nesting burrows for *Osmia lignaria* Say. *J. econ. Ent.* **50**, 506–7.

LEWIS, T., 1959. A comparison of water traps, cylindrical sticky traps and suction traps for sampling Thysanopteran populations at different levels. *Entomologia exp. appl.* **2**, 204–15.

LEWIS, T. and MACAULAY, E. D. M., 1976. Design and elevation of sex-attractant traps for pea moth, *Cydia nigricana* (Steph.) and the effect of plume shape on catches. *Ecol. Ent.* **1**, 175–87.

LEWIS, W. J., SNOW, J. W. and JONES, R. L., 1971. A pheromone trap for studying populations of *Cardiochiles nigricaps*, a parasite of *Heliothis virescens*. *J. econ. Ent.* **64**, 1417–21.

LEWIS, T. and TAYLOR, L. R., 1965. Diurnal periodicity of flight by insects. *Trans. R. ent. Soc. Lond.* **116**, 393–469.

LINSLEY, E. G., MACSWAIN, J. W. and SMITH, R. F., 1952. Outline for ecological life histories of solitary and semi-social bees. *Ecology* **33**, 558–67.

LOOMIS, E. C., 1959a. A method for more accurate determination of air volume displacement of light traps. *J. econ. Ent.* **52**, 343–5.

LOOMIS, E. C., 1959b. Selective response of *Aedes nigronaculis* (Ludlow) to the Minnesota light trap. *Mosquito News* **19**, 260–3.

LOOMIS, E. C. and SHERMAN, E. J., 1959. Comparison of artificial shelter and light traps for measurements of *Culex tarsalis* and *Anopheles freeborni* populations. *Mosquito News* **19**, 232–7.

LOVE, G. J. and SMITH, W. W., 1957. Preliminary observations on the relation of light trap collections to mechanical sweep net collections in sampling mosquito populations. *Mosquito News* **17**, 9–14.

LUCZAK, J. and WIERZBOWSKA, T., 1959. Analysis of likelihood in relation to the length of a series in the sweep method. *Bull. Acad. pol. Sci., Ser. Sci. Biol.* **7**, 313–18.

LUFF, M. L., 1968. Some effects of formalin on the numbers of Coleoptera caught in pitfall traps. *Ent. mon. Mag.* **104**, 115–6.

LUFF, M. L., 1973. The annual activity pattern and life cycle of *Pterostichus madidus* (F.) (Col. Carabidae). *Ent. scand.* **4**, 259–73.

LUFF, M. L., 1975. Some features influencing the efficiency of pitfall traps. *Oecologia* **19**, 345–57.

LUMSDEN, W. H. R., 1958. A trap for insects biting small vertebrates. *Nature* **181**, 819–20.

MACAN, T. T., 1958. Methods of sampling the bottom fauna in stony streams. *Mitt. int. Ver. Limnol.* **8**, 1–21.

McCLELLAND, G. A. H. and CONWAY, G. R., 1971. Frequency of blood-feeding in the mosquito *Aedes aegypti*. *Nature* **232**, 485–6.

MACLEOD, J. and DONNELLY, J., 1956. Methods for the study of Blowfly popultions. I. Bait trapping. Significance of limits for comparative sampling. *Ann. appl. Biol.* **44**, 80–104.

McPHAIL, M., 1939. Protein lures for fruit flies. *J. econ. Ent.* **32**, 758–61.

MALAISE, R., 1937. A new insect-trap. *Ent. Tidskr.* **58**, 148–60.

MANN, K. H., 1955. the ecology of the British freshwater leeches. *J. Anim. Ecol.* **24**, 98–119.

MASAKI, J., 1959. Studies on rice crane fly (*Tipula aino* Alexander, Tipulidae, Diptera) with special reference to the ecology and its protection. [In Japanese.] *J. Kanto-Tosan agric. Exp. Sta.* **13**, 195 pp.

MASON, H. C., 1963. Baited traps for sampling *Drosophila* populations in tomato field plots. *J. econ. Ent.* **56**, 897–8.

MAXWELL, C. W., 1965. Tanglefoot traps as indicators of apple maggot fly activities. *Can. Ent.* **97**, 110.

MAYER, K., 1961. Untersuchungen über das Wahlverhalten der Fritfliege (*Oscinella frit* L.) beim Anflug von Kulturpflanzen im Feldversuch mit der Fangschalen-methode. *Mitt. biol. Bund. Anst. Ld- u. Forstw., Berlin* **106**, 1–47.

MEDLER, J. T. and FYE, R. E., 1956. Biology of *Ancistrocerus antilope* (Panzer) in trap nests in Wisconsin. *Ann. ent. Soc. Am.* **49**, 97–102.

MEDLER, J. T., 1964. Biology of *Rygchium foraminatum* in trap-nests in Wisconsin (Hymbroptera: Vespidae). *Ann. ent. Soc. Am.* **57**, 56–60.

MELLANBY, K., 1962. Sticky traps for the study of animals inhabiting the soil surface. *In* Murphy, P. W. (ed.), *Progress in Soil Zoology* Butterworths, London. 226–7.

MENHINICK, E. F., 1963. Estimation of insect population density in herbaceous vegetation with emphasis on removal sweeping. *Ecology* **44**, 617–21.

MERZHEEVSKAYA, O. I. and GERASTEVICH, E. A., 1962. A method of collecting living insects at light. [In Russian.] *Zool. Zhur.* **41**, 1741–3.

MICHELBACHER, A. E. and MIDDLEKAUFF, W. W., 1954. Vinegar fly investigations in Northern California. *J. econ. Ent.* **47**, 917–22.

MIKKOLA, K., 1972. Behavioural and electrophysiological responses of night-flying insects, especially *Lepidoptera*, to near-ultraviolet visible light. *Ann. zool. Fennici* **9**, 225–54.

MILLER, T. A., STRYKER, R.G., WILKINSON, R. N. and ESAH, S., 1970. The influence of moonlight and other environmental factors on the abundance of certain mosquito species in light-trap collections in Thailand. *J. med. Ent.* **7**, 555–61.

MINTER, D. M., 1961. A modified Lumsden suction-trap for biting insects. *Bull. ent. Res.* **52**, 233–8.

MITCHELL, B., 1963. Ecology of two carabid beetles, *Bembidion lampros* (Herbst.) and *Trechus quadristriatus* (Schrank). II. *J. Anim. Ecol.* **32**, 377–92.

MOERICKE, V., 1950. Über den Farbensinn der Pfirsichblattlaus *Myzodes persicae Sulz.* *Z. Tierpsychol.* **7**, 265–74.

MORAN, P. A. P., 1951. A mathematical theory of animal trapping. *Biometrika* **38**, 307–11.

MORELAND, C., 1954. A wind frame for trapping insects in flight. *J. econ. Ent.* **47**, 944.

MORGAN, C. V. G. and ANDERSON, N. H., 1958. Techniques for biological studies of tetranychid mites, especially *Bryobia arborea* M. & A. and *B. praetiosa* Koch. (Acarina: Tetranychidae). *Can. Ent.* **90**, 212–15.

MORRIS, K. R. S., 1960. Trapping as a means of studying the game tsetse, *Glossina pallidipes* Aust. *Bull. ent. Res.* **51**, 533–57.

MORRIS, K. R. S., 1961. Effectiveness of traps in tsetse surveys in the Liberian rain forest. *Am. J. trop. Med. hyg.* **10**, 905–13.

MORRIS, K. R. S. and MORRIS, M. G., 1949. The use of traps against tsetse in West Africa. *Bull. ent. Res.* **39**, 491–523.

MUIRHEAD-THOMSON, R. C. (ed.), 1963. *Practical entomology in Malaria eradication.* WHO (MHO/PA/62.63) Part 1, Chapter 1 (mimeographed).

MULHERN, T. D., 1942. New Jersey mechanical trap for mosquito surveys. *N.J. Agric. Exp. Sta. Circ.* **421**, 1–8.

MULHERN, T. D., 1953. Better results with mosquito light traps through standardizing mechanical performance. *Mosquito News* **13**, 130–3.

MULLA, M. S., GEORGHIOU, G. P. and DORNER, R. W., 1960. Effect of ageing and concentration on the attractancy of proteinaceous materials to *Hippelates* gnats. *Ann. ent. Soc. Am.* **53**, 835–41.

MUNDIE, J. H., 1964. A sampler for catching emerging insects and drifting materials in streams. *Limnol. Oceanogr.* **9**, 456–9.

MURRAY, W. D., 1963. Measuring adult populations of the pasture mosquito, *Aedes migromaculis* (Ludlow). *Proc. 27th Conf. Calif. Mosq. Contr. Ass.* **1959**, 67–71.

MYERS, K., 1956. Methods of sampling winged insects feeding on the rabbit *Oryctolagus cuniculus* (L.). *Aust. C.S.I.R.O. Wildl. Res.* **1**, 45–58.

NAKAMURA, K., ITO, Y., MIYASHITA, K. and TAKAI, A., 1967. The estimation of population density of the green rice leafhopper, *Nephotettix cincticeps* Uhler. In spring field by the capture–recapture method. *Res. Pop. Ecol.* **9**, 113–29.

NEILSON, W. T. A., 1960. Field tests of some hydrolyzed proteins as lures for the apple maggot, *Rhagoletis pomonella* (Walsh). *Can. Ent.* **92**, 464–7.

NICKERSON, B., 1963. An experimental study of the effect of changing light intensity on the activity of adult locusts. *Ent. mon. Mag.* **99**, 139–40.

NIELSEN, B. OVERGAARD, 1963. The biting midges of *Lyngby aamose* (Culicoides: Ceratopogonidae). *Natura jutl.* **10**, 48 pp.

NIELSEN, E. T., 1960. A note on stationary nets. *Ecology* **41**, 375–6.

NIJHOLT, W. W. and CHAPMAN, J. A., 1968. A flight trap for collecting living insects. *Can. Ent.* **100**, 1151–53.

ODINTSOV, V. S., 1960. (Air catch of insects as a method of study upon entomofauna of vast territories.) [In Russian.] *Ent. Obozr.* **39**, 227–30.

OTAKE, A. and KONO, T., 1970. Regional characteristics in population trends of smaller brown planthopper, *Laodelphax striatellus* (Fallen) (Hemiptera: Delphacidae), a vector of rice stripe disease: an analytical study of light trap records. *Bull. Shikoku Agric. exp. Sta.* **21**, 127–47.

OTAKE, A., TAKIGAWA, N. and OYAMA, M., 1974. A device to trap and detect the presence of *Spodoptera litura* F. (Lepidoptera: Noctuidae) by simple virgin female traps. *Bull. Shikoku Agric. exp. Sta.* **28**, 59–64.

PARIS, O. H., 1965. The vagility of P³²-labelled Isopods in grassland. *Ecology* **46**, 635–48.

PARKER, A. H., 1949. Observations on the seasonal and daily incidence of certain biting midges (*Culicoides* Latreille – Diptera, Ceratopogonidae) in Scotland. *Trans. R. ent. Soc. Lond.* **100**, 179–90.

PAVIOUR-SMITH, K., 1964. The life history of *Tetratoma fungorum* F. (Col., Tetratomidae) in relation to habitat requirements, with an account of eggs and larval stages. *Ent. mon. Mag.* **100**, 118–34.

PETRUSKA, F., 1968. Members of the group Silhini as a component part of the insects fauna of sugar beet fields in the Unicov Plain. *Acta Univ. Palackianae Olomucenscs Fac. Rerum Natur.* **38**, 189–200.

PETRUSKA, F., 1969. On the possibility of escape of various components of the epigeic fauna of the fields from the pitfall traps containing formalin (Coeoptera) (in Czech, English summary). *Acta Univ. palackianae Olomucensis Fac. Rerum Natur.* **31**, 99–124.

PIECZYNSKI, E., 1961. The trap method of capturing water mites (Hydracarina). *Ekol. Polska B* **7** (2), 111–15.

PRESCOTT, H. W. and NEWTON, R. C., 1963. Flight Study of the clover root Curculio. *J. econ. Ent.* **56**, 368–70.

PROKOPY, R. J., 1968. Sticky spheres for estimating Apple Maggot adult abundance. *J. econ. Ent.* **61**, 1082–5.

PROKOPY, R. J., 1977. Host plant influences on the reproductive biology of Tephritidae. *In* Labeyrie, V. (ed.), *Comportement des Insectes et Milieu Trophique.* Coll. Int. C.N.R.S. **265**, 305–334.

PROVOST, M. W., 1960. The dispersal Of *Aedes taeniorhynchus*. III. Study methods for migratory exodus. *Mosquito News* **20**, 148–61.

RACE, S. R., 1960. A comparison of two sampling techniques for lygus bugs and stink bugs on cotton. *J. econ. Ent.* **53**, 689–90.

RENNISON, B. D. and ROBERTSON, D. H. H., 1959. The use of carbon dioxide as an attractant for catching tsetse. *Rep. E. Afr. Trypan. Res. Organ.* **1958**, 26.

RILEY, J. R., 1974. Radar observations of individual desert locusts (*Schistocerca gregaria* (Forsk) (Orthoptera Locustidae)). *Bull. ent. Res.* **64**, 19–32.

RILEY, J. R., 1975. Collective orientation in night-flying insects. *Nature* **253**, 113–4.

RIOUX, J. A. and GOLVAN, Y. J., 1969. Epidemiologie des Leishmanioses dans le sud de la France. *Monog. Inst. Nat. Sante Recherche Med.* **37**, 220 pp.

RIVARD, I., 1962. Un piège à fosse amélioré pour la capture d'insectes actifs à la surface du sol. *Can. Ent.* **94**, 270–1.

ROBERTS, R. H., 1970. Color of Malaise trap and collection of Tabanidae. *Mosquito News* **30**, 567–71.

ROBERTS, R. H., 1972. The effectiveness of several types of Malaise traps for the collection of Tabanidae and Culicidae. *Mosquito News*, **32**, 542–7.

ROBERTS, R. H., 1976. The comparative efficiency of six trap types for the collection of Tabanidae (Diptera). *Mosquito News* **36**, 530–7.

ROBERTS, R. J and SMITH, T. J. R., 1972. A plough technique for sampling soil insects. *J. appl. Ecol.* **9**, 427–30.

ROBINSON, H. S. and ROBINSON, P. J. M., 1950. Some notes on the observed behaviour of Lepidoptera in flight in the vicinity of light-sources together with a description of a light-trap designed to take entomological samples. *Ent. Gaz.* **1**, 3–15.

ROELOFS, W. L., 1975. Insect communication – chemical. *In* Pimental, D. (ed.) *Insects, Science & Society*, 79–99. Academic Press, New York.

ROESLER, R., 1953. Über eine Methode zur Feststellung der Flugzeit schädlicher Fliegenarten (Kirschfliege, Kohlfliege, Zwiebelfliege). *Mitt. biol. Reichsant. (Zent Anst.)Ld- u. Forstw.*, Berlin **75**, 97–9.

ROGERS, D. J and SMITH, D. T., 1977. A new electric trap for tsetse flies. *Bull. ent. Res.* **67**, 153–9.

ROMNEY, V. E., 1945. The effect of physical factors upon catch of the beet leafhopper (*Eutettix tenellus* (Bak.)) by a cylinder and two sweep-net methods. *Ecology* **26**, 135–48.

ROOS, T., 1957. Studies on upstream migration in adult stream-dwelling insects. *Inst. Freshw. Res. Drottningholm Rep.* **38**, 167–93.

ROTHSCHILD, M., 1961. Observations and speculations concerning the flea vector of myxomatosis in Britain. *Ent. mon. Mag.* **96**, 106–9.

RUDD, W. G. and JENSEN, R. L., 1977. Sweep net and ground cloth sampling for insects in soybeans. *J. econ. Ent.* **70**, 301–4.

RUDINSKY, J. A., 1977. Olfactory and auditory signals mediating behavioural patterns of bark beetles. *In* Labeyrie, V. (ed.), *Comportement des Insectes et Milieu Trophique*. Coll. int. CNRS. **265**, 188–95.

RUFFINELLI, A., ORLANDO, A. and BIGGI, E., 1962. Novos ensaios com Substâncias atrativas para as 'môscas das frutas' – *Ceratitis capitata* (Wied.) e *Anastrepha mombinpraeoptans* Sein. *Arq. Inst. biol. São Paulo* **27** (**1960**), 1–9.

SALITERNIK, Z., 1960. A mosquito light trap for use on cesspits. *Mosquito News* **20**, 295–6.

SAUGSTAD, E. S., BRAM, R. A. and NYQUIST, W. E., 1967. Factors influencing sweep-net sampling of alfalfa. *J. econ. Ent.* **60**, 421–6.

SAUNDERS, D. S., 1964. The effect of site and sampling and method on the size and composition of catches of tsetse flies (*Glossina*) and Tabanidae (Diptera). *Bull. ent. Res.* **55**, 483.

SCHAEFER, G. W., 1976. 8. Radar observations of insect flight. *In* Rainey, R. C. (ed.) *Insect Flight; Symp. R. ent. Soc. Lond.* **7**, 157–97.

SCHEEPERS, C. C. and GUNN, D. L., 1958. Enumerating populations of adults of the red locust, *Nomadacris septemfasciata* (Serville), in its outbreak areas in East and Central Africa. *Bull. ent. Res.* **49**, 273–85.

SCHNEIDER, D., LANGE, R., SCHWARZ, F., BEROZA, M. and BIERLE, B. A., 1974. Attraction of male Gypsy and Nun moths to disparlure and some of its chemical analogues. *Oecologia* **14**, 19–36.

SEBER, G. A. F., 1973. *The Estimation of Animal Abundance.* 506 pp., Griffin, London.

SEBER, G. A. F. and LE CREN, E. D., 1967. Estimating population parameters from catches large relative to the population. *J. Anim. Ecol.* **36**, 631–43.

SEBER, G. A. F. and WHALE, J. F., The removal method for two and three samples. *Biometrics* **26** (3), 393–400.

SEN, A. R., TOURIGNY, J. and SMITH, G. E. J., 1974. On the line transect sampling method. *Biometrics* **30** (2), 329–40.

SERVICE, M. W., 1963. The ecology of the mosquitoes of the northern Guinea savannah of Nigeria. *Bull. ent. Res.* **54**, 601–32.

SERVICE, M. W., 1976. *Mosquito Ecology – Field Sampling Methods.* Applied Science Publishers, London.

SEVASTPULO, D. G., 1963. Field notes from East Africa – Part XI. *Entomologist* **96**, 162–5.

SHEMANCHUK, J. A., 1959. Mosquitoes (Diptera: Culicidae) in irrigated areas of southern Alberta and their seasonal changes in abundance and distribution. *Can. J. Zool.* **37**, 899–912.

SHOREY, H. H., 1973. Behavioural responses to insect pheromones. *Ann. Rev. Ent.* **18**, 349–80.

SIDDORN, J. W. and BROWN, E. S., 1971. A Robinson light trap modified for segregating samples at predetermined time intervals, with notes on the effect of moonlight on the periodicity of catches of insects. *J. appl. Ecol.* **8**, 69–75.

SIMPSON R. G. and BERRY, R. E., 1973. A twenty-four hour directional aphid trap. *J. econ. Ent.* **66**, 291–2.

SKELLAM, J. G., 1958. The mathematical foundations underlying the use of line transects in animal ecology. *Biometrics* **14**, 385–400.

SMITH, B. D., 1962. The behaviour and control of the blackcurrant gall mite *Phytoptus ribis* (Nal.). *Ann. appl. Biol.* **50**, 327–34.

SMITH, I. M. and RENNISON, B. D., 1961. Studies of the sampling of *Glossina pallidipes* Aust. I, II. *Bull. ent. Res.* **52**, 165–89; III and IV. *ibid.* **52**, 601–19.

SMITH, P. W., TAYLOR, J. G. and APPLE, J. W., 1959. A comparison of insect traps equipped with 6- and 15-watt black light lamps. *J. econ. Ent.* **52**, 1212–14.

SNOW, W. E., PICKARD, E. and SPARKMAN, R. E., 1960. A fan trap for collecting biting insects attacking avian hosts. *Mosquito News* **20**, 315–16.

SOUTHWOOD, T. R. E., 1960. The flight activity of Heteroptera. *Trans. R. ent. Soc. Lond.* **112** (8), 173–220.

SOUTHWOOD, T. R. E., 1962. Migration of terrestrial arthropods in relation to habitat. *Biol. Rev.* **37**, 171–214.

SOUTHWOOD, T. R. E., JEPSON, W. F. and VAN EMDEN, H. F., 1961. Studies on the behaviour of *Oscinella frit* L. (Diptera) adults of the panicle generation. *Entomologia exp. appl.* **4**, 196–210.

STAGE, H. H., GJULLIN, C. M. and YATES, W. W., 1952. Mosquitoes of the Northwestern States. *U.S.D.A., Agric. Handbk* **46**, 95 pp.

STAHL, C., 1954. Trapping hornworm moths. *J. econ. Ent.* **47**, 879–82.

STAMMER, H. J., 1949. Die Bedeutung der Äthylenglycolfallen für tierökologische und phänologische Untersuchungen. *Verh. dtsch. Zoologen. Kiel* (**1948**), 387–91.

STANLEY, J. M. and DOMINICK, C. B., 1957. Response of tobacco and tomato hornworm moths to black light. *J. econ. Ent.* **51**, 78–80.

STANLEY, J. M. and DOMINICK, C. B., 1970. Funnel size and lamp wattage influence on light-trap performance. *J. econ. Ent.* **63**, 1423–6.

STAPLES, R. and ALLINGTON, W. B., 1959. The efficiency of sticky traps in sampling epidemic populations of the Eriophyid mite *Aceria tulipae* (K.), vector of wheat streak mosaic virus. *Ann. ent. Soc. Am.* **52**, 159–64.

STARR, D. F. and SHAW, J. G., 1944. Pyridine as an attractant for the Mexican fruit fly. *J. econ. Ent.* **37**, 760–3.

STEIGEN, A. L., 1973. Sampling invertebrates active below a snow cover. *Oikos* **24**, 373–6.

STEINER, P., WENZEL, F. and BAUMERT, D., 1963. Zur Beeinflussung der Arthropodenfauna nordwestdeutscher Kartoffelfelder durch die Anwendung synthetischer Kontaktinsektizide. *Mitt. biol. BundAnst. Ld- u. Forstw., Berlin* **109**, 38 pp.

SUDIA, W. D. and CHAMBERLAIN, R. W., 1962. Battery-operated light trap, an improved model. *Mosquito News.* **22**, 126–9.

SYMMONS, P. M., DEAN, G. J. W. and STORTENBERER, C. W., 1963. The assessment of the size of populations of adults of the red locust, *Nomadacris septemfasciata* (Serville), in an outbreak area. *Bull. ent. Res.* **54**, 549–69.

TAFT, H. M. and JERNIGAN, C. E., 1964. Elevated screens for collecting boll weevils flying between hibernation sites and cottonfields. *J. econ. Ent.* **57**, 773–5.

TAYLOR, L. R., 1962. The efficiency of cylindrical sticky insect traps and suspended nets. *Ann. appl. Biol.* **50**, 681–5.

TAYLOR, L. R., 1963. Analysis of the effect of temperature on insects in flight. *J. Anim. Ecol.* **32**, 99–112.

TAYLOR, L. R. and BROWN, E. S., 1972. Effects of light-trap design and illumination on samples of moths in the Kenya highlands. *Bull. ent. Res.* **62**, 91–112.

TAYLOR, L. R. and CARTER, C. I., 1961. The analysis of numbers and distribution in an aerial population of Macrolepidoptera. *Trans. R. ent. Soc. Lond.* **113**, 369–86.

TAYLOR, L. R. and FRENCH, R. A., 1974. Effects of light-trap design and illumination on samples of moths in an English woodland. *Bull. ent. Res.* **63**, 583–94.

TAYLOR, L. R. and PALMER, J. M. P., 1972. Aerial sampling. *In* van Emden, H. F. (ed.) *Aphid Technology*. Academic Press, London.

TAYLOR, L. R. and TAYLOR, R., 1977. Aggregation, migration and population mechanics. *Nature* **265**, 415–21.

TEDDERS, W. L. and EDWARDS, G. W., 1972. Effects of black light trap design and placement on catch of adult Hickory Shuckworms. *J. econ. Ent.* **65** (6), 1624–7.

THORSTEINSON, A. J., BRACKEN, G. K. and HANEC, W., 1965. The orientation behaviour of horse flies and deer flies (Tabanidae, Diptera). III. The use of traps in the study of orientation of Tabanids in the field. *Entomologia exp. appl.* **8**, 189–92.

TOWNES, H., 1962. Design for a Malaise trap. *Proc. ent. Soc. Wash.* **64**, 253–62.

TRETZEL, E., 1955. Technik und Bedeutung des Fallenfanges für ökologische Untersuchungen. *Zool. Anz.* **155**, 276–87.

TSHERNYSHEV, W. B., 1961. (Comparison of field responses of insects to the light of a mercury-quartz lamp and clear ultra-violet radiation of the same lamp.) [In Russian.] *Ent. Obozar.* **40**, 568–70 (transl. *Ent. Rev.* **40**, 308–9).

TURNER, E. C., 1972. An animal-baited trap for the collection of *Culicoides* spp. (Diptera: Ceratopogonidae). *Mosquito News* **32**, 527–30.

TURNER, F. B., 1962. Some sampling characteristics of plants and arthropods of the Arizona desert. *Ecology* **43**, 567–71.

UETZ, G. W., 1977. Co-existence in a guild of wandering spiders. *J. Anim. Ecol.* **46**, 531–41.

UETZ, G. W. and UNZICKER, J. D., 1976. Pitfall trapping in ecological studies of wandering spiders. *J. Arachnology* **3**, 101–11.

USINGER, R. L. (ed.), 1963. *Aquatic insects of California.* 508 pp., University of California Press, Berkeley and Los Angeles.

USINGER, R. L. and KELLEN, W. R., 1955. The role of insects in sewage disposal beds. *Hilgardia* **23**, 263–321.

VALE, G. A., 1971. Artificial refuges for tsetse flies (*Glossina* spp.). *Bull. ent. Res.* **61**, 331–50.

VALE, G. A., 1974. The responses of tsetse flies (Diptera, Glossinidae) to mobile and stationary baits. *Bull. ent. Res.* **64**, 545–88.

VANDERPLANK, F. L., 1960. The availability of the coconut bug, *Pseudotheraptus wayi* Brown (Coreidae). *Bull. ent. Res.* **51**, 57–60.

VAVRA, J. R., CARESTIA, R. R., FROMMER, R. L. and GERBERG, E. J., 1974. Field evaluation of alternative light sources as mosquito attractants in the Panama Canal Zone. *Mosquito News* **34**, 382–4.

VERHEIJEN, F. J., 1960. The mechanisms of the trapping effect of artificial light sources upon animals. *Arch. Néerland. Zool.* **13**, 1–107.

VITÉ, J. P. and GARA, R. I., 1962. Volatile attractants from Ponderosa pine attacked by bark beetles (Coleoptera: Scolytidae). *Contr. Boyce Thompson Inst.* **21**, 251–73.

VLIJM, L. and KESSLER-GESCHIERE, A. M., 1967. The phenology and habitat of *Pardosa monticola, P. nigriceps* and *P. pullata* (Araneae, Lycosidae). *J. Anim. Ecol.* **36**, 31–56.

WADA, Y., 1962. Studies on the population estimation for insects of medical importance. I. A method of estimating the population size of mosquito larvae in a fertilizer pit. *Endemic Dis. Bull. Nagasaki Univ.* **4**, 22–30. II. A method of estimating the population size of larvae of *Aedes togoi* in the tide-water rock pool. *ibid.* **4**, 141–56.

WAKERLEY, S. B., 1963. Weather and behaviour in carrot fly (*Psila rosae* Fab. Dipt. Psilidae) with particular reference to oviposition. *Entomologia exp. appl.* **6**, 268–78.

WALKER, A. R. and BOREHAM, P. F. L., 1976. Saline, as a collecting medium for *Culicoides* (Diptera: Ceratopogonidae) in blood-feeding and other studies. *Mosquito News* **36**(1), 18–20.

WALL, C., GREENWAY, A. R. and BURT, P. E., 1976. Electroantennographic and field responses of the Lea moth, *Cydia nigricana*, to synthetic sex attractants and related compounds. *Physiol. Ent.* **1**, 151–7.

WALOFF, N. and BAKKER, K., 1963. The flight activity of Miridae (Heteroptera) living on broom, *Sarothamnus scoparius* (L.) Wimm. *J. Anim. Ecol.* **32**, 461–80.

WALSH, G. B., 1933. Studies in the British necrophagous Coleoptera. II. The attractive powers of various natural baits. *Ent. mon. Mag.* **69**, 28–32.

WATERS, T. F., 1962. Diurnal periodicity in the drift of stream invertebrates. *Ecology* **43**, 316–20.

WATERS, T. F., 1969. Invertebrate drift-ecology and significance to stream fishes. *In Symp. on Salmon and Trout in Streams*, 121–34, Univ. B. C., Vancouver.

WELCH, R. C., 1964. A simple method of collecting insects from rabbit burrows. *Ent. mon. Mag.* **100**, 99–100.

WENK, P. and SCHLÖRER, G., 1963. Wirtsorientierung und Kopulation bei blutsaugenden Simuliiden (Diptera). *Zeit. Trop. Med. Parasit.* **14**, 177–91.

WHARTON, R. H., EYLES, D. E. and WARREN, MCW., 1963. The development of methods for trapping the vectors of Monkey Malaria. *Ann. Trop. Med. Parasit., Liverpool* **57**, 32–46.

WHITE, E. G., 1964. A design for the effective killing of insects caught in light traps. *N.Z. Entomologist* **3**, 25–7.

WILDE, W. H. A., 1962. Bionomics of the pear psylla, *Psylla pyricola* Foerster in pear orchards of the Kootenay valley of British Columbia, 1960. *Can. Ent.* **94**, 845–9.

WILKINSON, P. R., 1961. The use of sampling methods in studies of the distribution of larvae of *Boophilus microplus* on pastures. *Aust. J. Zool.* **9**, 752–83.

WILLIAMS, C. B., 1935. The times of activity of certain nocturnal insects, chiefly Lepidoptera as indicated by a light-trap. *Trans. R. ent. Soc. Lond.* **83**, 523–55.

WILLIAMS, C. B., 1939. An analysis of four years captures of insects in a light trap. Part I. General survey; sex proportion; phenology and time of flight. *Trans. R. ent. Soc. Lond.* **89**, 79–132.

WILLIAMS, C. B., 1940. An analysis of four years captures of insects in a light trap. Part II. The effect of weather conditions on insect activity; and the estimation and forecasting of changes in the insect population. *Trans. R. ent. Soc. Lond.* **90**, 228–306.

WILLIAMS, C. B., 1948. The Rothamsted light trap. *Proc. R. ent Soc. Lond. A* **23**, 80–5.

WILLIAMS, C. B., 1951. Comparing the efficiency of insect traps. *Bull. ent. Res.* **42**, 513–17.

WILLIAMS, C. B., FRENCH, R. A. and HOSNI, M. M., 1955. A second experiment on testing the relative efficiency of insect traps. *Bull. ent. Res.* **46**, 193–204.

WILLIAMS, C. B., SINGH, B. P. and EL ZIADY, S., 1956. An investigation into the possible effects of moonlight on the activity of insects in the field. *Proc. R. ent. Soc. Lond. A* **31**, 135–44.

WILLIAMS, D. F., 1973. Sticky traps for sampling populations of *Stomoxys calcitrans*. *J. econ. Ent.* **66**, 1279–80.

WILLIAMS, G., 1958. Mechanical time-sorting of pitfall captures. *J. Anim. Ecol.* **27**, 27–35.

WILSON, B. H. and RICHARDSON, C. G., 1970. Attraction of deer flies (*Chrysops*) (Diptera: Tabanidae) to traps baited with dry ice under field conditions in Louisiana. *J. med. Ent.* **7**, 625.

WILTON, D. P. and FAY, R. W., 1972. Air-flow direction and velocity in light trap design. *Entomologia exp. appl.* **15**, 377–86.

WOOD, D. L., STARK, R. W., SILVERSTEIN, R. M. and RODIN, J. O., 1967. Unique synergistic effects produced by the principal sex attractant compounds of *Ips confusus* (LeConte) (Coleoptera: Scolytidae). *Nature* **215**, 206.

WORTH, C. B. and JONKERS, A. H., 1962. Two traps for mosquitoes attracted to small vertebrate animals. *Mosquito News* **22**, 18–21.

YAPP, W. B., 1956. The theory of line transects. *Bird Study* **3**, 93–104.

YEO, D. and FOSTER, R., 1958. Preliminary note on a method for the direct estimation of populations of *Pseudotheraptus wayi* Brown on coconut palms. *Bull. ent. Res.* **49**, 585–90.

YOSHIMOTO, C. M. and GRESSITT, J. L., 1959. Trapping of airborne insects on ships on the Pacific (Part II). *Proc. Hawaiian Ent. Soc.* **17** (1), 150–5.

YOSHIMOTO, C. M., GRESSITT, J. L. and WOLFF, T., 1962. Airborne insects from the Galathea expedition. *Pacific Insects* **4**, 269–91.

ZHOGOLEV, D. T., 1959. Light-traps as a method for collecting and studying the insect vectors of disease organisms. [In Russian.] *Ent. Obozr.* **38**, 766–73.

ZIMMERMAN, J. R., 1960. Seasonal population changes and habitat preferences in the genus *Laccophilus* (Coleoptera: Dytiscidae). *Ecology* **41**, 141–52.

ZIPPIN, C., 1956. An evaluation of the removal method of estimating animal populations. *Biometrics* **12**, 163–89.

ZIPPIN, C., 1958. The removal method of population estimation. *J. Wildl. Mgmt* **22**, 82–90.

8

Estimates based on Products and Effects of Insects

Measures of the size of populations based on the magnitude of their products or effects are often referred to as population indices. The relationship of these indices to the absolute population varies from equivalence, when the number of exuviae are counted, to no more than an approximate correlation, when the index is obtained from general measures of damage.

8.1 Products

8.1.1 Exuviae

The larval or pupal exuviae of insects with aquatic larval stages are often left in conspicuous positions around the edges of water bodies, and where it is possible to gather these they will provide a measure of the emergence rate and of the absolute population of newly emerged adults. The method is most easily applied to large insects such as dragonflies (Corbet, 1957). The exuviae of the last immature, subterranean stage of some insects are also often conspicuous and with cicadas Strandine (1940) and Dybas & Lloyd (1962, 1974) found that the number of larval cases per unit area was correlated with the number of emergence holes and provided a useful index of population; different species could be recognized from the exuviae. Pupal exuviae on the branches and trunks of peach trees were used to provide relative estimates of moth, *Synanthedon*; these accorded with estimates made with pheromone traps (Yonce *et al.* 1977).

Paramonov (1959) drew attention to the possibility of obtaining an index of the population of arboreal insects from their exuviae, more particularly from the head capsules of lepidopterous larvae that may be collected in the same way as frass (see below). He found that the head capsules could be separated from the other debris by flotation. In most species the stage of the larva can be determined from the size of the capsule and if all the capsules could be collected an elegant measurement of absolute population would be obtained. However, a significant, and probably fluctuating, number fail to be recovered and this variable needs to be carefully investigated in a given situation before the method is employed to measure either absolute population or even an index of it.

8.1.2 Frass

The faeces of insects are generally referred to as frass; as this is based on the incorrect use of a German word, the term feculae has been suggested (Frost, 1928), but seldom used. (The German words *frass* = insect damage or food, whilst *kot* = faeces.) The frass-drop, the number of frass pellets falling to the ground, was first used as an index of both population and insect damage by a number of forest entomologists in Germany (Rhumbler, 1929; Gösswald 1935). The falling frass is collected in cloth or wooden trays or funnels under the trees (Fig. 8.1). In order for such collections to be of maximum value for population estimation one should be able to identify the species and the developmental stage, information is also required on the quantity of frass produced per individual per unit of time and the proportion of this that falls to the ground and is collected.

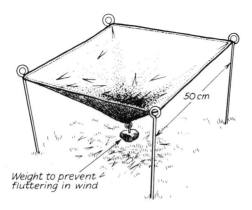

50 cm

Weight to prevent
fluttering in wind

Fig. 8.1 Frass-collector, constructed of cloth (after Tinbergen, 1960).

Identification

Although the frass of early instars of several species may be confused, in the later instars it is generally distinctive and keys for the identification of frass pellets have been prepared (Nolte, 1939; Morris, 1942; Weiss & Boyd, 1950, 1952; Hodson & Brooks, 1956; Solomon, 1977). With the nun moth (*Lymantria monacha*) Eckstein (1938) concluded the frass size was a more reliable indicator of larval instar than the width of the head capsule; in the sawfly *Diprion hercyniae* (Morris, 1949), and the armyworm (*Pseudaletia unipuncta*) (Pond, 1961) instar can easily be determined from pellet size. Water-soluble constituents are washed out when the pellets are exposed to rain and so pellet weight may fall by as much as 30 %; however, in these species the volume is not affected. Bean (1959) considered that in the spruce budworm (*Choristoneura fumiferana*), the width of the pellet rather than its volume was closely correlated with larval instar; furthermore this author also found that larvae, feeding on pollen, ejected frass pellets that were considerably larger than those produced by the same instar larvae feeding on foliage.

The rate of frass production

This can be measured in terms of dry weight or number of pellets, but the two may not vary in direct proportion (Fig. 8.2*a*). Whatever units are used the rate of frass production is affected by many factors: the phase and generation (Fridén, 1958; Iwao, 1962), the temperature and humidity (Gösswald, 1935; Morris, 1949; Green & de Freitas, 1955; Pond, 1961), the available food, both the plant species and its condition (Morris, 1949; Green & de Freitas 1955; Pond 1961; Dadd, 1960; Waldbauer, 1964), the developmental rate as determined by the phenology of the season (Tinbergen, 1960) and the presence of adult parasites (Green & de Freitas 1955) (Fig. 8.2). Therefore it is not possible to assume that the number of pellets produced in different areas or different seasons reflects very precisely changes in the actual populations. Direct comparisons should be made with other measurements of population; such a comparison was made by Tinbergen (1960), who found that in six out of seven years frass-drop provided a reliable index of the absolute population. When a difference is indicated this must be investigated further to determine whether the absolute method has become inaccurate or if the frass-drop rate per individual has changed. The latter can be measured by the use of a coprometer, that collects the frass voided each hour into a separate compartment (Green & Henson, 1953; Green & de Freitas, 1955).

Efficiency of collection

Another variable is the proportion of the frass that actually reaches the ground and is collected. As Morris (1949) points out, a greater proportion of the frass will be retained, on the foliage, in calm weather than under more windy conditions, and where the young larvae produce webbing the frass will tend to be caught up with this and its fall delayed (Bean, 1959).

Most of the above work has referred to Lepidoptera and sawflies; the frass of Coleoptera may also be identified (Eckstein, 1939) and Campbell (1960) used frass-drop as an index of population in studies of stick insects, Phasmida.

It can be concluded that frass-drop measurements may with caution be used as a second method to check trends established by another technique (see p. 4); if comparisons are limited to a certain area where the insect has but a single host plant the chances that a good relative estimate will be obtained are maximal.

8.1.3 Other products

Populations of web-building spiders can be estimated by counting the number of webs and the visibility of the webs increased by dusting with lycopodium

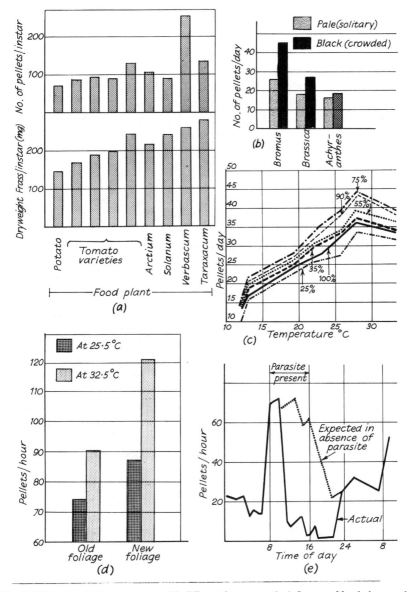

Fig. 8.2 Variations in frass-drop rate with different factors: *a.* the influence of food plant on the weight and number of pellets per instar from *Protoparce sexta* (Sphingidae) (data from Waldbauer, 1964); *b.* the influence of phase and foodplant on the number of pellets per day from *Leucania separata* (Noctuidae) (data from Iwao, 1962); *c.* the influence of temperature and humidity on the number of pellets per day from *Dendrolimus pini* (Lasiocampidae) (from Gösswald, 1935); *d.* the influence of temperature and condition of the foliage on the number of pellets per hour from 20 larvae of the sawfly, *Neodiprion lecontei* (Diprionidae) (data from Green & de Freitas, 1955); *e.* the influence of the presence of a dipterous parasite on the number of pellets per hour from 20 larvae of the sawfly, *N. lecontei* (modified from Green & de Freitas, 1955).

powder (Cherrett, 1964), or by spraying with a fine mist of water from a knapsack sprayer. I have found that the webs over a large area can be rapidly assessed using the latter method.

Indices of the populations of colonial nest-building caterpillars have been obtained by counting the number of nests, rather than the caterpillars; the great advantage of this approach lies in extensive work, as it enables the population level to be measured over many acres (Tothill, 1922; Legner & Oatman, 1962; Morris & Bennett, 1967; Ito *et al.*, 1970). Some mites also produce webbing and this may be used as an index (Newcomer, 1943). Cercopid larvae may be estimated by counting the spittle masses, but this is less accurate for the first instars (Whittaker, 1971).

8.2 Effects

8.2.1 Effects due to an individual insect

These are of great value to the ecologist, as they are immediately convertible into an absolute estimate. When, at a given stage of the life-cycle, each animal has some unique effect (for example, the commencement of a leaf mine), a count of these, after all the animals have passed this stage, will provide a precise measure of the total number passing through that stage, whilst the counts of the actual animals would need to be integrated to provide such a total (see p. 358). These effects are therefore measures of the number of animals entering a stage, and as such could be considered in the next chapter.

Cicada larvae construct conspicuous and distinctive turreted emergence holes which may be counted to measure the total emerging population (Dybas & Davis, 1962). Populations of solitary Hymenoptera can be assessed from the number of nest holes (Bohart & Lieberman, 1949); this will give the number of reproducing females, but in some species a female will construct more than one nest.

Measures of absolute population can be obtained for plant feeders when they enter the plant. Leaf-miners allow particularly elegant studies, as at the end of a generation it is possible to determine the number of larvae that commenced mining, the number that completed development and often some indication of the age and cause of death in those that failed. (Askew, 1968; Pottinger & Le Roux, 1971; Leroi, 1972, 1974; Payne *et al.*, 1972). When the larva cuts characteristic cases at different ages even more information may be gathered (De Gryse, 1934) and, because of the 'historical' record of spatial position precise studies of interspecific competition are possible (Murai, 1974).

Galls are easily counted, but many harbour a variable number of insects and it is usually impossible to determine externally whether these have been parasitized or not (Bess & Haramoto, 1959; Nakamura *et al.*, 1964; Howse & Dimond, 1965; Redfern, 1968).

Some stem-borers cause the growing shoot to die; when multiple invasion is sufficiently rare to be overlooked estimates of these 'deadhearts' may be taken as equivalent to the total number of larvae invading, and the same approach can be applied to insects in grains or seeds (Jepson & Southwood, 1958; Gomez & Bemardo, 1974). Oviposition punctures have been found to provide an accurate index of the number of eggs per plant unit for the weevil, *Hypera postica* (Harcourt, Mukerji & Guppy, 1974).

8.2.2 General effects – damage

The effect of an insect population on a plant stand is the product of two opposing processes: the eating rate of the insects and the growing rate of the plant. The eating rate of the insects will depend on their age, the temperature conditions and the other factors that affect frass production (see above), as well as on population size. An example of the complex interaction of these processes for various cruciferous plants and two different pests is given by Taylor & Bardner (1968*a* & *b*). Therefore measures of damage are only approximate indices of population size, although they may have advantages over other methods (e.g. Johnson & Burge, 1971). Damage *per se* is, of course, an important parameter for many entomologists; an outline of its assessment will be given.

CRITERIA

Economic damage

This measure is the one least related to insect population and describes the effects of the insects in economic terms; it is of considerable importance to the agriculturalist and forester as it indicates the need for control measures, which become desirable once the economic threshold is reached. The economic threshold is defined by Stern *et al.* (1959) as 'the density at which control measures should be determined to prevent an increasing pest population from reaching the economic injury level'. Changing economic conditions will alter the economic injury level from season to season and its relation to the size of the insect population will vary greatly from species to species. Pests that cause blemishes to fruits or destroy the leading shoots of growing trees cause damage whose economic level is very high relative to the amount of plant material destroyed (Smith, 1969; Strickland, 1970; Southwood & Norton, 1972; Stern, 1973).

Loss of yield

This is a measure of the extent to which the weight, volume or number of the marketable parts of the plant is changed; normally reduced, but occasionally increased (Harris, 1974).

Although more of a biological measure than economic damage, variations in yield may be caused by many factors other than pest numbers (Möllerström, 1963). Furthermore, the same number of insects attacking a plant can have very different effects depending on the timing of the attack in relation to the growth stages of the plant. For example, the larva of the frit fly invades and kills a young shoot of the oat plant; early in the season the plant will have time to develop another tiller that will contribute towards the yield of grain, and so the latter is not affected. Later in the season any tillers destroyed would have contributed towards the crop, but there is insufficient growing time for the tillers that replace them to ripen grain; this is the time of the maximum effect of attack on yield. At a still later time any tillers that are young enough to be susceptible to the frit fly are so far behind the bulk of the crop that their contribution to the eventual yield would be insignificant. If the frit fly attacks such tillers yield will not be reduced; it might even be increased. Gough (1947) describes a similar situation with the wheat bulb fly (*Leptohylemyia coarctata*). Prasad (1961) studied the reduction in weight of cabbage heads following the artificial colonization by newly emerged larvae of the small cabbage butterfly (*Pieris rapae*) at various dates after transplantation. It was found (Fig. 8.3) that the greatest reduction in yield occurred when the larvae were introduced seven weeks after transplantation, as the leaves damaged at this time contribute to the head. On the other hand attacks soon after transplanting affect the growth of the plant.

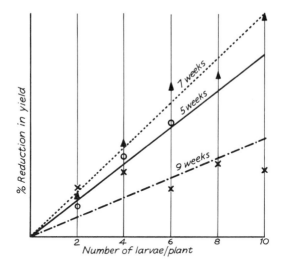

Fig. 8.3 The relationship of damage to the age of the plant; the reduction in the yield of cabbage plants when infested with different numbers of *Pieris rapae* larvae 5, 7 and 9 weeks after transplantation (after Prasad, 1961).

The amount of plant consumed

This is perhaps the most meaningful biological measure and could be expressed precisely in kilocalories as the amount of the primary production consumed by the insect (see Chapter 14); (Andrzejewska & Wojcik, 1970; Taylor & Bardner, 1970). The actual amount of damage may be assessed in terms of dry weight reduction (e.g. Ortman & Painter, 1960) or leaf area destroyed. The area of the leaf may be measured electronically (Hatfield *et al.* 1976), 'xeroxing' provides a rapid record, whilst the area of leaf blotching due to the removal of cell contents (by leafhoppers and mirids) may be estimated by photography with diazo process film (Fry & Henneberry, 1977). It is often easier to measure some index of the amount of plant destroyed rather than the actual quantity. For example, Coaker (1957) used various degrees of tattering of the leaf margin of cotton as an index of mirid damage; Dimond & Allen (1974) found the length of pine needles closely correlated with the numbers of first instar adelgid (*Pineus*) larvae; and Coombs (1963) assessed the level of damage in stored grain from the weight of fine dust present in a given volume of grain.

Estimates of the quantity of leaf consumed may be confused by the subsequent enlargement of the hole by the growth of the leaf; Reichle *et al.* (1973) allowed for this by studying the expansion of standard holes punched in young leaves.

Aerial photography, including infra-red remote sensing, is increasingly providing a method of surveying insect damage over large areas, generally in forests (Franz & Karafiat, 1958; Waters *et al.*, 1958; Aldrich *et al.*, 1959; Wear *et al.*, 1964; Klein, 1973), but also in orchards and crops (Chiang, Latham & Meyer, 1973; Hart & Ingle, 1969; Hart *et al.*, 1973; Stern, 1973; Harris *et al.*, 1976; Wallen *et al.*, 1976).

Insect defoliation of trees affects the growth rate and is reflected in the annual growth rings (Mott *et al.*, 1957; Varley & Gradwell, 1962; Lessard & Buffan, 1976; Creber, 1977), whilst aphids may also effect the time of leaf fall and other characters (Dixon, 1971). Considering more particularly the effects of defoliators on trees, Henson & Stark (1959) proposed that insect populations could be defined as:

(1) Tolerable – Populations that do not utilize the entire excess biological productivity of the host; that is, the insect and the host plant populations could continue at this level indefinitely.

(2) Critical – 'Populations that utilize more than the excess biological productivity of the hosts, but less than the total productivity'. Such population levels cannot be continued indefinitely and Henson & Stark (1959) (Fig. 8.4), and Churchill *et al.* (1964) show the long-term effects of insect attack on perennial hosts.

(3) Intolerable – 'Populations that are depleting the host at a rate greater than the current rate of production'.

The condition of the host tree (expressed as the number of leaf needles per branch tip) and the history of attack in previous years will interact with the intensity of larval population (larvae per tip) in the determination of the appropriate description of population in these terms (Fig. 8.4).

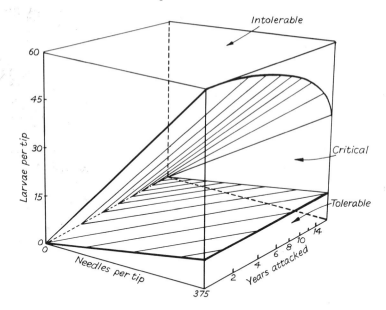

Fig. 8.4 The relationship of tree condition, insect population intensity and previous history to intolerable, critical and tolerable populations (after Henson & Stark, 1959).

DETERMINING THE RELATIONSHIP BETWEEN DAMAGE AND INSECT POPULATIONS

As has been indicated above, the level of damage a plant suffers is influenced by many factors, such as soil, climate, age and health of the plant, as well as the size of the insect population. Therefore attempts to relate insect numbers to damage by the correlation of these two variables measured in different areas and seasons are seldom satisfactory (Lockwood, 1924; Strickland, 1956), although it may be possible to separate the roles of the different factors by a carefully planned sampling programme emphasizing the comparison of nearby fields (Sen & Chakrabarty, 1964). The comparisons will be even more exact if they are made in the same field, as then climatic and soil effects will be identical. Theoretically there are three techniques for making such comparisons:

(1) The introduction of a known number of animals on to pest-free plants (e.g. Neiswander & Herr, 1930; Bowling, 1963; Andrzejewska and Wojcik, 1970; Ogunlana & Pedigo, 1974).

(2) The exclusion of the animals from certain plots by mechanical barriers.

(3) The exclusion of the animals or the reduction of their populations in some plots by the use of pesticides.

(4) A comparison of the growth of individual plants attacked by the pest compared with unattached plants growing in the same crop (Bardner, 1968; Bardner *et al.*, 1969).

Methods 1 and 2 both suffer from the fact that the barriers erected to exclude or retain the animals will severely modify the climate and growth conditions of the plant; the exclusion technique is of course widely used in studies on the effects of grazing by vertebrates.

The third method has been used successfully by Strickland (1956), Dahms & Wood (1957) and Jepson (1959). Working on the frit fly (*Oscinella frit*) the last named author found that a detailed knowledge of the periods of oviposition of the pest allowed frequent 'blanket spraying' to be replaced by a limited number of timed sprays or granular applications (Jepson & Mathias, 1960).

REFERENCES

ALDRICH, R. C., BAILEY, W. F. and HELLER, R. C., 1959. Large scale 70 mm color photography techniques and equipment and their application to a forest sampling problem. *Photogrammetric Eng.* **25**, 747–54.

ANDRZEJEWSKA, L. and WOJCIK, Z., 1970. The influence of Acridoidea on the primary production of a meadow (field experiment). *Ekol. Polska* **18**, 89–109.

ASKEW, R. R. 1968. A survey of leaf-miners and their parasites on laburnum. *Trans. R. ent. Soc. Lond.* **120**, 1–37.

BARDNER, R., 1968. Wheat bulb fly, *Leptohylemyia coarctata* Fall. and its effect on the growth and yield of wheat. *Ann. appl. Biol.* **61**, 1–11.

BARDNER, R., FLETCHER, K. E. and HUSTON, P., 1969. Recent work on wheat bulb flies. Proc. 5th Br. Insecticide Fungicide Conf. 500–4.

BEAN, J. L., 1959. Frass size as an indicator of spruce budworm larval instars. *Ann. ent. Soc. Am* **52**, 605–8.

BESS, H. A. and HARAMOTO, F. H., 1959. Biological control of Pamakani, *Eupatorium adenophorum*, in Hawaii by a tephritid gall fly, *Procecidochares utilis*. 2. Population studies of the weed, the fly and the parasites of the fly. *Ecology* **40**, 244–9.

BOHART, G. E. and LIEBERMAN, F. V., 1949. Effect of an experimental field application of DDT dust on *Nomia melanderi*. *J. econ. Ent.* **42**, 519–22.

BOWLING, C. C., 1963. Cage tests to evaluate stink bug damage to rice. *J. econ. Ent.* **56**, 197–200.

CAMPBELL, K. G., 1960. Preliminary studies on population estimation of two species of stick insects (Phasmatidae Phasmatodea) occurring in plague numbers in Highland Forest areas of south-eastern Australia. *Proc. Linn. Soc. N.S.W.* **85**, 121–37.

CHERRETT, J. M., 1964. The distribution of spiders on the Moor House National Nature Reserve, Westmorland. *J. Anim. Ecol.* **33**, 27–48.

CHIANG, H. C., LATHAM, R. and MEYER, M. P., 1973. Aerial photography: use in detecting simulated insect defoliation in corn. *J. econ. Ent.* **66** (3), 779–84.

CHURCHILL, G. B., JOHN, H. H., DUNCAN, D. P. and HODSON, A. C., 1964. Long-term effects of defoliation of aspen by the forest tent caterpillar. *Ecology* **45**, 630–3.

COAKER, T. H., 1957. Studies of crop loss following insect attack on cotton in East Africa. II. Further experiments in Uganda. *Bull. ent. Res.* **48**, 851–66.

COOMBS, C. W., 1963. A method of assessing the physical condition of insect-damaged grain and its application to a faunistic survey. *Bull. ent. Res.* **54**, 23–35.

CORBET, P. S., 1957. The life-history of the Emperor dragonfly, *Anax imperator* Leach (Odonata: Aeshnidae). *J. Anim. Ecol.* **26**, 1–69.

CREBER, G. T., 1977. Tree rings: a natural data storage system. *Biol. Rev.* **52**, 349–82.

DADD, R. H., 1960. Observations on the palatability and utilisation of food by locusts, with particular reference to the interpretation of performances in growth trials using synthetic diets. *Entomologia exp. Appl.* **3**, 283–304.

DAHMS, R. G. and WOOD, E. D., 1957. Evaluation of green bug damage to small grains. *J. econ. Ent.* **50**, 443–6.

DE GRYSE, J. J., 1934. Quantitative methods in the study of forest insects. *Sci. Agr.* **14**, 477–95.

DIMOND, J. B. and ALLEN, D. C., 1974. Sampling populations of pine leaf chermid *Pineus pinifoliae* (Homoptera: Chermidae). III. Neosisterites on white pine. *Can. Ent.* **106**, 509–18.

DIXON, A. F. G., 1971. The role of aphids in wood formation II. The effect of the lime aphid, *Eucallipterus tiliae* L. (Aphididae), on the growth of lime, *Tilia* × *vulgaris* Hayne. *J. appl. Ecol.* **8**, 393–9.

DYBAS, H. S. and DAVIS, D. D., 1962. A population census of seventeen-year periodical Cicadas (Homoptera: Cicadidae: Magicicada). *Ecology* **43**, 432–44.

DYBAS, H. S. and LLOYD, M., 1962. Isolation by habitat in two synchronized species of periodical cicadas (Homoptera: Cicadidae: Magicicada). *Ecology* **43**, 444–59.

DYBAS, H. S. and LLOYD, M., 1974. The habitats of 17-year periodical cicadas (Homoptera: Cicadidae: Magicicada spp.). *Ecol. Monogr.* **44** (3), 279–324.

ECKSTEIN, K., 1938. Die Bewertung des Kotes der Nonnenraupe, *Psilura monacha* L., als Grundlage für die Festellung ihres Auftretens und zu ergreifenden Massregeln. *Allgem. Forst. Jagdztg.* **114**, 132–48.

ECKSTEIN, K., 1939. Das Bohrmehl des Waldgärtners, *Myelophilus pimperda* L., nebst Bemerkungen über den 'Frass' der Borkenkäfer und anderen Insekten. *Arb. physiol. angew. Ent.* **6**, 32–41.

FRANZ, J. and KARAFIAT, H., 1958. Eigen sich Kartierung und Serienphotographie von Tannenläusen für Massenwechsel-studien? *Z. angew. Ent.* **43**, 100–12.

FRIDÉN. F., 1958. *Frass-drop frequency in Lepidoptera.* Uppsala (Almqvist & Wiksells Boktryckeri), 59 pp.

FROST, S. W., 1928. Insect scatology. *Ann. ent. Soc. Am.* **21**, 35–46.

FRY, K. E., and HENNEBERRY, T. J., 1977. Measuring leaf damage by the cotton leaf perforator. *J. econ. Ent.* **70**, 141–2.

GOMEZ, K. A. and BERNARDO, R. C., 1974. Estimation of stem borer damage in rice fields. *J. econ. Ent.* **67** (4), 509–16.

GÖSSWALD, K., 1935. Über die Frasstätigkeit von Forstschädlingen unter verschiedener Temperatur und Luftfeuch tigkeit und ihre praktische und physiologische Bedeutung. I. *Z. angew. Ent.* **21**, 183–7.

GOUGH, H. C., 1947. Studies on wheat bulb fly, *Leptohylemyia coarctata*, Fall. II. Numbers in relation to crop damage. *Bull. ent. Res.* **37**, 439–54.

GREEN, G. W. and DE FREITAS, A. S., 1955. Frass drop studies of larvae of *Neodiprion americanus banksianae*, Roh. and *Neodiprion lecontii*, Fitch. *Can. Ent.* **87**, 427–40.

GREEN, G. W. and HENSON, W. R., 1953. A new type of coprometer for laboratory and field use. *Can. Ent.* **85**, 227–30.

HARCOURT, D. G., MUKERJI, M. K. and GUPPY, J. C., 1974. Estimation of egg populations of the alfalfa weevil, *Hypera postica* (Coleoptera: Curculionidae). *Can. Ent.* **106**, 337–47.

HARRIS, M. K., HART, W. G., DAVIS, M. R., INGLE, S. J. and VAN CLEAVE, H. W., 1976. Aerial photography shows caterpillar infestation. *Pecan Quarterly.* **10** (2); 12–8.

HARRIS, P., 1974. A possible explanation of plant yield increases following insect damage. *Agro-ecosystems* **1**, 219–25.

HART, W. G. and INGLE, S. J., 1969. Detection of arthropod activity on citrus foliage with aerial infrared color photography. *Proc. Workshop Aerial Color Photography Pl. Sci.*, 85–8, Florida.

HART, W. G., INGLE, S. J., DAVIS, M. R. and MANGUM, C., 1973. Aerial photography with infra-red colour film as a method of surveying for Citrus Blackfly. *J. econ. Ent.* **66** (1), 190–4.

HATLFIELD, J. L., STANLEY, C. D. and CARLTON, R. E., 1976. Evaluation of an electronic foliometer to measure leaf area in corn and soybeans. *Agron. J.* **68**, 434–6.

HENSON, W. R. and STARK, R. W., 1959. The description of insect numbers. *J. econ. Ent.* **52**, 847–50.

HODSON, A. C. and BROOKS, M. A., 1956. The frass of certain defoliators of forest trees in the north central United States and Canada. *Can. Ent.* **88**, 62–8.

HOWSE, G. M. and DIMOND, J. B., 1965. Sampling populations of Pine leaf adelgid, *Pineus pinifoliae* (Fitch). I. The gall and associated insects. *Can. Ent.* **97**, 952–61.

ITO, Y., SHIBAZAKI, A and IWAHASHI, O., 1970. Biology of *Hyphantria cunea* Drury (Lepidoptera: Arctiidae) in Japan. XI. Results of road-survey. *Appl. Ent. Zool.* **5**, 133–44.

IWAO, S., 1962. Studies on the phase variation and related phenomena in some Lepidopterous insects. *Mem. Coll. Agric. Kyoto Univ.* (Ent. no. 12) **84**, 1–80.

JEPSON, W. F., 1959. The effects of spray treatments on the infestation of the oat crop by the frit fly (*Oscinella frit* L.). *Ann. appl. Biol.* **47**, 463–74.

JEPSON, W. F. and MATHIAS, P., 1960. The control of frit fly, *Oscinella frit* (L.) in sweet corn (*Zea mays*) by Thimet (*O, O-Diethyl S*-ethyethio-methyl phosphorodithioate). *Bull. ent. Res.* **51**, 427–33.

JEPSON, W. F. and SOUTHWOOD, T. R. E., 1958. Population studies on *Oscinella frit* L. *Ann. appl. Biol.* **46**, 465–74.

JOHNSON, C. G. and BURGE, G. A., 1971. Field trials of anti-capsid insecticides on farmers cocoa in Ghana, 1956–60. 2. Effects of different insecticides compared by counting capsids, and capsid counting compared with counting the percentage of newly damaged trees. *Ghana. J. agric. Sci.* **4**, 33–8.

KLEIN, W. H., 1973. Beetle killed pine estimates, *Photogrammetric Engineering* **39**, 385–8.

LEGNER, E. F. and OATMAN, E. R., 1962. Sampling and distribution of summer eye-spotted bud moth *Spilonota ocellana* (D. & S.) larvae and nests on apple trees. *Can. Ent.* **94**, 1187–9.

LEROI, B., 1972. A study of natural populations of the celery leaf-miner *Philophylla heraclei* L. (Diptera Tephritidae). I. Methods of counting of larval populations. *Res. Popul. Ecol.* **13**, 201–15.

LEROI, B., 1974. A study of natural populations of the celery leaf-miner, *Philophylla heraclei* L. (Diptera, Tephritidae). II. Importance of changes of mines for larval populations. *Res. Popul. Ecol.* **15**, 163–82.

LESSARD, G. and BUFFAM, P. E., 1976. Effects of *Rhyacionia neomexicana* on height and radial growth in Ponderose Pine reproduction. *J. econ. Ent.* **69**, 755–60.

LOCKWOOD, S., 1924. Estimating the abundance of, and damage done by grasshoppers. *J. econ. Ent.* **17**, 197–202.

MÖLLERSTRÖM, G., 1963. Different kinds of injury to leaves of the sugar beets and their effect on yield. *Medd. Växtskyddsanst.* **12**, 299–309.

MORRIS, R. F., 1942. The use of frass in the identification of forest insect damage. *Can. Ent.* **74**, 164–7.

MORRIS, R. F., 1949. Frass-drop measurement in studies of the European spruce sawfly. *Univ. Michigan Sch. Forestry and Conserv. Bull.* **12**, 58 pp.

MORRIS, R. F. and BENNETT, C. W., 1967. Seasonal population trends and extensive census methods for *Hyphantria cunea. Can. Ent.* **99**, 9–17.

MOTT, D. G., NAIRN, L. O. and COOK, J. F., 1957. Radial growth in forest trees and effects of insect defoliation. *For. Sci.* **3** (3), 286–304.

MURAI, M., 1974. Studies on the interference among larvae of the citrus leaf miner, *Phyllocristis citrella* Stainton (Lepidoptera: Phyllocristidae). *Res. Popul. Ecol.* **16**, 80–111.

NAKAMURA, M., KONDO, M., ITO, Y., MIYASHITA, K. and NAKAMURA, K., 1964. Population dynamics of the chestnut gall-wasp, *Dryocosmus buriphilus*. 1. Description of the survey station and the life histories of the gall wasp and its parasites. *Jap. J. appl. Ent. Zool.* **8**, 149–58.

NEISWANDER, C. R. and HERR, E. A., 1930. Correlation of corn borer populations with degree of damage. *J. econ. Ent.* **23**, 938–45.

NEWCOMER, E. J., 1943. Apparent control of the Pacific mite with xanthone. *J. econ. Ent.* **36**, 344–5.

NOLTE, H. W., 1939. Über den Kot von Fichten- und Kieferninsekten. *Tharandter forstl. Jahrb.* **90**, 740–61.

OGUNLANA, M. O. and PEDIGO, P., 1974. Economic injury levels of the potato leafhopper on Soybeans in Iowa. *J. econ. Ent.* **67**(1), 29–32.

ORTMAN, E. E. and PAINTER, R. H., 1960. Quantitative measurements of damage by the greenbug, *Toxoptera graminum* to four wheat varieties. *J. econ. Ent.* **53**, 798–802.

PARAMONOV, A., 1959. A possible method of estimating larval numbers in tree crowns. *Ent. mon. Mag.* **95**, 82–3.

PAYNE, J. A., TEDDERS, W. L., COSGROVE, G. E. and FOARD, D., 1972. Larval mine characteristics of four species of leaf-mining Lepidoptera in Pecan. *Ann. ent. Soc. Am.* **65**, 74–81.

POND, D. D., 1961. Frass studies of the armyworm, *Pseudaletia unipuncta. Ann. ent. Soc. Am.* **54**, 133–40.

POTTINGER, R. P. and LE ROUX, E. J., 1971. The biology and dynamics of *Lithocolletis blancardella* (Lepidoptera: Gracillariidae) on apple in Quebec. *Mem. Ent. Soc. Canada.* **77**, 437 pp.

PRASAD, S. K., 1961, (1962). Quantitative estimation of damage to cabbage by cabbage worm, *Pieris rapae* (Linn.). *Indian J. Ent.* **23**, 54–61.

REDFERN, M., 1968. The natural history of spear thistle-heads. *Field Studies* **2**(5), 669–717.

REICHLE, D. E., GOLDSTEIN, R. A., VAN HOOK, R. I. and DOBSON, G. J., 1973. Analysis of insect consumption in a forest canopy. *Ecology* **54**, 1076–84.

RHUMBLER, L., 1929. Zur Begiftung des Kiefernspanners (*Bupalus piniarius* L.) in der Oberförsterei Hersfeld–Ost 1926. *Z. angew. Ent.* **15**, 137–58.

SEN, A. R. and CHAKRABARTY, R. P., 1964. Estimation of loss of crop from pests and diseases of tea from sample surveys *Biometrics* **20**, 492–504.

SMITH, R. F., 1969. The importance of economic injury levels in the development of integrated pest control programs. *Qual. Plant. Mater. Veg.* **17**, 81–92.

SOLOMON, J. D., 1977. Frass characteristics for identifying insect borers (Lepidoptera: Cossidae and Sesiidae; Coleoptera: Cerambyciidae) in living hardwoods. *Can. Ent.* **109**, 295–303.

SOUTHWOOD, T. R. E. and NORTON, G. A., 1973. Economic aspects of pest management strategies and decisions. In Geier, P. W. *et al.* (eds.) *Insects: studies in population management. Mem. Ecol. Soc. Australia* **1**, 168–95.

STERN, V. M., 1973. Economic thresholds. *Ann. Rev. Ent.* **18**, 259–80.

STERN, V. M., SMITH, R. F., VAN DEN BOSCH, R. and HAGEN, K. S., 1959. The integrated control concept. *Hilgardia* **29**(2), 81–101.

STRANDINE, E. J., 1940. A quantitative study of the periodical cicada with respect to soil of three forests. *Am. Midl. Nat.* **24**, 177–83.

STRICKLAND, A. H., 1956. Agricultural pest assessment. I – The problem. *N.A.A.S. quart. Rev.* (H.M.S.O., London) **33**, 112–18. II – A partial solution. *Ibid.* **34**, 156–62.

STRICKLAND, A. H., 1970. Some attempts to predict yield losses in England from estimates of pest populations. *E.P.P.O. Public. Ser. A. No. 57*, 147–58.

TAYLOR, W. E. and BARDNER, R., 1968a. Effects of feeding by larvae of *Phaedon cochleariae* (F.) and *Plutella maulipennis* (Curt.) on the yield of radish and turnip plants. *Ann. appl. Biol.* **62**, 249–54.

TAYLOR, W. E. and BARDNER, R., 1968b. Leaf injury and food consumption by larvae of *Phaedon cochleariae* (Col. Chrysomelidae) and *Plutella maculipennis* (Lep. Plutellidae) feeding on turnip and radish. *Entomologia exp. appl.* **11**, 177–184.

TAYLOR, W. E. and BARDNER, R., 1970. Energy relationships between the larvae of *Phaedon cochleariae* (Coleoptera, Chrysomelidae) or *Plutella maculipennis* (Lepidoptera, Plutellidae) and radish and turnip plants. *Entomologia exp. appl.* **13**, 403–6.

TINBERGEN, L., 1960. The natural control of insects in pinewoods. 1. Factors influencing the intensity of predation by song birds. *Arch. Neérl. Zool.* **13**, 266–343.

TOTHILL, J. D., 1922. The natural control of the fall webworm (*Hyphentia cunea* Drury) in Canada. *Bull. Can. Dept. Agric.* **3**(n.s) (*Ent. Bull.* **19**), 1–107.

VARLEY, G. C. and GRADWELL, G. R., 1962. The effect of partial defoliation by caterpillars on the timber production of oak trees in England. *Proc. XI int. Congr. Ent.* **2**, 211–14.

WALDBAUER, G. P., 1964. Quantitative relationships between the numbers of fecal pellets, fecal weights and the weight of food eaten by tobacco hornworms, *Protoparce sexta* (Johan.) (Lepidoptera: Sphingidae). *Entomologia exp. appl.* **7**, 310–14.

WALLEN, V. R., JACKSON, H. R. and MACDIARMID, S. W., 1976. Remote sensing of corn aphid infestation, 1974 (Hemiptera: Aphididae) *Can. Ent.* **108**, 751–4.

WATERS, W. E., HELLER, R. C. and BEAN, J. L., 1958. Aerial appraisal of damage by the spruce budworm. *J. Forest.* **56**(4), 269–76.

WEAR, J. F., POPE, R. B. and LAUTERBACH, P. G., 1964. Estimating beetle-killed Douglas fir by photo and field plots. *J. Forest.* **62**(5), 309–15.

WEISS, H. B. and BOYD, W. M., 1950, 1952. Insect feculae I. *J. New York Ent. Soc.* **58**, 154–68. Insect feculae II. *ibid.* **60**, 25–30.

WHITTAKER, J. B., 1971. Population changes in *Neophilaenus lineatus* (L.) (Homoptera: Cercopidae) in different parts of its range. *J. Anim. Ecol.* **40**, 425–43.

YONCE, C. E., GENTRY, C. R., TUMLINSON, J. H., DOOLITTLE, R. E., MITCHELL, E. R. and McLAUGHLIN, J. R., 1977. Seasonal distribution of the lesser peach tree borer in Central Georgia as monitored by pupal skin counts and pheromone trapping techniques. *Environ. Ent.* **6**, 203–6.

9

Observational and Experimental Methods for the Estimation of Natality, Mortality and Dispersal

Values for the 'pathways' through which population size changes may also be obtained by the subtraction or integration of census figures in a budget: methods of caloulation and of analysis of budgets are discussed in the next chapter, but there is no hard and fast distinction between the contents of the two chapters. As is indicated below, in the appropriate sections, *the terms 'natality' and 'dispersal' are used in their widest sense.*

9.1 Natality

Natality is the number of births, that is strictly speaking the number of living eggs laid; however, from the practical point of view of constructing a population budget the number of individuals entering a post-ovarian stage, i.e. larval instar, pupa or adult, can be considered as the 'natality' of that stage.

9.1.1 Fertility

Fertility is the number of viable eggs laid by a female and fecundity is a measure of the total egg production; the latter is often easier to measure. In those insects where all the eggs are mature on emergence the *total potential fecundity* may be *estimated by examining the ovaries* as Davidson (1956) did with sub-imagines of a mayfly, *Cloeon*. The ovaries were removed and lightly stained in methylene blue, the eggs were separated by sieving through bolting silk and then counted in a Sedwick—Rafter plankton-counting cell.

More often eggs are matured throughout much of adult life and fecundity is *measured directly by keeping females caged* under as natural conditions as possible and recording the total number of eggs laid (e.g. Huffaker & Spitzer, 1950; Spiller, 1964); if viable eggs can be distinguished from non-viable ones, usually by the onset of development, then fertility may be measured (e.g. Fewkes, 1964). In some insects egg cannibalism may interfere with the estimation of fecundity; the influence of this behaviour may be calculated from the decimation of a known number of marked eggs (Rich, 1956).

It has been found in a wide range of insects that the *fecundity of the female is proportional to her weight* (e.g. Prebble, 1941; Richards & Waloff, 1954;

Waloff & Richards, 1958; Colless & Chellapah, 1960; Lozinsky, 1961; Murdie, 1969; Taylor, 1975). Female weight is directly related to size, so that measures of size such as wing-length (Gregor, 1960) or pupal length (Miller, 1957) or both (Hard, 1976) may be substituted for weight. Once the relation between size and fecundity has been established (regression analysis is a convenient method) it may be applied to estimate the natality in field populations by measuring wild females or female pupae – the length of the empty pupal cases is a particularly suitable criterion (Miller, 1957).

Within a given population the rate of oviposition will be influenced by temperature and the extent to which the potential fecundity is realized is influenced by the longevity of the females. The influence of temperature can be studied in the laboratory and incorporated into the regression equation; the observations of Richards & Waloff (1954) show that it may be justified to apply data from the laboratory to populations in the field. Information on longevity may be obtained from wild populations by marking and recapture. An equation incorporating all these variables will be in the form (Richards & Waloff, 1954):

No. of eggs = regression coeff. × weight ± reg. coeff. × temp. + reg. coeff. × longevity + constant

Such equations cannot, however, be applied outside the populations for which they were derived; females that have developed under other conditions, more especially of nutrition and of crowding, will have different potential fecundities (Blais, 1953; Clark, 1963); therefore in practice it is necessary to have different equations for different stages in a pest outbreak (Miller, 1957). Larval density of course affects size and longevity directly (Miller & Thomas, 1958) but the fact that new equations are necessary shows that the effects on fecundity are not just reflections of changes in these properties. Evidence is accumulating that changes may occur in the genetic constitution of fluctuating populations and the fecundity of different forms of a species may be very different (Wellington, 1964). Therefore laboratory measurements of fertility should be continually checked, for it is important in any population study to establish not only the potential fertility, but any variation in it, for this may be the essence of a population regulation mechanism (p. 377).

When large numbers of eggs are laid together in a group the weight of the individual female will fall sharply after each oviposition and slowly rise again until the next group is deposited. Where individual animals can be marked it may be possible by frequently recapturing and weighing to establish the time of oviposition of actual egg batches; Richards & Waloff (1954) were able to do this with a field population of grasshoppers. It may be possible with some insects to mark them with a radionucleid (e.g. ^{65}Zn) and then estimate egg production by measuring the loss of isotope (Mason & McGraw, 1973). The actual number of eggs per egg mass may vary with size or age of the female (Richards & Waloff, 1954) or from generation to generation (Iwao, 1956).

9.1.2 Numbers entering a stage

When an animal can be trapped as it passes from one part of its habitat to another part at a specific stage of its life-history the summation of the catches provides a convenient measure of the total populations entering the next stage. Such traps are commonly referred to as *emergence traps*. The particular design will depend on the insect and its habitat; a number are described by Peterson (1934), Nicholls (1963) and Ives *et al.* (1968). Basically the traps consist of a metal or cloth box that covers a known area of soil and glass collecting vials with 'lobster-pot' type baffles are inserted in some of the corners (Fig. 9.1). A newly emerged insect being positively phototactic will make its way into the collection vials, which are emptied regularly; Turnock

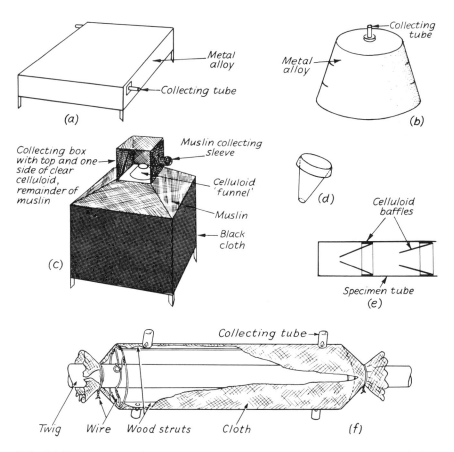

*Fig. 9.1 Emergence traps for terrestrial habitats: *a.* and *b.* metal box and tub respectively; *c.* cloth-covered, Calnaido type; *d.* celluloid baffle of collecting tube; *e.* sectional view of collecting tube with pair of baffles designed to separate, partially at least, large and small insects; *f.* cloth trap for twigs (*a, b* and *c* modified from Southwood and Siddorn, 1965; *f.* from Glen, 1976).

(1957) used an adhesive resin* in the collecting containers of his trap so that they could be emptied at less frequent intervals. The efficiency of a trap for a particular insect can be tested by releasing a known number of newly emerged individuals into it. The construction of the trap will influence its effect on the microclimate; all traps tend to reduce the daily temperature fluctuations (Fig. 9.2) and the deeper they are (the greater the insulating layer of air) the smaller the fluctuations. Cloth-covered traps lose the heat less quickly than metal ones. Although the daytime deficit may be approximately balanced by the greater warmth at night (Fig. 9.2), over a period of several weeks these small daily excesses or deficits will accumulate to levels where they might influence the development rate of pupae (Southwood & Siddorn, 1965). For this reason caution should be exercised in using emergence trap data for phenology. Rice & Reynolds (1971) collected twice as many pink bollworms (*Platyedra*) from metal-screened emergence cages as from plastic-screened ones. Special traps have been described by many workers, including one for fleas (Bates, 1962) and others for ceratopogonid midges (Campbell & Pelham-Clinton, 1960; Neilsen, 1963; Davies, 1966; Braverman, 1970), for anthomyiid flies (Dinther, 1953), for cecidomyiid midges (Speyer & Waede, 1956; Nijveldt, 1959; Guennelon & Audemard, 1963) and beetles (Richards & Waloff, 1961;

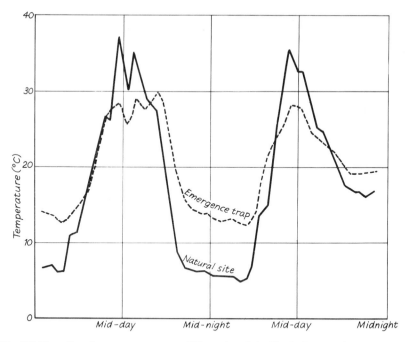

Fig. 9.2 The soil surface temperature over 48 hours in a Calnaido cloth-covered emergence trap compared with that from a natural site (after Southwood & Siddorn, 1965).

* See footnote p. 249

Polles & Payne, 1972; Boethel *et al.*, 1976) and for lepidopterous larvae emerging from maize 'ears' (Straub *et al.* 1973).

A number of special traps have been designed for bark-dwelling insects. For bark-beetles and other insects occurring at high densities it may even be desirable to subdivide the trap to record the dispersion pattern (Reid, 1963). Simpler traps have been developed by Nord & Lewis (1970) and Glen (1976), essentially consisting of a wire and wood frame, covered with black cloth (Fig. 8.5): Glen found mean temperature was hardly affected.

Emergence traps have been used extensively in aquatic environments for measuring the quality, quantity and biomass of insects emerging from the habitat (Illies, 1971; Speir & Anderson, 1974). In shallow water, amongst emergent vegetation and in sheltered situations floating box traps have been utilized (Adamstone & Harkness, 1923; Macan, 1949; Vallentyne, 1952; Sommerman, Sailer & Esselbaugh, 1955; Judge, 1957; Morgan & Waddell, 1961). In exposed situations submerged funnel traps have been used as these avoid damage from wind and rain (Grandilewskaja-Decksbach, 1935; Brundin, 1949; Jónasson, 1954; Palmén, 1955, 1962; Mundie, 1956; Darby, 1962; Mulla *et al.*, 1974; McCauley, 1976). Careful comparisons of these two types of trap has been made by Morgan, Waddell & Hall (1963) and Kimerle & Anderson (1967), from which the following conclusions may be drawn. Funnel traps are generally much less efficient than floating box traps. A number of factors probably contribute to this: one is the decomposition of the catch in the small air space of the collecting jar of the funnel trap; losses due to this cause will be proportionally greater the denser the population of emerging insects and the longer the intervals between emptying the trap, which should therefore be done daily. The ascending pupae and larvae are strongly phototactic and the slight shade produced by the gauze of the funnel trap will cause it to be avoided; furthermore some insects, e.g. Odonata, will crawl out again. It is impossible to exclude small predatory insects from the traps and they may consume much of the catch. Lastly the effective trapping area is reduced if the trap becomes tipped. However, in very exposed situations these traps must be used and Morgan *et al.* (1963) recommend a design, similar to that in Fig. 9.3, to overcome as many of these faults as possible. The collecting jar was originally devised by Borutsky (1955); it is very important that the 'Perspex' be kept clean for algal growth will soon render it opaque.

The most important causes of loss of efficiency of floating box traps are waves that will swamp the catch, and the shading effect of the trap itself on the ascending larvae and pupae. Morgan *et al.* (1963) point out that as it is probably only those animals that ascend near the edges of the trap, that can take successful avoiding reaction, the larger the area of the trap the smaller this edge-effect relative to the size of the catch. The trap designed by Morgan *et al.* (Fig. 9.3a) is constructed mainly from clear 'Perspex';* the apron

---

* Plexiglass is a similar material

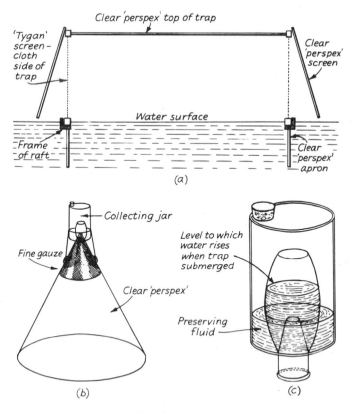

Fig. 9.3 Emergence traps for aquatic habitats: *a.* floating box trap for use under exposed situations; the screen may be omitted in sheltered situations (after Morgan, Waddell & Hall, 1963); *b.* submerged funnel trap; *c.* collecting jar (after Borutsky, 1955).

projecting in the water retains floating exuviae within the trap and helps to reduce wave damage; the latter function is also served by the lateral screens. The wooden frame should be as narrow as possible and painted white to minimize shadow; wire netting stretched under the frame reduces predation by fish. The trap needs to be emptied every two days and this is a difficult process: a floating tray is inserted under the trap and the whole towed to a boathouse or similar site where the unidirectional light source ensures that the edge on the darkest side may be lifted, to allow the entrance of an entomological pooter (= aspirator, Fig. 7.5), without any of the insects escaping. On dull days or at night, the trap may be tilted towards a paraffin lamp during emptying. In view of the time-consuming nature of this operation it would seem worthwhile to determine whether shrouding the traps for a short while in the day would drive a large proportion of the catch into a collecting tube, inserted in the roof; which is the method of collection from tents (p. 154).

Kimerle & Anderson (1967) had a sliding base in their trap that greatly facilitated the retrieval of the whole catch, and Corbet (1966) arranged for the traps to be closed and emptied in the laboratory. Carlson (1971) increased catch size (and interpretation difficulties regarding the catchment area) by incorporating a battery-operated light (see p. 261).

Traps with floats were designed by Langford & Daffern (1975) for work in large fast-flowing rivers: they found these robust, but there was some evidence that the numbers of Ephemeroptera, but not Trichoptera, caught were not affected by the presence of floats.

In very shallow fast-running streams it may be necessary to use a tent trap (Ide 1940; Anderson & Wold, 1972); basically it resembles a gauze tent that is attached to the substrate. A rather unsatisfactory feature of the design is that the observer has to enter the trap to remove the insects; not only, as Ide mentions, is he exposed to noxious insects (e.g. Simuliidae), but the resultant trampling of the substrate must affect future emergences. Possibly a smaller tent should be used – one that can be emptied from the exterior.

Chironomids were sampled from very shallow rock pools by Lindeberg (1958) using a trap similar in appearance to the submerged funnel type, but the funnel was made of glass and the whole upper part of the trap was airtight so that it could be kept full of water, even though it projected well above the level of water in the pool. Cook & Horn (1968) developed a sturdy trap for damselflies and similar insects emerging from shallow rush-growing water, frequented by cattle. It consisted of fencing wire, covered with fibre-glass window screening.

Material containing the resting stage of the animal may be collected and exposed under field conditions or in the laboratory. Bark beetle emergence from logs may be measured under natural conditions by placing the logs in screened cages (McMullen & Atkins, 1959; Clark & Osgood, 1964) and emergence of flies from heads of grass and corn determined by placing these in muslin bags in the field (Southwood *et al.*, 1961). The emergence of mites from overwintered eggs on twigs or bark was measured by Morgan & Anderson (1958) by attaching the substrate to the centre of a white card and surrounding it with a circle of a fruit-tree banding resin.* As the mites crawled out they were caught in the resin and being red, were easily counted.

Unseasonal emergence may be forced in the laboratory (Wilbur & Fritz, 1939; Terrell, 1959; van Emden, 1962; McKnight, 1969; Gruber & Prieto, 1976). Commercial containers such as shoeboxes, ice-cream cartons and biscuit tins have been found useful in such studies (Fig. 9.4*a*).

Insects that are arboreal for part of their life may often be trapped on their upward or downward journeys. Wingless female moths and others moving up tree trunks may be trapped in sackcloth and other trap-bands (DeBach, 1949; Reiff, 1955; Otvos, 1974) or in inverted funnel traps (Varley & Gradwell, 1963; Ives *et al.*, 1968; Agassiz, 1977). Larvae descending to pupate may be caught in funnel traps (Fig. 9.4*b*) (Ohnesorge, 1957; Pilon *et al.*, 1964).

* See footnote p. 249

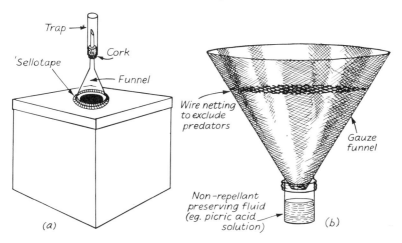

Fig. 9.4 *a*. Emergence tin to determine in the laboratory the emergence of insects from samples (after van Emden, 1962). *b*. Funnel trap for the collection of descending arboreal larvae.

9.1.3 The birth-rate from mark and recapture data

The details of marking animals and of the equations for estimating the birth-rate are given in Chapter 3. This birth-rate includes immigration and it is often, more correctly, referred to as the dilution rate.

9.2 Mortality

9.2.1 Total

Measurements of total mortality are also commonly obtained by the subtraction of population estimates for successive stages (e.g. Miller, 1958; Cook & Kettlewell, 1960; see Chapter 10).

i. Successive observations on the same cohort. When it is possible to make these, mortality may be measured directly. Completely natural cohorts can in general only be followed in sedentary or relatively immobile animals (e.g. the egg and pupal stages, scale insects, some aphids). Colonies present in the field in accessible positions may be repeatedly examined, each colony being identified by labelling the plant (e.g. MacLellan, 1962; van Emden, 1963); and with an animal as large as the hornworm (Sphingidae) Lawson (1959) was able to follow the larvae as well as eggs. In other situations it may be necessary to delimit the sample population or even to 'plant' it. Examples of the former are Ives & Prentice's (1959) and Turnock & Ives' (1962) studies where they placed pupal-free moss-filled trays or blocks of peat with screen bottoms and sides under the host trees of certain sawflies at pupation. Later in the season these 'natural traps' may be removed and the pupae classified. More mobile animals will need to be retained by a more extensive enclosure (Vlijm *et al.*, 1968;

Ashby, 1974) or 'field cage' (Dobson, Stephenson & Lofty, 1958; see p. 25). It is important to remember, however, that the screens and cages are bound to alter, albeit only slightly, the situation one is trying to assess. The planting of known populations in the field and their subsequent re-examination gives a measure of the level of mortality; sometimes it is possible to recover the remains of the dead individuals in such experiments and then further information may be gained about the cause of death (Graham, 1928; Morris, 1949; Buckner, 1959; Pavlov, 1961).

When an insect leaves a mark of its presence, as many that bore into plant tissue do, a single count at, for example, the pupal stage enables the still living and the dead to be distinguished (see also p. 292). Furthermore the latter may be divided into those whose burrows have been opened by predators and those killed by parasites and disease. Such a study, which of course is in effect population estimates of two successive stages (i.e. total young larvae = total burrows, and pupae), was made by Gibb (1958) for the Eucosmid moth, *Ernarmonia conicolana*, in pine cones.

ii The recovery of dead or unhealthy individuals. This gives another measure of mortality. Although Gary (1960) has devised a trap for collecting dead and unhealthy honey-bees, it is unusual to be able to recover the non-survivors of a mobile stage. However, unhatched eggs can often be examined to ascertain the cause of their death (e.g. Bess, 1961; Way & Banks, 1964).

9.2.2 The death-rate from mark and recapture data
Details of these models and the calculation of the death-rate, which includes emigration, are given in Chapter 3 (see also p. 339).

9.2.3 Climatic factors
Apart from direct observations on a known cohort, the main method of establishing the role of climate in the total mortality has been experimental. A known number of individuals may be exposed to field conditions (e.g. Lejeune, Fell & Burbidge, 1955) or predictions may be made from laboratory experiments (MacPhee, 1961, 1964; Green, 1962; Sullivan & Green, 1964; Kensler, 1967; Carter, 1972; Abdel Rahman, 1974; Neuenschwander, 1975); in the latter case it is important to allow for the effect of acclimatization. Cold hardiness may be influenced by the presence of free water in the surroundings and of food in the gut (Eguagie, 1974).

9.2.4 Biotic factors
As with climatic factors a knowledge of the role of various biotic ones is often obtained from successive observations on the same cohort (see above); this section is restricted to techniques for the recognition of the role of individual factors. It is important to remember that the effect of a parasite will vary with the host, as has been demonstrated by Loan & Holdaway (1961) with the braconid parasite of a weevil, which has no effect on the adult male, but causes the egg production of the female to fall off quickly. The quantitative

evaluation of natural enemies has been reviewed by Kiritani & Dempster (1973): there are four methods of direct assessment, together with the indirect approach by laboratory experiment.

EXAMINATION OF THE PREY (HOST)

Parasitic insects may often be detected in their hosts by dissection (e.g. Miller, 1955; Hafez, 1961; Evenhuis, 1962) or by breeding out the parasites from a sample of hosts (e.g. Richards, 1940; Sasaba & Kiritani, 1972). As the rate of parasitism varies throughout a generation of the host, a single assessment will not give an adequate degree of precision unless the hosts are closely synchronized. Parasites may sometimes be detected within their hosts without dissection by the use of soft X-rays (wavelength $> 0.25°$) and fine-grain film (Holling, 1958). It is possible that elutriation (p. 182) could also be used to diagnose parasitism. The proportion of hosts attacked by a parasite is referred to as the '*apparent parasitism*'. The problems involved in the combination of a series of apparent mortalities and its interpretation are discussed in the next chapter.

Sometimes it is possible to find a fair proportion of the corpses of insects killed by a predator, generally when the predator collects the food together in one place, e.g. spiders' webs (Turnbull, 1960), hunting wasps' nests (Rau & Rau, 1916; Richards & Hamm, 1939; Evans & Yoshimoto, 1962), thrushes' anvils (Goodhart, 1958), shrikes' larders. Ants provide unusual opportunities for this type of assessment because in many species the foragers return to the nest with virtually intact prey, along trails; special apparatus has been designed to sample returning foragers and their prey (Chauvin, 1966; Finnegan, 1969). The frequency of beak-marks on butterflies, arising from unsuccessful attacks by birds, has often been recorded; its interpretation is difficult, but it may be used as evidence for seasonal variation in predation intensity (Shapiro, 1974).

A semi-natural assessment of the role of predators may be made by 'planting' a known number of prey in natural situations; Buckner (1958) describes how sawfly cocoons were exposed in this way and the type of predator determined by the markings left on the opened cocoons.

The diagnosis and determination of the extent of infection by pathogens in an insect population is a complex and specialized problem outside the scope of this book; reference should be made to works such as Steinhaus (1963), Wittig (1963), Smirnoff (1967), Weiser (1969), Weiser & Briggs (1971) and Thomas (1974).* Good examples of the application of these methods to insect populations in the field are provided by Tanada (1961), Neilson (1963), Neilson & Morris (1964), Young (1974) and Entwistle (1977). It should be noted that some methods of collection (e.g. sweep net) may lead to an overestimate of the proportion of diseased insects in the population (Newman & Carner, 1975).

---

* see also Poinar, G. O. and Thomas, G. M. (1978). *Diagnostic manual for the identification of insect pathogens.* Plenum Press, New York.

EXAMINATION OF THE PREDATOR

As the young of many birds remain in the nest it is possible to record their food. The simplest method is to observe feeding through the glass side of a nesting box (Tinbergen, 1960); however, identification and counting of the prey is often difficult and Promptow & Lukina (1938) and Betts (1954, 1958) have shown that parent birds may be induced to put the whole, or at least part, of the food for the young into an artificial gape, from which it may be removed. The young of some birds will regurgitate the food if their necks are manipulated shortly after feeding (Errington, 1932; Lack & Owen, 1955), with others a neck ring needs to be used to prevent swallowing. Emetics may sometimes be administered to recover stomach contents (Radke & Fryden-dall, 1974). The remains of insects and other prey in the pellets of owls provides a further source of evidence of the role of these predators (Hartley, 1948; Miles, 1952; Southern, 1954); and it is possible that if the prey were marked in some way this might be made quantitative. The relationship between the analysis of stomach contents or pellets and the actual food is not completely reliable, however, because of different digestion rates (Hartley, 1948; Custer & Pitelka, 1975).

Other methods involve the slaughter of the predator, and although the removal of a small number of invertebrate predators may not have a significant effect on the population being studied, it is clearly undesirable (quite apart from conservational and legal considerations) to kill large numbers of vertebrates. When practical, therefore, the role of adult birds is best determined by exclusion techniques (see below), with only qualitative confirmation from gizzard analysis. Much information on the possible role of vertebrates as insect predators has been obtained by the examination of the gizzard and stomach contents (Hartley, 1948; Kennedy, 1950; Betts, 1955; Mook, 1963; Elliott, 1973) and the excreta of mammals (Chapman, Romer & Stark, 1955) and lizards (Fuseini & Kumar, 1975). Differential digestion is one of the problems in interpreting data from stomach and faecal contents (Buckner, 1966).

Occasionally prey remains can be detected and identified from the guts of large predatory insects (James, 1961), but in general more refined methods of detection are needed; these are*:

(1) the precipitin test in which the specific proteins of the prey are identified by their reaction with specific antisera prepared in an immunized animal (usually a rabbit).

(2) the use of labelled prey so that the labels may subsequently be detected in the predator.

Serological methods

Serological techniques, originally developed by vertebrate immunologists, have been used in entomology for the investigation of systematic relationships

---

* A novel, third type of method, based on the electrophoretic detection of prey enzymes, has been developed by Murray, R. A. and Solomon, M. G. (1978). *Ann. appl. Biol.* **90**(1), 7–10.

and metamorphosis, the identification of blood meals of biting flies and predator/prey relations: here we are only concerned with the last named aspect and relevant studies include those of Brooke & Proske (1946), Dempster, Richards & Waloff (1959), Loughton, Derry & West (1963), Dempster (1964), Rothschild (1966, 1970, 1971), Frank (1967), Davies (1969), Pickavance (1970, 1971), Service (1973, 1976), Healey & Cross (1975) and Vickermann & Sunderland (1975). For the population ecologist the value of serological tests may be limited by the fact that, unless certain assumptions are made, they cannot be quantified. The other important feature is that the specificity of the antisera must be well known so that the correct interpretation of the results may be made, taking into account the species present in the study area. The techniques given below are described with particular reference to the ecologist. Full details of the immunological aspects can be found in Weir (1973) and Hudson & Hay (1976).

i. *Preparation of antigen*

Numbers of prey are starved to empty the gut, killed with cyanide or by freezing and, with large insects, the legs and wings removed. They are then crushed in a pestle and mortar with buffered normal saline (0.85 % NaCl, pH 7.0) and maintained at 4°C overnight with constant stirring. Particulate matter is removed by centrifugation and a second extraction may be carried out. It is prefereble to extract insects with a large volume of buffered saline and concentrate the solution later in order to obtain the maximum amount of antigenic material. The pooled extracts are freeze-dried, reconstituted in a small volume of buffered saline, dialysed for 24 h at 4°C against buffered saline filtered through a Seitz EKS pad. The antigen may be stored at $-20°C$ or freeze dried in small aliquots. Working with the beetle *Phytodecta olivacea* Dempster (1960) found it necessary to add M/1000 potassium cyanide to prevent the deposition of melanin.

ii. *Production of antiserum*

The important feature of serological tests, to identify prey/predator interactions, is the preparation of high-titred specific antisera. A large number of different regimens has been used to prepare antisera. The two most suitable methods are: injection of alum precipitated antigen intramuscularly into rabbits (Weitz 1952) and injection of an emulsion of antigen and Freund's complete adjuvant into the lymph nodes of rabbits (Boreham & Gill 1973). In the former method 0.1 ml of 10 % potash alum is added to 2.5 ml. of antigen and after adjusting the pH to 6.8 with 1 N sodium hydroxide and washing the precipitate 3 times with buffered saline; the resulting suspension is injected intramuscularly into the hind limbs of rabbits. In the second method equal parts of Freund's complete adjuvant and the antigen are mixed to form an emulsion by drawing up the mixture into a syringe and expelling it through a fine gauge needle. The emulsion is divided into 4 aliquots and injected into the 2 axillary and 2 inguinal lymph nodes of rabbits. With both methods a second injection is given 7–10 days after the first and approximately ten days later a blood sample is taken to test the antibody titre.

If the titre is satisfactory, (the antiserum reacts with a dilution of the antigen at a dilution of 1 in 5000 or more) 50 ml of blood can be taken from the rabbit and serum separated. If the titre is not satisfactory a third or even fourth injection may be given to boost the titre. The advantage of the lymph node method is that only about ten per cent of the antigen is required than is for the alum precipitated method. In addition, the antisera tend to be more specific.

iii. *Collection of predator meals*

The predator may be killed with cyanide and either the whole animal (small species) or the gut (larger ones) smeared on to Whatman No. 1 filter paper, labelled, dried and stored in a desiccator, or the whole predator may be stored below $-20°C$.

iv. *Predator testing*

The filter paper containing a predator is soaked in a small volume of saline (0.2–0.5 ml depending upon the size) overnight, or the whole predator crushed in a similar volume of saline. After centrifugation, the extract is ready for testing. Various modifications of the basic precipitin test have been developed to meet particular needs and increase sensitivity. The basic test consists of allowing the antigen and antibody to interact when under appropriate conditions a precipitate is formed. Three main test procedures have been described:

a. *The capillary ring test* (Weitz 1956, Dempster 1960, Boreham, 1975). The predator extract is overlaid on the antiserum in a glass capillary tube and a positive result is obtained by the formation of a visible ring of precipitate at the interface of the two liquids. The main advantages are that it is adaptable to large-scale use and gives a rapid result (2 h). However, both liquids must be completely clear and this can present difficulties.

b. *Agar gel double diffusion* (Ouchterlony 1948, Pickavance 1970). Antigen and antisera are allowed to diffuse towards each other through 0.6 % agar gel, and a line of precipitation is formed where the reactants. meet at equivalent proportions. This technique has the advantage that a single predator can be tested against several antisera simultaneously and that cloudy predator extracts can be used. It may be further improved by staining with a protein stain such as thiazine red (Pickavance 1970). A modification of this basic test has been used by Davies (1969).

c. *Crossover electrophoresis* (Immunoelectro-osmophoresis, countercurrent electrophoresis, Healy & Cross 1975). This test is basically similar to the Ouchterlony technique except that the two reactants are brought together rapidly by electrophoresis. Results may be obtained in under 30 min.

d. *Latex agglutination.* A new development in serological testing of predator meals is the use of a direct latex agglutination test (Ohiagu & Boreham, 1978). This test has been adapted from a similar technique designed for bloodmeal identification under field conditions (Boorman *et al.* 1977). It consists of separating immunoglobulin G (IgG) from the antisera by caprylic acid (Steinbuch & Audran, 1969) and allowing this to attach to polystyrene latex particles. When this sensitized latex is mixed with the homologous predator meal extract, agglutination will occur. The great advantage of this test is that it can be done in the field, the only equipment required are a microscope slide and pipettes. The reagents are known to be stable for at least 3 months at 4°C and the test is at least as sensitive as the precipitin test.

e. *Passive haemagglutination inhibition.* This technique allows discrimination between prey in related genera (Greenstone, 1977). It depends on attaching the antigen to red blood cells (commonly sheep red blood cells); placed with the appropriate antibody these red blood cells would then agglutinate. The procedure involves the preparation of both the specific antibody and sensitized red blood cells (i.e. with the antigen attached). The extracts from the gut of the predator are added to antibody and the mixture incubated before adding the sensitized red blood cells. When the specific prey is present in the predators' gut, the antibody will combine with it leaving no free antibodies to combine with the sensitized red cells, so these will settle to form a small spot at the bottom of the tube. When the specific prey is absent and in the controls (with the antibody alone) the sensitized red blood cells will agglutinate, i.e. the prey antigen is not present to inhibit the reaction. The method involves a significant measure of skill and experience and a ready supply of antigens; each assay takes about two and a half days, although some 150 tests may be handled in the same batch. Full details of this sophisticated method are given in Greenstone (1977).

v. *Disadvantages of the precipitin test* (methods *a–e* above)
a. Cross reaction: The precipitin test cannot be satisfactorily used to analyse prey species which are taxonomically closely related since the antisera will tend to cross react. Cross reactions can often be removed by absorption (Weitz 1952, Dempster 1960, Service 1976), but for closely related species, if this is attempted, the titre is reduced. However, if the

fauna of the study area is accurately known this problem may not be important. For example, the antisera prepared by Dempster (1960) for *Phytodecta* cross reacted with Chrysomelidae, Cassididae and Coccinellidae. Chrysomelidae and Cassididae were considered unimportant in the study area and reaction with ladybirds was removed by absorption. In his study of predators of rice borers Rothschild (1971) was unable to distinguish between predation on *Chilo suppressalis* and *Tryporyza incertulas* because of cross reactions.

b. Specificity of the antiserum: The second limitation is that the antiserum may not react equally well with all stages of the insect used as antigen, for example, Spencer & Boreham (unpublished) found that an antiserum prepared against adult *Locusta migratoria* reacted poorly with eggs, first and second instars. Their table also shows cross reactions occurring with closely related Orthoptera species. Similar results were obtained with the spruce budworm *Choristoneura* (Loughton *et al.* 1963).

Ashby (1974) working with *Pieris rapae* found that antiserum prepared from pupae was more sensitive than that prepared from larvae. It is therefore preferable to use all stages of the insect as antigen when preparing an antiserum or at least to test carefully the completed product.

c. The precipitin test gives no indication whether dead or live prey have been eaten (Sutton 1970, Tod 1973). This criticism will also apply to microscopical examination of the predator meal.

d. It is not possible to determine whether more than one prey has been eaten by a predator and unless this assumption can be made, quantification of predation cannot be undertaken. The rate of digestion of the predator depends upon intrinsic factors as well as extrinsic ones such as temperature. Measurement of the rate of digestion gives an indication of the length of time a meal can be detected. This is partly a matter of the relative size of the predator and the meal but the data of Hall *et al.* (1953), Dempster (1960) and Loughton *et al.* (1963) for a number of predators (ranging in size from mites and mirids to earwigs and reduviid bugs) with lepidopterous and coleopterous prey, are reasonably consistent: eggs were detectable for 18–48 hours after feeding and large larvae for about five days. For precise population studies it is desirable to know not only that a given predator has eaten the prey within the last x hours, but also whether it has eaten one or more individuals. By studying the dispersion of the prey in the field and the rate of movement of the predator, Dempster (1960) was able to argue, supported by laboratory feeding trials, that the chance of a positive representing more than one feed was very small. Therefore it was possible to estimate the number of prey consumed by the predator in the total population during the digestion period; e.g. if in a sample of 200 predators, 6 were shown to contain meals of the prey, which remain detectable for 1 day, we may say that the total number of prey destroyed by that predator during the previous 24 h is $= 6 \times$ total population of predator $+ 200$ (Dempster 1960).

The majority of studies undertaken to date using the precipitin test have been concerned only with determining the range of predators which are preying on a particular species. It is an especially useful technique for predators that suck the juices of their prey rather than take the whole insect.

Labelled prey

The prey may be labelled with a dye, a rare element or a radioactive isotope; it is important that the label does not modify the behaviour of the predator. The egg predators of a moth were detected by spraying the eggs with an alcoholic suspension of a powdered fluorescent dye. The dye could be detected, under UV light, in the gut of the predators; this was particularly easy if the gut was macerated and dried (Hawkes, R. B., 1972).

Feeding can also easily be established if the prey are marked with a rare element (see p. 77) or a radioactive isotope and the label is subsequently detected in the predator. Ito *et al.* (1972) used europium for this purpose, but found it to be rapidly excreted by insects. If each prey carried a similar burden of rare element or radionuclide, the level of these labels in the predator would be a measure of the prey consumed.

Theoretically, therefore, this method could be of value with voracious predators that consume prey at frequent intervals, a situation in which the precipitin test is unsuitable for quantitative studies. However, in practice it is difficult to arrange for all the prey to carry an equal burden of label: van Dinther & Mensink (1971) found this impossible to achieve with fly eggs whether they were labelled directly or via the female.

Most studies using this technique have involved prey tagged with ^{32}phosphorus (see Chapter 3 for marking methods) and have had as their objective the identification of the predators (Jenkins & Hassett, 1950; Fredeen *et al.*, 1953; Pendleton & Grundmann, 1954; Baldwin, James & Welch, 1955; James, 1961, 1966; Jenkins, 1963). An approximate measure of the number of prey taken may be obtained from the level of activity in the predator. However, Baldwin *et al.*, (1955) showed, in laboratory tests, that the same level of radioactivity (in counts per minute) could result from the consumption of different numbers of prey; this may be due to the variable radionuclide burdens of the prey, but could arise from different assimilation rates on the part of the predator (p. 316). If a large number of results were obtained and the radioactivity of the prey were normally distributed, it would be justifiable to calculate the number of prey eaten from the mean counts for 1, 2, ... x prey. In essence this was the approach of Pendleton & Grundmann (1954), who made a thistle plant radioactive by placing 1 μc ^{32}P in 3 cm^3 of water in a hollow in the pith; the aphids feeding on the plant gave an average of 250 ct/min and the predators that ate *whole* aphids gave counts roughly in multiples of 250 ct/min; those such as spiders that consume only part of the aphid gave lower counts and so the lowest group of counts (120–180) was taken to represent the killing of a single aphid and approximate multiples of this, the killing of two or more.

Greater precision for more continuous studies might be introduced by using a modification of the approach of Crossley (1963) which was developed for measuring the consumption of vegetation by insects. He points out that if the biological half-life of the nuclide is long relative to the life-span of the insect, the amount in the insect (Q_t) may be given by:

$$Q_t = \frac{ra}{K}(1 - e^{-Kt}) \tag{9.1}$$

where r = rate of ingestion, a = proportion of ingested nuclide assimilated, K = average elimination constant and t = time. (This approach could not be used in a situation when the radionuclide in the predator reaches equilibrium with that of its food, when the relationship $Q_e = ra/K$ holds; this is discussed on p. 466). Now if the isotope used had a long biological half-life (e.g. ^{45}Ca, ^{65}Zn) for any predator Q_t could be measured, and if a and K were known from laboratory studies, it should be possible to calculate r which would give a measure of the total number of isotope marked prey eaten. If the proportion of these in the total population was known (it would probably have to be fairly high), it could be assumed that the marked and unmarked prey were taken equally and the total prey consumption easily estimated. Gamma emitters and the use of a scintillation counter would be particularly suitable for measuring radioactivity in these studies, since it would eliminate problems of counting geometry associated with whole body counts.

Moulder & Reichle (1972) were able to assess the role of spider predation in a forest at Oak Ridge which had been tagged with ^{137}cesium: they found the cesium concentrations of potential prey reasonably similar, although there were seasonal variations. Allowances had to be made for this, for the portion of the prey discarded by the spider and for the different size categories of spiders and prey. This was, of course, a rather special situation where the whole environment was labelled (i.e. every prey) with an isotope with a long half-life.

Extreme caution should be exercised in all work of this type to ensure that the predator's intake of radionuclide came only from the prey. Baldwin *et al.* (1955) point out that if mosquito larvae are returned to a pond directly after marking by immersion in a solution of ^{32}P, their radioactive excreta contaminates the whole pond. They overcome this by washing the larvae twice and keeping them for two days in freshwater before release. If the prey are marked through the host plant it must be remembered that some predators also feed, to a greater or lesser extent, on the plant. Van der Meijden (1973) determined the fate of lepidopterous larvae by labelling them with paint containing an isotope (see p. 80), the labels could be recovered and it could be determined whether they were close to a pupa or isolated when the larvae was assumed to have been killed by ants (or other causes).

PREDATOR OR PARASITE EXCLUSION TECHNIQUES
These methods demonstrate the effects of predators or parasites by artificially excluding them and measuring the increase in the prey's population under the new conditions; this increase represents the action of the predator (or parasite) in the exposed situation. Conceptually the method is very attractive but its weakness is that the techniques of excluding the predators often affect the microclimate and other aspects of the habitat, so that it is difficult to be certain that the observed effects are entirely due to the predator (Fleschner, 1958).

Mechanical or other barriers
Mammals and birds are relatively easily excluded by wire netting or nets, and it is unlikely that such methods have marked side-effects on the prey species, unless of course they also exclude herbivorous mammals whose activities have an important role on the vegetation. Buckner (1959) used mammal exclusion cages in his study of the predation of larch sawfly, *Pristiphora*, cocoons; in this case both the protected and exposed cocoons were 'planted' and therefore there was the risk that the resulting artificially high density might have acted as a bait to the mammals.

Predatory insects need to be excluded by smaller muslin cages of the sleevecage type (Smith & DeBach, 1942), but as Fleschner (1958) has shown such cages may affect the population directly and therefore, even if some cages are left open for predator access, as the cages affect the prey's rate of increase the natural situation cannot be measured (De Bach & Huffaker, 1971).

Non-flying insects can be more easily excluded by mechanical or insecticidal barriers. Wright, Hughes & Worrall (1960) and Coaker (1965) demonstrated the effects of carabid beetles on populations of the cabbage root fly (*Erioischia brassicae*) by means of trenches round the plots, filled with straw soaked with an insecticide (Fig. 9.5). Spiders (Clarke & Grant, 1968) and ants can often be

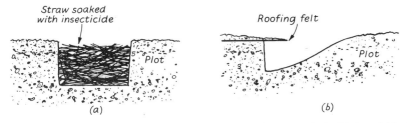

Fig. 9.5 Barriers to carabid beetles: *a.* exclusion barrier of straw soaked with insecticide; *b.* emigration barrier that allows beetles to enter plot but not to leave (based on Wright, Hughes & Worrall, 1960).

eliminated by similar barriers or bands on trees (p. 249).

Various combinations of exclusion cages may be used to separate the roles of vertebrates (wire cages), ground dwelling invertebrates (mechanical barriers) and flying invertebrates (net cages – variations in net size can produce further separation) (Ashby, 1974; Eickwort, 1977).

Elimination of predator or parasite
Vertebrates may be shot (Dowden, Jaynes & Carolin, 1953), or trapped (Buckner, 1959), until their numbers are extremely low; as this must be done over a fairly large area to be effective it will lead to major disturbance in the habitat; this is undesirable in ecological studies. Fleschner (1958) found hand-picking of invertebrate predators from a part of the tree was both a feasible and reliable method.

As some predators are very susceptible to certain insecticides to which the prey are resistant, it is possible to assess their effects by an *insecticidal check method* (DeBach, 1946; DeBach & Huffaker, 1971). It is very important to ensure that the pesticide has no side-effects on the prey species. Although this method was found reliable for studies on predators of *Aonidiella* scale insects (DeBach, 1955), it was not satisfactory for various plant-feeding tetranycid mites (Fleschner, 1958).

A *biological check method* has also been described in which large populations of ants are built up; these attack and often prevent predator and parasite action (Fleschner, 1958). Although such observations do give a measure of predator importance, there are so many other factors involved that they cannot be considered as equivalent to the quantitative estimation of predation.

DIRECT OBSERVATION
When a single parasite larva emerges from each individual of an arboreal host and drops to the ground to pupate, these may be collected in trays or cone traps (see p. 308) to give an absolute measure of their own population, which is equivalent to the number of hosts they have killed. Such conditions are fulfilled, for example, in many tachinid parasites of Lepidoptera. When the percentage parasitism of the host is known *accurately* such figures may also be used to calculate the actual host population (Dowden, Jaynes & Carolin, 1953; Bean, 1958):

$$\text{Total host population} = \frac{\text{Total parasites} \times 100}{\% \text{ parasitism}}$$

Sometimes it is possible to estimate predation on the basis of sightings of predators consuming prey (Kiritani *et al.*, 1972; Kiritani & Dempster, 1973). The observed frequency of feeding (F) is determined by routine counts, together with the length of time captured prey are retained (R) and the diurnal rhythm of feeding to give the proportion of feeding (C) that occurs during the time interval when the observations are made. Now if the probability of observing feeding (P_f) is retention time in hours (R) divided by 24, then an estimate of the total number taken is:

$$n = FC/P_f \qquad (9.2)$$

A series of values of n may be plotted against time and the area under the graph gives the total prey killed in that habitat. This 'sight-count' method was devised and used by Kiritani *et al.* (1972) in studying spider predation of leafhoppers. It depends on a high accuracy in observing all instances of predation at a given time and on the values of C and R being fairly constant.

EXPERIMENTAL ASSESSMENT OF NATURAL ENEMIES

There is now an extensive theoretical background, supported by field and laboratory studies, on the interactions of prey with their natural enemies; the subject is reviewed by Murdoch & Oaten (1975), Beddington *et al.* (1976), Hassell *et al.* (1976), Hassell (1976*a* & *b*), May (1976) and Hassell (1978). The interactions may be viewed in relation to two population variables (Hassell, 1976*a*; Hassell *et al.* 1976; Beddington *et al.* 1976):

(1) the death rate of the prey
(2) the rate of increase of the predator

The two are, of course, interrelated; the prey death rate depends on the numbers of predators and on their searching efficiency. To some extent this classification parallels that between functional and numerical responses (Solomon, 1949; Holling, 1959*a,b*, 1961, 1965, 1966), but is somewhat broader in that prey density is now only one of the independent variables against which predator efficiency or reproductive rate is viewed. Although sufficient laboratory and field studies have been made to allow some general comparisons and insights into predator–prey relationships (e.g. Murdoch & Oaten, 1975; Hassell *et al.* 1976; Beddington *et al.* 1976), there is a considerable need, both general and particular, for the comparative assessment of various components of natural enemy action in the laboratory and in the field, especially with the same natural enemy.

The major relationships are summarized in Table 9.1. The following sections indicate some approaches to the measurement of these relationships. Once this has been done the best procedure, as almost invariably in ecology, is to plot the data; when this has been completed it will in many cases be possible to select the most appropriate model and determine precisely the value of the major component parameters. For some other relationships, clear models have yet to be developed. The uses and methods of modelling are discussed in Chapter 12.

Death rate of prey/prey density

This is most elegantly measured in the laboratory with different prey densities

Table 9.1 Some relationships whose measurement is important in assessing the role of natural enemies (other variables: A = age of predator (developmental response), B = age of other predators of same species, C = climate and other externals, D = other prey (switching), E = other animals (competition and mutualism))

Component	In relation to	Other variables	Affecting
attack rate a handling time T_h	prey density 'functional response'	A,C,D,E.	death rate of prey
interference constant m quest constant q or time lost through encounters bt_w	predator density	A,B,C,D,E.	
predator dispersion P_i (number/ith area) T_i (time/ith area)	prey dispersion N_i 'aggregative component of numerical response'	A,C,D,E.	
predator fecundity survival rate of predator of different stages duration of different stages of predator	prey density 'breeding component of numerical response'	B,C,D,E.	rate of increase of predator

per predator and the numbers of prey killed or hosts parasitized (for insect parasitoids) recorded.

The numbers of prey eaten per predator at different prey densities can also be obtained from field data, although it is rare that an adequate estimate of the density of searching predators is available. Such results, however, should only be interpreted as a functional response *sensu strictu* if the density is relatively constant. Wide fluctuations in predator density may confuse the relationship by different levels of predator interference occurring (see below).

There are two essential components to a functional response: an instantaneous attack rate (a) and a saturation term, often called the 'handling time' (T_h) but also including any effects of satiation. In addition the speed of movement of the predator (Glen, 1975), and its reactive distance (both of which may be influenced by habitat characteristics) and the proportion of attacks that are successful (Nelmes, 1974) and of prey that are fully consumed are further variables. The number of prey dying and the number eaten may not be the same, either because the predator kills more than it consumes (Buckner, 1966; Kruuk, 1972; Toth & Chew, 1972) or because it disturbs the prey, driving them from the host plant, etc., and exposing it to other mortalities, as Nakasuji *et al.* (1973) showed for spiders with lepidopterous larvae and W. Milne (unpublished) for birds with aphids. The type of habitat will influence predator/prey interactions through its effect on speed of moving and reactive distance, as well as through dispersion (Martin, 1969).

Many functional responses have been found to be of the 'Holling Type II form' (Fig. 9.6a). This has been traditionally described by Hollings disc equation:

$$N_a = \frac{a\,N\,T\,P}{1 + a\,T_h N} \tag{9.3}$$

where N_a = total number of prey attacked, a = attack rate, N = total number of prey, P = predator density, usually one in experimental systems, T_h = handling time and T = total time (searching time $+ T_h N_a$, but excluding any regular sleeping time, i.e. sleeping not related to the amount of feeding).

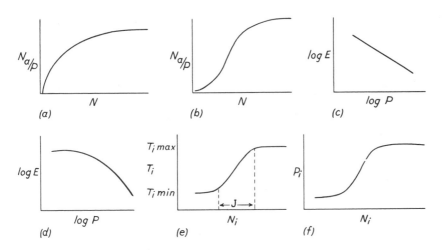

Fig. 9.6 Some characteristic forms of predator-prey relationships: *a.* Holling type II functional response; *b.* Holling type III functional response; *c.* interference – Hassell and Varley model; *d.* interference – Rogers and Hassell or Beddington model; *e.* and *f.* aggregative response measured in time and numbers.

The method of estimating a and T_h from functional response data using the above equation is described in Holling (1959b). This technique, however, does not allow for the removal of prey during the experiment (i.e. it assumes prey replaced as eaten or a systematically searching predator) and will lead to incorrect estimates if prey depletion is significant (Royama, 1971; Rogers, 1972) (c.f. removal trapping, p. 230). A more satisfactory method for attaining a and T_h is discussed by Rogers (1972), which assumes random exploitation of the available prey. Using this technique, it is necessary to distinguish between predators (that remove prey as eaten) and parasitoids (that can re-encounter hosts, thus involving more handling time):

$$\text{Random predator:} \quad N_a = N[1 - \exp\{-a(PT - N_a T_h)\}] \qquad (9.4)$$

$$\text{Random parasitoid:} \quad N_a = N[1 - \exp\{-TaP/(1 + a\,T_h\,N)\}] \qquad (9.5)$$

where P = number of predators (parasitoids).

Although the type II response of Holling is usually regarded as the form typical of invertebrate predators and parasitoids, there is increasing evidence that sigmoid type III responses are also widespread, at least amongst insects (Murdoch & Oaten, 1975; Hassell, 1978). Experimental evidence suggests that these responses result from some component of a, or the time initially available for search (T), now being a *rising* function of prey density. The increased number of parameters involved in describing such responses makes their adequate estimation less straightforward than for type II responses, and requires a non-linear, least squares technique.

Death rate of prey/predator density
The presence of other predators may enhance (co-operation) or reduce (interference) the prey death rate due to a single predator. Instances of co-operation are particularly common amongst vertebrates and range from the group-hunting tactics of carnivorous mammals to the mere confusion of prey by several birds fishing in the same region. The defence mechanisms of, for example, a colony of aphids would probably be more easily disrupted by a number of coccinellid larvae than by a singleton, but this does not seem to have been quantitatively demonstrated; the more general result with invertebrates is interference (Hassell *et al.* 1976). It is important to remember that in laboratory experiments, predators or parasitoids are often prevented from dispersing when crowded, therefore unreal levels of interference and superparasitism may be observed (Hassell *et al.* 1976). Interference effects, when plotted in a log scale, are commonly linear (Fig. 9.6c) or curvilinear (Fig. 9.6d). When the relationship is linear then the component coefficients may be extracted using the model of Hassell & Varley (1969):

$$\log E = \log Q - m \log P \qquad (9.6)$$

where P = predator density, m = interference constant, Q = quest constant ($= E$ when $P = 1$) and E = searching efficiency of the predator. In general E may be defined (Hassell, 1976b):—

$$E = \frac{1}{P} \log_e \left[\frac{N}{N - N_{ha}} \right] \qquad (9.7)$$

when N_{ha} = prey actually eaten or parasitized. Under the special circumstances when every prey encountered is actually eaten and the number attacked is small compared with the total prey population, then

$$E = N_a/NP \qquad (9.8)$$

The curvilinear relationships were discussed and modelled by Rogers & Hassell (1974) and also by Beddington (1975), who provided an expression for searching efficiency (E):

$$E = a\, T/[1 + bt_w(P-1)] \qquad (9.9)$$

where a = attack rate = $N_a/NT_s\, P'$ (where T_s = total searching time and P' = predator number), P = predator density, b = rate of encounter between parasites and t_w = time wasted *per encounter*.

Predator dispersion/prey dispersion

Predators may aggregate in a region of high prey density. This aggregative or behavioural component of the numerical response has considerable significance in the stability of predator–prey systems (Hassell & May, 1974; Murdoch & Oaten, 1975). Observations may be made in the laboratory or in the field, measuring the number of predators found in regions of different prey density; when this can be done sequentially (e.g. by time lapse photography) so that the series of numbers may be used to calculate the proportion of time spent in each area. More rarely continuous observations can be made to give data on actual searching time per area (Murdie & Hassell, 1973).

The results may be plotted as numbers of predators per unit area (P_i) or the time spent by predators per unit area (T_i) against prey density per unit area (N_i) (Fig. 9.6e and f). In their analysis of the significance of the form of the aggregative response for stability, Hassell & May (1974) showed that the greater the difference between the maximum and minimum times per unit area and the closer the region J (Fig. 9.6e) corresponded to the average range of the prey's density, the more stabilizing the aggregative response.

Fecundity, developmental and survival rates of predators/prey density

The importance of these relationships is reviewed by Beddington *et al.* (1976) who suggest certain expressions for them, but stress that the lack of documentation for all the components for one species 'represents a major gap in experimental ecological work'. The methodology for such studies is, of course, basically similar to that used for any animal (e.g. Turnbull, 1962).

The role of other prey

'Other prey' are often important in allowing the survival of natural enemies over a period when the prey species being considered is sparse. However, the few quantitative studies in this area have been concerned with "switching", a sudden change in the predators' preference between various prey: switching is reviewed by Murdoch & Oaten (1975). The idea arose from L. Tinbergen's theory of a 'search image'; some supporting evidence is given by Murton (1971). Royama (1970) suggested however that the predator maximizes the 'profitability' of its searching: the profitability of feeding on a particular prey is compounded of size, density and ease of capture. Royama's own work together with that of Bryant (1973), in which the prey of insectivorous birds were compared with the available prey, tends to support his concept, larger prey items tending to be relatively favoured: a finding that also accords with Hespenheide's (1975). However, even in these studies the prey spectrum was not markedly different from that available. No significant preference or switching has been discerned for trout (*Salmo trutta*) (Elliott, 1970), coccinellid larvae (Murdoch & Marks, 1973), mites (Santos, 1976), or generally for predatory littoral snails (Murdoch, 1969).

A number of different indices have been used to detect preference, of which V. S. Ivlev's 'elective index' and various forage ratios are examples (Lawton *et al.* 1974; Jacobs, 1974; Mustafa, 1976). These have been reviewed by Cock (1978), who follows Lawton *et al.* (1974) in using the random equations (9.4 and 9.5) to predict the numbers of each prey type (subscripts 1 and 2 in the notation) eaten:

$$N_{a1} = N_1[1 - \exp.\{-a_1 P(T - T_{h1} N_{a1} - T_{h2} N_{a2})\}]$$
$$N_{a2} = N_2[1 - \exp.\{-a_2 P(T - T_{h2} N_{a2} - T_{h1} N_{a1})\}] \tag{9.10}$$

when the ratio of the two prey types becomes:

$$\frac{N_{a1}}{N_{a2}} = \frac{N_1[1 - \exp.(-a_1 P T_s)]}{N_2[1 - \exp.(-a_2 P T_s)]} \tag{9.11}$$

where $T_s = (T - T_{h1} N_{a1} - T_{h2} N_{a2})$, T = total time available for searching and feeding and T_h = handling time. A similar procedure based on the random parasitoid equation (9.5) should be used with parasitoids.

Cock (1978) (see also Hassell, 1978) recommends the following procedure for detecting preference:

(1) Carry out function response experiments using each prey separately.
(2) Estimate a_1, a_2, T_{h1} and T_{h2} from the 'random equations' (9.4 and 9.5) as appropriate (i.e. predator or parasitoid).
(3) Any preference resulting from differences in the functional response parameters (i.e. $a_1 \neq a_2$ and/or $T_{h1} \neq T_{h2}$) can now be conveniently displayed in terms of N_{a1}/N_{a2} (calculated from 9.11) plotted against N_1/N_2 or, alternatively, as the proportion of one of the species in the total diet against the proportion available (e.g. $N_{a1}/(N_{a1} + N_{a2})$ against $N_1/(N_1 + N_2)$). Such innate preference will then be detected as a deviation from a slope of unity passing through the origin.

(4) Carry out a further experiment in which various ratios of the two prey types are presented together, and so contrast predicted and observed results. Ideally, this procedure should then be repeated for a range of total prey densities that encompass those used in the functional response experiments (1). Any difference between the predicted preference from (3) and that observed will now be due either to an active rejection of one of the prey or some change on a_1, a_2, T_{h1} or T_{h2} as a result of the predator experiencing the two prey types together.

In the field situation the position may be further complicated because polyphagous predators will often have ·wider trivial ranges than their prey whose micro-habitat preferences may be more restricted: thus if a predator modifies its hunting pattern from one microhabitat to another in response to relative differences in prey availability, then the 'switching effect' might well be greater than that estimated from preference along when prey are randomly mixed. Jolicoeur & Brunel's (1966) observations on the cod (*Gadus morhus*) may be an illustration of this.

Changes during the development of the predator
This concept was formalized by Murdoch (1971) and termed the developmental response. Clearly many components (T_h, a, m, etc.) will change with the age of the predator; Hassell *et al.* (1976) show how age structure of predator and prey may alter the parameter values of the functional response. Such data are most easily gained from laboratory experiments (Turnbull, 1962; Thompson, 1975; Evans, 1976; Wratten, 1976).

Changes due to other animals
Competition between predators is often recorded (e.g. between parasitoids, leading to multiparasitism). This has yet to be considered in detail against the general theory of competition between two populations (May, 1976). Mutualism between two predators or between a non-predator (buffalo) and a predator (egret) has not been related to the theoretical framework for this relationship now being developed (May, 1973, 1976). The important variables for measurement would be prey killed, prey density, density and survival of both types of predator.

Climatic and similar effects
The appropriate experimental methods are referred to on p. 310.

9.3 Dispersal

The term dispersal covers any movement away from an aggregation or a population and may refer to the movement of newly hatched larvae away from their egg mass, a secondary dispersive process (Henson, 1959) (= interspersal of MacLeod & Donnelly, 1963) or the migration of adults away from their population territory (Southwood, 1962, 1971; Johnson, 1969). Taylor & Taylor (1976) refer to all such processes as migration, restricting immigration and emigration to movements in and out of the population's habitat. Flying insects and birds, detected by radar and other methods as appropriate (Chapters 4 and 7), provide data on dispersing or migrating populations. The interpretation of such information is often facilitated by some of the methods described below which measure dispersal directly. They may be grouped according to the type of experiment: laboratory, field with marked animals or field with unmarked populations.

9.3.1 Laboratory assessments

Useful measurements of the potential for movement of insects may be obtained from laboratory observations, either by tethered flight (e.g. Dingle, 1966; Dingle & Arora, 1973) or more realistically, but with greater complexity of equipment in a flight chamber (wind tunnel) (Kennedy & Booth, 1963; Kring, 1966; Kennedy & Ludlow, 1974). An automatic device for measuring tethered flight is described by Cullis & Hargrove (1972). Readiness and frequency of take-off and the condition of the wings and musculature are other indicators of dispersive potential that may be assessed in the laboratory. Such laboratory studies may underestimate dispersive distance with airborne animals because the effects of gliding and of wind can be overlooked; any density dependent effects will also be missed. Animals that have been kept in laboratory cultures for some generations may well provide aberrant data, because of the unusual selective forces to which they have been exposed.

9.3.2 The use of marked or introduced animals

The animals are marked using the various methods described in Chapter 3. When they are released where they were found, or directly marked in the field, (Greenberg & Bornstein, 1964) their subsequent movements may be followed to determine their home range or territory (if this exists). It is also possible to release unmarked animals on an 'empty' habitat. Some workers have used laboratory-bred material for these releases; although this may be acceptable (MacLeod & Donnelly, 1956; Fletcher & Economopoulos, 1976), for the reasons outlined above it should not be adopted without a check.

The use of insects captured in the wild also poses problems: the level of migratory activity is commonly related to the insect's age (Johnson, 1969) and the process of marking may so disturb them that dispersal is exceptional during the first few days after release (Clark, 1962; Greenslade, 1964). Dean (1973) studied the movements of aphids in an uninfested crop by placing pots of cereals, with dense colonies of aphids, at various points.

THE MEASUREMENT AND DESCRIPTION OF DISPERSAL

The animals are released at a known point and are subsequently recaptured.

Long distance migrants, mainly vertebrates, but including some butterflies are eventually recaptured at random; providing information on distance and direction of movement (e.g. Urquhart, 1960). Other experiments involve the release of marked individuals at a point and their recapture in traps placed at increasing distances (in annuli or in the form of a cross) from the release point (e.g. Doane, 1963). The influence of irregularities in environmental favour-ableness around different traps may be reduced by using the ratio of unmarked to marked animals in each trap (Gilmour, Waterhouse & McIntyre, 1946; MacLeod & Donnelly, 1963).

Firstly it should be determined if there is drift or non-randomness in the direction of dispersal. One approach to this is to superimpose upon the map of the release and recapture sites a horizontal and vertical grid, and the mean and the variance are calculated for each day in terms of these grids (Clark, 1962). If the means differ significantly, then there was markedly more dispersal in one direction than another–drift. If the means are similar, but the variances differ, then movement was non-random. Clark (1962) describes a method for testing whether the resulting spatial distribution is circular or elliptical.

Another approach is that of Paris (1965), who, in a series of experiments, determined the number of woodlice in different radii and compared the results for eight radii using Friedman's (1940) analysis of variance by ranks test, which is non-parametric.

Both the above methods are limited to situations when a group of marked animals are released from a central point. When the members of a natural population have been marked individually it may be possible to use the approach of Frank (1964), who demonstrated that individual limpets (*Acmaea digitalis*) were not moving randomly and hence concluded that they had home ranges. The habitat is regarded as a grid of identical squares; if the probability of movement from one square to the next remains constant, then, knowing the movement in one time period, it is possible to calculate, by applying a Markov process, the transition matrix at the end of a given number of time periods for the probabilities of movement from one square to another. The actual movements may then be compared with the expected using the χ^2 test. Recent work on bird navigation has revealed some ability to orientate in a time-compensated manner to sun, moon and certain constellations and perhaps magnetic field; attention has therefore been directed at analysing the effect of a small bias on an otherwise random movement (Matthews, 1974; Kendall, 1976).

Provided the drift in one direction is not excessive the dispersal of a marked population from a point source may be represented by Fig. 9.7, the shaded panels represent the successive density/distance profiles (Inoue *et al.*, 1973). Five interrelated variables may be calculated to determine the properties of a dispersing population as revealed by such an experiment:

(1) The extent to which the population is heterogeneous with regard to dispersal, i.e. some dispersing, others being virtually sedentary.

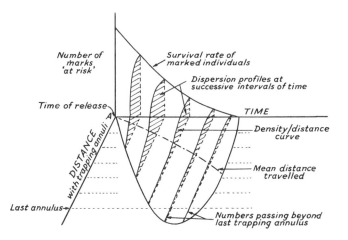

Fig. 9.7 A schematic representation of the dispersion of marked individuals released at a fixed point *A*, showing some of the variables that may be estimated (modified from Inoue *et al.*, 1973).

(2) The numbers or relative densities at successive distances from the release point.

(3) The fall-off of density with distance; i.e. the form of the leading edge of the shaded segments in Fig. 9.7.

(4) The mean distance travelled (by the emigrating component if some flies are stationary) (Fig. 9.7).

(5) The proportion or number of marked individuals that have passed beyond the area.

Account must also be taken of the density at the start because these variables may well be density modified (Taylor & Taylor, 1976). Furthermore the tail of the distribution, representing the most mobile individuals, is likely to be unrepresented in small samples.

The detection of heterogeneity, with respect to the rate of dispersal, in the individuals of the population

In an experiment on the dispersal of marked *Drosophila* from a central point of a cross of traps, Dobzhansky & Wright (1943) demonstrated that the flies were heterogeneous with respect to the distance they travelled. This was detected by determining the departure from the normal curve of the frequency curve of numbers with distance from the point of release on a given day. The departure or kurtosis is given by the formula:

$$Ku = \frac{R \sum_{i=1}^{y} x_i^4 n_i}{\left(\sum_{i=1}^{y} x_i^2 n_i \right)^2} \tag{9.12}$$

where R = total animals caught (recaptured) in all traps, x_i = distance of the recapture point (i) from the point of release, n_i = total number of animals caught in traps at the same distance (x_i) from the release point and y = the last equidistant set of recapture points (the first point will be at the centre i.e. $x_i = 0$).

Shaw (1970) has shown how even alate aphids may differ greatly in their potential for dispersal.

The numbers at various distances from the release point
The approach with the fewest assumptions is that of Fletcher (1974) who studied the Queensland fruit fly (*Dacus tryoni*) and calculated the *proportion* of the migrant flies that were in a particular annulus (F_i). The density of flies in the ith annulus is proportional to:

total number of marked flies trapped (recaptured)/no. of traps = n_i/g_i

Then

$$\hat{F}_i = \frac{n_i}{g_i}\left(x_{i+1}^2 - x_i^2 \right) \div \sum_{i-1}^{y} \frac{n_i}{g_i}\left(x_{i+1}^2 - x_i^2 \right) \qquad (9.13)$$

where x_i = the distance the inner radius of the ith annulus is from the central release point, therefore x_{i+1} = the distance of the outer radius of the same annulus from the central point and y = total number of annuli.

This method cannot be used when significant numbers have migrated beyond the outer annulus: proportions are used because the catchment area of the traps could not be ascertained.

Crumpacker & Williams (1973) in their study of *Drosophila* assumed that the zone of attraction of one trap would overlap the next but the number of marked flies (M) in the annulus could be estimated for a particular day:

$$\hat{M}_i = r_i\left(\frac{U_i}{u_i} \right)\left(\frac{A_i}{\alpha_i} \right) \qquad (9.14)$$

where r_i = marked flies recaptured on the particular day in the ith annulus, u_i = the unmarked flies captured, U_i = the total unmarked flies, A_i = the total area of the annulus and α_i = attractive area of the annulus. On their assumption $A_i/\alpha_i = 1$; clearly M_i and U_i are both unknown, but these could easily be computed from a series of equations.

The probability that a marked fly reached the ith annulus is estimated by:

$$\hat{q}_i = \frac{M_i}{\sum\limits_{i=1}^{y} M_i} \qquad (9.15)$$

Given the assumption that $A_i = \alpha_i$, this may be expressed as:

$$\hat{q} = \frac{A_i r_i / u_i}{\displaystyle\sum_{i=1}^{y} \frac{A_i r_i}{u_i}} \tag{9.16}$$

where the first annulus is number 1 and the last y.

Of course these probabilities will be influenced by heterogeneity in the population discussed above, (see also p. 339 for a method assuming random movement.)

The fall-off of density with distance

This topic has recently been comprehensively reviewed by Freeman (1977) and Taylor (1978); the latter shows that the equations belong to two 'families' (Table 9.2). The shapes of the curves generated vary significantly (Fig. 9.8) and both Taylor (1978) and Freeman (1977) found that the equations in family II give the best fits to most field data for insects; this implies that the lengths of individual movements are not random: such dispersal would be better fitted by equations in family I. The general equation for family II is:

$$N = \exp(\eta + bX^c) \tag{9.17*}$$

Table 9.2 Equations for the dispersal of organisms from a release point following the classification of Taylor (1978), where N = number dispersing to distance X and η, b and c are constants.

	General form for family	Equations	Author
I.	$N = \eta + b\,f(X)$	$N = \eta + c/X$	Paris, 1965
		$N = \eta + b.\log_e X$	Wolfenbarger, 1946
		$N = \eta + b.\log_e X + c/X$	Wolfenbarger, 1946
II.	$N = \exp(\eta + bX^c)$	$N = \exp(\eta + c/X)$	Taylor, 1978
		$N = \exp(\eta + b.\log_e X)$	MacLeod & Donnelly, 1963
		$N = \exp(\eta + b\sqrt{X}$	Hawkes, C., 1972
		$N = \exp(\eta + bX)$	Gregory & Read, 1949
		$N = \exp(\eta + bX^2)$	Dobzhansky & Wright, 1943

This may be taken as indicating non-random movement, and it seems probable that, as Taylor (1978) suggests, the parameter c is a measure of non-randomness, representing randomness only when $c \sim 2$; when $c < 2$ there is a very hollow curve that could indicate some tendency to aggregation (around

* *Note* the constant $\eta = a$ of Taylor (1978), but in this chapter this symbol is used for the attack-rate (e.g. equation 9.3) or, in accordance with chapter 3, the number of marked animals released (e.g. equation 9.21).

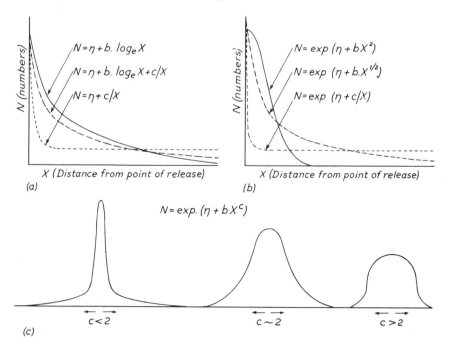

Fig. 9.8 Various forms of dispersal – distance curves. *a.* and *b.* Half distributions for various equations (Table 9.2) (after Taylor (1978) based on Dobzhansky and Wright's (1943) data); *a.* Family I type curves; *b.* Family II type curves that imply a density relationship; *c.* representations of the whole distributions for Family II type curves with different values of *c* in the general equation.

the point of release), whilst $c > 2$ may be a reflection of repulsion leading to regularity (see Fig. 2.5 and p. 47). Taylor & Taylor (1976, 1978) have taken this concept of density-dependence further in their general model (see below).

A simple test for random movement – a random-walk, Brownian motion or diffusion model – is to compare the fall-off of numbers with distance with half a normal distribution (Fletcher, 1974). Considered as a diffusion phenomenon, the density of marked insects per unit area (m') at a particular point is given by (Scotter *et al.*, 1971; Alkman & Hewitt, 1972):

$$m' = \frac{M}{4\pi\psi t} \exp\left(\frac{-x^2}{4\psi t}\right) \tag{9.18}$$

where M = effective (i.e. surviving and moving) number of marked insects released, x = the radial distance from the point of release, t = time, $\pi = 3.14$ and ψ = dispersal rate coefficient. This equation may be solved by iteration, but ψ will be found to have two values, one as the density of insects at a particular distance builds up and one as it gradually falls as they disperse even further (as shown by Clark, 1962 in his study of grasshoppers). The density at which the switch from congregation to dispersion occurs has no significance in this purely physical model, but in a more realistically biological situation, it could be most illuminating (see 9.28 below).

It should be noted that the magnitude of the maximum x in relation to the scale of the animals habitat will strongly influence the type of frequency distribution found with distance: many species will show a different pattern within the population habitat and outside it.

The mean-distance travelled and rate of dispersal
This is most easily calculated from data on the proportional frequency (F_i) (Fletcher, 1974) or actual numbers $(\hat{M}_i)$ (Crumpacker & Williams, 1973) of marked individuals estimated for successive annuli (see above). The estimate is:

$$\hat{\hat{d}} = \sum_{i=0}^{y} F_i \tfrac{1}{2}(x_{i+1} + x_i) \qquad (9.19)$$

where x_i is the distance the inner radius of the ith annulus is from the central point, x_{i+1} is the same for the outer radius and y is the outermost annulus. If numbers have been estimated, $\hat{M}_i$ is substituted for F_1 in the above equation. These estimates are reliable only so long as no animals pass beyond the sampling annuli.

The rate of dispersal may be estimated by the dispersal rate coefficient (ψ) of Scotter *et al.* (1971), described above. Its use is based on the assumption of random walk movement. In the experiment with *Drosophila* referred to above, Dobzhansky & Wright (1943, 1947) determined the speed of dispersal by comparing the change in variance on successive days. If the speed of dispersal is constant the variance should change by a constant amount from day to day. They suggested that as the curve of numbers on distance was not normal, variance should be estimated:

$$s^2 = \frac{\pi \sum_{i=2}^{y} x_i^3 F_i}{\left(\sum_{i=2}^{y.} x_i \bar{r}_i \right) + r_1} \qquad (9.20)$$

where x_i = distance of recapture point from central release point, $\bar{r}_i$ = the mean number of animals captured per trap at a given distance (x_i) from the release point, y = the equidistant set of traps furthest from the release point, r_i = the number of recaptures in the central trap (i.e. where $x = 0$) and $\pi = 3.14$. If animals are released at a central point it is possible, making the assumptions of random flight movements and constant speed, to calculate theoretical curve for the fall-off in numbers of marked individuals with distance. By relating this curve to actual data of the ratio of marked to unmarked flies Gilmour, Waterhouse & McIntyre (1946) were able to calculate the number of marked blowflies beyond the last ring of traps.

The number of marked animals that have left an area

Scotter et al. (1971) developed this approach, but maintaining the assumption of random walk. They used their dispersal rate coefficient (ψ) calculated as above or from the numbers of marked flies (M_i) in an annulus (i) assuming trapping or searching methods 100% efficient (i.e. $M_i = r_i$):

$$\hat{M}_i = a^1 \left[\exp\left(\frac{-x_i^2}{4\psi t}\right) - \exp\left(\frac{-x_{i+1}^2}{4\psi t}\right) \right] \tag{9.21}$$

where (as above) a^1 = number of marked flies released, t = time, x_i = distance of the inner radius of the ith annulus from the central point and x_{i+1} = distance of outer radius from ith annulus from point. Where $x_i = 0$, i.e. the innermost radius, there is a unique solution to ψ, which is expressed as unit of area/unit time. Then the probability (P) that a particular insect travels more than a certain distance beyond the radius of the outer annulus (y) is:

$$P = \exp\left(\frac{-y^2}{4\psi t}\right) \tag{9.22}$$

The actual number of animals migrating into or out of a population may also be determined from mark and recapture analysis (see Chapter 3). If there are no births or deaths, then the loss and dilution rates may be taken as equivalent to migration rates; alternatively the two components may be separated by measuring natality and mortality separately (see pp. 97–113).

The simple Lincoln Index may be used to calculate the proportion of a population that has migrated if the total population is known from some other method and there are no births or deaths. A number of marked individuals are released; knowing this number, the total population and the size of subsequent samples, expected recapture values may be calculated. The proportion of the marked insects that have emigrated (a_e) is estimated by the ratio of actual to expected recaptures:—

$$a_e = r/r_e \tag{9.23}$$

where r = actual recaptures, r_e = expected recaptures = an/N (where a = number of marked animals released, n = total size of sample and N = total population known from some other estimate). Such an approach is of particular value for a highly mobile animal.

THE RATE OF POPULATION INTERCHANGE BETWEEN TWO AREAS

Richards & Waloff (1954) described a method for studying the movement between two grasshopper colonies; their basic assumption was that the survival rates in the two colonies were similar. Their method has been further developed by Iwao (1963), who has derived equations that are applicable to populations where both the survival rates and sampling ratios differ. A series of three sets of observations on days 1 (t_1), 2(t_2) and 3(t_3) are necessary as with

Bailey's triple catch (p. 104). A number $(_x a_1)$ of animals are marked and released in-both areas on day 1; on day 2 a sample (n_2) is taken and the number of already marked individuals recorded, all the individuals $(_x a_2)$ are then given a distinctive mark and released. On the third day samples are again taken in both areas and the number of already marked individuals recorded together with the details of their marks.

Thus the estimate of the emigration rate from area x to area y during the time interval from day 1 to day 2 is given by:

$$_{xy}\hat{e}_1 = \frac{\left(\dfrac{_{yy}r_{31}\; _{y}a_2 + _{yy}r_{21}\; _{yy}r_{32}}{_{yy}r_{32}\; _{y}a_1}\right)_{y}a_1\; _{xy}r_{21}}{_{x}a_1\; _{yy}r_{21}} \quad (9.24)$$

where the notation has been adapted to conform with that in Chapter 3, the anterior subscripts being added to denote the areas; with both the anterior and posterior subscripts the symbol nearest the character represents the actual condition and that furthest away its previous history, thus $_{xy}r_{21}$ represents the recaptures in area y on day 2 that were marked in area x on day 1. To recapitulate the notation for the above equation:

$_{x}a_1 =$ no. of marked individuals released in area x on day 1
$_{y}a_1 =$,,   ,,   ,,   ,,   ,, y ,,   1
$_{y}a_2 =$,,   • ,,   ,,   ,,   ,, y ,,   2

$_{yy}r_{21} =$ recaptures in area y on day 2 marked in area y on day 1
$_{yy}r_{31} =$,,   ,, y ,,   3   ,, y 1
$_{yy}r_{32} =$,,   ,, y ,,   3   ,, y 2

The equivalent equation for the estimation of the emigration rate from y to x is:

$$_{yx}\hat{e}_1 = \frac{\left(\dfrac{_{xx}r_{31}\; _{x}a_2 + _{xx}r_{21}\; _{xx}r_{32}}{_{xx}r_{32}\; _{x}a_1}\right)_{x}a_1\; _{yx}r_{21}}{_{y}a_1\; _{xx}r_{21}} \quad (9.25)$$

If the total population (N) has been estimated by capture–recapture or some other ways, then the actual numbers that are estimated to have emigrated $(\hat{E})$ are:

$$_{xy}\hat{E} = {}_{x}N_1 \times {}_{xy}\hat{e}_1 \quad (9.26)$$

and

$$_{yx}\hat{E} = {}_{y}N_1 \times {}_{yx}\hat{e}_1$$

Survival rates (ϕ) may also be calculated:

$$_{x}\phi_1 = {}_{xx}\mu_1 + {}_{xy}\hat{e}_1 \quad (9.27)$$

where

$$_{xx}\mu_1 = \frac{_{yy}r_{31}\; _{y}a_2 + _{yy}r_{21}\; _{yy}r_{32}}{_{yy}r_{32}\; _{y}a_1}$$

i.e. the bracket term in equation 9.24 above, and

$$_y\phi_1 = {}_{yy}\mu_1 + {}_{yx}\hat{e}_1$$

where $_{yy}\mu_1$ = the bracket term in equation 9.25 above

THE DESCRIPTION OF POPULATION DISPLACEMENT IN RELATION TO ITS DISPERSION

Migration has long been recognized as one of the three pathways of population change; however, its formal incorporation into population mechanisms is very recent. Taylor & Taylor (1976, 1978) showed, on the basis of the extensive monitoring of adult Lepidoptera in Britain, that such populations must be regarded as spatially fluid. Emigration and immigration are linked as two opposing density-related behaviour sets such that the coefficient of displacement is:

$$\Delta = G\rho^p - H\rho^q \qquad (9.28)$$

where ρ = population density; p = the density-related moderator in relation to emigration and q that for congregation. This may be written:—

$$\Delta = \Gamma\left[\left(\frac{\rho}{\rho_0}\right)^p - \left(\frac{\rho}{\rho_0}\right)^q\right] \qquad (9.29)$$

where ρ_0 is the density at which emigration balances immigration and $\Gamma = {}^{(p-q)}\sqrt{H^p. g^{-q}}$ is a scale factor. The net result of displacement is therefore congregation when $\rho < \rho_0$ and emigration when $\rho > \rho_0$. Taylor & Taylor fitted (by interaction on the computer) the above expression to various sets of density/distance frequency data for animals and calculated values for the parameters E, ρ_0, p and q. The significance of ρ_0, the equilibrium density, is most easily grasped, although of course habitat, as well as behavioural characters will influence its magnitude. Taylor & Taylor give the following values of ρ_0 (per km²): *Drosophila* in California–6 000 000; man in Sweden–2000; blackbirds (*Turdus merula*) in Britain–225; or monarch butterflies (*Danaus*) migrating in N. America–54.

The underlying biological processes of the model are intuitively clear; it may be fitted to all the range of distributions illustrated in Fig. 9.8 (and more), and it gives some meaning to the equilibrium density, touched on in the model of Scotter *et al.* (1971) where it was unstable because the scale was so local that the population could be conceived of as expanding indefinitely. Taylor & Taylor's provides not only a description of displacement, but with its link to spatial pattern (dispersion), provides a dynamic description of the changes of the population of an animal with time. This new concept will undoubtedly be explored and developed (see also Taylor & Taylor, 1978).

THE MEASUREMENT AND DESCRIPTION OF HOME RANGE AND TERRITORY

The determination of the home range or territory of an individual or, for social animals, a colony, is of value in the analysis of competition and density effects, the assessment of resources and similar problems (Brown & Onans, 1970). Although many studies have been made on the territories of vertebrates, little work has been done with insects except for crickets (Alexander, 1961), dragonflies (Borror, 1934; Moore, 1952, 1957), some grasshoppers (Clark 1962), and ants (Elton, 1932). Some insects may be marked and recaptured on a number of occasions (e.g. Borror, 1934; Green & Pointing, 1962; Greenslade, 1964); this type of experiment gives data that would be suitable for the computation of the home range, using the methods of vertebrate ecologists. Ten recaptures are usually taken as sufficient to calculate the home range, but considerably more data would be required to use the method of Frank (1964) (p. 329) and ensure that one was not measuring an artifact.

The term territory implies the exclusion of at least certain other individuals, as occurs in crickets and dragonflies (Alexander, 1961; Moore, 1952, 1957); it is probable that this is the exception rather than the rule in insects; nevertheless most insects will have a home range, over which they forage and search for mates and where they rest; that is the area over which they engage in trivial movement (*sensu* Southwood, 1962). As Moore (1957) points out, the home range, especially for insects, should not be regarded as a hard and fast geographical area, but provided it is measured in a strictly comparative way it should be possible to detect changes due to inter- or intra-specific competition, to variation in the available resources or to changes in the behaviour of the animals themselves (e.g. Wellington, 1964). The calculation of the home range of vertebrates is important to the student of arthropod ectoparasites (e.g. Mohr & Stumpf, 1964*a* and *b*). There are a number of methods of calculating home range of vertebrates; these have been reviewed by Sanderson (1966) and Jennrich & Turner (1969). The conclusion reached by the latter authors was that in a homogeneous environment the most statistically stable graphical technique is the minimum area method, but the calculation of the determinant of the covariance matrix is to be preferred. They provide a table that allows the comparison of home ranges compared by different methods.

Minimum area method

The points of recapture are mapped and the outermost points joined up to enclose the area, which may be measured (Mohr, 1947; Odum & Kuenzler, 1955). This method has the advantage that no assumption is made about the shape of the range and it has been shown that in some animals, at least, they may be linear (Stumpf & Mohr, 1962). However, with such a method the worker may feel it necessary to exclude 'incidental forays outside the area' (Jorgensen & Tanner, 1963). All re-entrant angles must be avoided for the depth to which they are drawn will depend on individual judgement and thus different results may be obtained from the same set of data.

The matrix index
Developed by Jennrich & Turner (1969) this index I_m is

$$I_m = 6\pi |S|^{\frac{1}{2}} \tag{9.30}$$

Here $|S|$ is the determinant of the capture point covariance matrix

$$S = \begin{pmatrix} S_{xx} & S_{xy} \\ S_{yx} & S_{yy} \end{pmatrix} \tag{9.31}$$

defined by the equations

$$S_{xx} = \frac{1}{n-2} \sum_{i=1}^{n} (x_i - \bar{x})^2, \quad S_{yy} = \frac{1}{n-2} \sum_{i=1}^{n} (y - \bar{y})^2,$$

$$S_{xy} = S_{yx} = \frac{1}{n-2} \sum_{i=1}^{n} (x_i - \bar{x})(y_i - \bar{y}), \tag{9.32}$$

$$\bar{x} = \frac{1}{n} \sum_{i=1}^{n} x_i, \quad \bar{y} = \frac{1}{n} \sum_{i=1}^{n} y_i.$$

where (x_i, y_i) is the ith ordered capture point (defined on two coordinates, x and y) from a total of n capture points.

Where an animal's habitat is not homogeneous, where it spans two types of ecosystem, the above methods are not appropriate for the calculation of home-range. Van Winkle *et al.* (1973) developed a method for frogs living on the edge of a pond which could be applied to other animals in similar ecotones.

9.3.3 Direct field measurements

THE ELIMINATION OF EMIGRATION
With a very few insects, such as apterous ground beetles, it may be possible to allow immigration, but prevent emigration by a barrier of some type (e.g. Fig. 9.5*b*).

THE USE OF QUADRAT COUNTS OF UNMARKED INDIVIDUALS
Where the habitat is uniform and it is possible to assume random diffusion, movement may be separated from mortality by using the method of Dempster (1957), which is based on random diffusion theory. The changes in the numbers of insects (f) with time will be represented (Skellam, 1951) by:

$$\frac{df}{dt} = \alpha \left(\frac{d^2f}{dx^2} + \frac{d^2f}{dy^2} \right) - \mu f \tag{9.33}$$

where $\alpha =$ the mobility of the insects; $d^2f/dx^2 + d^2f/dy^2 =$ a measure of the density gradient along the two axes of the quadrat in which the insects occur and μ = mortality rate.

Dempster shows that a representation of density gradient expression is given:

$$\frac{d^2f}{dx^2} + \frac{d^2f}{dy^2} = \frac{1}{3}(3\,\Sigma x^2 f'_{(x,\,y)} + 3\,\Sigma\,y^2 f'_{(x,\,y)} - 4\,\Sigma\,f'_{(x,\,y)}) \qquad (9.34)$$

Now if we consider a block of 9 quadrats (Fig. 9.9) and take quadrat f_{11}, its

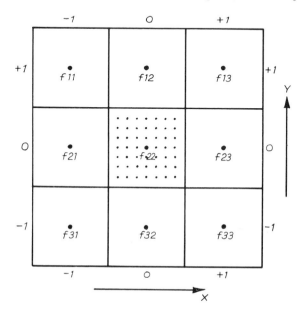

Fig. 9.9 The populations $(f_{11}, f_{12} \ldots)$ along the two axes (x and y) across a central quadrat surrounded by quadrats of equal area (from Dempster, 1957).

numerical value in this expression would be

$$3 \times (-1)^2 + 3 \times (1)^2 - 4 = 2$$

Similarly for all the other squares in turn, so that:

$$\frac{d^2f}{dx^2} + \frac{d^2f}{dy^2} = \frac{1}{3}(2f_{11} - 1f_{12} + 2f_{13} - 1f_{21} - 4f_{22} - 1f_{23} + 2f_{31} - f_{32} + 2f_{33})$$

$$(9.35)$$

Thus for a central quadrat surrounded by 8 other quadrats the best estimator for $d^2f/dx^2 + d^2f/dy^2$ is given by equation 9.35 and this value may be substituted in equation 9.33, the only unknowns then are α and μ. A series of simultaneous equations can therefore be developed for a number of central squares; as these must be more than the number of unknowns (2 in this case), the absolute minimum number of equal sized quadrats would be 15 arranged in a block of 3 × 5. A larger number would be preferable and Dempster used 18. Equations for the four middle squares can be calculated. The df/dt value being the changes in numbers normally decrease, from one instar to the next (or whatever the period of time). For the values of the estimator of $d^2f/dx^2 + d^2f/dy^2$, equation 9.35, the estimates of the populations in the quadrats used are those for the commencement of the period in question. The equations are then solved by normal mathematical procedures. The value for μ will be fractional and that for α should be positive; when mobility is extremely small negative values of α may occur by chance.

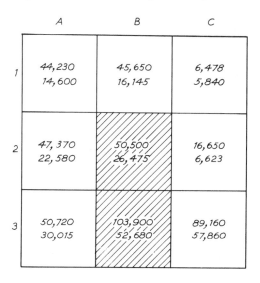

Fig. 9.10 The estimated number of locusts entering the first (upper figure) and second instar (lower figure) in a number of quadrats (from Dempster, 1957).

EXAMPLE (from Dempster, 1957): Consider square B_2 in Fig. 9.10. The number of insects in this square has changed from 50 500 to 26 475.

$$\therefore \frac{df}{dt} = -24\,025$$

and
$$\left(\frac{d^2f}{dx^2} + \frac{d^2f}{dy^2}\right) = \frac{1}{3}\left(\begin{array}{l} 2 \times 44\,230 - 45\,650 + 2 \times 6478 - \\ 47\,370 - 4 \times 50\,500 - 16\,650 + 2 \times \\ 50\,720 - 103\,900 + 2 \times 89\,160 \end{array}\right) = -11\,465$$

Substituting in equation 9.33

we get
$$-24\,205 = -11\,465\alpha \times -50\,500\mu$$

At least 3, preferably more, similar equations are obtained. Then the normal equation for μ is found by multiplying each equation by the coefficient of μ in it and adding all equations. The normal equation for α is found in the same way. These two equations are then solved.

THE RATE OF COLONIZATION OF A NEW HABITAT

This property, which is related to migration, may be measured by planting virgin artificial habitats (e.g. new plants in a field or stones in a stream) and determining the rate of colonization (Breymeyer & Pieczynski, 1963) (see also pp. 210 and 262).

THE DIRECTION OF MIGRATION

There are a number of traps (p. 244) that enable one to determine the direction an animal was flying at the time of capture. It is possible that the proportionality values obtained from a series of such traps round a habitat (Sylven, 1970) might be used in conjunction with measures of either the net change in population or a measurement of migration to determine values for immigration and emigration. Furthermore it has been pointed out how the dilution and 'loss' rates obtained from multiple capture–recapture analysis are compounded of birth and immigration and death and emigration respectively. Sometimes it is possible to separate the two components of the dilution rate from a knowledge of natality or the numbers entering the stage, when 'births' are known, the proportion of emigration to immigration determined from a 'directional trap', might be used to give an indication of the amount of immigration and hence allow the partitioning of the death-rate into mortality and immigration. Caution would need to be used in such an approach, particularly to ensure an adequate distribution of traps around the habitat and difficulty would be experienced in determining the numbers of animals leaving and arriving in vertical air currents, i.e. only crossing the habitat boundaries at a considerable height.

REFERENCES

ABDEL RAHMAN, I., 1974. The effect of extreme temperatures on Californian red scale, *Aonidiella aurantii* (Mask.) (Hemiptera: Diaspididae) and its natural enemies. *Aust. J. Zool.* **22**, 203–12.

ADAMSTONE, F. B. and HARKNESS, W. J. K., 1923. The bottom organisms of Lake Nipigon. *Univ. Toronto Stud. Biol.* **22**,121–70.

AGASSIZ, D., 1977. A trap for wingless female moths. *Proc. Brit. ent. nat. hist. Soc.* **10**, 69–70.

AIKMAN, D. and HEWITT, G.,1972. An experimental investigation of the rate and form of dispersal in grasshoppers. *J. appl. Ecol.* **9**, 807–17.

ALEXANDER, R. D.,1961. Aggressiveness, territoriality and sexual behaviour in field crickets (Orthoptera: Gryllidae). *Behaviour* **17**, 130–223.

ANDERSON, N. H. and WOLD, J. L.,1972. Emergence trap collections of Trichoptera from an Oregon stream. *Can. Ent.* **104**, 189–201.

ASHBY, J. W., 1974. A study of arthropod predation of *Pieris rapae* L. using serological and exclusion techniques. *J. appl. Ecol.* **11**, 419–25.

BALDWIN, W. F., JAMES, H. G. and WELCH, H. E., 1955. A study of predators of mosquito larvae and pupae with a radio-active tracer. *Can. Ent.* **87**, 350–6.

BATES, J. K.,1962. Field studies on the behaviour of bird fleas. 1. Behaviour of the adults of three species of bird flea in the field. *Parasitology* **52**, 113–32.

BEAN, J. L., 1958. The use of larvaevorid maggot drop in measuring trends in spruce budworm populations. *Ann. ent. Soc. Am.* **51**, 400–3.

BEDDINGTON. J. R., 1975. Mutual interference between parasites or predators and its effect on searching efficiency. *J. Anim. Ecol.* **44**, 331–40.

BEDDINGTON, J. R., HASSELL, M. P. and LAWTON, J. H., 1976. The components of arthropod predation II. The predator rate of increase. *J. Anim. Ecol.* **45**, 165–85.

BESS, H. A., 1961. Population ecology of the gipsy moth, *Porthetria dispar* L. (Lepidoptera: Lymantridae). *Bull. Conn. agric. Exp. Sta.* **646**, 43 pp.

BETTS, M. M., 1954. Experiments with an artificial nestling. *Brit. Birds* **47**, 229–31.

BETTS, M. M., 1955. The food of titmice in oak woodlands. *J. Anim. Ecol.* **24**, 282–323.

BETTS, M. M., 1958. Further experiments with an artificial nestling gape. *Brit. Birds* **49**, 213–5.

BLAIS, J. R., 1953. The effects of the destruction of the current year's foliage of balsam fir on the fecundity and habits of flight of the spruce budworm. *Can. Ent.* **85**, 446–8.

BOETHEL, D. J., MORRISON, R. D. and EIKENBARY, R. D., 1976. Pecan weevil *Curculio caryae* (Coleoptera: Curculionidae). 2. Estimation of adult populations. *Can. Ent.* **108**, 19–22.

BOORMAN, J., MELLOR, P. S., BOREHAM, P. F. L. and HEWETT, R.S., 1977. A latex agglutination test for the identification of blood-meals of *Culicoides* (Diptera: Ceratopogonidae). *Bull. ent. Res.* **67**, 305–11.

BOREHAM, P. F. L., 1975. Some applications of bloodmeal identifications in relation to the epidemiology of vector-borne tropical disease. *J. trop. Med. Hyg.* **78**, 83–91.

BOREHAM, P. F. L. and GILL, G. S., 1973. Serological identification of reptile feeds of *Glossina. Acta trop.* **30**, 356–65.

BORROR, D. J., 1934. Ecological studies of *Agria moesta* Hogen (Odonata: Coenagrionidae) by means of marking, *Ohio J. Sci.* **34**, 97–108.

BORUTSKY, E. V., 1955. [A new trap for the quantitative estimation of emerging chironomids.] [In Russian.] *Trudý vses. gidrobiol. Obshch.* **6**, 223–6.

BRAVERMAN, Y., 1970. An improved emergence trap for Culicoides. *J. econ. Ent.* **63**, 1674–5.

BREYMEYER, A. and PIECZYNSKI, E., 1963. Review of methods used in the Institute of Ecology, Polish Academy of Sciences, for investigating migration. [In Polish.] *Ekol. Polska B* **9**, 129–44.

BROOKE, M. M. and PROSKE, H. O., 1946. Precipitin test for determining natural insect predators of immature mosquitoes. *J. nat. Malaria Soc.* **5**, 45-56.

BROWN, J. L. and ORIANS, G. H., 1970. Spacing patterns in mobile animals. *Ann. Rev. Ecol. Syst.* **1**, 239–62.

BRUNDIN, L., 1949. Chironomiden und andere Bodentiere der Südschwedischen Urgebirgsseen. *Rep. Inst. Freshw. Res. Drottringholm* **20**, 915 pp.

BRYANT, D. M., 1973. The factors influencing the selection of food by the House Martin (*Delichon urbica* (L.)). *J. Anim. Ecol.* **42**, 539–64.

BUCKNER, C. H., 1958. Mammalian predators of the larch sawfly in eastern Manitoba. *Proc. X int. Congr. Ent.* **4**, 353–61.

BUCKNER, C. H., 1959. The assessment of larch sawfly cocoon predation by small mammals. *Can. Ent.* **91**, 275–82.

BUCKNER, C. H., 1966. The role of vertebrate predators in the biological control of forest insects. *Ann. Rev. Ent.* **11**, 449–70.

CAMPBELL, J. A. and PELHAM-CLINTON, E. C., 1960. A taxonomic review of the British species of 'Culicoides' Latreille (Diptera, Ceratopogonidae). *Proc. R. Soc. Edin. B* **67** (3), 181–302.

CARLSON, D., 1971. A method for sampling larval and emerging insects using an aquatic black light trap. *Can. Ent.* **103**, 1365–9.

CARTER, C. L., 1972. Winter temperature and survival of the green spruce aphid *Elatobium abietinum*. *Forestry Commission. Forest Record* no. **84**, 10 pp.

CHAPMAN, J. A., ROMER, J. I. and STARK, J., 1955. Ladybird beetles and army cut-worm adults as food for grizzly bears in Montana. *Ecology* **36**, 156–8.

CHAUVIN, R., 1966. Un Procédé pour recolter automatiquement les proies que les *Formica polyctena* rapportent au rid. *Ins. Soc.* **13**, 59–67.

CLARK, D. P., 1962. An analysis of dispersal and movement in *Phaulacridium vittatum* (Sjöst.) (Acrididae). *Aust. J. Zool.* **10**, 382–99.

CLARK., E. W. and OSGOOD, E. A., 1964. An emergence container for recovering southern pine beetles from infested bolts. *J. econ. Ent.* **57**, 783–4.

CLARK, L. P., 1963. The influence of population density on the number of eggs laid by females of *Cardiaspina albitextura* (Psyllidae). *Aust. J. Zool.* **11**, 190–201.

CLARKE, R. D. and GRANT, P. R., 1968. An experimental study of the role of spiders as predators in a forest litter community. Part 1. *Ecology* **49**, (6), 1152–4.

COAKER, T. H., 1965. Further experiments on the effect of beetle predators on the numbers of the cabbage root fly, *Erioischia brassicae* (Bouché), attacking brassica crops. *Ann. appl. Biol.* **56**, 7–20.

COCK, M. J. W., 1978. The assessment of preference. *J. Anim. Ecol.* (in press).

COLLESS, D. H. and CHELLAPAH, W. T., 1960. Effects of body weight and size of blood-meal upon egg production in *Aëdes aegypti* (Linnaeus) (Diptera, Culicidae). *Ann. trop. Med. Parasit.* **54**, 475–82.

COOK, L. M. and KETTLEWELL, H. B. D., 1960. Radioactive labelling of lepidopterous larvae: a method of estimating late larval and pupal mortality in the wild. *Nature* **187**, 301–2.

COOK, P. P. and HORN, H. S., 1968. A sturdy trap for sampling emergent Odonata. *Ann. ent. Soc. Am.* **61** (6), 1506–7.

CORBET, P. S., 1966. Diel periodicities of emergence and oviposition in riverine Trichoptera. *Can. Ent.* **98**, 1025–34.

CROSSLEY, D. A., 1963. Consumption of vegetation by insects. *In* Schultz, V. & Klement, A. W. (eds.), *Radioecology,* 431–40. Rheinhold, New York.

CRUMPACKER, D. W. and WILLIAMS, J. S., 1973. Density, dispersion and population structure in *Drosophila pseudoobscura*. *Ecol. Monogr.* **43** (4), 499–538.

CULLIS, N. A. and HARGROVE, J. W., 1972. An automatic device for the study of tethered flight in insects. *Bull. ent. Res.* **61**, 533–7.

CUSTER, T. W. and PITELKA, F. A., 1974. Correction factors for digestion rates for prey taken by snow buntings (*Plectrophenax nivalis*). *Condor* **77**, 210–12.

DARBY, R. E., 1962. Midges associated with California rice fields, with special reference to their ecology (Diptera: Chironomidae). *Hilgardia* **32**, 1–206.

DAVIDSON, A., 1956. A method of counting Ephemeropteran eggs. *Ent. mon. Mag.* **92**, 109.

DAVIES, J. B., 1966. An evaluation of the emergence or box trap for estimating sandfly (*Culicoides* spp. Heleidae) populations. *Mosquito News* **26**, 170–2

DAVIES, R. W., 1969. The production of antisera for detecting specific triclad antigens in the gut contents of predators. *Oikos* **20**, 248–60.

DEAN, G. T., 1973. Aphid colonization of spring cereals. *Ann. appl. Biol.* **75**, 183–93.

DEBACH, P., 1946. An insecticidal check method for measuring the efficacy of entomophagous parasites. *J. econ. Ent.* **39**, 695–7.

DEBACH, P., 1949. Population studies of the long-tailed mealy bug and its natural enemies on citrus trees in Southern California, 1946. *Ecology* **30**, 14–25.

DEBACH, P., 1955. Validity of the insecticidal check method as a measure of the effectiveness of natural enemies of Diaspine scale insects. *J. econ. Ent.* **48**, 584–8.

DEBACH, P., and HUFFAKER, C. B., 1971. Experimental techniques for evaluation of the effectiveness of natural enemies, in Huffaker, C. B. (ed.) *Biological Control* 113–40. Plenum Press, New York.

DEMPSTER, J. P., 1957. The population dynamics of the Moroccan locust (*Dociostaurus maroccanus* Thunberg) in Cyprus. *Anti-Locust Bull.* **27**, 1–60.

DEMPSTER, J. P., 1960. A quantitative study of the predators on the eggs and larvae of the broom beetle, *Phytodecta olivacea* Forster, using the precipitin test. *J. Anim. Ecol.* **29**, 149–67.

DEMPSTER, J. P., 1964. The feeding habits of the Miridae (Heteroptera) living on broom (*Sarothamnus scoparius* (L.) Wimm.). *Entomologia. exp. appl.* **7**, 149–54.

DEMPSTER, J. P., RICHARDS, O. W. and WALOFF, N., 1959. Carabidae as predators on the pupal stage of the Chrysomelid beetle, *Phytodecta olivacea* (Forster). *Oikos* **10**, 65–70.

DINGLE, H., 1966. Some factors affecting flight activity in individual milkweed bugs. (*Oncopeltus*). *J. exp. Biol.* **44**, 335–43.

DINGLE, H. and ARORA, G., 1973. Experimental studies of migration in bugs of the genus *Dysdercus*. *Oecologia* **12**, 119–40.

DINTHER, J. B. M. VAN, 1953. Details about some flytraps and their application to biological research. *Ent. Ber.* **14**, 201–4.

DINTHER, J. B. M. VAN and MENSINK, F. T., 1971. Use of radioactive phosphorus in studying egg predation by carabids in cauliflower fields. *Meded. Fak. Landb. Wetenschappen, Gent.* **36**, 283–93.

DOANE, J. F., 1963. Dispersion on the soil surface of marked adult *Ctenicera destructor* and *Hypolithus bicolor* (Coleoptera: Elateridae), with notes on fight. *Ann. ent. Soc. Am.* **56**, 340–5.

DOBSON, R. M., STEPHENSON, J. W. and LOFTY, J. R., 1958. A quantitative study of a population of wheat bulb fly, *Leptohylemyia coarctata* (Fall.), in the field. *Bull. ent. Res* **49**, 95–111.

DOBZHANSKY, T. and WRIGHT, S., 1943. Genetics of natural populations: X. Dispersion rates in *Drosophila pseudoobscura*. *Genetics* **28**, 304–40.

DOBZHANSKY, T. and WRIGHT, S., 1947. Genetics of natural populations. XV. Rate of diffusion of a mutant gene through a population of *Drosophila pseudoobscura*. *Genetics* **32**, 303–24.

DOWDEN, P. B., JAYNES, H. A. and CAROLIN, V. M., 1953. The role of birds in a spruce budworm outbreak in Maine. *J. econ. Ent.* **46**, 307–12.

EGUAGIE, W. E., 1974. Cold hardiness of *Tingis ampliata* (Heteroptera: Tingidae). *Entomologia exp. appl.* **17**, 204–14.

EICKWORT, K. R., 1977. Population dynamics of a relatively rare species of milkweed beetle, (*Labidomera*). *Ecology* **58**, 527–38.

ELLIOTT, J. M., 1970. Diel changes in invertebrate drift and the food of trout (*Salmo trutta* L.) *J. Fish. Biol.* **2**, 161–5.

ELLIOTT, J. M., 1973. The food of Brown and Rainbow Trout (*Salmo trutta* and *S. gairdneri*) in relation to the abundance of drifting invertebrates in a Mountain Stream. *Oecologia* **12**, 329–47.

ELTON, C., 1932. Territory among wood ants (*Formica rufa* L.) at Picket Hill. *J. Anim. Ecol.* **1**, 69–76.

EMDEN, H. F. VAN, 1962. A preliminary study of insect numbers in field and hedgerow. *Ent. mon. Mag.* **98**, 255–9.

EMDEN, H. F. VAN, 1963. A field technique for comparing the intensity of mortality factors acting on the cabbage aphid, *Brevicoryne brassicae* (L) (Hem., Aphididae) in different areas of a crop. *Entomologia exp. appl.* **6**, 53–62.

ENTWISTLE, P. F., 1977. The development of an epizootic of a nuclear polyhedrosis virus disease in European spruce sawfly, *Gilpinia hercyaniae*. *Proc. int. Colloq. Invert.*

Path., Kingston, Canada, 1976.

ERRINGTON, P., 1932. Technique of raptor food habits study. *Condor* **34,** 75–86.

EVANS, H. E. and YOSHIMOTO, C. M., 1962. The ecology and nesting behaviour of the Pompilidae (Hymenoptera) of the Northeastern United States. *Misc. Pub. ent. Soc. Am.* **3** (3), 65–119.

EVANS, H. F., 1976. The role of predator–prey size ratio in determining the efficiency of capture by *Anthrocoris nemorum* and the escape reactions of its prey, *Acyrthosiphon pisum. Ecol. Ent.* **1,** 85–90.

EVENHUIS, H. H., 1962. Methods to investigate the population dynamics of aphids and aphid parasites in orchards. *Entomophaga* **7,** 213–20.

FEWKES, D. W., 1964. The fecundity and fertility of the Trinidad sugar-cane froghopper, *Aeneolamia varia saccharina* (Homoptera, Cercopidae). *Trop. Agriculture, Trin.* **41,** 165–8.

FINNEGAN, R. J., 1969. Assessing predation by ants on insects. *Insectes Sociaux* **16,** 61–5.

FLESCHNER, C. A., 1950. Studies on searching capacity of the larvae of three predators of the citrus red mite. *Hilgardia* **20** (13), 233–65.

FLETCHER, B. S., 1974. The ecology of a natural population of the Queensland Fruit Fly, *Dacus tryoni* V. The dispersal of adults. *Aust. J. Zool.* **22,** 189–202.

FLETCHER, B. S. and ECONOMOPOULOS, A. P., 1976. Dispersal of normal and irradiated laboratory strains and wild strains of the olive fly *Dacus oleae* in an olive grove. *Entomologia. exp. appl.* **20,** 183–9.

FRANK, J. H., 1967. The insect predators of the pupal stage of the winter moth, *Operophtera boumata* (L.) (Lepidoptera: Hydriomenidae).*J. Anim. Ecol.* **36,** 375.

FRANK, P. W., 1964. On home range of limpets. *Am. Nat.* **98,** 99–1.

FREDEEN, F. J. H., SPINKS, J. W. T., ANDERSON, J. R., ARNASON, A. P. and REMPEL, J. G., 1953. Mass tagging of blackflies (Diptera: Simuliidae) with radio-phosphorus. *Can. J. Zool.* **31,** 1–15.

FREEMAN, G. H., 1977. A model relating numbers of dispersing insects to distance and time. *J. appl. Ecol.* **14,** 477–87.

FRIEDMAN, M., 1940. A comparison of alternative tests of significance for the problem of *m* rankings. *Ann. math. Stat.* **11,** 86–92.

FUSEINI, B. A. and KUMAR, R., 1975. Ecology of cotton stainers (Heteroptera: Pyrrhocoridae) in southern Ghana. *Biol. J. Linn. Soc.* **7,** 113–46.

GARY, N. E., 1960. A trap to quantitatively recover dead and abnormal honey bees from the hive. *J. econ. Ent.* **53,** 782–5.

GIBB, J. A., 1958. Predation by tits and squirrels on the Eucosmid *Ernarmonia conicolana* (Heyl.). *J. Anim. Ecol.* **27,** 375–96.

GILMOUR, D., WATERHOUSE, D. F. and MCINTYRE, G. A., 1946. An account of experiments undertaken to determine the natural population density of the sheep blowfly, *Lucilia cuprina* Wied. *Bull, Coun. sci. indust. Res. Aust.* **195,** 1–39.

GLEN, D. M., 1975. Searching behaviour and prey-density requirements of *Blephandopterus angulatus* (Fall.) (Heteroptera: Miridae) as a predator of the Lime Aphid, *Eucallipterus tiliae* (L.) and Leafhopper, *Alnetoidea alneti* (Dahlbom). *J. Anim. Ecol.* **44,** 116–34.

GLEN, D. M., 1976. An emergence trap for barkdwelling insects, its efficiency and effects on temperature. *Ecol. Ent.* **1,** 91–4.

GOODHART, C. B., 1958. Thrush predation on the snail *Cepaea hortensis. J. Anim. Ecol.* **27,** 47–57.

GRAHAM, S. A., 1928. The influence of small mammals and other factors upon larch sawfly survival. *J. econ. Ent.* **21,** 301–10.

GRANDILEWSKAJA-DECKSBACH, M. L., 1935. Materialien zur Chironomidenbiologie verschiedener Becken. Zur Frage über die Schwankungen der Anzahl und der Biomasse der Chironomidenlarven. *Trudy limnol. Sta. Kosine* **19**, 145–82.

GREEN, G. W., 1962. Low winter temperatures and the European pine shoot moth, *Rhyacionia buoliana* (Schiff.) in Ontario. *Can. Ent.* **94**, 314–36.

GREEN, G. W., and POINTING, P. J., 1962. Flight and dispersal of the european pine shoot moth, *Rhyacionia buoliana* (Schiff.) II. Natural dispersal of egg-laden females. *Can. Ent.* **94**, 299–314.

GREENBERG, B. and BORNSTEIN, A. A., 1964. Fly dispersion from a rural mexican slaughter house. *Am. J. trop. Med. Hyg.* **13** (6), 881–6.

GREENSLADE, P. J. M., 1964. The distribution, dispersal and size of a population of *Nebria brevicollis* (F.) with comparative studies on three other carabidae. *J. Anim. Ecol.* **33**, 311–33.

GREENSTONE, M. H., 1977. A passive haemagglutination inhibition assay for the identification of stomach contents of invertebrate predators. *J. appl. Ecol.* **14**, 457–64.

GREGOR, F., 1960. Zur Eiproduktion des Eichenwicklers (*Tortrix viridana* L.). *Zool. Listy.* **9**, 11–18.

GREGORY, P. H., and READ, D. R., 1949. The spatial distribution of insect-borne plant virus diseases *Ann. appl. Biol.* **36**, 475–82.

GRUBER, F. and PRIETO, C. A., 1976. A collecting chamber suitable for recovery of insects from large quantities of host plant material. *Environ. Ent.* **5**, 343–4.

GUENNELON, G. and AUDEMARD, M. H., 1963. Enseignements écologiques donnés par la méthode de captures par cuisses-éclosion de la cécidomyie des lavandes (*Thomasmiana lavandulae* Barnes). Critique de la méthode. Conclusions pratiques. *Ann. Epiphyt. C* **14**, 35–48.

HAFEZ, M., 1961. Seasonal fluctuations of population density of the cabbage aphid, *Brevicoryne brassicae* (L.) in the Netherlands, and the role of its parasite, *Aphidius* (*Diaeretiella*) *rapae* (Curtis). *Tijdschr. PlZiekt.* **67**, 445–548.

HALL, R. R., DOWNE, A. E. R., MACLELLAN, C. R. and WEST, A. S., 1953. Evaluation of insect predator–prey relationships by precipitin test studies. *Mosquito News* **13**, 199–204.

HARD, J. S., 1976. Estimation of Hemlock sawfly (Hymenoptera: Diprionidae) fecundity. *Can. Ent.* **108**, 961–6.

HARTLEY, P. H. T., 1948. The assessment of the food of birds. *Ibis* **90**, 361–81.

HASSELL, M. P., 1976a. Arthropod predator–prey systems. *In* May, R. M. (ed.) *Theoretical Ecology, Principles and Applications.* pp. 71–93.

HASSELL, M. P., 1976b. *The Dynamics of Competition and Predation.* Edward Arnold, London.

HASSELL, M. P., 1978. *The dynamics of arthropod predator–prey relationships.* Princeton Monograph on Population Biology **13**, 245pp., Princeton, New Jersey.

HASSELL, M. P., LAWTON, J. H. and BEDDINGTON, J. R., 1976. The components of arthropod predation. I. The prey death rate. *J. Anim. Ecol.* **45**, 135–64.

HASSELL, M. P. and MAY, R. M., 1974. Aggregation of predators and insect parasites and its effect on stability. *J. Anim. Ecol.* **43**, 567–94.

HASSELL, M. P., and VARLEY G. C., 1969. A new inductive population model for insect parasites and its bearing on biological control. *Nature* **223**, 1133–7.

HAWKES, C., 1972. The estimation of the dispersal rate of the adult cabbage root fly (*Erioischia brassicae* (Bouché) in the presence of a brassica crop. *J. appl. Ecol.* **9**, 617–32.

HAWKES, R. B., 1972. A fluorescent dye technique for marking insect eggs in predation studies. *J. econ. Ent.* **65**, 1477–8.

HEALEY, J. A. and CROSS, T. F., 1975. Immunoelectroosmophoresis for serological identification of predators of the sheep-tick *Ixodes ricinus. Oikos* **26**, 97–101.

HENSON, W. R., 1959. Some effects of secondary dispersive processes on distribution. *Am. Nat.* **93**, 315–20.

HESPENHEIDE, H. A., 1975. Prey characteristics and predator niche width. *In* Cody, M. L. & Diamond, J. M. (eds). *Ecology and Evolution of Communities* 158–80. Harvard University Press, Cambridge, Mass.

HOLLING, C. S., 1958. A radiographic technique to identify healthy, parasitised and diseased sawfly prepupae within cocoons. *Can. Ent.* **90**, 59–61.

HOLLING, C. S., 1959a. The components of predation as revealed by a study of small mammal predation of the european pine sawfly. *Can. Ent.* **91**, 293–320.

HOLLING, C. S., 1959b. Some characteristics of simple types of predation and parasitism. *Can. Ent.* **91**, 385–98.

HOLLING, C. S., 1961. Principles of insect predation. *Ann. Rev. Ent.* **6**, 163–82.

HOLLING, C. S., 1965. The functional response of predators to prey density and its role in mimicry and population regulation. *Mem. ent. Soc. Can.* **No. 45**, 60pp.

HOLLING, C. S., 1966. The functional response of invertebrate predators to prey density. *Mem. ent. Soc. Can.*, **No. 48**, 86pp.

HUDSON, L. and HAY, S. C., 1976. *Practical Immunology*, 298 pp., Oxford University Press, Oxford.

HUFFAKER, C. B. and SPITZER, C. H., 1950. Some factors affecting red mite populations on pears in California. *J. econ. Ent.* **43**, 819–31.

IDE, F. P., 1940. Quantitative determination of the insect fauna of rapid water. *Univ. Toronto Stud. Biol. Ser.* **47** (*Publ. Ontario Fish. Res. Lab. 59*), 20 pp.

ILLIES, J., 1971. Emergenz 1969 im Breitenbach. *Arch. Hydrobiol.* **69**, 14–59.

INOUE, T., KAMIMMURA, K. and WATANABE, M., 1973. A quantitative analysis of dispersal in a horse-fly *Tabanus iyoensis* Shiraki and its application to estimate the population size. *Res. Popul. Ecol.* **14**, 209–33.

ITO, Y., YAMANAKA, H., NAKASUJI, F. and KIRITANI, K., 1972. Determination of predator-prey relationship with an activable tracer, Europium-151. *Kontyu* **40**, 278–83.

IVES, W. G. H. and PRENTICE, R. M., 1959. Estimation of parasitism of larch sawfly cocoons by *Bessa harveyi* Tnsd. in survey collections. *Can. Ent.* **91**, 496–500.

IVES, W. G. H., TURNOCK, W. J., BUCKNER, J., H., HERON, R. J. and MULDREW, J. A., 1968. Larch sawfly population dynamics: techniques. *Manitoba Entomologist* **2**, 5–36.

IWAO, S., 1956. On the number of eggs per egg-mass of the paddy rice borer, *Schoenobius incertellus* Walker and the percentage of their parasitization. [In Japanese.] *Gensei* (Kochi Konchu Dokokai) **5**, 45–9.

IWAO, S., 1963. On a method for estimating the rate of population interchange between two areas. *Res. Popul. Ecol.* **5**, 44–50.

JACOBS, J., 1974. 'Quantitative measurement of food selection: a modification of the forage ratio and Ivlev's selectivity index. *Oecologia* **14**, 413–7.

JAMES, H. G., 1961. Some predators of *Aedes stimulans* (Walk) and *Aedes trichurus* (Dyar) (Diptera: Culicidae) in woodland pools. *Can. J. Zool.* **39**, 533–40.

JAMES, H. G., 1966. Location of univoltine Aedes eggs in woodland pool areas and experimental exposure to predators. *Can. Ent.* **98**, 550–5.

JENKINS, D. W., 1963. Use of radionuclides in ecological studies of insects. *In* Schultz, V. and Klement, A. W. (eds), *Radioecology* 431–40. Rheinhold, New York.

JENKINS, D. W. and HASSETT, C. C., 1950. Radioisotopes in entomology. *Nucleonics* **6** (3), 5–14.

JENNRICH, R. I. and TURNER, F. B., 1969. Measurement of non-circular home range. *J. Theor. Biol.* **22**, 227–37.

JOHNSON, C. G., 1969. *Migration and Dispersal of Insects by Flight.* 763pp., Methuen, London.

JOLICOEUR, P. and BRUNEL, D., 1966. Application du diagramme hexagonal a l'étude de la sélection de ses proies par la morue. *Vie et Milieu* **17**, 419–33.

JÓNASSON, P. M., 1954. An improved funnel trap for capturing emerging aquatic insects, with some preliminary results. *Oikos* **5**, 179–88.

JORGENSEN, C. D. and TANNER, W. W., 1963. The application of the density probability function to determine the home ranges of *Uta stansburiana stansburiana* and *Cnemidophorus tigris tigris*. *Herpetologica* **19**, 105–15.

JUDGE, W. W., 1957. A study of the population of emerging and littoral insects trapped as adults from tributary waters of the Thames River at London, Ontario. *Am. midl. Nat.* **58**, 394–412.

KENDALL, D. G., 1974. Pole seeking, Brownian motion and bird navigation. *J. R. Stat. Soc. (B)* **36**, 365–417.

KENNEDY, C. H., 1950. The relation of American dragonfly-eating birds to their prey. *Ecol. Monogr.* **20**, 103–42.

KENNEDY, J. S. and BOOTH, C. O., 1963. Free flight of aphids in the laboratory. *J. exp. Biol.* **40**, 67–85.

KENNEDY, J. S. and LUDLOW, A. R., 1974. Co-ordination of two kinds of flight activity in an aphid. *J. exp. Biol.* **61**, 173–96.

KENSLER, C. B., 1967. Desiccation resistance of intertidal crevice species as a factor in their zonation. *J. Anim. Ecol.* **36**, 391–406.

KIMERLE, R. A. and ANDERSON, N. H., 1967. Evaluation of aquatic insect emergence traps. *J. Econ. Ent.* **60**, 1255–9.

KIRITANI, K. and DEMPSTER, J. P., 1973. Different approaches to the quantitative evaluation of natural enemies. *J. appl. Ecol.* **10**, 323–30.

KIRITANI, K., KAWAHARA, S., SASABA, T. and NAKASUJI, F., and 1972. Quantitative evaluation of predation by spiders on the green rice leafhopper, *Nephotettix cincticeps* Uhler, by a sight-count method. *Res. Popul. Ecol.* **13**, 187–200.

KRING, J. B., 1966. An aphid flight chamber: construction and operation. *J. econ. Ent.* **59**, 1518–20.

KRUUK, H., 1972. Surplus killing by carnivores. *J. Zool.* **166**, 233–44.

LACK, D. and OWEN, D. F., 1955. The food of the swift. *J. Anim. Ecol.* **24**, 120–36.

LANGFORD, T. E. and DAFFERN, J. R., 1975. The emergence of insects from a British river warmed by power station cooling-water. Part. I. *Hydrobiologia* **46**, 71–114.

LAWSON, F. R., 1959. The natural enemies of the hornworms on tobacco (Lepidoptera: Sphingidae). *Ann. ent. Soc. Am.* **52**, 741–55.

LAWTON, J. H., BEDDINGTON, J. R. and BONSER, R., 1974. Switching in invertebrate predators. *In* Usher, M. B. & Williamson, M. H. (ed.) *Ecological Stability*, pp. 141–158.

LEJEUNE, R. R., FELL, W. H. and BURBIDGE, D. P., 1955. The effect of flooding on development and survival of the larch sawfly *Pristiphora erichsonii* (Tenthredinidae). *Ecology* **36**, 63–70.

LINDEBERG, B., 1958. A new trap for collecting emerging insects from small rockpools, with some examples of the results obtained. *Suom. hyönt. Aikak. (Ann. ent. fenn.* **24** 186–91.

LOAN, C. and HOLDAWAY, F. G., 1961. *Microctonus aethiops* (Nees) auctt. and *Perilitus rutilus* (Nees) (Hymenoptera: Braconidae). European parasites of *Sitona* weevils (Coleoptera: Curculionidae). *Can. Ent.* **93**, 1057–78.

LOUGHTON, B. G., DERRY, C. and WEST, A. S. 1963. Spiders and the spruce budworm. *Mem. ent. Soc. Can.* **31**, 249–68.

LOZINSKY, V. A., 1961. On the correlation existing between the weight of pupae and the number and weight of eggs of *Lymantria dispar* L. [In Russian.] *Zool. Zh.* **40**, 1571–3.

MACAN, T. T., 1949. Survey of a moorland fishpond. *J. Anim. Ecol.* **18**, 160–86.

MCAULEY, V. J. E., 1976. Efficiency of a trap for catching and retaining insects emerging from standing water. *Oikos* **27**, 339–345.

MCKNIGHT, M. E., 1969. Distribution of hibernating larvae of the Western Budworm, *Choristoneura occidentalis* on Douglas Fir in Colarado. *J. econ. Ent.* **62**, 139–42.

MACLELLAN, C. R., 1962. Mortality of codling moth eggs and young larvae in an integrated control orchard. *Can. Ent.* **94**, 655–66.

MACLEOD, J. and DONNELLY, J., 1956. Methods for the study of blowfly populations. II. The use of laboratory-bred material. *Ann. appl. Biol.* **44**, 643–8.

MACLEOD, J. and DONNELLY, J., 1963. Dispersal and interspersal of blowfly populations. *J. Anim. Ecol.* **32**, 1–32.

MCMULLEN, L. H. and ATKINS, M. D., 1959. A portable tent-cage for entomological field studies. *Proc. ent. Soc. B. C.* **56**, 67–8.

MACPHEE, A. W., 1961. Mortality of winter eggs of the european red mite *Panonychus ulmi* (Koch), at low temperatures, and its ecological significance. *Can. J. Zool.* **39**, 229–43.

MACPHEE, A. W., 1964. Cold-hardiness, habitat and winter survival of some orchard Arthropods in Nova Scotia. *Can. Ent.* **96**, 617–36.

MARTIN, F. J., 1969. Searching success of predators in artificial leaf litter. *Am. Midl. Nat.* **81**, 218–27

MASON, W. H. and MCGRAW, K. A., 1973. Relationship of ^{65}Zn excretion and egg production in *Trichoplusia ni* (Hubner). *Ecology* **54**, 214–6.

MATTHEWS, G. V. T., 1974. On bird navigation with some statistical undertones. *J. R. Stat. Soc. (B)* **36**, 349–64.

MAY, R. M., 1973. Qualitative stability in model ecosystems. *Ecology* **54**, 638–41.

MAY, R. M. 1976. Models for two interacting populations. *In* May, R. M. (ed.) *Theoretical Ecology, Principles and Applications* pp. 49–70.

MEIJDEN, E. VAN DER., 1973. Experiments on dispersal, late-larval predation, and pupation in the Cinnabar moth (*Tyria jacobaeae* L.) with a radioactive label (^{192}Ir.). *Netherlands J. Zool.* **23**, 430–45.

MILES, P. M., 1952. Entomology of Bird Pellets. *Amat. Ent. Soc. Leaflet* **24**, 8 pp.

MILLER, C. A., 1955. A technique for assessing larval mortality caused by parasites. *Can. J. Zool.* **33**, 5–17.

MILLER, C. A., 1957. A technique for estimating the fecundity of natural populations of the spruce budworm. *Can. J. Zool.* **35**, 1–13.

MILLER, C. A., 1958. The measurement of spruce budworm populations and mortality during the first and second larval instars. *Can. J. Zool.* **36**, 409–22.

MILLER, R. S. and THOMAS, J. L., 1958. The effects of larval crowding and body size on the longevity of adult *Drosophila melanogaster*. *Ecology* **39**, 118–25.

MOHR, C. O., 1947. Table of equivalent populations of North American small mammals. *Am. midl. Nat.* **37**, 223–49.

MOHR, C. O. and STUMPF, W. A., 1964a. Relation of tick and chigger infestations to home areas of California meadow mice. *J. med. Ent.* **1** (1), 73–7.

MOHR, C. O. and STUMPF, W. A., 1964b. Louse and chigger infestations as related to host size and home range of small mammals. *Trans. 29th N. Am. Wildl. nat. Res. Confr.* 181–95.

MOOK, L. J., Birds and the spruce budworm. *Mem. ent. Soc. Canada.* **31**, 268–71.

MOORE, N. W., 1952. On the so-called 'territories' of dragonflies (Odonata – Anisoptera). *Behaviour* **4**, 85–100.

MOORE, N. W., 1957. Territory in dragonflies and birds. *Bird study* **4**, 125–30.

MORGAN, C. V. G. and ANDERSON, N. H., 1958. Techniques for biological studies of Tetranychid mites, especially *Bryobia arborea* M. and A. and B. *praetiosa* Koch (Acarina: Tetranychidae). *Can. Ent.* **90**, 212–15.

MORGAN, N. C. and WADDELL, A. B., 1961. Insect emergence from a small trout loch, and its bearing on the food supply of fish. *Sci.Invest. Freshw. Fish. Scot.* **25**, 1–39.

MORGAN, N. C., WADDELL, A. B. and HALL, W. B., 1963. A comparison of the catches of emerging aquatic insects in floating box and submerged funnel traps. *J. Anim. Ecol.* **32**, 203–19.

MORRIS, R. F., 1949. Differentiation by small mammals between sound and empty cocoons of the European spruce sawfly. *Can. Ent.* **81**, 114–20.

MOULDER, B. C. and REICHLE, D. E., 1972. Significance of spider predation in the energy dynamics of forest-floor arthropod communities. *Ecol. Monogr.* **42** (4), 473–98.

MULLA, M. S., NORLAND, R., IKESHOJI, T. and KRAMER, W. L., 1974. Insect growth regulators for the control of aquatic midges. *J. econ. Ent.* **67** (2), 165–70.

MUNDIE, J.H. 1956. Emergence traps for aquatic insects. *Mitt. int. Verein. theor. angew. Limnol.* **7**, 1–13.

MURDIE, G., 1969. The biological consequences of decreased size caused by crowding or rearing temperatures in apterae of the pea aphid, *Acyrthosiphon pisum* Harris. *Trans. R. ent. Soc. Lond.* **121**, 443–55.

MURDIE, G. and HASSELL, M. P., 1973. Food distribution, searching success and predator-prey models. *In* Hiorns, R. W. (ed.) *The Mathematical Theory of the Dynamics of Biological Populations*, pp. 87–101, Academic Press, London.

MURDOCH, W. W., 1969. Switching in general predators: experiments on predators specificity and stability of prey populations. *Ecol. Monogr.* **39**, 335–54.

MURDOCH, W. W., 1971. The developmental response of predators to changes in prey density. *Ecology* **52**, 132–7.

MURDOCH, W. W. and MARKS, J. P., 1973. Predation by coccinellid beetles: experiments on switching. *Ecology* **54**, 160–7.

MURDOCH, W. W. and OATEN, A., 1975. Predation and population stability. *Adv. ecol. Res.* **9**, 1–131.

MURTON, R. K., 1971. The significance of a specific search image in the feeding behaviour of the wood pigeon. *Behaviour* **40**, 10–42.

MUSTAFA, S., 1976. Selective feeding behaviour of the common carp, *Esomus danricus* (Ham.) in its natural habitat. *Biol. J. Linn. Soc.* **8**, 279–84.

NAKASUJI, R., YAMANAKA, H. and KIRITANI, K., 1973. The disturbing effect of microphantid spiders on the larval aggregation of the tobacco cutworm, *Spodoptera litura* (Lepidoptera: Noctuidae). *Kontyu* **41**, 220–227.

NEILSON, M. M., 1963. Disease and the spruce budworm. *Mem. ent. Soc. Can.* **31**, 272–88.

NEILSON, M. M. and MORRIS, R. F., 1964. The regulation of European spruce sawfly numbers in the Maritime Provinces of Canada from 1937–1963. *Can. Ent.* **96**, 773–84.

NELMES, A. J., 1974. Evaluation of the feeding behaviour of *Prionchulus punctatus* (Corb.) a nematode predator. *J. Anim. Ecol.* **43**, 553–65.

NEUENSCHWANDER, P., 1975. Influence of temperature and humidity on the immature stages of *Hemerobius pacificus*. *Environ. Ent.* **4** (2), 215–20.

NEWMAN, G. G. and CARNER, G. R., 1975. Disease incidence in soy bean loopers collected by two sampling methods. *Environ. Ent.* **4**, 231–2.

NICHOLLS, C. F., 1963. Some entomological equipment. *Res. Inst. Can. Dept. Agric. Belleville, Inf. Bull.* **2**, 85 pp.

NIJVELDT, W., 1959. Overhet gebruik van vangekegels bij het galmugonderzoek. *Tijdschr. PlZiekt.* **65**, 56–59.

NORD, J. C. and LEWIS, W. C. 1970. Two emergence traps for wood-boring insects. *J. Ga. ent. Soc.* **5**, 155–7.

ODUM, E. P. and KUENZLER, E. J., 1955. Measurement of territory and home range size in birds. *Auk.* **72**, 128–37.

OHIAGU, C. E. and BOREHAM, P. F. L., 1978. A simple field test for evaluating insect prey-predator relationships. *Entemologia exp. appl.* **23**, 40–7.

OHNESORGE, B., 1957. Untersuchungen über die Populationsdynamik der kleinen Fichtenblattwespe, *Pristiphora abietina* (Christ) (Hym. Tenthr.). I. Teil. Fertilität und Mortalität. *Z. angew. Ent.* **40**, 443–93.

OTVOS, I. S., 1974. A collecting method for pupae of *Lambderia fiscellaria fiscellaria* (Lepidoptera: Geometridae). *Can. Ent.* **106**, 329–31.

OUCHTERLONY, O., 1948. In vitro method for testing the toxin producing capacity of diphtheria bacteria. *Acta path. microbiol. Scand.* **25**, 186–91.

PALMEN, E., 1955. Diel periodicity of pupal emergence in natural populations of some chironomids (Diptera). *Ann. zool. Soc. Vanamo* **17** (3), 1–30.

PALMÉN, E., 1962. Studies on the ecology and phenology of the Chironomids (Dipt.) of the Northern Baltic. *Ann. ent. fenn.* **28** (4) 137–68.

PARIS, O. H., 1965. The vagility of P^{32}-labelled Isopods in grassland. *Ecology* **46**, 635–48.

PAVLOV, I. F., 1961. Ecology of the stem moth *Ochsenheimeria vaculella* F.-R. (Lepidoptera Tineoidea). [In Russian.] *Ent. Obozr.* **40**, 818–27 (transl. *Ent. Rev.* **40**, 461–6).

PENDLETON, R. C. and GRUNDMANN, A. W., 1954. Use of P^{32} in tracing some insect-plant relationships of the thistle, *Cirsium undulatum. Ecology* **35**, 187–91.

PETERSON, A., 1934. *A manual of entomological equipment and methods.* Pt 1. Edwards Bros. Inc. Ann Arbor.

PICKAVANCE, J. R., 1970. A new approach to the immunological analysis of invertebrate diets. *J. Anim. Ecol.* **39**, 715–24.

PICKAVANCE, J. R., 1971. The diet of the immigrant planarian *Dugesia tigrina* (Girard). II. Food in the wild and comparison with some British species. *J. Anim. Ecol.* **40**, 637–50.

PILON, J. G., TRIPP, H. A., McLEOD, J. M. and ILNITZKEY, S. L., 1964. Influence of temperature on prespinning eonymphs of the Swaine jack-pine sawfly, *Neodiprion swainei* Midd. (Hymenoptera: Diprionidae). *Can. Ent.* **96**, 1450–7.

POLLES, S. G. and PAYNE, J. A., 1972. An improved emergence trap for adult Pecan weevils. *J. econ. Ent.* **65**, 1529.

PREBBLE, M. L., 1941. The diapause and related phenomena in *Gilpinia polytoma* (Hartig). IV. Influence of food and diapause on reproductive capacity. *Can. J. Res. D* **19**, 417–36.

PROMPTOW, A. N. and LUKINA, E. W., 1938. Die Experimente beim biologischen Studium und die Ernährung der Kohlmeise (*Parus major* L.) in der Brutperiode. [In Polish.] *Zool. Zh.* **17**, 777–82.

RADKE, W. J. and FRYDENDALL, M. J., 1974. A survey of emetics for use in stomach contents recovery in the house sparrow. *Am. Midl. Nat.* **92**, 164–72.

RAU, P. and RAU, N., 1916. The biology of the mud-daubing wasps as revealed by the contents of their nests. *J. Anim. Behavior* **6**, 27–63.

REID, R. W., 1963. Biology of the mountain pine beetle, *Dendroctonus monticolae* Hopkins, in the East Kootenay Region of British Columbia. III. Interaction between the beetle and its host, with emphasis on brood mortality and survival. *Can. Ent.* **95**, 225–38.

REIFF, M., 1955. Untersuchungen zum Lebenszyklus der Frostspanner *Cheimatobia (Operophthera) brumata* L. und *Hibernia defoliaria* Ch. *Mitt. schweiz. ent. Ges.* **26**, 129–44.

RICE, R. E. and REYNOLDS, H. T., 1971. Seasonal emergence and population development of the Pink Bollworm in Southern California. *J. econ. Ent.* **64**, 1429–32.

RICH, E. R., 1956. Egg cannibalism and fecundity in *Tribolium*. *Ecology* **37**, 109–20.

RICHARDS, O. W., 1940. The biology of the small white butterfly (*Pieris rapae*), with special reference to the factors controlling its abundance. *J. Anim. Ecol.* **9**, 243–88.

RICHARDS, O. W. and HAMM, A. H., 1939. The biology of the British Pompilidae (Hymenoptera). *Trans. Soc. Brit. Ent.* **6**, 51–114.

RICHARDS, O. W. and WALOFF, N., 1954. Studies on the biology and population dynamics of British grasshoppers. *Anti-Locust Bull.* **17**, 184 pp.

RICHARDS, O. W. and WALOFF, N., 1961. A study of a natural population of *Phytodecta olivacea* (Forster) (Coleoptera: Chrysomelidae). *Phil. Trans. B* **244**, 205–57.

ROGERS, D. J., 1972. Random search and insect population models. *J. Anim. Ecol.* **41**, 369–83.

ROGERS, D. J. and HASSEL, M. P., 1974. General models for insect parasite and predator searching behaviour: interference. *J. Anim. Ecol.* **43**, 239–53.

ROTHSCHILD, G. H. L., 1966. A study of a natural population of *Conomelus anceps* (Germar) (Homoptera: Delphacidae) including observations on predation using the precipitin test. *J. Anim. Ecol.* **35**, 413–34.

ROTHSCHILD, G. H. L., 1970. Observations on the ecology of the rice ear bug, *Leptocoris oratorius* (F.) (Hemiptera: Alydidae) in Sarawak (Malaysian Borneo). *J. appl. Ecol.* **7**, 147–67.

ROTHSCHILD, G. H. L., 1971. The biology and ecology of rice-stem borers in Sarawak (Malaysian Borneo). *J. appl. Ecol.* **8**, 287–322.

ROYAMA, T. 1970. Factors governing the hunting behaviour and selection of food by the great tit (*Parus major* L.). *J. Anim. Ecol.* **39**, 619–68.

ROYAMA, T., 1971. A comparative study of models for predation and parasitism. *Res. Pop. Ecol. Supp. 1* 90 pp.

SANDERSON, G. C., 1966. The study of mammal movements—a review. *J. Wildl. Manag.* **30**, 215–35.

SANTOS, M. A., 1976. Prey selectivity and switching response of *Zetzellia maki*. *Ecology* **57**, 390–4.

SASABA, T. and KIRITANI, K., 1972. Evaluation of mortality factors with special reference to parasitism of the Green rice leafhopper, *Nephotettix cinticeps* Uhler (Hemiptera: Deltocephalidae). *Appl. Ent. Zool.* **7**, 83–93.

SCOTTER, D. R., LAMB, K. P. and HASSAN, E., 1971. An insect dispersal parameter. *Ecology* **52**, 174–7.

SERVICE, M. W., 1973. Study of the natural predators of *Aedes cantaris* (Meigen) using the precipitin test. *J. med. Ent.* **10**, 503–10.

SERVICE, M., 1976. *Mosquito Ecology: field sampling techniques*, Applied Science Publishers, London.

SHAPIRO, A. M., 1974. Beak-mark frequency as an index of seasonal predation intensity on common butterflies. *Am. Nat.* **108**, 229–32.

SHAW, M. J. P., 1970. Effects of population density on alienicolae of *Aphis fabae* Scop. *Ann. appl. Biol.* **65**, 191–4, 197–203, 205–212.

SKELLAM, J. G., 1951. Random dispersal in theoretical population. *Biometrika* **38**, 196–218.

SMIRNOFF, W. A., 1967. A method for detecting viral infection in populations of *Neodiprion swainei* by examination of pupae and adults. *Can. Ent.* **99**, 214–6.

SMITH, H. S. and DEBACH, P. 1942. The measurement of the effect of entomophagous insects on population densities of the host. *J. econ. Ent.* **35**, 845–9.

SOLOMON, M. E., 1949. The natural control of animal population *J. Anim. Ecol.* **18**, 1–35.

SOMMERMAN, K. M., SAILER, R. I. and ESSELBAUGH, C. O., 1955. Biology of Alaskan black flies (Simuliidae, Diptera). *Ecol. Monogr.* **25**, 345–85.

SOUTHERN, H. N., 1954. Tawny owls and their prey. *Ibis* **96**, 384–410.

SOUTHWOOD, T. R. E., 1962. Migration of terrestrial arthropods in relation to habitat. *Biol. Rev.* **37**, 171–214.

SOUTHWOOD, T. R. E., 1971. The role and measurement of migration in the population system of an insect pest. *Trop. Sci.* **13**, 275–8.

SOUTHWOOD, T. R. E., JEPSON, W. F. and EMDEN, H. F. VAN, 1961. Studies on the behaviour of *Oscinella frit* L. (Diptera) adults of the panicle generation. *Entomologia exp. appl.* **4**, 196–210.

SOUTHWOOD, T. R. E. and SIDDORN, J. W., 1965. The temperature beneath insect emergence traps of various types. *J. Anim. Ecol.* **34**, 581–5.

SPEIR, J. A. and ANDERSON, N. H., 1974. Use of emergence data for estimating annual production of aquatic insects. *Limnol. Oceanogr.* **79** (1), 154–6.

SPEYER, W. and WAEDE, M., 1956. Eine Methode zur Vorhersage des Weizengallmücken-fluges. *Nachr bl. dt. PflSchDienst. Stuttg.* **8**, 113–21.

SPILLER, D., 1964. Numbers of eggs laid by *Anobium punctatum* (Degeer). *Bull. ent. Res.* **55**, 305–11.

STEINBUCH, M. and AUDRAN, R., 1969. The isolation of Ig Cr from mammalian sera with the aid of caprylic acid. *Arch. Biochem. Biophys.* **134**, 279–284.

STEINHAUS, E. A., 1963. Background for the diagnosis of insect diseases. In Steinhaus, E. A. (ed.), *Insect pathology, an advanced treatise* **2**, 549–89, Academic Press, New York and London.

STRAUB, R. W., FAIRCHILD, L. M. and KEASTER, A. J., 1973. Corn earworm: use of larval traps on corn ears as a method of evaluating corn lines of resistance. *J. econ. Ent.* **66** (4), 989–90.

STUMPF, W. A. and MOHR, C. O., 1962. Linearity of home ranges of California mice and other animals. *J. Wildl. Mgmt.* **26**, 149–54.

SULLIVAN, C. R. and GREEN, G. W., 1964. Freezing point determination in immature stages of insects. *Can. Ent.* **96**, 158.

SUTTON, S. L., 1970. Predation on woodlice: an investigation using the precipitin test. *Entomologia exp. Appl.* **13**, 279–85.

SYLVEN, E., 1970 Field movement of radioactively labelled adults of *Dasyneura brassicae* Winn. (Dipt., Cecidomyiidae). *Ent. Scan.* **1**, 161–87.

TANADA, Y., 1961. The epizootiology of virus disease in field populations of the armyworm, *Pseudaletia unipuncta* (Hamorth). *J. Insect Path.* **3**, 310–23.

TAYLOR, L. R., 1975. Longevity, fecundity and size: control of reproductive potential in a polymorphic migrant, *Aphis fabae* Scop. *J. Anim. Ecol.* **44**, 135–63.

TAYLOR, L. R and TAYLOR, R. A. J., 1976. Aggregation, migration and population mechanics. *Nature* **265**, 415–21.

TAYLOR, L. R. and TAYLOR, R. A. J., 1978. Dynamics of spatial behaviour. *In Population control by social behaviour*, (*Inst. Biol. Symp.*). 181–212.

TAYLOR, R. A. J. 1978. The relation between density and distance of dispersing insects. *Ecol. Ent.* **3**, 63–70.

TERRELL, T. T., 1959. Sampling populations of overwintering spruce budworm in the Northern Rocky Mountain region. *Res. note Intermountain Forest Range Exp. Sta., Ogden, Utah* **61**, 8 pp.

THOMAS, G. M., 1974. Diagnostic techniques. *In* Cantwell, G. E. (ed.) *Insect Diseases* 1. Marcel Dekker, New York.

THOMPSON, D. J., 1975. Towards a predator–prey model incorporating age structure, *J. Anim. Ecol.* **44**, 907–16.

TINBERGEN, L., 1960. The natural control of insects in pinewoods. 1. Factors influencing the intensity of predation by songbirds. *Arch. Néerland. Zool.* **13**, 266–343.

TOD, M. E., 1973. Notes on beetle predators of molluscs. *Entomologist* **106**, 196–201.

TOTH, R. S. and CHEW, R. M., 1972. Development and energetics of *Notonecta undulata* during predation on *Culex tarsalis*. *Ann. ent. Soc. Am.* **5**, 1270–9.

TURNBULL, A. L., 1960. The prey of the spider *Linyphia triangularis* (Clerck) (Araneae: Linyphiidae). *Can. J. Zool.* **38**, 859–73.

TURNBULL, A. L., 1962. Quantitative studies of the food of *Linyphia triangularis* Clerck (Araneae: Linyphiidae). *Can. Ent.* **94**, 1233–49.

TURNOCK, W. J., 1957. A trap for insects emerging from the soil. *Can. Ent.* **89**, 455–6.

TURNOCK, W. J. and IVES, W. G. H., 1962. Evaluation of mortality during cocoon stage of the larch sawfly, *Pristiphora erichsonii* (Htg.). *Can. Ent.* **94**, 897–902.

VALLENTYNE, J. R., 1952. Insect removal of nitrogen and phosphorus compounds from lakes. *Ecology* **33**, 573–7.

VARLEY, G. C. and GRADWELL, G. R., 1963. The interpretation of insect population change. *Proc. Ceylon Assn Adv. Sci. (D)* (1962), **18**, 142–56.

VICKERMAN, G. P. and SUNDERLAND, K. D., 1975. Arthropods in cereal crops: Nocturnal activity, vertical distribution and aphid predation. *J. appl. Ecol* **12**, 755–66.

VLIJM, L., VAN DIJCK, T. S. and WIJMANS, S. Y., 1968. Ecological studies on Carabid beetles. III. Winter mortality in adult *Calathus melanocephalus* (Linn.) egg production and locomotory activity of the population which has hibernated. *Oecologia* **1**, 304–14.

WALOFF, N. and RICHARDS, O. W., 1958. The biology of the Chrysomelid beetle, *Phytodecta olivacea* (Forster) (Coleoptera: Chrysomelidae). *Trans. R. ent. Soc. Lond.* **110**, 99–116.

WAY, M. J. and BANKS, C. J., 1964. Natural mortality of eggs of the black bean aphid, *Aphis fabae* (Scop.), on the spindle tree, *Euonymus europaeus*. *Ann. Appl. Biol.* **54**, 255–67.

WEIR, D. M., 1973. *Handbood of Experimental Immunology* 2nd Ed., Blackwells, Oxford.

WEISER, J., 1969. *An Atlas of Insect Diseases*. 292 pp. Irish Univ. Press, Shannon and Academia, Prague.

WEISER, J. and BRIGGS, J. D., 1971. Identification of pathogens. *In* Burges, H. D. & Hussey, N. W. (eds.) *Microbial Control of Insects and Mites*. pp. 13–66, Academic Press, London and New York.

WEITZ, B., 1952. The antigenicity of sera of man and animals in relation to the preparation of specific precipitating antisera. *J. Hyg. Camb.* **50**, 275–94.

WEITZ, B., 1956. Identification of blood-meals of blood-sucking arthropods. *Bull. Wld. Hlth. Org.* **15**, 473–90.

WELLINGTON, W. G., 1964. Qualitative changes in populations in unstable environments. *Can. Ent.* **96**, 436–51,

WILBUR, D. A. and FRITZ, R., 1939. Use of shoebox emergence cages in the collection of insects inhabiting grasses. *J. econ. Ent.* **32**, 571–3.

WINKLE, W. VAN., MARTIN, D. C. and SEBETICH, M. J., 1973. A home-range model for animals inhabiting an ecotone. *Ecology* **54**, 205–9.

WITTIG, G., 1963. Techniques in insect pathology. *In* Steinhaus, E. A. (ed.), *Insect pathology, an advanced treatise* **2**, 591–636. Academic Press, London and New York.

WOLFENBARGER, D. O., 1946. Dispersion of small organisms. *Am. midl. Nat.* **35**, 1–152.

WRATTEN, S. D., 1976. Searching by *Adatea bipunctata* (L.) (Coleoptera: Coccinellidae) and escape behaviour of its aphid and cicadellid prey on lime (*Tilia x vulgaris* Hayne). *Ecol. Ent.* **1**, 139–42.

WRIGHT, D. W., HUGHES, R. D. and WORRALL, J., 1960. The effect of certain predators on the numbers of cabbage root fly (*Erioischia brassicae* (Bouché)) and on the subsequent damage caused by the pest. *Ann. appl. Biol.* **48**, 756–63.

YOUNG, E. C., 1974. The epizootiology of two pathogens of the coconut palm rhinoceros beetle. *J. Invert. Path.* **24**, 82–92.

10

The Construction, Description and Analysis of Age-specific Life-tables

10.1 Types of life-table and the budget

The construction of a number of life-tables is an important component in the understanding of the population dynamics of a species. Although some animal ecologists, such as Richards (1940), had expressed their results showing the successive reductions in the population of an insect throughout a single generation, Deevey (1947) was really the first to focus attention on the importance of this approach. Life-tables have long been used by actuaries for determining the expectation of life of an applicant for insurance and thus the column indicating the expectation of life at a given age (the e_x column) is an essential feature of human life-tables. However, the fundamental interests of the ecologist and, even more so, of the economic entomologist are essentially different from those of the actuary and it is a mistake to believe that these approaches and parameters of primary interest in the study of human populations are also those of greatest significance to the animal ecologist. Because many insects have discrete generations and their populations are not stationary, the age-specific life-table is more widely applicable than the time-specific life-table. The differences between these two types are as follows:

An age-specific (or horizontal) life-table is based on the fate of a real cohort; conveniently the members of a population belonging to a single generation. The population may be stationary or fluctuating.

A time-specific (or vertical) life-table is based on the fate of an imaginary cohort found by determining the age structure, at a point of time, of a sample of individuals from what is assumed to be a stationary population with considerable overlapping of generations, i.e. a multi-stage population. Age determination is a prerequisite for time-specific life-tables (Chapter 11). A modification of this approach is the *variable life-table* of Gilbert *et al.* (1976), which is an inductive strategic computer model of the population: this is varied until it provides a reasonable description of the population (see Chapters 11 and 12).

The data in age-specific life-tables may be corrected so as to start with a fixed number of individuals, e.g. 1000; however, this practice, which simplifies the calculation of life expectancy, causes the very important information on actual population size to be lost. It will be seen in this chapter that the variations in population size, from generation to generation, provide the frame of reference against which the roles of the various factors are analysed; therefore, in much work on insect populations, the type of table required lists the actual absolute populations at different stages and records the action of mortality factors where these are known (see Table 10.5, p. 377). Such a table, giving just the observed data, is well described by the term *budget* proposed by Richards (1961), which also has the advantage that it emphasizes the distinction between this approach and that of the actuary.

10.2 The construction of a budget

The ideal budget will contain absolute estimates of the total population of as many stages as possible. At some points in the generation it may be possible to determine the total number entering a stage directly, as described in Chapter 9 (see also p. 357). For other stages there will be a series of estimates, made on successive sampling days, using methods based on the numbers per unit area (Chapters 4–6) or from mark and recapture (Chapter 3), nearest neighbour (p. 358) or removal trapping and line transect (pp. 228–236) techniques. The problem now arises as to how to determine from these estimates the total number of individuals that pass through a particular stage in one generation.

The degree of synchronization in the life-cycle is an important factor affecting the ease or difficulty of this step. The ideal situation is when there is a point of time when all the individuals of a generation are in a given stage; a census at this time will provide a 'peak estimate' that may be used in a life-table. As the overlap in time of successive stages increases, it will be necessary to integrate a number of estimates to obtain the total population; special techniques for doing this are described below. When there is complete overlap of all stages, methods based on the age structure of the population are most appropriate (see Chapter 11). Once the series of estimates of total population at each stage have been developed it can be assumed that the differences between these represent mortality and/or dispersal. It may be possible to check this assumption by direct measurement of these factors (Chapter 9).

A budget is to some extent self-checking; erroneous population measurements may be exposed by increases in numbers at a stage when immigration is impossible or by other inconsistencies. Thus, the more terms in a budget the greater the confidence that can be placed in it (Richards, 1959); this confidence is additional to that obtained from the statistical confidence limits of the individual estimates. The latter are based solely on the information gathered by the given method for that particular stage and mean that the true value is likely (to the extent of the chosen probability level) to lie within them. It is reasonable to claim that when there is agreement with other estimates, by other methods made simultaneously (p. 4) or sequentially, this substantially increases the probability that the true value lies close to the estimate.

There are several different graphical and statistical techniques available for estimating the numbers entering a stage or at the mid-point of the stage from a series of population samples. These methods vary in their assumptions and requirements (Table 10.1). In determining which to use, the first consideration must be the extent to which the assumptions are valid and the special requirements can be met; the availability of computing facilities will be a second factor. The graphical method is probably the simplest and most robust (Ruesink, 1975); that of Kiritani & Nakasuji (1967) as modified by Manly (1976) is also relatively simple and reliable (Manly, 1974c). Birley's (1977) new method looks most promising because of the non-restrictive nature of its assumptions; however, it requires computer facilities. Ashford *et al.* (1970) have described a mathematically complex method based on the general theory of stochastic branching processes and using maximum likelihood estimates. As Manly (1974c) indicates, its procedure is somewhat opaque.

10.2.1 Graphical method

This is the crudest and simplest of the methods of integration. It gives reasonable estimates of the numbers entering a stage if mortality, which may be heavy, occurs entirely at the end of the stage, or if the mortality is light, and occurs at constant rate throughout the stage which is of relatively long duration (Birley, unpub.). If the mortality rate is constant, then the estimate is an approximation (and an overestimation) of the numbers at the medium age of the stage. Successive estimates are plotted on graph paper, most conveniently allowing one square per individual and per day (Fig. 10.1). The points are joined up and the number of squares under the line counted; this total is then

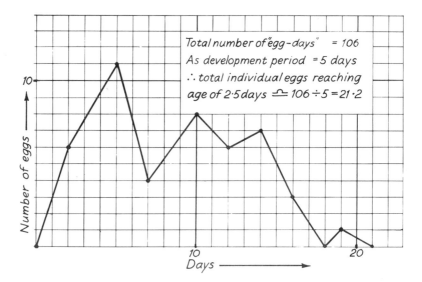

Fig. 10.1 The determination of the total number of individuals in a stage from series of estimates by graphical summation (hypothetical example with unrealistically small numbers for simplicity).

Table 10.1 Comparison of assumptions and additional requirements of methods for estimating stage recruitment and/or survival [partly after Manly (1974c) and Birley (unpublished)].

Method	Assumptions						
	Recruitment	Survival	Sampling efficiencies for different stages	Nos. recruited	Duration of stage sampled	Sampling intervals
Richards & Waloff First	All at about the same time	Constant all stages	Equal	–	–	–
Manly	Normal distribution	Constant all stages	Equal	–	–	–
Birley	Variable	Variable	Variable	Index	–	–
Graphical	Variable	Constant within stage or at end of stage	Variable	–	Yes	–
Ruesink	Variable	At end of stage	Equal	–	–	–
Dempster	Variable	Constant within stage	Equal	Yes	–	–
Richards & Waloff Second	Variable	Constant within stage	Equal	Yes	Yes	Equal
Kiritani, Nakasuji & Manly	Variable	Constant all stages	Equal	–	–	Equal

divided by the mean developmental time under field conditions. This method was used by Southwood & Jepson (1962), who compared it with another method in which the total time, during which the stage was found, was divided arbitrarily into periods each corresponding to the developmental time of an individual; the mean population was calculated for each of these arbitrary periods and these means summed. The two methods are essentially similar and give similar results. This technique has been found to be robust and has been quite widely used (e.g. Helgeson & Haynes, 1972).

10.2.2 Richards & Waloff's first method

This method (Richards & Waloff, 1954) is applicable to a stage with a well-marked peak and an approximately steady mortality rate. It assumes a single impulse of recruitment and consists of plotting the regression of the fall-off of numbers with time after the peak and then extending this back to the time when the stage was first found (t_i) (Fig. 10.2). The population corresponding to t_0 will be the total number of individuals alive just before the stage is seen for the first time (Manly, 1975), hence this method slightly overestimates recruitment.

A population exposed to a steady mortality may be expressed:

$$Y_t = N_0 \phi^t \qquad (10.1)$$

where Y_t = population on day t, N_0 is the peak population (ideally the number hatched) and ϕ = the fraction of the population which survives to the end of a unit of time (e.g. a day), the survival rate per unit time. Hence:

$$\log Y_t = \log N_0 + t \log \phi \qquad (10.2)$$

and after the peak this will describe a straight line which is conveniently obtained from the regression of $\log Y_t$ on t: the regression coefficient will be $\log \phi$, the logarithm of the average, and supposedly constant, survival rate. The number entering the stage will then be given by the value of Y_t, found by inserting into the equation a value of t corresponding to the start of the stage. Alternatively, if t is numbered from the start of the stage in question, $\log N_0$ will equal the number entering the stage.

More reliable estimates are obtained if Y_t is taken as the accumulated population. That is, after the peak of a particular instar, the values of Y_t are the total of that instar, plus any individuals that have passed through that instar and are in subsequent ones. For example, Y_{tx} for the second instar would consist of the populations of second instar, third instar and fourth instar, etc. as estimated for day tx, but Y_{tx} for the third instar would consist only of the populations of the third and subsequent instars. Dempster (1956) found that if Y_t was based solely on the numbers of the particular instar the estimates of population were too high, the slope of the line being too steep due to the confusion of moulting with mortality; the use of accumulated totals avoids this.

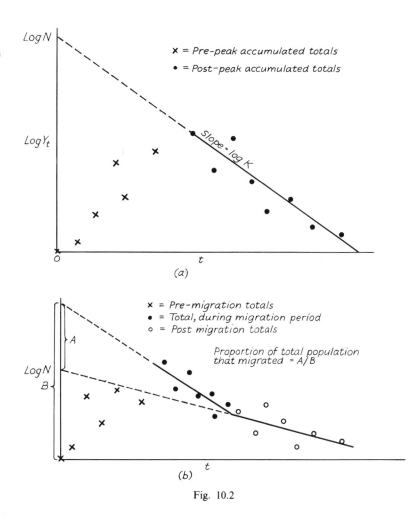

Fig. 10.2

Waloff & Bakker (1963) used a modification of this method to determine the total change in population due to migration. Basically, separate regressions were calculated for the migration and post-migration period and the slopes ($\log \phi$) compared. The value of $\log \phi$ for the migration period represents migration and mortality, for the post-migration period just mortality. Therefore, if the difference between these is expressed as a percentage of the larger, this will be the percentage of the initial population that was lost during the migratory period.

10.2.3 Manly's method
Manly's (1974 a & b) method is essentially a modification of Richards &

Waloff's First Method, but it avoids the need for a distinct peak. Recruitment is assumed to be normal, but Manly considered that so long as the entry was unimodal, reasonable estimates would be provided. Simulation studies suggest that significant biases will arise if recruitment to the first stage is 'far from normal' (Manly, 1974c). The basic equation is:

$$Y_t = N_0 \int_{-\infty}^{t} \exp[-\theta(t-x)] f(x)(dx) \qquad (10.3)$$

where Y_t = the number of animals in a particular stage and later stages at time t, N_0 = total number entering the stage, $\exp(-\theta(t-x))$ = the probability that an animal that enters the population at time x is still alive at time t (θ being the constant age specific death rate) and $f(x)$ is the frequency of the stage at time x. Because of Manly's assumption of normality, the frequency function for this distribution can be inserted and equation 10.3 becomes:

$$Y_t = N_0^* \exp(-\theta t) \int_{-\infty}^{t-\mu^*/\sigma} (2\pi)^{-\frac{1}{2}} \exp(-\tfrac{1}{2}x^2) dx \qquad (10.4)$$

where $N_0^* = \exp[\theta(\mu + \tfrac{1}{2}\theta\sigma^2)]N_0$ and $\mu^* = \mu + \theta\sigma^2$, where μ = the mean time of entry to the stage and σ = its standard deviation.

The four unknowns, N_0 (the numbers recruited), μ (the mean time of recruitment), σ (the standard deviation of μ) and θ (the age specific death rate) may be found from a series of equations derived from at least four different sampling occasions (i.e. Y_t's with different values of t). Computer programmes may be developed for this purpose. Manly (1974a) also describes how the values can be estimated from graphical methods.

10.2.4 Birley's Method
This method makes the fewest assumptions, but like the two preceding, belongs to a class of mathematical models termed transfer functions or convolution models (Birley, 1977); for one is considering two processes (recruitment and survival) whose time-scales are parallel but opposed. This may be seen from the basic form of the equation for the numbers at time t:

$$Y_t = f(t)\phi_0 + f(t-1)\phi_1 + f(t-2)\phi_2 \ldots f(t-d)\phi_d$$

or

$$Y_t = \sum_{j=0}^{d} f(t-j)\phi_j \qquad (10.5)$$

where Y_t = observed stage density at time t, $f(t)$ = stage recruitment at time t, $f(t)\phi_0$ = the recruits at time t exposed to the survival rate (ϕ) for zero time, and $f(t-d)\phi_d$ = the recruits which have survived d days, where d is the maximum stage duration.

Now $f(t) = N_0 \cdot g(t)$ where N_0 = total stage recruitment and $g(t)$ = the fractional stage recruitment at time t the sum becomes:

$$Y_t = N_0 \sum_{j=0}^{d} g(t-j)\phi_j \qquad (10.6)$$

The field data must provide the series of Y_t's, the numbers of animals in a given instar at time t and also, independently, the fraction of recruitment at different times (g_t's), simply an index of recruitment with time, the shape of the 'hatching curve'. The objective is to find N_0 and the survival rate (ϕ's). Equation 10.6 may be written in the expanded general form:

$$Y_t = N_0\phi_0 g(t) + N_0\phi_1 g(t-1) + \ldots N_0\phi_d g(t-d) \qquad (10.7)$$

As Birley points out this may be viewed as a multiple regression with the $N_0\phi$'s values regarded as a series of regression coefficients. These coefficients may be extracted by computer, but there will be $d+1$ variables and if the number of data points is less than d, then the model is said to be overparameterized. However, as emphasized above (p. 347), in order to conform to biological reality life-tables must be internally consistent and in this instance Birley places the constraint:

$$N_0\phi_0 > N_0\phi_1 > N_0\phi_2 > N_0\phi_3 \ldots > 0$$

which simply indicates that once dead, animals cannot revive and that one cannot have negative numbers, the cohort size cannot fall below zero!

The parameters for such a constrained multiple linear regression may be estimated using the Quadratic Programming Algorithm (Birley, 1977; Judge & Takayama, 1966) and the constraints, using *a priori* information in a Bayesian sense, may be considered to improve the estimates over the unconstrained estimator. These estimators will then give cohort survival directly, without depending on the subtractive processes of the life budget.

It should be noted that this method allows for a wide variety of stage survival and recruitment patterns and it places constraints on the estimates to conform to biological reality. There is, however, no general theory of estimation that would allow confidence limits to be attached to these estimates. As has been stressed elsewhere in this book (pp. 4, 98), the ecologist must work with techniques that allow the incorporation of additional data, sometimes merely indices, into the estimation. Such approaches will in the long run prove more useful than methods whose statistical propriety may mask their underlying ecological indecency, and more profitable than, as R. F. Morris has expressed it, 'the pessimistic contemplation of individual standard errors'!

10.2.5 Ruesink's method
Like Birley's method Ruesink's (1975) allows the calculation of the stage survival rate and its does this by a comparison of the rate of recruitment into the stage in question with the rate of recruitment to the next stage. The

calculations are most easily performed on a digital computer: the basic equation is:

$$\phi_{jj+1} = \frac{C_{j+1}(t+n) - C_{j+1}(t-n)}{D_j(t+n) - D_j(t-n)} \qquad (10.7)$$

where $\phi_{jj+1} =$ the survival between the jth stage and the next stage, $C_{j+1} =$ the total number of individuals that have entered stage $j+1$ in the periods $(t+n)$ and $(t-n)$, likewise $D_j =$ the total number of individuals that have left stage j, $n =$ an arbitary time period: Ruesink (1975) points out that if there are large sampling errors, n must be large to provide meaningful estimates. The values of C_{j+1} and D_j are obtained either graphically or by computer. In the former method a continuous line is drawn through the points from field data for the numbers of that stage: this is in essence the approach of Richards & Waloff's First Method.

10.2.6. Dempster's method

This method requires an independent estimate of the total natality of the first stage, together with a series of population estimates, at least two more than the numbers of stages (Dempster, 1961). If the mortality rate (μ) is assumed to be a constant, then:

$$Y_0 - Y_t = N\alpha_{(0 \to t)} - \frac{(I_0 + I_t)}{2}t\mu_1 - \frac{(II_0 + II_t)}{2}t\mu_2 \ldots - \frac{(Ad_0 + Ad_t)}{2}t\mu_a$$

$$(10.8)$$

where Y_0 and $Y_t =$ total populations at days 0 and t (successive sampling dates), $N =$ the total number hatching (or emerging) which is found independently; $\alpha_{(0 \to t)} =$ the proportion of the total hatch that occurs between days 0 and t; I_0 and $I_t =$ the total numbers of first instar larvae on the first and second sampling dates respectively, II_0 and II_t are the numbers of the second instar larvae; further terms are inserted, as appropriate, until the adult (Ad_0 and Ad_t) stage is reached; t is the time internal between day 0 and day t (the two sampling occasions and $\mu_1, \mu_2 \ldots \mu_a$ are the average daily mortality of first and second stage larvae and adults.

The values $\mu_1 \ldots \mu_a$ are unknown, but may be found from a series of simultaneous equations, provided there is one more equation than there are unknowns. These equations may be regarded as a series of multiple regressions and solved by computer.

Where there is migration a further term needs to be inserted into the equation; let $M =$ the net result of migration in terms of increase or decrease of population. A measure of the amount of migration may be obtained from one or more suction, rotary or interception traps placed close to the population and from this data the statistic x calculated:

$$x = \frac{\text{no. trapped between days 0 and } t}{\text{total number trapped}}$$

x is thus the migratory analogue of α. The following term is then inserted in the equation: $Mx_{(0 \to t)}$ the value of M and its sign being unknown; if M is positive migration has resulted in a gain (immigration), if negative in a loss (emigration). Estimates obtained in this way should, where possible, be compared with others obtained from directional interception traps and other approaches (pp. 244, 342).

This is a method capable of wide application; however, in order to obtain reliable estimates, a large number of accurate population samples are required, that is, it is not a 'robust' method (Dempster, pers. comm.; Manly, 1974c).

10.2.7 Richards & Waloff's Second Method

This method may be used when recruitment and mortality overlap widely. The total number (N_i) of the ith stage taken in all the daily samples will be given by:

$$N_i = N_0 \int_0^a \phi^t \, dt = \frac{N_0(\phi^a - 1)}{\log_e \phi} \tag{10.9}$$

where N_0 = the total number entering the stage, ϕ = the survival rate (for a unit of time) and a = the duration of the stage (Richards, 1959; Richards, Waloff & Spradbery, 1960; Richards & Waloff, 1961). Now if the number of eggs laid (N_0) and duration of the egg stage (a) are known and N_i obtained from samples, then the equation can be solved for ϕ. The percentage mortality is $100(1 - \phi^a)$, and this may be used to calculate the number surviving the egg stage and entering the first larval instar. Thus N_0, a and N_i will again be known values in the equation for the first instar so that the percentage mortality for this instar and the numbers entering the next may be found. The process can be repeated to the end of the life-cycle.

In order to apply this method one must have an accurate estimate for the initial N_0; Richards & Waloff (1961) obtained this from the population of adult females assessed independently and the oviposition rate measured under the field conditions. It is also necessary to determine the duration of the stage (a) experimentally and, as Richards *et al.* (1960) pointed out, differences, of as little as half a day, in the estimate of the duration of an instar can make a large difference in the estimated mortality. It is clear, therefore, that inaccuracies in these independent estimates of a and N_0, or discrepancies between the actual and assumed field conditions, which will influence the values assigned to a and N_0, are potential sources of error with this method.

10.2.8 Kiritani, Nakasuji & Manly's method

As originally developed (Kiritani & Nakasuji, 1967), this method required that samples be taken at regular intervals throughout the generation and, unless the number entering stage 1 are known independently, only survival rates can be estimated. Manly (1976) has modified the method to remove these restrictions. Samples are taken at irregular intervals (h_1, h_2 . . .), the area under the frequency trend curve is estimated by the trapezoidal rule:

$$\hat{A}_i = \frac{1}{2} \sum_{l=1}^{n} (h_l + h_{l+1}) \hat{f}_{il} \tag{10.10}$$

where $\hat{f}_{il}$ = the number of the ith instar estimated from the samples taken on the lth occasion, which is at the end of the sampling intervals h_l, there are $n + 1$ sampling intervals, the last interval extending from the last occasion when the stage was present to the next sampling occasion (when it was found to be absent). It will be noted that these f_i's correspond to $I, II \dots Ad$ of Dempster's method but whereas in his method two numbers are averaged over one sampling interval, in this method the numbers are 'spread' into the sampling intervals on either side and then the sum divided by two.

Then the survival rate of the ith stage is estimated:

$$\hat{\phi}_i = |-A_i| \sum_{j=1}^{q} A_j \tag{10.11}$$

where j = the next stage and q = the last stage.

Manly (1976) shows how further estimates can be obtained if the median stage (instar value, which will not, of course, normally be a whole number) is plotted against time so that the time for a 'median' animal to pass from stage to another can be read off. Now the daily survival rate raised to the power of the duration of the stage should equal the stage survival rate as determined above:

$$\hat{\phi}_i = \hat{\phi}_d{}^{\hat{a}_i} \tag{10.12}$$

where ϕ_d = daily survival rate and a_i = the duration of the ith stage as determined above. (This calculation can be done for two or more instars at once, which may give a better estimate of duration). Thus:

$$\log \hat{\phi}_d = \log \hat{\phi}_i / \hat{a}_i$$

If on the other hand time is taken over several instars (i.e. the divisor is, say, $a_2 + a_3 + a_4$), then having found $\hat{\phi}_i$ and $\hat{\phi}_d$, the appropriate stage durations (a_i's) can be found. Then the numbers entering each stage may be found:

$$\hat{N}_{0i} = -\log_e \phi_d \left[\sum_{j=i}^{q} A_j \right] \tag{10.13}$$

where N_{0i} = the number entering the ith stage and the others as above.

It will be apparent that this method depends on the assumption of a constant daily survival rate for all stages.

10.3 The description of budgets and life-tables

10.3.1 Survivorship curves
The simplest description of a budget is the graphical representation of the fall-

off of numbers with time – the survivorship curve. The number living at a given age (l_x) are plotted against the age (x); the shape of curve will describe the distribution of mortality with age. Slobodkin (1962) shows four basic types of curve (Fig. 10.3); in type I mortality acts most heavily on the old individuals, in type II (a straight line when the l_x scale is arithmetic) a constant number die per unit of time; in type III (a straight line when the l_x scale is logarithmic) the mortality rate is constant and in type IV mortality acts most heavily on the young stages. Deevey (1947) also drew attention to these different types, but only recognized three and, in order to avoid confusion between his classification and Slobodkin's, it should be noted that Deevey plotted survivors (l_x) on a log scale and thus type II of Slobodkin was not recognized by him; type II of Deevey = type III of Slobodkin and type III of Deevey = type IV of Slobodkin.

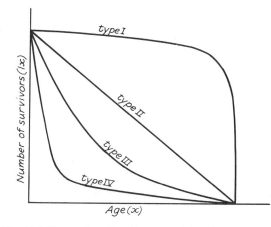

Fig. 10.3 Types of survivorship curves (after Slobodkin, 1962).

In insects mortality often occurs in distinct stages so that the survivorship curves show a number of distinct steps (Itô, 1961).

10.3.2 The life-table and life expectancy

For the further description of the data collected in the form of a budget or any type of life-table (Chapter 11) it is most convenient if these are corrected so as to commence with a fixed number, usually 1000.

A table may then be constructed with the following columns (Deevey, 1947):

x the pivotal age for the age class in units of time (days, weeks, etc.)
l_x the number surviving at the beginning of age class x (out of a thousand originally born)
d_x the number dying during the age interval x
e_x the expectation of life remaining for individuals of age x

In practice the table may have two further columns (Table 10.2) to facilitate the calculation of the expectation of life as follows:

Table 10.2 A life-table (hypothetical)

x	l_x	d_x	L_x	T_x	e_x	$1000q_x$
1	1000	300	850	2180	2.18	300
2	700	200	600	1330	1.90	286
3	500	200	400	730	1.46	400
4	300	200	200	330	1.10	667
5	100	50	75	130	1.30	500
6	50	30	35	55	1.10	600
7	20	10	15	20	1.00	500
8	10	10	5	5	0.50	1000

(*i*) The number of animals alive between age x and $x + 1$ is found. Precisely this is:

$$L_x = \int_x^{x+1} l_x d_x \tag{10.14}$$

but if the age intervals are reasonably small it may be found from:

$$L_x = \frac{l_x + l_{x+1}}{2} \tag{10.15}$$

(*ii*) The total number of animal x age units beyond the age x, which is given by:

$$T_x = L_x + L_{x+1} + L_{x+2} \ldots L_w \tag{10.16}$$

where w = the last age. In practice it is found by summing the L_x column from the bottom upwards.

(*iii*) The expectation of life which is theoretically:

$$e_x = \frac{\int_x^w l_x d_x}{l_x} \tag{10.17}$$

and is therefore given by:

$$e_x = \frac{T_x}{l_x} \tag{10.18}$$

When the survivorship curve is of type I (Fig. 10.3) e_x will decrease with age; it will be constant for type II, and it will vary for types III and IV

A further column is sometimes added to life-tables: the mortality rate per age interval (q_x), usually expressed as the rate per thousand alive at the start of that interval:

$$1000 q_x = 1000 \frac{d_x}{l_x} \qquad (10.19)$$

10.3.3 Life and fertility tables and the net reproductive rate

In the two sections above we have been concerned solely with the description of one of the pathways of population change – mortality. In this section and the next methods of describing natality and its interaction with mortality in the population will be discussed.

A life and fertility table (or fecundity schedule) may be constructed by preparing a life-table with x and l_x columns as before, except that the l_x column refers entirely to females and should represent the number of females alive, during a given age interval, as a fraction of an initial population of one. Or, expressed another way, the life expectancy at birth to age x as a fraction of one (Birch, 1948) (Table 10.3).

Table 10.3 Life and fertility table for the beetle, *Phyllopertha horticola* (modified from Laughlin, 1965)

x (in weeks)	l_x	m_x	$l_x m_x$ ($= V_x$)
0	1.00	—	
49	0.46	—	Immature stages
50	0.45	—	
51	0.42	1.0	0.42
52	0.31	6.9	2.13
53	0.05	7.5	0.38
54	0.01	0.9	0.01

A new column is then added on the basis of observations, this is the m_x or age-specific fertility* column that records the number of living females born per female in each age interval. In practice it is frequently necessary to assume a 50:50 sex ratio whem $m_x = N_x/2$, N_x being the total natality per female of age x.

Columns l_x and m_x are then multiplied together to give the total number of female births (female eggs laid) in each age interval (the pivotal age being x); this is $l_x m_x$ column.

The number of times a population will multiply per generation is described

---

* This is often termed the fecundity column, but as it refers to live births, 'fertility' is a more appropriate term.

by the net reproductive rate R_0, which is:

$$R_0 = \int_0^\infty l_x m_x d_x$$

$$R_0 = \Sigma l_x m_x \tag{10.20}$$

Thus from table 10.3, $R_0 = 2.94$; this net reproductive rate may be expressed in another way as the ratio of individuals in a population at the start of one generation to the numbers at the beginning of the previous generation. Thus,

$$R_0 = \frac{N_{t+\tau}}{N_t} \tag{10.21}$$

where $\tau = $ generation time.

Clearly, values of R_0 in excess of one imply an increasing population, of less than one a decreasing population; when $R_0 = 1$ the population will be stationary.

Where the generation limits are obscrued the value of R_0 as a description is limited and this led A. J. Lotka, a student of human demography, to propose the consideration of the growth rates of populations.

10.3.4 Population growth rates

As A. J. Lotka pointed out, the growth rate of a population is r in the equation:

$$\frac{dN}{dt} = rN \tag{10.22}$$

where N is the number of individuals at any given time (t); which may be expressed as:

$$N_t = N_0 e^{rt} \tag{10.23}$$

where $e = $ the base of natural logarithms. The parameter r in this equation describes population growth. Under conditions of an unlimited environment and with a stable age distribution this parameter becomes a constant.

The maximum value of the parameter r that is possible for the species under the given physical and biotic environment is denoted as r_m and is variously termed the intrinsic rate of natural increase, the Malthusian parameter, the innate capacity for increase or various combinations of these (Leslie & Ranson, 1940; Birch, 1948; Caughley & Birch, 1971), whilst $e^{r_m} = \lambda$ is the finite capacity for increase. It must be stressed that this value is the maximum under various *natural* conditions, that is it allows for some mortality; the theoretical potential r_p with maximal fertility and zero mortality will be higher (Southwood, 1969). Under optimal physical conditions r_m will approach r_p; at the limits of a species environmental range r_m will cease to be positive and become zero; beyond the boundary of the range r_m will be negative. If the population is constrained by its environment (perhaps through intraspecific effects), another parameter r_s, may be determined which is the rate at which the population would change (it may be positive or negative) if the age distribution was stable (Caughley & Birch, 1971). As environmental constraints (due to population density) approach zero, so r_s comes to equal r_m. The parameter r_s is of particular interest to students of vertebrate populations.

The intrinsic rate of natural increase (r_m) is of value as a means of describing the growth potential of a population under given climatic and food conditions (Messenger, 1964; Watson, 1964). It is an important parameter in inductive strategic and management models for insect pest populations (Chapters 11 and 12), and in fisheries work (Jensen, 1975).

When a stable age distribution has been achieved, but the population is still growing in an unlimited environment, and given $l_x m_x$ values from a life table, r_m may be calculated (see below) from the expression:

$$\sum_x e^{-r_m x} l_x m_x = 1 \tag{10.24}$$

However, ecologists are interested in r_m or the 'time r' in a broader context, its relationship to life history parameters such as generation size and fecundity (Smith, 1954; Fenchel, 1974; Southwood *et al.*, 1974), whilst the establishment of a stable age distribution in an unlimited environment may be virtually impossible. Sometimes the instantaneous birth rate per individual ($\sim r_p$) has been used instead of the instantaneous rate of natural increase per individual, thereby eliminating the mortality pathway.

An approximate r_m, the capacity for increase (r_c) has long been used by insect ecologists, and Laughlin (1965) presented a graph that showed its error over r_m in relation to the relative length of the reproductive period and the net reproductive rate. It is defined:

$$r_c = \log_e R_0 / T_c \tag{10.25}$$

where R_0 is as defined in equation 10.19 and 10.20 and T_c = cohort generation time, the mean age of the females in the cohort at the birth of female offspring or the pivotal age where $l_x m_x = 0.5 R_0$ (Bengstron, 1969).

The simple and general relationship between r_m and r_c (and between τ and T_c) has been elucidated by May (1976).

$$r_c = r_m \left(1 - \frac{r\sigma^2}{2T_c} + \ldots \right) \tag{10.26}$$

where σ^2 = the variance of the $l_x m_x$ distribution defined as:

$$\sigma^2 = (\Sigma x^2 \, l_x m_x / R_0) - T_c^2 \tag{10.27}$$

The further correction terms, omitted from equation 10.26, involve higher order moments (skewness, etc.): if we can assume that $l_x m_x$ is normal (Gaussian) then the equation is exact. May (1976) rearranged the equation for

r_c to give an expression of the relative error involved in calculating r_c, rather than r_m; it is:

$$r_m - r_c/r_c \sim \frac{1}{2}(\log_e R_0)(CV)^2 \tag{10.28}$$

where $CV = \sigma/T_c$, the variation in the distribution of age of reproduction within a cohort. A consideration of equation 10.28 immediately shows that, provided the $l_x m_x$ distribution is approximately normal, the relative error of r_c compared with r_m will be small, provided:

(1) $R_0 \sim 1$, that is population essentially just replaces itself e.g. most human populations, other vertebrates with low fecundities (e.g. albatross, whales).

(2) CV is small, that is reproduction tends to occur in all individuals at about the same age and is not spread out over a long reproductive life. This is true of many arthropods, but not of long-lived, virtually iteroparous species such as the tropical rain forest butterflies, *Heliconius*.

$CV (= \sigma/T_c)$ is clearly related to Laughlin's (1965) comparison of reproductive period to total period (to oldest reproductive age).

Another parameter that is involved in expression of life table data is generation time (τ) which is defined:

$$\tau = \log_e R_0/r_m \tag{10.29}$$

and has been considered to have little biological significance; however, May (1976) shows that:

$$\tau = T_c\left[1 - \frac{r_m \sigma^2}{2T_c} + \ldots\right] \tag{10.30}$$

Thus the difference is significant only if the $l_x m_x$ distribution has a high variance, particularly if this implies (as it generally will) a significant departure from the normal (i.e. additional terms are required in the above). Lefkovitch (1963) suggests that variation in the stored products beetle, *Lasioderma serricome*, is not normally distributed; Messenger (1964) also obtained various anomalous results, possibly for the same reason. The subject would repay further investigation in the light of May's (1976) paper, but such errors are not among the most substantial facing ecologists!

10.3.5 The calculation of r

Where a life-table can be constructed for a stable age distribution in an unlimited environment then r_m may be extracted from equation 10.24, i.e.:

$$\sum_x e^{-r_m x} l_x m_x = 1$$

The advent of computers has made the extraction of r_m from equation 10.24 a matter of routine; even the use of an electronic calculator (with appropriate exponent keys) is sufficient to avoid the labour of the elaborate graphical and iterative techniques described by Birch (1948) and Watson (1964). The collection of the required $l_x m_x$ data is, however, time consuming (e.g. Hardman, 1976; Evans, 1977).

For many insect species, where the conditions listed above are met, r_c is adequate and may be calculated from equation 10.25.

Several species of aphid and spider mites have been found by Wyatt & White (1977) to have steep cumulative $l_x m_x$ curves early in adult life that flatten out at a time period that happens to equal the period of pre-reproductive life, i.e. the development time (ξ) from birth to reproduction (Fig. 10.4). The value on the $\Sigma l_x m_x$ curve at a particular time represents the total births until that time per number of original females. Wyatt & White (1977) point out that it is much easier to determine (1) the pre-reproductive development period (ξ) and (2) the total reproduction/original female (R_ξ) in the first part of the reproductive

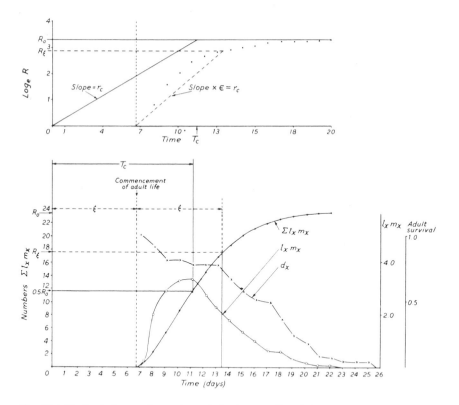

Fig. 10.4 The intrinsic rate of natural population increase. Population growth data for the mite, *Tetranychus urticae* (from Bengstron, 1969) displayed to illustrate the values used to make approximate estimates following the methods of Laughlin (1965) and Bengstron (1969) (r_c) and Wyatt and White (1977)(r_c/ϵ) (see text for further explanation).

period of length $= \xi$, than the $l_x m_x$ values. The slope given by these values will be slightly steeper than r_c (Fig. 10.4) so that

$$r = \varepsilon(\log_e R_\xi)/\xi \qquad (10.31)$$

where ε is a constant that corrects the slope of this line to correspond with r_c; Wyatt & White (1977) found $\varepsilon = 0.74$ to hold for many aphids and mites, but when a whole series of determinations are to be made under different climatic or biotic conditions the value of the constant should be confirmed, because it is very dependent on the form of the $\Sigma l_x m_x$ curve. This method additionally assumes, of course, that CV, as defined for equation 10.28, is small.

10.4 The analysis of life-table data

In the previous section we were concerned with the description of a life-table by a single parameter, r_m; the present section will discuss the methods of 'taking the life-table apart' so as to determine the role of each factor. Several life-tables are necessary; ideally these will be a series for a number of generations of the same population, but some information can be obtained from life-tables from different populations or by measuring density, mortality and the associated factors in different parts of the habitat. Such analysis is not only of considerable theoretical interest, but will eventually provide a rational and predictive basis for pest control: enabling both the forecasting of climatically induced outbreaks, and also the making of prognoses of the effects of changes in cultural or control practices.

10.4.1 The comparison of mortality factors within a generation [Table 10.4]

APPARENT MORTALITY
This is the measured mortality, the numbers dying as a percentage of the numbers entering that stage, i.e. d_x as a % of l_x (see also p. 358). Its main value is for simultaneous comparison either with independent factors or with the same factor in different parts of the habitat.

REAL MORTALITY
This is calculated on the basis of the population density at the beginning of the generation, i.e. $100 \times d_i/l_c =$ the deaths in the ith age interval and l_c the size of the cohort at the commencement of the generation. The real mortality row in Table 10.4 is the only % row that is additive and is useful for comparing the role of population factors within the same generation.

INDISPENSABLE (OR IRREPLACEABLE) MORTALITY
This is that part of the generation mortality that would not occur, should the mortality factor in question be removed from the life system, after allowance is made for the action of subsequent mortality factors. It is often assumed that

Table 10.4 Various measures for the comparison of mortality factors

Measure	Eggs		Larvae		Pupae		Adults
l_x	1000		500		300		30
d_x		500		200		270	
% apparent mortality	50		40		90		
% real mortality	50		20		27		
% indispensible mortality	3		2		27		
Mortality/ survivor ratio	1.00		0.66		9.00		
log population	3.00		2.70		2.48		1.48
k-values		0.30		0.22		1.00	

these will still destroy the same percentage independent of the change in prey density; clearly this assumption will not always be justified. To take an example of the calculation of the indispensible mortality from the table, consider the egg stage mortality: if there is no egg mortality 1000 individuals enter larval stage where a 40 % mortality leaves 600 survivors to pupate; in the pupal stage a 90 % mortality leaves 60 survivors, that is 30 more than when egg mortality occurs, and thus its indispensible mortality

$$= 30/1000 \times 100 = 3\%.$$

When it is known that the subsequent mortalities are unrelated to density the indispensable component of a factor may be used for assessing its value in control programmes (e.g. Huffaker & Kennet, 1965). If the exact density relationship of subsequent factors is known then a corrected, and more realistic, indispensable mortality can be calculated.

MORTALITY–SURVIVOR RATIO
Introduced by Bess (1945) this measure represents the increase in population that would have occurred if the factor in question had been absent. If the final population is multiplied by this ratio then the resulting value represents, in individuals, the indispensable mortality due to that factor.

10.4.2 The simple statistical relationship of population size to a factor

If the size of the population of a certain stage is measured over a number of generations, straightforward statistical correlation or regression analysis methods may be used to test the relationship of the variations in population with some other factor, e.g. climate, the numbers of natural enemies, the quantity of food available. Williams (1961) used this method extensively to demonstrate relationships between light-trap catch and immediate and previous climate; he was able to demonstrate that the relationships held, over many years and in two different areas, and that it would be possible to forecast the size of the catch if the weather conditions could be forecast. It is of course an elementary statistical principle that correlation is not causation. Where there is extensive data, as in Williams's work, it seems likely that there is a causal link, but the mechanism through which previous weather influences the catch remains obscure (Varley, 1963). The strongest evidence for the validity of a relationship established in this way will come from direct experiments; often, however, the ecologist has to be content with an apparently sound biological link that holds under a variety of conditions, which are such that possible 'common causes' have varied independently and so been eliminated.

10.4.3 Survival and life budget analysis
Very significant advances in the analysis of population census data were made by Watt (1961, 1963, 1964), R. F. Morris (1959, 1963a & b) and their co-workers. Watt developed the mixed deductive-inductive model that blended field data and mathematical analysis in a way that allowed 'feed-back' and correction in the model's development (Conway, 1973). Morris introduced the concept of key-factor analysis that aimed to determine the factor (or factors) that were of the greatest predictive value in forecasting future population trends; he also sought to estimate the magnitude of density dependence in the system. Subsequent studies (Hassell & Huffaker, 1969; Southwood, 1967; Maelzer, 1970; St. Aman, 1970; Luck, 1971; Brockelman & Fagen, 1972; Benson, 1973) have shown that many difficulties arise in the interpretation of results from the Morris method; the method of Varley & Gradwell (1960), the development of which was stimulated by Morris's original paper, is to be preferred.

VARLEY & GRADWELL'S METHOD
This method developed by Varley & Gradwell (1960, 1963a and b, 1965; Varley, 1963; Varley, Gradwell & Hassell, 1973) demands the data from a series of successive life-tables. The whole generation is considered; thus it is immediately apparent in which age interval the density dependent and key factors lie, and it provides a direct method of testing the role of changes in natality from generation to generation.

Varley & Gradwell's method may be outlined as follows:

(1) The maximum potential natality is found by multiplying the number of females of reproductive age by the maximum mean fecundity/female and this figure is entered in the budget (Table 10.5).

(2) The values in the budget are converted to logarithms.

(3) A convenient generation basis in this method is adult to adult; the base chosen will affect the recognition of key factors (Varley & Gradwell, 1965). The total generation 'mortality' is given by subtracting the log of the

Table 10.5 A budget prepared for Varley & Gradwell's analysis

	Nos/10 m² (observed, unless marked*)	Log nos/10 m²	k's
Maximum potential natality	× 30 × 100	3.176	
(no. reprod. ♀ × maximum natality)	= 1500		
k^0 (variation in natality)			0.076
Eggs laid	1260	3.100	
k^1 (egg loss)			0.171
Eggs hatching	850	2.929	
k^2 (predation, etc.)			0.498
3rd stage larvae	270	2.431	
k^3 (apparent parasitism of larvae)	100		0.201
[Larvae surviving parasitism]	170*	2.230	
k^4 (predation and other larval mortality)			0.276
Pupae	90	1.954	
k^5 (overwintering loss)			0.301
Adults emerging	45	1.653	
k^6 (dispersal, etc.)			0.352
Adults reproducing	20	1.301	
		K	= 1.875

population of adults entering the reproductive stage from the log maximum potential natality of the previous generation – this value is referred to as K^* (Table 10.5).

(4) The series of age-specific mortalities are calculated by subtracting each log population from the previous one (Table 10.5); these are referred to as k's, so that:

$$K = k_0 + k_1 + k_2 + \ldots k_i \qquad (10.32)$$

Where precise estimates of mortality and migration (see Chapter 9) are available these are also incorporated into the equation. These series of k's – one series for each generation – provide a complete picture of population change. In the subsequent steps of this analysis the role of each k factor is examined separately, but it must be remembered that sampling errors are 'hidden' in each k and may be responsible for spurious results (Kuno, 1971). The identification of the k value with a specific factor presents no problem when it is based on an apparent mortality, but when it is the difference of successive estimates it should, strictly, be referred to as 'overwintering loss' or 'loss of young adults'. It may be possible from other knowledge to indicate the major components of these losses; such assumptions could be checked by

* This K should not be confused with the more widespread use of the symbol for the carrying capacity of the habitat as defined by the logistic equation.

testing the correlation between the k value and an independent measure of the loss factor, e.g. the abundance of a predator, over a number of generations. k_0, the difference between the log maximum potential natality and log actual natality, has a special significance, for it does not represent mortality in the strict sense, but the variation in natality. This is compounded of two separate causes: death of the reproducing females before the end of the reproductive life and variations in the fertility of the females. The two could be separated by determining the form of survival curve of the reproducing female (Chapter 11).

(5) The next step involves the recognition of the key factor for the index of population trend from adult to adult. An assessment may be made by visual correlation, K and k_0 to k_i are plotted against generation and it may easily be seen which k is most closely correlated with K (Fig. 10.5). A quantitative evaluation of the roles of each k is provided by Podoler & Rogers' (1975)

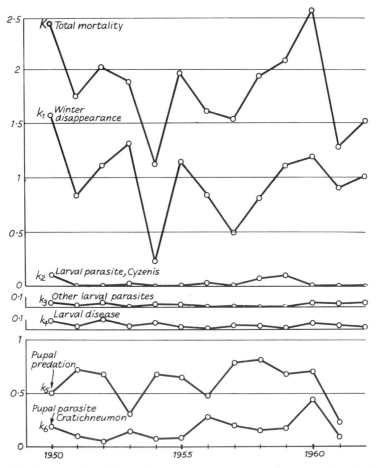

Fig. 10.5 The recognition of the key factor by the visual correlation of various k's with K (after Varley & Gradwell, 1960).

method in which the individual k-values are plotted (on the y-axis) against total mortality (K on the x axis) and the regression calculated. The relative importance of each factor will be proportional to its regression coefficient. Because sampling errors incorporated in k will also appear in K this is not a precise statistical test for the importance of each mortality, but for the role of the estimates of each mortality in contributing to changes in the value of K.

(6) The various k's are then tested for direct density-dependence. Each k is plotted against the number (N_t) entering the stage (age interval) on which it acts (Fig. 10.6) and if the regression is significant then density-dependence may

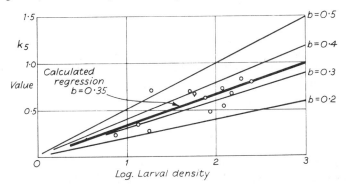

Fig. 10.6 The recognition of density dependence of a k factor by plotting the value of this factor against the log density at the start of the stage on which it acts (after Varley & Gradwell, 1963a).

be suspected. But the principal difficulty now arises: the variables are not independent (Varley & Gradwell, 1960; Bulmer, 1975), the regression is:

$$k = \log \alpha + b \log N_t \qquad (10.33)$$

where b = the slope of the density-dependence (see also May *et al.*, 1974) and α a constant, but

$$k = \log N_t - \log N_{t+1}$$

Thus the regression could be spurious, due to sampling erros. The second step therefore is to plot the log numbers entering the stage ($\log N_t$) against the log number of survivors ($\log N_{t+1}$). The regressions of $\log N_t$ on $\log N_{t+1}$ and of $\log N_{t+1}$ on $\log N_t$ should be calculated, and if both the regression coefficients depart significantly from 1.0, then the density-dependence may be taken as real.* Bulmer (1975) has devised a more precise test based on the ratio:

$$R = V/U \qquad (10.34)$$

where
$$V = \sum_{t=1}^{n} (\log N_t - \log \overline{N})^2 \qquad (10.35)$$

* For an alternative regression technique in which errors are equally apportioned between each axis see Bartlett, M. S. (1949), *Biometrics* **5**, 207–12 (expounded in Sokal, R. R. & Rohlf, F. J. (1969) *Biometry*, pp. 483–6).

and
$$U = \sum_{t=1}^{n-1} (\log N_{t+1} - \log N_t)^2 = \sum_{t=1}^{n-1} (k)^2 \qquad (10.36)$$

where $\overline{N}$ = the mean population level at time t over the years studied; R is the reciprocal of von Neumann's ratio which is known to have a certain distribution depending on the value of N; Bulmer (1975) shows by simulation that a value of R smaller than the lower limit of R (due to random variation as predicted from tables of von Neumann's ratio) is an indication of density-dependence. However, as he points out, this test will eliminate any trend (i.e. delayed density-dependence) and it is not strictly valid if the estimates of N_t and N_{t+1} contain sampling errors (as they normally do). Another ratio, that provides a less powerful test, may be used to allow for these. However, Bulmer concludes 'a series of 62 observations is only likely to reveal the existence of rather strong density-dependence and its existence may be masked by temporal trends (i.e. delayed effects) in the data'; so the practical utility of this test is unfortunately minimal (unless the organisms have very short generation times so that a series of about 100 generations may be accumulated). Likewise, Southwood & Reader (1976) from a relatively long (12-year) series of insect census data could not show density-dependence by the methods given here. At present one must conclude that the demonstration of density-dependence from census data is fraught with difficulties: in particular failure to detect it in no way proves its absence. Precise studies on individual cohorts (p. 5) and experimental work with particular components of the population system (pp. 309–37) would seem to be profitable methods of investigating its role.

(7) If density dependence is believed to be real, attention may now be refocused on the plot of k_x against the numbers entering the stage (Fig. 10.6). The slope of the line, the regression coefficient, should be determined as this will give a measure of how the factor will act; the closer the regression coefficient is to 1.0, the greater the stabilizing effect of that regulatory factor. If the coefficient is exactly 1.0 the factor will compensate completely for any changes in density; if the coefficient is less than 1.0 the factor will be unable to compensate completely for the changes in density caused by other disturbing factors; whilst a coefficient of more than 1.0 implies overcompensation.

Using equation 10.33, Stubbs (1977) was able to detect density dependence in forty-six sets of life-budget data and draw significant conclusions. Hassell (1975) developed a more general model for density-dependence., for single species populations:

$$k = b \log(1 + aN_t) \qquad (10.37)$$

where k is as defined in 10.33, b is the measure of density dependence and:

$$a = 1/N_c \qquad (10.38)$$

where N_c = a critical density for the organism at which scramble competition starts to operate. Hassell (1975) was able to estimate the level of density-dependence in populations of several species by fitting his model to published population budgets.

(8) The detection of temporal trends (*sensu* Bulmer, 1975) or of delay in density dependence may be undertaken by plotting the k value against log initial density (as in Fig. 10.6) and then joining the points up in a time sequence plot (Varley, 1947, 1953; Morris, 1959; Varley & Gradwell, 1965). The different types of factor will trace different patterns (Fig. 10.7): direct density-dependent factors will trace a more or less straight line or narrow band of points, delayed

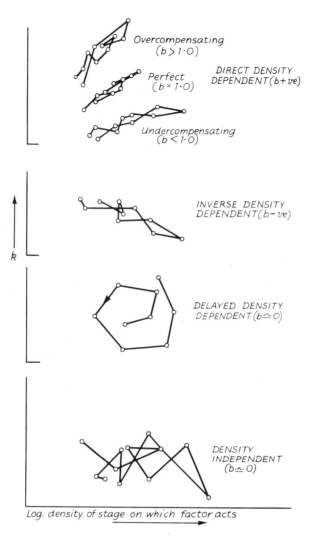

Fig. 10.7 Time sequence plots showing how the different density relationships may be recognized from the patterns produced.

density factors circles or spirals; density-independent factors irregular or zigzag plots, whose amplitude reflects the extent to which they fluctuate. Solomon (1964) discusses how the order in which various factors act will influence the magnitude of their effect. This type of plot, the linking of consecutive points serially, is of particular value in investigating what Hassell (1966) has appropriately termed 'intergeneration relationships', and it is the only method by which a delayed relationship (a delayed density-dependent factor) can be recognised.

In conclusion, Varley & Gradwell's method allows:

(1) The recognition of the key factor (or factors) or the period in which it acts.

(2) The investigation of the role of natality in population dynamics.

(3) The consideration of the role of mortality factors at every stage of the life-cycle, the recognition of the different density relationships of these factors and an indication of their mode of operation: direct density-dependent factors tend to stabilize, delayed density-dependent factors lead to oscillations, density-independent factors lead to fluctuations and inverse density-dependent factors will tend to accentuate the fluctuations.

Many life tables have now been analysed by these methods and the results are beginning to allow some more general understanding of population dynamics (Podoler & Rogers, 1975; Southwood, 1975) and comparative evaluations of the role of density-dependent and density-independent factors have been made by Hassell (1975) and Stubbs (1977).

THE INVESTIGATION OF THE ROLES OF DIFFERENT FACTORS IN SINGLE-GENERATION BUDGETS

The above methods can be used with continuous populations for which a series of estimates have been made in successive generations, but with many pests of arable crops or extremely migratory insects continuous populations do not exist and therefore it is not possible to calculate a meaningful population trend from generation to generation. It is, however, possible to obtain a number of incomplete budgets from egg to late immature stage or to the emerging adult. Such data may be used to determine the relative importance of natality compared with mortality in determining the size of the final or economic population (Southwood 1967). In other words to answer the question of whether the size of the population of a caterpillar in a field is determined mostly by the number of eggs laid, or by the level of egg parasitism or by the amount of early larval mortality, the data may be examined initially by graphical methods (2 below), or statistical analysis alone may be used (3 below).

The procedure is as follows:

(1) The various population estimates are converted to logarithms: if N_E = number (absolute density of eggs laid and N_R = number of the resulting population (late larvae, pupae or emerging adults), then

$$\log N_E - \log N_R = \kappa \qquad (10.39)$$

κ being the total mortality over that part of the life-cycle studied. (The symbol κ (kappa) is used to distinguish this from K, which is the total mortality including variation in natality over the whole life-cycle.)

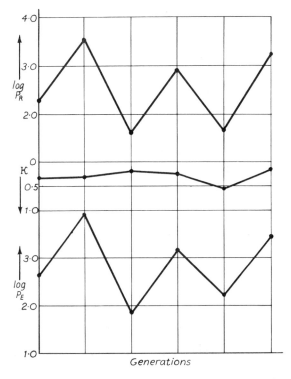

Fig. 10.8 The graphical investigation of the roles of natality (N_E) and mortality (κ) in determining the size of the resulting population (N_R). Data on the frit fly from Southwood & Jepson (1962) showing the dominant role of natality.

If apparent parasitism or other mortality factors are known, appropriate κ values are calculated as in Varley & Gradwell's method (p. 376). Then a residual mortality K_r is found:

$$K_r = \kappa - (k_1 + k_2 + \ldots k_n) \tag{10.40}$$

(2) The correlation of the different values may be examined by a graphical method (Fig. 10.8). N_E and N_R are plotted normally, but κ or the series of k's are plotted on an inverted scale so that high levels of mortality will be represented by dips that will correspond with low values in N_E. In some cases it will be found that fluctuations in N_R are obviously accounted for almost entirely by variations in natality (N_E). As N_E and N_R are statistically independent this conclusion need not be questioned. However, reliance cannot

be placed on the correlation of κ with N_R as these are not independent (see Watt, 1964).

(3) A more precise evaluation of the role of natality may be obtained by calculating the value of the coefficient of determination (r^2) for N_R plotted against N_E. This is a measure, as a percentage, of the amount of the variance of N_R accounted for by N_E, i.e. of the role of natality.

This method allows the relative importance of different pathways of population change to be assessed in crop insects and others where there is a 'gap' in the budget data for part or all of the adult stage and successive generations are not part of the same population. It differs from the other methods in that the comparative basis is not total mortality or the index of population trend, but the actual size of the population of a certain stage. It will probably be useful in ecological investigations, in connection with integrated pest control programmes, to determine the role in population fluctuation of predator and parasite induced mortality compared with that due to variations in the number of eggs laid, which stems from the numbers of females invading the field and their survival and fertility (Metcalfe, 1972; Southwood, 1975).

The mode of action of individual mortality factors may be investigated by comparing the intensity of their action in different parts of the habitat where the host's (or prey's) densities differ (see p. 312).

REFERENCES

ASHFORD, J. R., READ, K. L. Q. and VICKERS, G. G., 1970. A system of stochastic models applicable to studies of animal population dynamics. *J. Anim. Ecol.* **39**, 29–50.

BENGSTRON, M., 1969. Estimating provisional values for intrinsic rate of natural increase in population growth studies. *Aust. J. Sci.* **32**, 24.

BENSON, J. F., 1973. Some problems of testing for density-dependence in animal populations. *Oecologia* **13**, 183–90.

BESS, H. A., 1945. A measure of the influence of natural mortality factors on insect survival. *Ann. ent. Soc. Am.* **38**, 472–82.

BIRCH, L. C., 1948. The intrinsic rate of natural increase of an insect population. *J. Anim. Ecol.* **17**, 15–26.

BIRLEY, M. H., 1977. The estimation of insect density and instar survivorship functions from census data. *J. Anim. Ecol.* **46**, 497–510.

BROCKELMAN, W. Y. and FAGEN, R. M., 1972. On modelling density-independent population change. *Ecology* **53**, 944–8.

BULMER, M. G., 1975. The statistical analysis of density-dependence. *Biometrics* **31**, 901–11.

CAUGHLEY, G. and BIRCH, L. C., 1971. Rate of increase. *J. Wldl. Manag.* **35**, 658–63.

CONWAY, G. R., 1975. Experience in insect pest modelling: a review of models, uses and future directions. *Mem. Aust. Ecol. Soc.* **1**, 103–30.

DEEVEY, E. S., 1947. Life tables for natural populations of animals. *Quart. Rev. Biol.* **22**, 283–314.

DEMPSTER, J. P., 1956. The estimation of the numbers of individuals entering each stage during the development of one generation of an insect population. *J. Anim. Ecol.* **25**, 1–5.

DEMPSTER, J. P., 1961. The analysis of data obtained by regular sampling of an insect population. *J. Anim. Ecol.* **30**, 429–32.

EVANS, D. E., 1977. The capacity for increase at a low temperature of several Australian populations of *Sitophilus oryzae* (L.) *Aust. J. Ecol.* **2**, 55–79.

FENCHEL, T. 1974. Intrinsic rate of natural increase: the relationaship with body size. *Oecologia* **14**, 317–26.

GILBERT, N., GUTIERREZ, A. R., FRAZER, B. D. and JONES, R. E., 1976. *Ecological Relationships*. 157pp., Addison-Wesley, Reading & San Francisco.

HARDMAN, J. M., 1976. Life-table data for use in deterministic and stochastic simulation models predicting the growth of insect populations in Malthusian conditions. *Can. Ent.* **108**, 897–906.

HASSELL, M. P., 1966. Evaluation of parasite or predator responses. *J. Anim. Ecol.* **35**, 65–75.

HASSELL, M. P., 1975. Density-dependence in single-species populations. *J. Anim. Ecol.* **44**, 283–95.

HASSELL, M. P. and HUFFAKER, C. B., 1969. The appraisal of delayed and direct density-dependence. *Can. Ent.* **101**, 353–61.

HELGESON, R. G. and HAYNES, D. L., 1972. Population dynamics of the cereal leaf beetle, *Aulema melanopus* (Coleoptera: Chrysomelidae): model for age-specific mortality. *Can. Ent.* **104**, 797–814.

HUFFAKER, C. B. and KENNET, C. E., 1965. Ecological aspects of control of olive scale *Parlatonia oleae* (Colvee) by natural enemies in California. *Proc. XII. int. Congr. Ent.*, 585–6.

ITÔ, Y., 1961. Factors that affect the fluctuations of animal numbers, with special reference to insect outbreaks. *Bull. Nat. Inst. Agric. Sci. C* **13**, 57–89.

JENSEN, A. L., 1975. Comparison of logistic equations for population growth. *Biometrics* **31**, 853–862.

JUDGE, G. G. and TAKAYAMA, T., 1966. Inequality restrictions in regression analysis. *J. Am. Statist. Ass.* **61**, 166–81.

KIRITANI, K. and NAKASUJI, F., 1967. Estimation of the stage-specific survival rate in the insect population with overlapping stages. *Res. Popul. Ecol.* **9**, 143–52.

KUNO, E., 1971. Sampling error as a misleading artifact in key factor analysis. *Res. Popul. Ecol.* **13**, 28–45.

LAUGHLIN, R., 1965. Capacity for increase: a useful population statistic. *J. Anim. Ecol.* **34**, 77–91.

LEFKOVITCH, L. P., 1963. Census studies on unrestricted populations of *Lasioderma serricorne* (F.) (Coleoptera: Anobiidae). *J. Anim. Ecol.* **32**, 221–31.

LESLIE, P. H. and RANSON, R. M., 1940. The mortality, fertility and rate of natural increase of the vole (*Microtus agrestis*) as observed in the laboratory. *J. Anim. Ecol.* **9**, 27–52.

LUCK, R. F., 1971. An appraisal of two methods of analyzing insect life tables. *Can. Ent.* **103**, 1261–71.

MAELZER, D. A., 1970. The regression of log N_{n+1} on log N_n as a test of density dependence: an exercise with computer-constructed density-independent populations. *Ecology* **51** (5), 810–22.

MANLY, B. F. J., 1974a. Estimation of stage-specific survival rates and other parameters for insect populations developing through several stages. *Oecologia* **15**, 277–85.

MANLY, B. F. J., 1974b. A note on the Richards, Waloff & Spradbery method for estimating stage-specific mortality rates in insect populations. *Biom. Z.* **17**, 77–83.

MANLY, B. F. J., 1974c. A comparison of methods for the analysis of insect stage-frequency data. *Oecologia* **17**, 335–48.

MANLY, B. F. J., 1976. Extensions to Kiritani and Nakasuji's method for analysing insect stage-frequency data. *Res. Popul. Ecol.* **17**, 191–9.

MAY, R. M., 1976. Estimating *r*: a pedagogical note. *Am. Nat.* **110**, 496–9.

MAY, R. M., CONWAY, G. R., HASSELL, M. P., and SOUTHWOOD, T. R. E., 1974. Time delays, density-dependence and single-species oscillations. *J. Anim. Ecol.* **43**, 747–70.

MESSENGER, P. S., 1964. Use of life tables in a bioclimatic study of an experimental aphid–braconid wasp host–parasite system. *Ecology* **45**, 119–31.

METCALFE, J. R., 1972. An analysis of the population dynamics of the Jamaican sugar-cane pest, *Saccharosydne saccharivore* (Westw.) (Hom. Delphacidae). *Bull. Entomol. Res.* **62**, 73–85.

MORRIS, R. F., 1959. Single-factor analysis in population dynamics. *Ecology* **40**, 580–8.

MORRIS, R. F. (ed.), 1963*a*. The dynamics of epidemic spruce budworm populations. *Mem. ent. Soc. Can.* **31**, 1–332.

MORRIS, R. F., 1963*b*. Chapters 6, 7 and 18 in 'The dynamics of epidemic spruce budworm populations'. *Mem. ent. Soc. Can.* **31**, 30–7, 116–29.

PODOLER, H. and ROGERS, D., 1975. A new method for the identification of key factors from life-table data. *J. Anim. Ecol.* **44**, 85–114.

RICHARDS, O. W., 1940. The biology of the small white butterfly (*Pieris rapae*), with special reference to the factors controlling abundance. *J. Anim. Ecol.* **9**, 243–88.

RICHARDS, O. W., 1959. The study of natural populations of insects. *Proc. R. ent. Soc. Lond. C* **23**, 75–9.

RICHARDS, O. W., 1961. The theoretical and practical study of natural insect pop-ulations. *Ann. Rev. Ent.* **6**, 147–62.

RICHARDS, O. W. and WALOFF, N., 1954. Studies on the biology and population dynamics of British grasshoppers. *Anti-Locust Bull.* **17**, 182pp.

RICHARDS, O. W. and WALOFF, N., 1961. A study of a natural population of *Phytodecta olivacea* (Forster) (Coleoptera: Chrysomeloidea). *Phil. Trans. B* **244**, 205–57.

RICHARDS, O. W., WALOFF, N. and SPRADBERY, J. P., 1960. The measurement of mortality in an insect population in which recruitment and mortality widely overlap. *Oikos* **11**, 306–10.

RUESINK, W. G., 1975. Estimating time-varying survival of arthropod life stages from population density. *Ecology* **56**, 244–7.

ST. AMANT, J. L. S., 1970. The detection of regulation in animal populations. *Ecology* **51**, 823–8.

SLOBODKIN, L. B., 1962. *Growth and regulation of animal populations.* 184 pp. Holt, Rinehart and Winston, New York.

SMITH, F. E., 1954. Quantitative aspects of population growth. *In* Boell, E. (ed.), *Dynamic of growth processes.* Princeton University Press, Princeton, New Jersey.

SOLOMON, M. E., 1964. Analysis of processes involved in the natural control of insects. *Adv. Ecol. Res.* **2**, 1–58.

SOUTHWOOD, T. R. E., 1967. The interpretation of population change. *J. Anim. Ecol.* **36**, 519–29.

SOUTHWOOD, T. R. E., 1969. Population studies of insects attacking sugar cane. *In* Williams, J. R. *et al.* (eds). *Pests of Sugar Cane*, 427–59. Elsevier, Amsterdam.

SOUTHWOOD, T. R. E., 1975. The dynamics of insect populations. *In* Pimentel, D. (ed.), *Insects, Science and Society*, pp. 151–99. Academic Press, New York.

SOUTHWOOD, T. R. E. and JEPSON, W. F., 1962. Studies on the populations of *Oscinella frit* L. (Dipt.: Chloropidae) in the oat crop. *J. Anim. Ecol.* **31**, 481–95.

SOUTHWOOD, T. R. E., MAY, R. M., HASSELL, M. P. and CONWAY, G. R. 1974. Ecological strategies and population parameters. *Am. Nat.* **108**, 791–804.

SOUTHWOOD, T. R. E. and READER, P. M., 1976. Population census data and key factor analysis for the Viburnum whitefly, *Aleurotrachelus jelinekii* (Frauenf.) on three bushes. *J. Anim. Ecol.* **45**, 313–25.

STUBBS, M., 1977. Density-dependence in the life-cycles of animals and its importance in *K* and *r*-strategies. *J. Anim. Ecol.* **46**, 677–88.

VARLEY, G. C., 1947. The natural control of population balance in the knapweed gallfly (*Urophora jaceana*). *J. Anim. Ecol.* **16**, 139–87.

VARLEY, G. C., 1953. Ecological aspects of population regulation. *Trans. IX int. Congr. Ent.* **2**, 210–14.

VARLEY, G. C., 1963. The interpretation of change and stability in insect populations. *Proc. R. ent. Soc. Lond. C* **27**, 52–7.

VARLEY, G. C. and GRADWELL, G. R., 1960. Key factors in population studies. *J. Anim. Ecol.* **29**, 399–401.

VARLEY, G. C. and GRADWELL, G. R., 1963a. The interpretation of insect population changes. *Proc. Ceylon Assoc. Adv. Sci.* **18** (D), 142–56.

VARLEY, G. C. and GRADWELL, G. R., 1963b. Predatory insects as density dependent mortality factors. *Proc. XVI int. zoo. Congr.* **1**, 240.

VARLEY, G. C. and GRADWELL, G. R., 1965. Interpreting winter moth population changes. *Proc. XII int. Congr. Ent.* 377–78.

VARLEY, G. C., GRADWELL, G. R. and HASSELL, M. P. 1973. *Insect Population Ecology, an analytical approach.* 212pp, Blackwells, Oxford.

WALOFF, N. and BAKKER, K., 1963. The flight activity of Miridae (Heteroptera) living on broom, *Sarothamnus scoparius* (L.) Wimn. *J. Anim. Ecol.* **32**, 461–80

WATSON, T. F., 1964. Influence of host plant condition on population increase of *Tetranychus telarius* (Linnaeus) (Acarina: Tetranychidae). *Hilgardia* **35** (11), 273–322.

WATT, K. E. F., 1961. Mathematical models for use in insect pest control. *Can. Ent.* **93**, suppl. 19, 62 pp.

WATT, K. E. F., 1963. Mathematical population models for five agricultural crop pests. *Mem. ent. Soc. Can.* **32**, 83–91.

WATT, K. E. F., 1964. Density dependence in population fluctuations. *Can. Ent.* **96**, 1147–8.

WILLIAMS, C. B., 1961. Studies in the effect of weather conditions on the activity and abundance of insect populations. *Phil. Trans. B* **244**, 331–78.

WYATT, I. J. and WHITE, P. F., 1977. Simple estimation of intrinsic increase rates for aphids and tetranychid mites. *J. appl. Ecol.* **14**, 757–66.

11

Age-grouping of Insects, Time-specific Life-tables and Predictive Population Models

This chapter is concerned with techniques for animals whose generations overlap widely; age-grouping is a prerequisite for these methods which have been most widely applied with vertebrate populations. Analysis is easiest with two extreme types of population – the stationary and the expanding. If the population can be assumed to be stationary, then the fall-off in numbers in successive age groups will reflect the survivorship curve and thus a time-specific (or vertical) life-table can be constructed on this basis (p. 393). If the population is expanding, unconstrained by its environment, the age structure may become stabilized and then mortality can be estimated from the difference between expected and actual growth rates (p. 370). Age-grouping may also provide useful information on the fertility or potential fertility of the population (see p. 302).

11.1 Age-grouping of insects*

The methods of age-grouping (or age-grading) insects will be referred to only in outline, as the details will vary from species to species. The immature stages of most insects are easily distinguished, larval instars being recognized by the diameter of the head, the length of the appendages and other structural features, although even so basic a feature as the width of the larval head capsule may be misleading (Kishi, 1971). It is possible to age-group within the instars of some Heteropterous larvae as the number of eye facets increases for some time after moulting; this was noted in lace-bugs (Tingidae) (Southwood & Scudder, 1956) and in Corixidae, where it is even more marked; the interocular distance decreases throughout the instar (E. C. Young, unpub.). In the last larval instar of Exopterygota the wing pads frequently darken shortly before the final moult. Noctuid larvae may be sexed (Hinks & Byers, 1973).

Different ages of the pupal stage may often be determined by dissection, particularly in the Diptera, where various categories based on pigmentation can be recognized: e.g. unpigmented, eyes of pharate adult pigmented, head

---

* Strictly it is not correct to equate age with stage; the age (in time units) of a given stage will vary with environmental condition and between individuals

pigmented, pharate adult fully pigmented (Schneider & Vogel, 1950; van Emden *et al.*, 1961). Using the falling phenolic content, measured by the reduction of a dye, Chai & Dixon (1971) were able to assess the age of sawfly cocoons.

Most ageing methods have, however, been concerned with the adult insect; some of the types of criteria used are indicated below. Age-grouping of mosquitoes is reviewed by Hamon *et al.* (1961) and by Muirhead-Thomson (1963) and for arthropods of medical importance as a whole by Detinova (1962, 1968); Russian workers have contributed greatly to this subject. Most methods rely on criteria that are indications of some physiological process, particularly reproduction and excretion, or on general wear and tear; therefore the precise chronological equivalent of a given condition may vary from habitat to habitat or even between individuals.

11.1.1 Cuticular bands
The demonstration by Neville (1963) that there were apparently daily growth layers present in some areas of the cuticle of certain insects opened up the possibility of an ageing criterion for insects, as precise as that available from fish scales.

Further studies have shown that such cuticular banding is widespread, but care is needed in its interpretation. In some Heteroptera food availability affects band deposition (Neville, 1970), but not in the blowfly, *Lucilia* (Tyndale-Biscoe & Kitching, 1974). The banding was thought to depend on a rhythm that was merely influenced by temperature, although in Coleoptera the number of bands may vary in different parts of the same animal (Neville, 1970); however, Tyndale-Biscoe & Kitching (1974) showed that in *Lucilia* they were dependent on temperature changes; precisely a rise of $3.5°C$ or more over a threshold of $15.5°C$. With the normal daily temperature rhythm one band (a dark layer + a light one, seen most clearly under polarized light) is laid down each day: but this commences at apolysis, so the number of bands formed in the pharate adult must be determined before they can be used to estimate post-ecdysal adult age. In the Exopterygota, where there seems to be more rhythmicity in their deposition, bands may be seen by sectioning the hind fibia, and have been demonstrated in most orders (Neville, 1970). Similar rings have been noted in honey-bees and beetles. In the Diptera, Schlein & Gratz (1972, 1973) showed that the skeletal apodemes exhibited more bands than other regions of the cuticle: they describe a staining technique, but Tyndale-Biscoe & Kitching (1974) could count banding in *Lucilia* without polarized light or staining. Band formation appears to be limited to the early part of adult life, up to 10–13 days in *Anopheles* mosquitoes (Schlein & Gratz, 1973) and as many as 63 have been observed in a large grasshopper (Neville, 1970). Growth rings in the shells of land snails may perhaps be used for age-grouping (Pollard, 1973), but not in marine snails (Fotheringham, 1971).

11.1.2 Sclerotization and colour changes in the cuticle and wings

In all insects the newly emerged adult is pale and the cuticle untanned – in most species the major part of tanning is completed during a short teneral period.

However, complete sclerotization may not occur for a considerable time, especially if a period of diapause or aestivation intervenes; e.g. Lagace & van den Bosch (1964) found in the weevil, *Hypera*, that the elytra remained thin and almost teneral for much of the summer diapause. During teneral development in the Corixidae and Notonectidae the cuticle is pigmented at different rates; the sequence of pigmentation of the mesotergum is remarkably constant, proceeding forwards, in a number of stages, from the posterior region (Young, 1965). The wings of young dragonflies are milky, becoming clear when mature (Corbet, 1962a and b).

Various excretory pigments are accumulated throughout adult life and especially during diapause; some of these occur in the cuticle and may cause progressive colour changes. In other instances cuticular colour change is a product of the sclerotization process. Dunn (1951) showed that the region of the discal cell and the associated veins of the hind wing of the Colorado beetle (*Leptinotarsa*) change from yellowish, in the newly emerged individual, to reddish after hibernation, and the green shieldbug, *Palomena prasina*, becomes redder, often a dark bronze, during hibernation (Schiemenz, 1953); in contrast, the pink colouring of another pentatomid, *Piezodorus lituratus*, disappears while overwintering. Many Miridae change colour during or after hibernation and yellows on the forewings of various Heteroptera often darken with age becoming orange or even reddish (Southwood & Leston, 1959). The coloration of the bodies of adult dragonflies changes with age (Corbet, 1962a and b).

11.1.3 Developmental changes in the male genitalia

The males of many insects have a period of immaturity during which they may be recognized by the condition of the male genitalia (e.g. in *Dacus tryoni* as described by Drew, 1969). In most Nematocera the male hypogydium rotates soon after emergence; Rosay (1961) found in mosquitoes that the time for rotation was similar in the species of *Aedes* and *Culex* studied, but was strongly influenced by temperature (at $17°C - 58$ hours, at $28°C - 12$ hours). The aedaegus of the weevil, *Hypera*, remains relatively unsclerotized throughout the period of aestivation lasting several months (Lagace & van den Bosch, 1964). In carabid beetles, comparable changes and the colour of the accessory glands may be used to determine age class (van Dijk, 1972).

11.1.4 Changes in the internal non-reproductive organs

The fat-body is perhaps the most variable of these; in some insects it is large at the start of adult life, gradually waning (e.g. in the moth *Argyroploce* (Waloff, 1958)), and in others it is slowly built up prior to diapause. The availability and quality of food affect the condition of the fat-body and other organs (Fedetov, 1947, 1955, 1960; Haydak, 1957), as does parasitism.

The deposition of coloured excretory products in various internal organs, especially the malpighian tubules, provides another measure of age. Haydak (1957) found in worker honey-bees that the malpighian tubules are clear for the first three days of life, milky from three to ten and in older bees generally (not always) yellowish-green. Female *Culicoides* acquire a burgundy-red pigment in the abdominal wall during the development of the ovarian follicles and as this remains parous midges can easily be recognised without dissection (Dyce, 1969).

In newly emerged mosquitoes, bees, moths and other endopterygote insects part of the gut is filled with the brightly coloured meconium; this is usually voided within a day or two, at the most (Haydak, 1957; Rosay, 1961).

The detailed studies of Haydak (1957) and others on the honey-bee have shown that a range of organs, especially various glands, change their appearance with age.

11.1.5 The condition of the ovaries and associated structures
Criteria associated with changes in the female reproductive system have been used widely in ageing studies; they are valuable for even when they do not provide a precise chronological age, they do give information on the extent of egg laying, which is for some purposes more useful. The principal characters that have been used may be summarized under the following headings.

i. Egg rudiments. In those insects that do not develop additional egg follicles during adult life, a count of the number of egg rudiments will indicate the potential fecundity; at the start of adult life the count will be high, gradually falling off, though a large number of rudiments may still remain at the end of life (Waloff, 1958; Corbet, 1962*b*).

ii. Follicular relics. After an egg has been laid, the relics of the follicle, which are often pigmented, will remain in bases of the ovariole pedicels or in the oviducts for a variable period enabling a parous female to be recognized (Lineva, 1953; Gillies, 1958; Lebied, 1959; Corbet, 1960, 1961; Hamon *et al.*, 1961; Saunders, 1962, 1964; Anderson, 1964; Vlijm & van Dijk, 1967; Lewis *et al.* 1970; van Dijk, 1972). In some insects the total amount of relics accumulates with an increasing number of cycles (Anderson, 1964), in others they are eventually lost. V. P. Polovodova demonstrated in *Anopheles* that each time an egg develops it causes local stretching of the stalk of the oviduct. These dilations contain the remains of the follicles and if the ovaries are carefully stretched and examined (see Giglioli, 1963) the number of dilations may be counted. In certain mosquitoes the maximum number of bead-like dilations in any ovariole can be taken as equal to the number of ovarian cycles; with other species difficulties have been encountered (Muirhead-Thomson, 1963).

iii. Ovarian tracheoles. The tracheoles supplying the ovaries of nulliparous females are tightly coiled ('tracheal skeins'); as the eggs mature these become stretched, so that they do not resume their previous form, even in interovular periods. These changes have been observed and used for ageing in mosquitoes (Hamon *et al.*, 1961; Detinova, 1962; Kardos & Bellamy, 1962), in dragonflies (Corbet, 1961) and in calypterate flies (Anderson, 1964). A standard method of dissection for age-grading mosquitoes, by this method, is given by Davies *et al.* (1971).

iv. Ovariole cycles and combined evidence. In many insects the ovaries pass through a series of cycles; the stage in any given cycle may be easily recognized by the size of the most mature egg rudiments. Kunitskaya (1960) recognized six stages in fleas; but five stages seem normal within the Diptera, as initially recognized in mosquitoes by Christophers (1960) (e.g. ceratopogonids (Linley, 1965)); simuliids (Le Berre, 1966), anthomyiids (Jones, 1971); by combining these with information on follicular relics, Tyndale-Biscoe & Hughes (1969) were able to recognise seventeen different stages in *Musca*, and Vogt *et al.* (1974) sixteen in *Lucilia*. Taken in conjunction with other evidence that indicates whether the female is nulliparous or parous, the age may be determined over two ovarian cycles. In viviparous insects and others (e.g. Schizopteridae) where one large egg is laid at a time, the ovaries develop alternatively and this may allow a larger number of reproductive cycles to be recognized (Saunders, 1962, 1964). In some insects the number of functional ovarioles decreases after the first two cycles and this may be used as an index of age.

11.1.6 Indices of copulation

Although frequency of copulation is not a reliable index of chronological age, it may be 'calibrated' for a particular population and gives *per se* information of ecological significance. The presence of sperms in the spermatheca will provide evidence of pairing; in some mosquitoes a gelatinous mating plug remains in the oviduct for a short period (Muirhead-Thomson, 1963). In insects where the sperms are deposited in a spermatophore the remains of these in the spermatheca will give a cumulative measure of copulation (Waloff, 1958).

In some insects the appendages of the male will leave characteristic 'copulation marks'. In dragonflies these are on the compound eyes or occipital triangle of the female; in Zygoptera some sticky secretion may remain on the sides of the female's thorax (Corbet, 1962a); in tsetse flies they are on the sixth sternum (Squire, 1952). In Cimicidae with a broad spermalege, every pairing will leave a characteristic groove (Usinger, 1966). The mated status of male calliphorine flies can be determined as traces of the accessory gland secretion remain in the lateral penis ducts for up to a week after mating (Pollock, 1969).

11.1.7 Changes in weight

Insects that feed little or not at all during adult life will become progressively lighter; Waloff (1958) showed that individual variation due to size differences could be reduced if a weight/length ratio was used. With Lepidoptera weight/wing length is a convenient measure and Waloff found in a number of species that this falls off strikingly with age, e.g. male *Argyroploce* with an expectation of life of 10 days had a ratio of 1.93, but when the expectation of life was 2.5 days the ratio was 1.13.

In other insects, the adult weight will fluctuate, reflecting ovarian cycles (Waloff, 1958; see also p. 303) or changes associated with diapause. In Crustacea, Mollusca and those few insects (e.g. Collembola) that continue to grow after the onset of reproduction, size and weight will be positively, but perhaps not precisely, correlated with age.

11.1.8 'Wear and tear'
As an adult insect becomes older its cuticle and appendages become damaged by contact with the environment and this 'wear and tear' may be used as an index of age. Kosminskii (1960) has found that fleas may be age-graded by the extent to which the ctenidal (genal comb) bristles are broken; whilst Michener *et al.* (1955) and Daly (1961) used the wear on the mandibles of wild bees. However, the most widely applicable index of this type is provided by the tearing or tattering of the wings, e.g. in Lepidoptera, Diptera (Saunders, 1962) and Hymenoptera (Michener *et al.*, 1955). Although apparently crude, the number of tears ('nicks') in the wings has been found to correlate well with other indices of ageing (Michener & Lange, 1959; Saunders, 1962). Phoretic water mites die or fall off mosquitoes, and possibly other insects that emerge from water, with time: under certain conditions (Corbet, 1970), they may be used to indicate age (Gillett, 1957; Corbet, 1963).

11.2 Time-specific life-table and survival rates

This method may be used if it is justified to assume that the population is steady. That is, that over the period of life of the individuals in the table, the recruitment rate has been constant and the mortality rate within each stage has been steady (although, of course, it may be different for each stage – the table will reveal this).

In contrast to the age-specific life-tables, where successive estimates need to be compared, the time-specific life-table is based solely on the age-grouping of the individuals collected at a single instant of time. Thus, a relative method (Chapter 7) may be used for sampling the population, provided that it samples at random with respect to the different age groups. The youngest age group may be equated with a convenient number* (e.g. 1; 100; 1000), and the other values corrected accordingly to give the l_x column. The remaining columns of

---

* There is no reason to retain the actual number, for this is entirely arbitrary, depending on sample size; in contrast in age-specific life-tables it represents the size or density of an actual cohort.

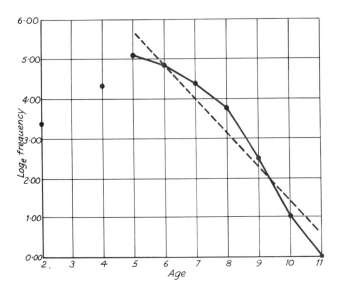

Fig. 11.1 A time-specific survivorship curve. The logarithms of the frequencies plotted against age groups in a season's sample of the fish, *Leucichthys sardinella* (after Wohlschlag, 1954).

the table can then be calculated as described in Chapter 10 (p. 367). An example of this type of life-table and its calculation is that given by Paris and Pitelka (1962) for the woodlouse, *Armadillidium*.

A number of time-specific life tables were constructed for the mosquito *Aedes aegypti* by Southwood *et al.* (1972); they were for limited parts of the yearly cycle when the assumption of steady recruitment appeared justified. Modified key factor may be applied to these life-tables; it will indicate differences between populations under different conditions (with *Aedes* in different breeding sites) at the period of steady recruitment, but it will not disclose anything about the periods of population change.

An extensive collection of age-grouping data is provided by the 'catch curves' obtained in fishing, and their analysis has been described by Ricker (1944, 1948), Wohlschlag (1954), Beverton & Holt (1957) Ebert (1973) and others. Another approach to the analysis is provided when log numbers are plotted against age, by the comparison of actual shape of the plot and the straight line that would result from a Slobodkin type III constant survival rate (Fig. 11.1). Variations in the rate of recruitment and in both the age- and time-specific survival rates will affect the shape of the curve. Age-specific survival rates may be obtained from mark and recapture data (Chapter 3) and a measure of recruitment can sometimes be obtained from another source, e.g. with insects from emergence traps. By obtaining information from other sources in this way the departures of the time-specific survivorship curve (Fig. 11.1) from linearity can often be interpreted; if recruitment and time-specific survival rates are constant then the fall-off with age in Fig. 11.1 implies increasing mortality with increasing age (Wohlschlag, 1954).

The hunting records of terrestrial vertebrates provide further extensive age-grouped data. Surveys of the proportions of various ages and sexes are made before and after hunting and the numbers of the different groups killed are also known; from this information it is possible to calculate natality (referred to as production), survival rates and absolute population. The methods of computation have been reviewed by Hanson (1963) and Seber (1973), and some of these might be combined with the 'removal-sampling' approach and used with insects. Some of the methods do not require hunting returns; of these the equation for the calculation of differential survival of age classes is of particular potential interest to insect ecologists (Hanson, 1963; Kelker & Hanson, 1964). It could be applied to a situation where generations overlapped, but each was comparatively well synchronized; from biological knowledge we must select three times: t_1 shortly before generation II becomes adult, t_2 just after generation II has become adult and t_3 a subsequent occasion. Assuming that the mortality of both sexes is equal, then the ratio of the survival of generation I to that of generation II during the period t_2 to t_3 is given by:

$$\frac{S_{II}}{S_I} = \frac{(\male t_2 - \female t_2)(\male t_3 \female t_1 - \male t_1 \female t_3)}{(\male t_3 - \female t_3)(\male t_2 \female t_1 - \male t_1 \female t_2)} \tag{11.1}$$

where the symbols represent the number of males and females in the samples on dates t_1, t_2 and t_3 as defined above (see also p. 114).

The use, in entomology, of age-grouping, time-specific life-tables and survivorship curve analysis of this type will probably prove most important in attempts to investigate the dynamics of the adult population. Detailed studies of this type will be necessary if the full meaning of k_0 (p. 377) is to be investigated.

11.3 Predictive population models

Key factor analysis of age-specific life tables or budgets (Chapter 10) and survival analysis of time-specific life tables provide a quantitative description of the population processes. From this predictions may be made of future population trends or of the response of the population to some change: a control measure or harvesting. A simple management model could be constructed (Chapter 12) to utilize the information obtained in this way. However, both these life-tables are inappropriate for many populations, but for these (continuously breeding or with overlapping generations) it is possible to tackle the problem 'the other way round': that is, to predict population growth under 'Malthusian conditions' (i.e. 'unlimited' resources, when $r_m = dN/Ndt$ (see p. 370) and to compare the actual field results with those from the predictions. In most cases the same model may be used to predict the impact of change on the population processes: because of the capacity of computers this may sometimes be undertaken directly, apparently without the intermediate step analagous to key-factor analysis. The approaches are more comparable than they may superficially appear and grade into the methods of life-table construction (p. 357). The use of models in ecology is reviewed in Chapter 12; these predictive models are either component or inductive strategic depending on the completeness with which the system is described.

11.3.1 Physiological time

These models predict the rate of population growth with time; population growth is dependent on generation time which depends on development rate. Development rate is influenced by temperature and therefore, in strongly seasonal regions where the same stage may be exposed to significantly different temperatures, either a correction must be made for the different development rates or 'physiological time' may be used (Hughes, 1962, 1963; Gilbert *et al.* 1976; Hardman, 1976a; Atkinson, 1977). Theoretically there is no reason why this should not always be used, however it seems advisable to restrict it to models where it is really necessary. As most biologists can only easily visualize their populations against calendar time, the use of physiological time makes more difficult an intuitive check on the output of the model. Physiological time is commonly expressed as day degrees ($D°$) (Hughes, 1962; Hardman, 1976a) or hour degrees ($h°$) (Atkinson, 1977), being the cumulative product of total time $\times$ temperature (above the threshold). Allen (1976) describes a sine wave method for computing day degrees, and Scopes and Biggerstaff (1977) show how a temperature integrator may be usefully used in the field as an effective measure of accumulated physiological temperature. Atkinson's (1977) studies suggest that the common practice of determining the physiological time for development at constant temperatures may be misleading when applied to fluctuating field conditions.

In most population models it is necessary to divide time, physiological or calendar, into intervals. The length of the interval is critical; if it is too large the changes between intervals are too sudden and the heterogeneity within the time-interval age class too great. Too small an interval will add to the complexities and costs of the work and the field data may be inadequate in quantity. Gilbert *et al.* (1976) tested a number of different time units and concluded that for their aphid a quarter instar period (a 'quip') was suitable: they make no claims for the generality of this unit, but it or a half-instar period seem to be of a reasonable order of magnitude to test initially. In their case, instar length was similar (in physiological time) for the first three instars; with other insects the instar-period might have to be based on the average or perhaps the shortest instar (excluding very short first instars) development time.

11.3.2 Life-table parameters

Predictive models require an input of information on the magnitude and variation of the development times of the different stages: a complete example of this type of data is given by Gomez *et al.* (1977) for *Culex*. When key-factor or similar analysis is to be undertaken, it is debatable whether the mortalities of the different stages detected under apparently 'Malthusian conditions' (e.g. Hardman, 1976a) should be included in the prediction or whether, as seems safer, the prediction should assume 100 % survival. Complete models, including the adult stage will require age-specific fecundity schedules.

11.3.3 Recruitment in the field

A reliable estimate of recruitment, depending on the form of the model either of adult emergence or daily oviposition or some combination of the two is essential. (see also Chapters 9 and 10). These models generally assume that age distribution within a stage, at any one time, is effectively uniform.

11.3.4 Empirical models

A graphical model was developed by Edmondson (1968) for the analysis of field data of the ratio of eggs/female (E/w) in plankton communities. The eggs per female per day (B) is given by:

$$B = E/Dw \qquad (11.2)$$

where $D =$ the mean development time of the eggs. Thus the basic inputs for this model are an observed recruitment value (or in this case its equivalent) and an experimentally determined life-table parameter. The birth rate is calculated and from this the population growth rate, in the absence of death, predicted (Fig. 11.2). The real growth rate, as expressed by the field population can be compared with this.

The model of Southwood *et al.* (1972) allows variable hatching and developmental times to be incorporated into the prediction (Fig. 11.3). The size of the egg cohort of a particular time period is known from field observation. It is also necessary to know the developmental spectra of the various stages (e.g. the shape of the hatching curve for eggs of the same age). If physiological time is not used, but developmental periods differ at different seasons, this variation may also be incorporated. The model calculates the contribution of each egg cohort to the numbers of subsequent stages and then sums the contributions from each cohort to predict the total population of that stage in that time interval (e.g. in Fig. 11.3, for adults emerging in time period $x + 2$ the estimated contributions of egg cohorts 2, 3, 4 and 5 would be summed). The whole operation is conceptually simple, but the summations are so numerous that a computer is required. These predictions may be compared with actual data from field sampling to estimate mortality levels: it is assumed that at a particular time the members of an age group suffer mortality at random, i.e. irrespective of their cohort origin. Vandermeer

Time-specific Life-tables

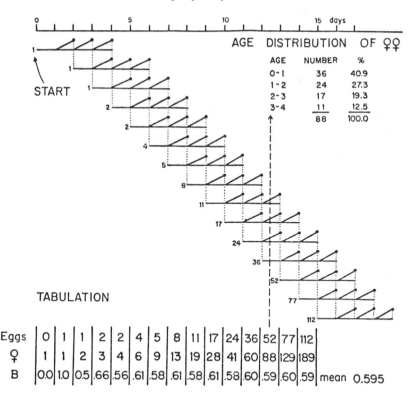

AGE DISTRIBUTION OF ♀♀

AGE	NUMBER	%
0-1	36	40.9
1-2	24	27.3
2-3	17	19.3
3-4	11	12.5
	88	100.0

START

TABULATION

Eggs	0	1	1	2	2	4	5	8	11	17	24	36	52	77	112	
♀	1	1	2	3	4	6	9	13	19	28	41	60	88	129	189	
B	0.0	1.0	0.5	.66	.56	.61	.58	.61	.58	.61	.58	.60	.59	.60	.59	mean 0.595

Fig. 11.2 A graphical model to predict the rate of population growth for a simple organism (after Edmondson, 1968). The population consists of parthenogenetic females that mature after one day and live a further three; on each day of maturity a female lays a single egg, these follow the same cycle. The 'foundress' therefore produces one egg at times 1.0, 2.0, and 3.0. The eggs take one day to develop (shown by the diagonal lines) and on hatching (represented by the 'pin heads' on the top of the diagonals) contribute the number of females shown at the end of the dotted lines descending from the 'pin heads'. The vertical column on any day give the age distribution; this is shown for day 12.5.

(1975) has queried the validity of making this assumption in the Lewis-Leslie matrix model, where it is also basic, if the age classes are too wide.

Shiyomi (1967) developed a model for the reproduction of aphids, recognizing that as aphids are not distributed randomly in space, reproduction per parent will be variable and will conform to the negative binomial. It enables a prediction to be made of the probability that a certain number of offspring will arise on a plant from a particular density of females. A noteworthy aspect of this model is that it emphasizes the spatial dimension in population dynamics; one that is frequently overlooked (Taylor, 1977).

11.3.5 Intrinsic rate models and variable life tables

Hughes (1962, 1963) laid the foundations of this approach when he showed

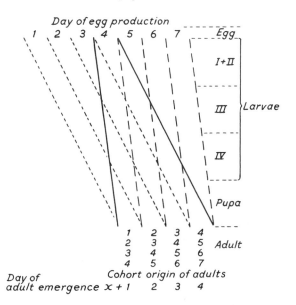

Fig. 11.3 The contribution of a single egg cohort (laid on day 4) (solid lines) to a population of over-lapping generations. The dotted lines represent the slowest developers and the dash lines the fastest developers, and hence the age extremes, of the members of other egg cohorts (after Southwood, Murdie, Yasuno, Tonn and Reader, 1972).

that stable age distributions often occured in expanding aphid populations; under these conditions the potential rate of population increase could be predicted by a model based on the intrinsic rate of increase in Lotka's population growth equation. The models have been extensively developed in collaboration with N. Gilbert (Hughes & Gilbert, 1968; Gilbert & Hughes, 1971) and colleagues (Gilbert & Gutierrez, 1973; Gilbert *et al.*, 1976). They have been termed 'variable life tables' by Gilbert *et al.* (1976), who give a detailed account of the basic philosophy and the construction of the models, including the facsimile reproduction of the print-out of the computer programmes: reference should be made to this work for details. The variable life table also owes much in its methodology to C. S. Holling's component analysis, for the approach is to identify the major components of population change, insert these in the model, test the model against the field data, examine the places where it does not fit (as Gilbert *et al.* (1976) stress, these are the most interesting results), then modify that part of the model in a biologically meaningful way and re-run, comparing the prediction against field data. The initial components of a variable life table might be expected to include the age fecundity schedule, the fecundity density relationship, the development schedule (normally on a physiological time-scale), the impact of density on development, and mortality schedules in relation to various natural enemies.

Gilbert & Hughes (1971) investigated the value of stochastic, as opposed to deterministic models and concluded that they do not provide more insights; Hardman's (1976*b*) tests with a flour beetle, *Tribolium*, confirm this endorsement of the prediction of May (1973) on theoretical grounds, that stochastic models would be necessary only with small populations (of some hundreds). Hardman's models could also be described as variable life-tables, and he illustrates their computer flow diagrams.

11.3.6 Lewis-Leslie matrices

During the 1940's, E. G. Lewis and P. H. Leslie both independently developed a deterministic model in matrix form for the age structure of a reproducing population. The general form of these models, in matrix notation, is:

$$\mathbf{M} \cdot a_t = a_{t+1} \tag{11.3}$$

where $\mathbf{M}$ is a square matrix that describes the transition of the population over one time period and thus contains terms describing age-specific fecundities and survival rates, a is a column vector of the age structure of the population at time t (the start) and a_{t+1}, is the column vector at time $t+1$. Set out in full:

$$
\begin{bmatrix}
f_0 & f_1 & f_2 & f_3 \\
p_0 & 0 & 0 & 0 \\
0 & p_1 & 0 & 0 \\
0 & 0 & p_2 & 0
\end{bmatrix}
\times
\begin{bmatrix}
a_0 \\
a_1 \\
a_2 \\
a_3
\end{bmatrix}
=
\begin{bmatrix}
f_0 a_0 & f_1 a_1 & f_2 a_2 & f_3 a_3 \\
p_0 a_0 & 0 & 0 & 0 \\
0 & p_1 a_1 & 0 & 0 \\
0 & 0 & p_2 a_2 & 0
\end{bmatrix}
\tag{11.4}
$$

where the f's represent the fecundity of each age class (0, 1, 2 and 3) as expressed in daughters born between t and $t+1$ and $p_0 \ldots p_2$ represents the probability of a female (of the age indicated by the subscript) surviving. It will be noted that these probabilities will always form a diagonal that commences at the left-hand end of the second row and ends (reaches the last row) in the penultimate column. In practice f_0 is normally zero. The vector column $(a_0 \ldots a_3)$ represents the number of animals at each age group $t = 0$ to $t = 3$ at time t. It will be seen that the element in the matrix in row i and the column j multiplies the jth element in the vector and puts it into the ith position, e.g. the element in the first row and the third column is f_2 and this multiplies the third element a_2 in the vector and places it in the first row, third column.

After a number of multiplications this matrix will give stable age proportions, this is when the fluctuations due to the initial column vector proportions have ceased and the population is growing at a constant rate (e^r). The column of this age distribution approximates to the dominant latent vector (or eigenvector) of the matrix and the increase (a dimensionless number — a scaler) is the dominant (or first) latent root (or eigenvalue) of the matrix:

$$\mathbf{M}a = \lambda a \tag{11.5}$$

where λ is the latent root and a the corresponding vector.

Because of the particular features of Lewis-Leslie matrices (the lack of negative values), there is only one latent root (effectively the largest) that has a latent vector having all its elements non-negative. The root may be found by a standard procedure; Davies (1971) gives a computer programme (No. 26). This root is the finite rate (or capacity) for increase (see p. 370):

$$\lambda = e^r \quad \text{or} \quad \log_e \lambda = r \tag{11.6}$$

Lucid accounts of the Lewis-Leslie matrix and its development are given by Williamson (1967, 1972) and Usher (1972).

The male sex was introduced into the model by Williamson (1959) by doubling the number of rows and columns, whilst Pennycuick *et al.* (1968) produced a computer model incorporating density-dependence. They made the elements in the matrix **M** dependent on the size of the population (the sum of the appropriate vector column) by multiplying the f values in the top row by a function F, and the p values (in the line below the diagonal) by another function S. F and S are expressions that modify fecundity and survival to give sigmoid curves; they combine the present size of the population, the equilibrium population and the degree of density-dependence in a rather complex manner (May *et al.* 1974).

A requirement of the Lewis-Leslie matrix is that the time intervals are exactly equal, yet insects are generally age-grouped by stages which are of unequal length. Lefkovitch (1965) considered this problem and describes a method of stage grouping taking account of development times and the influence of one stage on another. Many elements in the matrix will contain terms for more than one stage and it is clear that unless a stable age distribution is reached, the survival probabilities will change as the proportions of the stages in the elements changes. Vandermeer (1975) emphasizes the need to make age categories as brief as possible, otherwise the same criticisms could apply to the age class matrix if a stable age distribution has not been achieved, for the precise ages (and survival probabilities) of the individuals within an age category may then vary from occasion to occasion.

Longstaff (1977) has utilized these techniques to study the population dynamics of Collembola in culture. Besides deterministic models, he developed a stochastic model. The assumption was made that the parameter values were normally distributed around the mean and the actual value was selected by a random number generator. The accuracy of the deterministic and stochastic models varied, but the predictors were sufficiently close to the form of the actual populations for it to seem likely that the population parameters derived from the models (the density dependent function and the equilibrium population) are meaningful.

The Lewis-Leslie Matrix was found conceptually useful by Conway *et al.* (1975) in their study of the control of the sugar cane froghopper; by using age-intervals of a day they overcame some of the problems arising from metamorphosis (Lefkovitch, 1965) and from too wide an age category (Vandermeer, 1975).

REFERENCES

ALLEN, J. C., 1976. A modified sine wave method for calculating degree days. *Environ. Ent.* **5**, 388–96.

ANDERSON, J. R., 1964. Methods for distinguishing nulliparous from parous flies and for estimating the ages of *Fannia canicularis* and some other cyclorrophous Diptera. *Ann. ent. Soc. Am.* **57**, 226–36.

ATKINSON, P. R., 1977. Preliminary analysis of a field population of citrus red scale, *Aonidiella aurontii* (Maskell) and the measurement and expression of stage duration and reproduction for life tables. *Bull. ent. Res.* **67**, 65–87.

BEVERTON, R. J. H. and HOLT, S. J., 1957. On the dynamics of exploited fish populations. *Fishery investigations, ser. 2.* **19**, 533 pp. Min. Agric. Fish. Food Gt Britain, London, H.M.S.O.

CHAI, F. -C. and DIXON, S. E., 1971. A technique for ageing cocoons of the sawfly *Neodiprion sertifer* (Hymenoptera: Diprionidae). *Can. Ent.* **103**, 80–3.

CHRISTOPHERS, S. R., 1960. *Aedes aegypti (L.) the yellow fever mosquito.* 739 pp., Cambridge University Press, Cambridge.

CONWAY, G. R., NORTON, G. A., SMALL, N. J. and KING, A. B. S., 1975. A systems approach to the control of the sugar cane froghopper. *In* Dalton, G. E. (ed.) *Study of Agricultural Systems*, pp. 193–229. Applied Science Publishers, London.

CORBET, P. S., 1960. Recognition of nulliparous mosquitoes without dissection. *Nature* **187**, 525–6.

CORBET, P. S., 1961. The recognition of parous dragonflies (Odonata) by the presence of follicular relics. *Entomologist* **94**, 35–7.

CORBET, P. S. 1962a. Age-determination of adult dragonflies (Odonata). *Proc. XI int. Congr. Ent.* **3**, 287–9.

CORBET, P. S., 1962b. *A biology of dragonflies.* 247 pp., Witherby, London.

CORBET, P. S., 1963. Reliability of parasitic water mites (*Hydracarina*) as indicators of physiological age in mosquitoes (Diptera: Culicidae). *Entomologia exp. appl.* **6**, 215–33.

CORBET, P. S., 1970. The use of parasitic water-mites for age-grading female mosquitoes. *Mosquito News* **30**, 436–8.

DALY, H. V., 1961. Biological observations on *Hemihalictus lustrans*, with a description of the larva (Hymenoptera: Halictidae). *J. Kansas ent. Soc.* **34**, 134–41.

DAVIES, J. B., CORBET, P. S., GILLIES, M. T. and MCCRAE, A. W. R., 1971. Parous rates in some Amazonian mosquitoes collected by three different methods. *Bull. ent. Res.* **61**, 125–32.

DAVIES, R. G., 1971. *Computer Programming in Quantitative Biology*. Academic Press, London.

DETINOVA, T. S., 1962. Age-grouping methods in Diptera of medical importance with special reference to some vectors of malaria. *Monogr. Ser. World Hlth Org.* **47**, 216 pp.

DETINOVA, T. S., 1968. Age structure of insect populations of medical importance. *Ann. Rev. Ent.* **13**, 427–50.

DIJK, TH. S. VAN., 1972. The significance of the diversity in age composition of Calathus melanocephalus L. (Col. Carabidae) in space and time at Schiermannikoog. *Oecologia* (Berl.) **10**, 111–36.

DREW, R.A.L., 1969. Morphology of the reproductive system of *Strumeta tryoni* (Froggatt) (Diptera: Trypetidae) with a method of distinguishing sexually mature males. *J. Aust. ent. Soc.* **8**, 21–32.

DUNN, E., 1951. Wing coloration as a means of determining the age of the Colorado beetle (*Leptinotarsa decemlineta* Say). *Ann. appl. Biol.* **38**, 433–4.

DYCE, A. L., 1969. The recognition of multiparous and parous *Culicoides* without dissection. *J. Aust. Ent. Soc.* **8**, 11–5.

EBERT, T. A., 1973. Estimating growth and mortality rates from size data. *Oecologia (Berl.)* **11**, 281–98.

EDMONDSON, W. T., 1968. A graphical model for evaluating the use of the egg ratio for measuring birth and death rates. *Oecologia* **1**, 1–37.

EMDEN, H. F. VAN, JEPSON, W. F. and SOUTHWOOD, T.R.E., 1961. The occurrence of a partial fourth generation of *Oscinella frit* L. (Diptera: Chloropidae) in southern England. *Entomologia exp. appl.* **4**, 220–5.

FEDETOV, D. M., 1947, 1955, 1960. The noxious little tortoise, *Eurygaster integriceps* Put. [In Russian.] **1** (272 pp.) and **2** (1947); **3** (278 pp.) (1955); **4** (239 pp.) (1960).

FOTHERINGHAM, N., 1971. Life history of the littoral gastropods *Shaskyus festivus* (Hinds) and *Oceneboa poulsoni* Carpenter (Prosobranchia: Muricidae). *Ecology* **52**, 742–57.

GIGLIOLI, M. E. C., 1963. Aids to ovarian dissection for age determination in mosquitoes. *Mosquito News* **23**, 156–9.

GILBERT, N. and GUTIERREZ, A. P., 1973. A plant-aphid-parasite relationship. *J. Anim. Ecol.* **42**, 323–40.

GILBERT, N., GUTIERREZ, A. P., FRAZER, B. D. and JONES, R. E., 1976. *Ecological Relationships*. 157 pp., Addison-Wesley, Reading and San Francisco.

GILBERT, N. and HUGHES, R. D., 1971. A model of an aphid population–three adventures. *J. Anim. Ecol.* **40**, 525–34.

GILLETT, J. D., 1957. Age analysis of the biting cycle of the mosquito *Taeniorhynchus* (*Mansonioides*) *africana* Theobald, based on the presence of parasitic mites. *Ann. trop. Med. Hyg.* **51**, 151–8.

GILLIES, M. T., 1958. A review of some recent Russian publications on the technique of age determination in *Anopheles*. *Trop. Dis. Bull.* **55**, 713–21.

GOMEZ, C., RABINOVICH, J. E. and MACHADO-ALLISON, C. E., 1977. Population analysis of *Culex pipiens fatigans* Wielt. (Diptera: Culicidae) under laboratory conditions. *J. Med. Ent.* **13**, 453–63.

HAMON, J., GRJEBINE, A., ADAM, J. P., CHAUVET, G., COZ., J. and GRUCHET, H., 1961. Les méthodes d'évaluation de l'âge physiologique des moustiques. *Bull. Soc. ent. Fr.* **66**, 137–61.

HANSON, W. R., 1963. Calculation of productivity, survival, and abundance of selected vertebrates from sex and age ratios. *Wildl. Monogr.* **9**, 1–60.

HARDMAN, J. M., 1976a. Life table data for use in deterministic and stochastic simulation models predicting the growth of insect populations under Malthusian conditions. *Can. Ent.* **108**, 897–906.

HARDMAN, J. M., 1976b. Deterministic and stochastic models simulating the growth of insect populations over a range of temperatures under Malthusian conditions. *Can. Ent.* **108**, 907–24.

HAYDAK, M. H., 1957. Changes with age in the appearance of some internal organs of the honeybee. *Bee World* **38**, 197–207.

HINKS, C. F. and BYERS, J. R., 1973. Characters for determining the sex of cutworms and other noctuid larvae (Lepidoptera: Noctuidae). *Can. J. Zool.* **51**, 1235–41.

HUGHES, R. D., 1962. A method for estimating the effects of mortality on aphid populations, *J. Anim. Ecol.* **31**, 389–96.

HUGHES, R. D., 1963. Population dynamics of the cabbage aphid, *Brevicoryne brassicae* (L.). *J. Anim. Ecol.* **32**, 393–424.

HUGHES, R. D. and GILBERT, N., 1968. A model of an aphid population – a general statement. *J. Anim. Ecol.* **37**, 553–63.

JONES, M. G., 1971. Observations on changes in the female reproductive system of the wheat bulb fly, *Leptohylemyia coarctata* (Fall.). *Bull. ent. Res.* **61**, 55–68.

KARDOS, E. H. and BELLAMY, R. E., 1962. Distinguishing nulliparous from parous female *Culex tarsalis* by examination of the ovarian tracheation. *Ann. ent. Soc Am.* **54**, 448–51.

KELKER, G. H. and HANSON, W. R., 1964. Simplifying the calculation of differential survival of age-classes. *J. Wildl. Mgmt* **28** (2), 411.

KISHI, Y. 1971. Reconsideration of the method to measure the larval instars by use of the frequency distribution of head-capsule widths or lengths. *Can. Ent.* **103**, 1011–5.

KOSMINSKII, R. B., 1960. The method of determining the age of the fleas *Leptopsylla segnis* Schönh. 1811 and *L. taschenbergi* Wagn. 1898 (Suctoria-Aphaniptera) and an experiment on the age analysis of a population of *L. Segnis*. [In Russian.] *Med. Parazitol.* **29**, 590–4.

KUNITSKAYA, N. T., 1960. On the reproductive organs of female fleas and determination of their physiological age. [In Russian] *Med. Parazitol.* **29**, 688–701.

LAGACE, C. F. and VAN DEN BOSCH, R., 1964. Progressive sclerotization and melanization of certain structures in males of a field population of *Hypera brunneipennis* (Coleoptera: Curculionidae). *Ann. ent. Soc. Am.* **57**, 247–52.

LE BERRE, R., 1966. Contribution a l'étude biologique et écologique de *Simulium damnosium* (Diptera, Simuliidae). *Off. Rech. Sci. Tech. Outre-Mer*, 204 pp.

LEBIED, B., 1959. Détermination de l'âge physiologique des diptères. Nouvelle méthode basée sur la recherche des vestiges du processus de l'ovulation. *Riv. Parassit.* **20**, 91–106.

LEFKOVITCH, L. P., 1965. The study of population growth in organisms grouped by stages. *Biometrics* **21**, 1–18.

LEWIS, D. J., LAINSON, R. and SHAW, J. J., 1970. Determination of parous rates in phlebotomine sandflies with special reference to Amazonian species. *Bull. ent. Res.* **60**, 209–19.

LINEVA, V. A., 1953. Physiological age of females of *Musca domestica* L. (Diptera: Muscidae). [In Russian.] *Ent. Obozr.* **33**, 161–73.

LINLEY, J. R., 1965. The ovarian cycle and egg stage in *Leptoconops* (*Holoconops*) *becquaerti* (Kieff.) (Diptera, Ceratopogaoidae). *Bull. ent. Res.* **56**, 37–56.

LONGSTAFF, B. C., 1977. The dynamics of collembolan populations: a matrix model of single species population growth. *Can. J. Zool.* **55**, 314–24.

MAY, R. M., 1973. *Stability and complexity in model ecosystems.* Princeton University Press, Princeton, New Jersey.

MAY, R. M., CONWAY, G. R., HASSELL, M. P. and SOUTHWOOD, T. R. E., 1974. Time delays, density dependence and single-species oscillations. *J. Anim. Ecol.* **43**, 747–70.

MICHENER, C. D., CROSS, E. A., DALY, H. V., RETTENMEYER, C. W. and WILLE, A., 1955. Additional techniques for studying the behaviour of wild bees. *Insectes Sociaux* **2**, 237–46.

MICHENER, C.D. and LANGE, R.B., 1959. Observations on the behaviour of Brazilian Halictid bees (Hymenoptera, Apoidea). IV. *Augochloropsis*, with notes on extralimital forms. *Am. Mus. Novitates* **1924**, 1–41.

MUIRHEAD-THOMSON, R. C., 1963. *Practical entomology in malaria eradication.* WHO (MHO/PA/62.63). Part I: 64–71.

NEVILLE, A. C., 1963. Daily growth layers in locust rubber-like cuticle, influenced by an external rhythm. *J. Ins. Physiol.* **9**, 177–86.

NEVILLE, A. C. 1970. Cuticle ultrastructure in relation to the whole insect. *In* Neville, A. C. (ed.) *Insect Ultrastructure, Symp. R. ent. Soc. Lond.* **5**, 17–39.

PARIS, O. H. and PITELKA, A., 1962. Population characteristics of the terrestrial isopod *Armadillidium vulgare* in California grassland. *Ecology* **43**, 229–48.

PENNYCUICK, C. J., COMPTON, R. M. and BECKINGHAM, L., 1968. A computer model for simulating the growth of a poulation, or of two interacting populations. *J. Theort. Biol.* **18**, 316–29.

POLLARD, E., 1973. Growth classes in the adult Roman snail (*Helix pomatia* L.). *Oecologia (Berl.)* **12**, 209–12.

POLLOCK, J., 1969. Test for the mated status of male sheep blowflies. *Nature* **223**, 1287–8.

RICKER, W. E., 1944. Further notes on fishing mortality and effort. *Copeia* **1944**, 23–44.

RICKER, W. E., 1948. Methods of estimating vital statistics of fish populations. *Indiana Univ. Publ. Sci. Ser.* **15**, 101 pp.

ROSAY, B., 1961. Anatomical indicators for assessing the age of mosquitoes: the teneral adult (Diptera: Culicidae). *Ann. ent. Soc. Am.* **54**, 526–9.

SAUNDERS, D. S., 1962. Age determination for female tsetse flies and the age composition of samples of *Glossina pallidipes* Aust., *G. palpalis fuscipes* Newst. and *G. brevipalpis* Newst. *Bull. ent. Res.* **53**, 579–95.

SAUNDERS, D. S., 1964. Age-changes in the ovaries of the sheep ked, *Melophagus ovinus* (L.) (Diptera: Hippoboscidae). *Proc. R. ent. Soc. Lond. A* **39**, 68–72.

SCHIEMENZ, H., 1953. Zum Farbwechsel bei heimischen Heteropteren unter besonderer Berücksichtigung von *Palomena* Muls. & Rey. *Beitr. Ent.* **3**, 359–71.

SCHLEIN, J. and GRATZ, N. G., 1972. Age determination of some flies and mosquitoes by daily growth layers of skeletal apodemes. *Bull. Wld. Hlth. Org.* **47**, 71–6.

SCHLEIN, J. and GRATZ, N. G., 1973. Determination of the age of some anopheline mosquitoes by daily growth layers of skeletal apodemes. *Bull. Wld. Hlth Org.* **49**, 371–5.

SCHNEIDER, F. and VOGEL, W., 1950. Neuere Erfahrungen in der Bekämpfung der Kirschenfliege (*Rhagoletis cerasi*) Schweiz. *Z. Obst-weinbau* **59**, 37–47.

SCOPES, N. E. A. and BIGGERSTAFF, S.H., 1977. The use of a temperature integrator to predict the developmental period of the parasite *Aphidius matricariae* (Hal.). *J. appl. Ecol.* **14**, 1–4.

SEBER, G. A. F., 1973. *The Estimation of Animal Abundance and Related Parameters.* 506 pp., Griffin, London.

SHIYOMI, M., 1967. A statistical model of the reproduction of aphids. *Res. Popul. Ecol.* **9**, 167–76.

SOUTHWOOD, T. R. E. and LESTON, D., 1959. *Land and water bugs of the British Isles.* London, **436** pp.

SOUTHWOOD, T. R. E. and SCUDDER, G. G. E., 1956. The bionomics and immature stages of the thistle lace bugs (*Tingis ampliata* H-S. and *T. cardui* L.; Hem., Tingidae). *Trans. Soc. Brit. Ent.* **12**, 93–112.

SOUTHWOOD, T. R. E., MURDIE, G., YASUNO, M.,TONN, R. J. and READER, P. M., 1972. Studies on the life budget of *Aedes aegypti* in Wat Samphaya, Bangkok, Thailand. *Bull. Wld. Hlth. Org.* **46**, 211–26.

SQUIRE, F. A., 1952. Observations on mating scars in *Glossina palpalis* (R.-D.). *Bull. ent. Res.* **42**, 601–4.

TAYLOR, L. R., 1977. Migration and the spatial dynmaics of an aphid, *Myzus persicae.* *J. Anim. Ecol.* **46**, 411–24.

TYNDALE-BISCOE, M. and HUGHES, R. D., 1969. Changes in the female reproductive system as age indicators in the bushfly *Musca vetustissima* Wlk. *Bull. ent. Res.* **59**, 129–41.

TYNDALE-BISCOE, M. and KITCHING, R. L., 1974. Cuticular bands as age criteria in the sheep blowfly, *Lucillia cuprina* (Wied.) (Diptera, Calliphoridae). *Bull. ent. Res.* **64**, 161–74.

USHER, M. B., 1972. Developments in the Leslie Matrix Model. *In* Jeffers, J. N. R. (ed.) *Mathematical Models in Ecology., Symp. Brit. Ecol. Soc.* **12**, 29–60.

USINGER, R. L., 1966. *Monograph of Cimicidae*. The Thomas Say Foundation, Ent. Soc. Am.

VANDERMEER, J. H., 1975. On the construction of the population projection matrix for a population grouped in unequal stages. *Biometrics* **31**, 239–42.

VLIJM, L. and VAN DIJK, Th. S., 1967. Ecological studies on carabid beetles. II. General pattern of population structure in *Calathus melanocephalus* at Schiermonnikoog. *J. Morph. Okol. Tiere* **58**, 396–404.

VOGT, W. G., WOODBURN, T. L. and TYNDALE-BISCOE, M., 1974. A method of age determination in *Lucilia cuprina* (Wied) (Diptera, Calliphoridae) using cyclic changes in the female reproductive system *Bull. ent. Res.* **64**, 365–70.

WALOFF, N., 1958. Some methods of interpreting trends in field populations. *Proc. X int. Congr. Ent.* **2**, 675–6.

WILLIAMSON, M. H., 1959. Some extensions of the use of matrices in population theory. *Bull. Math. Biophys.* **21**, 13–17.

WILLIAMSON, M. H., 1967. Introducing students to the concepts of population dynamics. *In* Lambert, J. (ed.). *The Teaching of Ecology, Sym.Brit. Ecol. Soc.* **7**, 169–76.

WILLIAMSON, M. H., 1972. *The Analysis of Biological Populations*. 180 pp., Arnold, London.

WOHLSCHLAG, D. E., 1954. Mortality rates of whitefish in an arctic lake. *Ecology* **35**, 388–96.

YOUNG, E. C., 1965. Teneral development in British Corixidae. *Proc. R. ent. Soc. Lond. A.* **40**, 159–68.

12

Systems Analysis and Modelling in Ecology

As ecology is concerned with the quantitative interactions in complex systems, computers have proved very powerful tools. K.E.F. Watt (1961, 1966, 1968) and C.S. Holling (1963, 1964) were pioneers in their utilization and in combining their use with the disciplines of systems analysis and modelling. However, not all models utilize computers, nor does systems analysis always require the use of a computer.

A model is a representation, a mimic: its mimicry may be categorized according to the qualities of generality, realism and accuracy; a 'good model' is one in which these features are appropriate for the purposes for which it was built. This means that a clear definition of the objective is essential before the construction of the model can be attempted.

Systems analysis is a technique developed in engineering; it involves the mathematical characterization of the link between output and input in terms of the intervening system (Fig. 12.1).

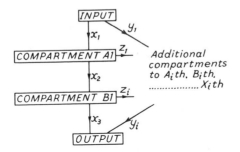

Fig. 12.1 The concept of systems analysis. The input consists of the 'state variables', $x_1\text{--}x_i$, $y_1\text{---}y_i$ and $z_1\text{----}z_i$ are transfer functions (functional relationships) that describe mathematically the process ('flow') between the compartments (or components, that contain quantities). Transfer functions relate to a time scale, so that the output is determined at time $t, t+1 \ldots t+i$.

Computers are of two types: the digital, that processes numerical information and handles discrete data; and the analog, which can handle continuous input. As Patten (1971) graphically expresses it, an analog computer is 'played' more like a musical instrument; the operator receives instantaneous response to a parameter change and can use it almost as an 'extension' to his brain, but accuracy and realism are sacrificed to speed of response.

407

The present chapter aims briefly to outline general approaches in this field and to provide an entry to the literature. In addition to the works cited below reference should be made to the volumes edited by Watt (1966), Van Dyne (1969), Patten (1971-1976), Jeffers (1972), Bartlett & Hiorns (1973) and to Streifer's (1974) review of the underlying mathematical concepts.

12.1 Types of systems model

Models may be classified in various ways. Because of the importance of determining the objective of model building before commencement, the

Model Type	System					Objectives
	Already fully defined	Described	Fully quantified experimentally	Some aspects simplified and quantified	Economic criteria added and quantified	
STATISTICAL	X					Information from data
DIORISTIC		X				Distinguish components Planning research and teaching Holistic description
COMPONENT		X	X			Complete quantification of system. Transfer functions experimentally determined
STRATEGIC		X		X		Analysis or simulation of complex systems.
MANAGEMENT		X		X	X	Decisions on management of complex systems.

Fig. 12.2 Outline classification of models by form of system and objective.

present classification is designed to reflect the function of the model and the complexity of the systems analysis involved (Fig. 12.2). Other modes of classification ('deterministic v. stochastic', 'simulation v. analytical' and 'differential v. difference' are discussed later).

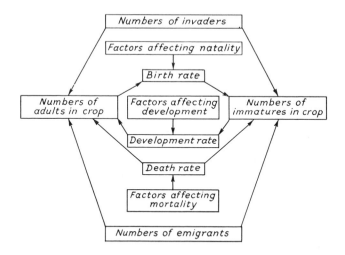

Fig. 12.3 A qualitative dioristic model: the population system of a pest (after Southwood and Way, 1970).

12.1.1 Dioristic* models

These serve to define the relationships under study and help the investigator to consider the processes in a logical and complete functional framework. In its simplest form such a model may be entirely qualitative and will describe the system in words (Fig. 14.5), this is the first step in systems analysis. The next step is to determine the quantititive relationship between the components of the system: fundamental biological knowledge may be adequate to suggest their basic form (additive, multiplicative, etc.) (Fig.12.4). Models of this type emphasize generality and realism and have an educational function. Their construction at an early stage in the design of ecological field work will often greatly facilitate the planning of the work, forcing the investigator to consider the various inputs and outputs, the system itself and its links. Southwood (1968), Klomp (1973) and Stark (1973) give other examples of dioristic models.

12.1.2 Component models

These models incorporate experimental data and serve as a test of the correctness (or otherwise) of our understanding of the components of the

* 'dioristic – serving to define' (Oxf. Eng. Dict.), from a Greek word meaning to distinguish or separate, as ordaining the division of a land by boundaries.

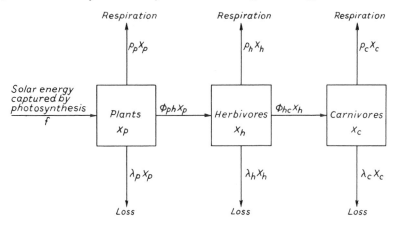

Fig. 12.4 A linear dioristic model with algebraically defined transfer functions: energy flow through three trophic levels (after Williams, 1971).

system, and the form and magnitude of the transfer functions. They may be viewed as the next step in complexity from the algebraically quantified dioristic model. An early and sound example of this type of model is provided by Holling's (1963, 1965, 1966) study on the response of predators to prey density (Fig.12.5). Experiments are established to measure the precise form and magnitude of the relationship (transfer function) of each component, the expression for each of these is in itself a model (e.g. Table 9.2). A computer programme is constructed on the model, the starting (initial, input) conditions (state variables) and various functional relationships are fed in, and the result can be compared with actual observations. If they agree the model is sound, if they disagree it is unsound: a relationship may be incorrect, a feed-back link missed or even a component overlooked. These errors must be sought. Kitching's (1971) model of the dispersal of animals in relation to habitat units is essentially a component model, as are Hardman's (1976*a* and *b*) models of population growth in an unlimited environment, Dill's (1973) avoidance learning model, and that of Coulson *et al.* (1976) on interspecific competition effects. Component models aim at completeness: thus the system has to be simple or the complexity becomes overwhelming.

12.1.3 Strategic models
These models serve to predict an outcome from a system given certain initial conditions. Component models are also predictive in the final test, but strategic models involve an additional (and difficult) step in their construction – the decision as to which links characterize the system; which are underlying its strategy and hence what are the 'key' transfer functions. In this they represent some measure of simplification: whereas component analysis aims to split the system down to every component, strategic models treat some parts as 'black

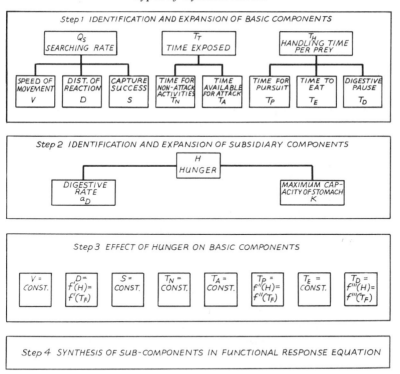

Fig. 12.5 The steps taken in the component analysis of the response of predators to prey density – the roles of successful search, time exposed, handling time and hunger (after Holling, 1963).

boxes'. There are basically three approaches: deductive ('building down'), mixed deductive–inductive and inductive ('building up').

DEDUCTIVE AND MIXED MODELS

The initial conditions are based on a general theory, like the Lotka-Volterra equations; then certain additional conditions that appear to accord with reality are applied, and the consequences determined by orthodox (though often very sophisticated) mathematical procedures e.g. stability analysis. Such studies then give general conclusions, e.g. 'the population damps exponentially (under-compensates) if the return time is greater than generation time' (May *et al.* 1974). Levins (1968), May (1973, 1976) and Maynard Smith (1974) provide many examples of these types of model. One of the most important and skilful parts of such modelling is to determine the significant relationships and eliminate third and fourth order effects: without such a process the models soon become so complex as to obscure any general conclusions. This type of modelling requires considerable mathematical dexterity, as well as biological intuition.

The combination of predictions from general theory with actual data from field observations leads to the construction of the mixed model. The data are used to provide evidence of the actual form and magnitude of certain of the function relationships. These mixed inductive–deductive models have proved to give valuable insights into the understanding of population processes. Examples of mixed models are those of Hassell & Rogers (1972), Hassell & May (1974), Hassell (1975), Lawton, Hassell & Beddington, (1975), Hassell, Lawton & May (1976). Southwood & Comins (1976) and Kiritani (1977) (see also p. 395).

INDUCTIVE MODELS
A particular system is studied, its principal relationships recognized and measured, and formulated into the model. These models derive from the component type of model, but the modeller recognizes that he is simplifying the system, combining certain components into a single compartment (Wiegert, 1973). The predictive population models of Gilbert and others are of this type and are discussed on p. 395. Examples for populations are provided in Hughes & Gilbert (1968), Gilbert & Gutierrez (1973), Meats (1974), Broadhead & Cheke (1975) and Gilbert *et al.* (1976), for bioenergetics by Hubbell (1971) and for ecosystems by Wiegert (1973) and Lugo *et al.* (1976). If the predictions accord with reality, then the modeller's selection of compartments is likely to have been correct, and conclusions may be drawn as to the relative significance of different factors in natural systems (e.g. Fraleigh & Wiegert, 1975).

These strategic models grade almost imperceptibly into the next type of model, as the range of prediction is increased and non-biological inputs are included.

12.1.4 Management models
These serve to provide guidance as to the outcome from the system of various changes that can be brought about by man (management). They differ in concept from the strategic type of model described above in that they often seek to predict the system's behaviour under conditions that may lie beyond the limits of the data from which it has been constructed, and they always include economic (or at least quasi-economic) criteria. Because of the latter the model will often be developed so as to balance certain components against others, the process of optimization.

There are different forms of management models and, as Conway & Murdie (1972) point out, the difference lies in the emphasis that is placed on the various characters, generality, realism and accuracy. Teaching models, such as that of Conway & Murdie, emphasize generality and are essentially dioristic models. In a wide-ranging review of applied models Conway (1977*b*) classifies them into:

(1) *Policy models* in which one might say the emphasis is on realism; they are designed to provide general guidelines for management and have been

extensively developed in fisheries from the classic studies of M. Graham, M.B. Schaefer and Beverton & Holt (1957), and in epidemiology from Ross and Macdonald's models (Conway 1977*b*).

(2) *Tactical models* where the emphasis is on accuracy, and which hope to provide day-to-day guidance on the management of a particular resource. As Conway (1977*b*) says, they have a formidable task and "few if any have shown practical success".

Although there is no fundamental difference between them, management models for individual populations (or a handful of related populations) will be considered separately from those for ecosystems (or biomes) because the massive complexity of the latter poses problems of system characterization of a different order of magnitude.

POPULATION MANAGEMENT MODELS.

The primary objectives of such models are to control population size, either to reduce it (pest control models), or to maximize it (harvest or yield models). Harvest models are sometimes concerned with maximizing yield, but because of the self-renewing properties of biological populations they are generally concerned with optimizing yield, the 'maximum sustainable yield': fisheries biology provides many examples of such models e.g. Paulik & Greenough (1966); some recent developments are reviewed by Conway (1977*b*). Epidemiological models for vector-borne diseases follow Macdonald's lead, and generally determine the level to which the vector population needs to be reduced to break transmission, or the appropriate combination of prophylaxis and vector control (Dietz *et al.*, 1974; Bradley & May, 1978).

Pest control models have two aspects: 'crop loss'—predicting the balance of economic loss and gain at different levels of pest control (Norton, 1976); and 'control'—predicting the effect of specified measures, commonly called 'control measures', that are designed to lower (or restrain) the population level. Many models on biological control (Huffaker *et al.*, 1977) or sterile male technique (reviewed by Conway, 1973) are of this latter type. Although still within the policy model definition, models that bring in crop loss and control features approach the tactical type. Examples are Murdie & Campion (1972) for the red bollworm and cotton, Conway *et al.* (1975) on the sugar cane froghopper and Regev *et al.* (1976) on the alfalfa weevil. Cotton pest management in general is considered by Blood *et al.* (1975) and Gutierrez *et al.* (1977): an even wider look at a pest's ecosystem has been made by Holling *et al.* (1977) in which they have investigated many aspects of the interactions between forest management policy and the spruce budworm. Additionally there is a 'family' of highly mathematical pest control models, that, although not yet providing practical insights (because their biological assumptions are unrealistic), suggest approaches to optimization (Becker, 1970; Mann, 1971; Chatterjee, 1973). Reviews of pest control models are given by Conway (1973, 1977*a*) and Ruesink (1976).

ECOSYSTEM (BIOME) MODELS:

The IBP, particularly its grassland and desert biome studies, has given a greater impetus to work on modelling these situations. The approach has been to classify the interactions into hierarchies, representing trophic levels. This inevitably involved pooling species, but nevertheless the models are highly complex (Goodall, 1972*a* & *b*; 1974; Van Dyne, 1972). Recently Walker *et al.* (1978) have proposed a somewhat different approach based on the identification of the 'key' ecological processes and the 'key' management questions: the relevant models are prepared for these questions. The key components are identified from a review of the existing biological (and climatological) data, the patterns of change and perturbation. This approach relies on dioristic modelling to give holism, the individual models are more tractable and accurate and hence provide results of direct value to decision makers.

12.1.5 Statistical models

As utilized these are quite distinct from the types 12.1.1–12.1.4, in all of which the investigator is concerned with the description of the system. In statistical and related models the system has been fully defined and the model serves to facilitate the handling by the computer of data for procedures such as multivariate analysis (p. 435), mark and recapture population estimation (p. 92) and the estimation of the numbers entering a stage (p. 357). When originally developed, these would be component models (e.g. equations in Table 9.2). Davies (1971) gives programmes for many standard procedures: more complex models are used in the analysis and description of vegetation data (see papers in Jeffers 1972).

12.2 Analysis and simulation in models

Analytical models provide general information about a system; they describe its characteristics in a handful of relationships and parameters, e.g. 'if the host has a high rate of increase, unstable population cycles will develop in a host-parasite system unless the mutual interference constant is greater than a half' (Hassell & May, 1973). The solution of such models seldom demands elaborate computing facilities and they may then be used as virtual statistical models to extract parameters (e.g. equation 10.33 as used by Stubbs (see p. 380)).

Simulation models provide information on the behaviour of the system under certain specific conditions and normally demand computer facilities, often with significant storage. Wiegert (1973; 1974) for example, simulates the growth of algal and 'brine' fly populations under various conditions, and Sollins *et al.* (1976) the effect of defoliation on deciduous forests. Theoretically, analog computers would provide a rapid means of gaining insights from simulations, but digital machines are usually used because of their availability and accuracy and the properties of simulation languages (Brennan *et al.* 1970).

Wiegert (1975) provides a thorough review of simulation, together with an introduction to the techniques of modelling; whilst a stochastic simulation model, particularly for insect populations, has been designed by Fujii (1975).

12.3 Deterministic and stochastic models

Deterministic models assume a fixed and unique outcome, e.g. 50 % mortality in a deterministic model means that in a population of 10 there would be 5 deaths, in one of 1000 there would be 500; but many outcomes are not fixed. When the ecological process can result in several possible outcomes then a distribution is involved and a probability can be defined for each event. A model that incorporates probabilities is 'stochastic' e.g. 50 % may be the mean mortality, but it may take values between 40 and 60 %, with equivalent range in the number of deaths realized.

Stochastic models therefore tend to be more complex than deterministic; Watt (1968) and May (1973) concluded that demographic stochasticity will only become manifest when populations are small, a theoretical insight confirmed experimentally by Gilbert & Hughes (1971) and Hardman (1976b) (see p. 399). A stochastic element may be inserted in a model when appropriate, by selecting the value of the variable (within an appropriate range, e.g. 10 % of the mean, two standard deviations) by a random number generator or by a Monte Carlo technique (Hammersby & Handscomb, 1964).

12.4 Difference and differential models

Traditionally, mathematical modelling is undertaken using the language of differential calculus, which assumes that the process is continuous. With some populations where there is either a stable age distribution (e.g. man) or very rapid reproduction (e.g. bacteria), this provides a reasonable description. However, in many populations, especially of temperate insects, reproduction is at discrete time intervals and the effects of these time delays (and of thresholds) are best illustrated by difference equations (Holling, 1964; May *et al.* 1974; Hassell, 1978). These equations (Goldberg, 1961) express the change in a variable (e.g. population size) in relation to its initial value and the time interval, i.e. the difference in magnitude over a certain interval of time. Complex models based on difference equations are more easily translated into computer language than those based on differential equations.

REFERENCES

BARTLETT, M. S. and HIORNS, R. W. (eds.), 1973. *The Mathematical Theory of the Dynamics of Biological Populations.* 347pp., Academic Press, London.
BECKER, N. G., 1970. Control of a pest population. *Biometrics* **26** (3), 365–75.

BEVERTON, R. and HOLT, R. J., 1957. On the dynamics of exploited fish populations. *Fisheries Investigations Ser.* **2**, (19), 533pp. *Min. Agric. Fish. Food*, HMSO, London.

BLOOD, P. R. B., LONGWORTH, J. W. and EVENSON, J. P., 1975. Management of the cotton agroecosystem in Southern Queensland: a preliminary modelling framework. *Proc. Ecol. Soc. Aust.* **3**, 230–49.

BRADLEY, D. J. and MAY, R. M., 1978. Consequences of helminth aggregation for the dynamics of Schistosomiasis. *Proc. R. Soc. trop. Med. Hyg.* **72**, (in press).

BRENNAN, R. D., DE WIT, C. T., WILLIAMS, W. A. and QUATTRIN, E. V., 1970. The utility of a digital simulation language for ecological modelling. *Oecologia* **4**, 113–32.

BROADHEAD, E. and CHEKE, R. A., 1975. Host spatial pattern, parasitoid interference and the modelling of the dynamics of *Alaptus fusculus* (Hym.: Mymaridae) a parasitoid of two Mesopsocus species (Psocoptera). *J. Anim. Ecol.* **44**, 767–93.

CHATTERJEE, S., 1973. A mathematical model for pest control. *Biometrics* **29** (4) 727–34.

CONWAY, G. R., 1973. Experience in insect pest modelling: a review of models, uses and future directions. *Mem Aust. Ecol. Soc.* **1**, 103–30.

CONWAY, G. R., 1977a. The utility of systems analysis techniques in pest management and crop production. *Proc. XV. Int. Congr. Ent.* 541–52.

CONWAY, G. R., 1977b. Mathematical models in applied ecology. *Nature* **269**, 291–7.

CONWAY, G. R. and MURDIE, G., 1972. Population models as a basis for pest control *In* Jeffers, J. N. R. (ed.) *Mathematical Models in Ecology. (Symp. Brit. Ecol. Soc.)* **12**, 195–213.

CONWAY, G. R., NORTON, G. A., SMALL, N. J. and KING, A. B. S., 1975. A systems approach to the control of the sugar cane froghopper. *In* Dalton, G. E., (ed.) *Study of Agricultural systems.* pp. 193–229.

COULSON, R. N., MAYYASI, A. M., FOLTZ, J. L. and HAIN, F. P., 1976. Interspecific competition between *Monochamus titillator* and *Dendroctonus frontalis*. *Environ. Ent.* **5**, 235–47.

DAVIES, R. G., 1971. *Computer Programming in Quantitative Biology.*, 492pp., Academic Press, London and New York.

DIETZ, K., MOLINEAUX, L. and THOMAS, A., 1974. A malaria model tested in the African savannah. *Bull. Wld. Health Org.* **50**, 347–57.

DILL, L. M., 1973. An avoidance-learning sub-model for a general predation model. *Oecologia* **13**, 291–312.

FRALEIGH, P. C. and WIEGERT, R. G., 1975. A model explaining successional change in standing crop of thermal blue-green algae. *Ecology* **56** (3), 656–64.

FUJII, K., 1975. A general simulation model for laboratory insect populations I. From cohort of eggs to adult emergences. *Res. Popul. Ecol.* **17**, 85–133.

GILBERT, N. and GUTIERREZ, A. P., 1973. An aphid-plant-parasite relationship. *J. Anim. Ecol.* **42**, 323–40.

GILBERT, N., GUTIERREZ, A. P., FRAZER, B. D. and JONES, R. E., 1976. *Ecological Relationships.* 157pp. Addison-Wesley, Reading & San Francisco.

GILBERT, N. and HUGHES, R. D., 1971. A model of an aphid population – three adventures. *J. Anim. Ecol.* **40**, 525–34.

GOLDBERG, S., 1961. *Introduction to Difference Equations*, Wiley, New York.

GOODALL, D. W., 1972a. Building and testing ecosystem models. *In* Jeffers, J. N. R. (ed.) *Mathematical Models in Ecology. Symp. Brit. Ecol. Soc.* **12**, 173–94.

GOODALL, D. W., 1972b. Potential applications of biome modelling. *Terre et la Vie* **No. 1**, 118–38.

GOODALL, D. W., 1974. The hierarchical approach to model building. *Proc. 1st int. Congr. Ecol., The Hague, 1974.*, 244–9.

GUTIERREZ, A. P., DE MICHELE, D. W. and WANG, Y., 1977. New systems technology for cotton production and pest management. *Proc. XV Int. Congr. Ent.* 553–9.

HAMMERSBY, S. M. and HANDSCOMB, D. C., 1964. *Monte Carlo Methods.*, Methuen, London.

HARDMAN, J. M., 1976a. Life table data for use in deterministic and stochastic simulation models predicting the growth of insect populations under Malthusian conditions. *Can. Ent.* **108**, 897–906.

HARDMAN, J. M., 1976b. Deterministic and stochastic models simulating the growth of insect populations over a range of temperatures under Malthusian conditions. *Can. Ent.* **108**, 907–24.

HASSELL, M. P., 1975. Density-dependence in single-species populations. *J. Anim. Ecol.* **44**, 283–95.

HASSELL, M. P., 1978. *The Dynamics of Arthropod Predator-Prey Systems.* Princeton Monograph in Population Biology, **13**, 245pp., Princeton, New Jersey.

HASSELL, M. P., LAWTON, J. H. and MAY, R. M., 1976. Patterns of dynamical behaviour in single-species populations. *J. Anim. Ecol.* **45**, 471–86.

HASSELL, M. P. and MAY, R. M., 1973. Stability in insect host-parasite models. *J. Anim. Ecol.* **42**, 693–726.

HASSELL, M. P. and MAY, R. M., 1974. Aggregation in predators and insect parasites and its effect on stability. *J. Anim. Ecol.* **43**, 567–94.

HASSELL, M. P. and ROGERS, D. J., 1972. Insect parasite responses in the development of population models. *J. Anim. Ecol.* **41**, 661–76.

HOLLING, C. S., 1963. An experimental component analysis of population processes. *Mem. Ent. Soc. Can.* **32**, 22–32.

HOLLING, C. S., 1964. The analysis of complex population processes. *Can. Ent.* **96**, 335–47.

HOLLING, C. S., 1965. The functional response of predators to prey density and its role in mimicry and population regulation. *Mem. ent. Soc. Can.* **45**, 5–60.

HOLLING, C. S., 1966. The functional response of invertebrate predators to prey density. *Mem. ent. Soc. Can.* **48**, 3–86.

HOLLING, C. S., JONES, D. D. and CLARK, W. C., 1977. Ecological policy design; a case study of forest and pest management. *In* Norton, G. A. & Holling, C. S. (eds.) *Proc. IIASA Pest Management Confr.* pp. 13–90, IIASA, Luxemburg.

HUBBELL, S. P., 1971. Of sowbugs and systems: the ecological bioenergetics of a terrestrial Isopod. *In* Patten, B. C. (ed.) *Systems Analysis & Simulation in Ecology* **1**, 269–324. Academic Press, London and New York.

HUFFAKER, C. B., LUCK, R. F. and MESSENGER, P. S., 1977. The ecological basis of Biological Control. *Proc. XV Int. Congr. Ent.*, 560–86.

HUGHES, R. D. and GILBERT, N., 1968. A model of an aphid population – a general statement. *J. Anim. Ecol.* **37**, 553–63.

JEFFERS, J. N. R. (ed.) 1972. *Mathematical Models in Ecology, Symp. Brit. Ecol. Soc.* **12**, 398pp., Blackwells, Oxford.

KIRITANI, K., 1977. Systems approach for management of rice pests. *Proc. XV Int. Congr. Ent.*, 591–8.

KITCHING, R., 1971. A simple simulation model of dispersal of animals among units of discrete habitats. *Oecologia* **7**, 95–116.

KLOMP, H., 1973. Population dynamics: a key to the understanding of integrated control. *Mem. Ecol. Soc. Aust.* **1**, 69–79.

LAWTON, J. H., HASSELL, M. P. and BEDDINGTON, J. R., 1975. Prey death rates and rate of increase of arthropod predator populations. *Nature* **255**, 60–2.

LEVINS, R., 1968. *Evolution in Changing Environments.* Princeton University Press, Princeton, New Jersey.

LUGO, A. E., SELL, M. and SNEDAKER, S. C., 1976. Mangrove Ecosystem Analysis. *In* Patten, B. C. (ed.) *Systems Analysis & Simulation in Ecology*. **4**, 113–45. Academic Press, London and New York.

MANN, S. H., 1971. Mathematical models for the control of pest populations. *Biometrics* **27**(2), 357–68.

MAY, R. M., 1973. On relationships between various types of population model. *Am. Nat.* **107**, 46–57.

MAY, R. M., 1976. *Theoritical Ecology*, Blackwell, Oxford, Chapters 2, 4 & 8.

MAY, R. M., CONWAY, G. R., HASSELL, M. P. and SOUTHWOOD, T. R. E., 1974. Time delays, density-dependence and single-species oscillations. *J. Anim. Ecol.* **43**, 747–70.

MAYNARD SMITH, J., 1974. *Models in Ecology*. Cambridge University Press, Cambridge.

MEATS, A., 1974. Simulation of population trends of *Tipula paluolosa* using a model fed with climatological data. *Oecologia* **16**, 139–47.

MURDIE, G. and CAMPION, D. G., 1972. Computer simulation of Red Bollworm populations in control programmes using sterile males and sex attractants. *Cotton. Grow. Rev.* **49**, 276–84.

NORTON, G. A., 1976. Analysis of decision-making in crop protection. *Agroecosystems* **3**, 27–44.

PATTEN, B. C., 1971. A primer for ecological modelling and simulation with analog and digital computers. *In* Patten, B. C. (ed.) *Systems Analysis and Simulation in Ecology* **1**, 3–121. Academic Press, London and New York.

PATTEN, B. C., 1971–76. *Systems Analysis and Simulation in Ecology*, Vols. **1–4**, Academic Press, London and New York.

PAULIK, G. J. and GREENOUGH, J. W., 1966. Management analysis for a Salmon Resource System. *In* Watt, K. E. F. (ed.) *Systems Analysis in Ecology*, 245–52. Academic Press, London and New York.

REGEV, U., GUTIERREZ, A. P. and FEDER, G., 1976. Pests as a common property resource: a case study of alfalfa weevil control. *J. Am. Agric. Econ.* **58**, 186–97.

RUESINK, W. G., 1976. Status of the systems approach to pest management. *Ann. Rev. Ent.* **21**, 27–44.

SOLLINS, P., HARRIS, W. F. and EDWARDS, N. T., 1976. Simulating the physiology of a temperate deciduous forest. *In* Pattern, B. C. (ed) *Systems Analysis and Simulation in Ecology* **4**, 173–220. Academic Press, London and New York.

SOUTHWOOD, T. R. E., 1968. The Abundance of Animals. *Inaug. Lect. Imp. Coll. Sci. Technol.* **8**, 1–16.

SOUTHWOOD, T. R. E. and COMINS, H. N., 1976. A Synoptic Population Model. *J. Anim. Ecol.* **45**, 949–65.

SOUTHWOOD, T. R. E. and WAY, M. J., 1970. Ecological background to pest management. *In* Rabb, R. L. & Guthrie, F. *Concepts of Pest Management*: 6–28. N. Carolina State Univ. Press.

STARK, R. W.; 1973. The systems approach to insect pest management – a developing program in the United States of America: the pine bark beetles. *Mem. Ecol. Soc. Aust.* **1**, 265–73.

STREIFER, W., 1974. Realistic models in Population Ecology. *Adv. Ecol. Res.* **8**, 200–66.

VAN DYNE, G. M. (ed.), 1969. *The Ecosystem Concept in Natural Resource Management*. 383pp., Academic Press, New York.

VAN DYNE, G. M., 1972. Organisation and management of an integrated ecological research program – with special emphasis on systems analysis, universities and scientific cooperation. *In* Jeffers, J. N. R. (ed.) *Mathematical Models in Ecology, Symp. Brit. Ecol. Soc.* **12**, 111–72.

WALKER, B. H., NORTON, G. A., CONWAY, G. R., COMINS, H. N. and BIRLEY, M., 1978. A procedure for multidisciplinary ecosystem research: with reference to the South

African Savanna Ecosystem Project. *J. appl. Ecol.* **15**.

WATT, K. E. F., 1961. Mathematical models for use in insect pest control. *Can. Ent.* **93** (suppl. 19), 62pp.

WATT, K. E. F. (ed.), 1966. *Systems Analysis in Ecology.* 267pp. Academic Press, London and New York.

WATT, K. E. F., 1968. *Ecology and Resource Management.* 450pp., McGraw-Hill, New York.

WIEGERT, R. G., 1973. A general ecological model and its use in simulating algal-fly energetics in a thermal spring community. *In* Geier, P. W., Clark, L. R., Anderson, D. J. & Nix, H. A. (ed.) *Insect: Studies in population management. Mem. Ecol. Soc. Aust.* **1**, 11–30.

WIEGERT, R. G., 1974. Simulation modelling of the algal fly components of a thermal ecosystem: effects of spatial heterogeneity, time delays and model condensation. *In* Patten, B. C. (ed.) *Systems Analysis and Simulation in Ecology* **3**, 157–81. Academic Press, London and New York.

WIEGERT, R. G., 1975. Simulation models of ecosystems. *Ann. Rev. Ecol. Syst.* **6**, 311–38.

WILLIAMS, R. B., 1971. Computer simulation of energy flow in Cedar Bog Lake, Minnesota based on the classical studies of Lindeman. *In* Patten, B. C. (ed.) *Systems Analysis and Simulation in Ecology* **1**, 543–82. Academic Press, London and New York.

13

Diversity, Species Packing and Habitat Description

A great deal of time and expertise has been expended on the compilation of faunal lists for particular habitats, but the consequent increase in our understanding of the structure and functioning of animal communities or the impact on them of natural or man-induced change has been meagre. In the last two decades a substantial body of quantitative theory on communities has been developed, especially by the 'Hutchinson-MacArthur school', that allows both the organization of data on communities and also poses a number of hypotheses that require confirmation, refinement or refutation by field data. It is axiomatic that the objectives and methods of analysis should be fully and carefully considered, before the field programme is undertaken. The methods of sampling are those outlined in Chapters 2 and 4 to 7, although the differential response of species to trapping is an indication for caution in respect of many of the methods in Chapter 7. The present chapter aims to describe the methods of handling and analysing data on animal communities (from guilds to continental faunas). Details of the theoretical studies and examples of field work are given in MacArthur and Wilson (1967), MacArthur (1972), May (1973, 1976*a* & *b*), Cody and Diamond (1975) and Pianka (1976*a* and *b*).

13.1 Diversity

Diversity, even restricted as it is (for the purposes of this chapter) to the variety of animal species, is one of those common-sense ideas that prove elusive and multifaceted when precise quantification is sought (Peet, 1974; Mound and Waloff, 1978). A useful classification, due to Whittaker (1972), is:

$\alpha - diversity$: the diversity of species within a community or habitat.

$\beta - diversity$: a measure of the rate and extent of change in species along a gradient, from one habitat to others.

$\gamma - diversity$: the richness in species of a range of habitats in a geographical area (e. g. island); it is a consequence of the alpha diversity of the habitats together with the extent of the beta diversity between them.

α and γ diversity are thus qualities that simply have magnitude and could, theoretically, be described entirely by a single number (a scalar). β diversity in contrast is analogous to a quality that requires both magnitude and direction

420

to characterize it fully (a vector). Their descriptions therefore require different approaches.

The restriction of the measurement of diversity to the consideration of the numbers of species, and the failure to take account of the variety, or otherwise, of form represented by these species has been criticized by Van Valen (1965), Hendrickson and Ehrlich (1971) and Findley (1973). The latter has suggested that taxonomic distance between the components of a fauna should be used to give a measure of its 'phenetic packing'.

13.1.1 Description of α and γ diversity

When a fauna (or a flora) is sampled it is found that a few species are represented by a lot of individuals and a large number of species by few individuals. These relative abundances must be considered to represent the basic pattern of niche utilization in the community (or area) and one approach to diversity has been to seek to expose the distributions that underlie the species patterns; others have been more pragmatic and sought the best description (e.g. Williams, 1964). The result has been an 'explosive speciation' of diversity indices (perhaps only rivalled in ecological methodology by 'new designs' for light traps*), that has brought confusion to the subject; to this, the ubiquitousness of some relationships and the apparent constancy of certain numerical values have added a measure of mystique. A penetrating analysis by May (1975a) of the underlying mathematical similarity of the assemblage of indices, and a careful systematic review by Whittaker (1972) should allow the subject to develop more rationally. Other useful reviews are by Cancela da Fonseca (1968, 1969) and Pielou (1975).

If diversity data are collected continuously (e.g. by trapping) a graph of the type shown in Fig. 13.1a may be prepared. Eventually the curve virtually flattens out at a value which is considered to represent the total species (S_T) for the habitat or area. This is the 'species equilibrium' (MacArthur and Wilson, 1967); if the survey time is prolonged it will be detected that the curve continues to 'creep' upwards as new species are added to the fauna by colonization (others may be lost by extinction, but this loss will not show on Fig. 13.1a); the rate of this change in the composition of the fauna will depend on the rate of change (the durational stability) of the habitat (Wilson, 1969; Southwood, 1977) (see p. 450). This number S_T is a straightforward measure of diversity, but it can only be found from a complete survey (which is virtually impossible for insects in most habitats); as may be seen from Fig. 13.1a, during the course of the survey the relationship between the cumulative sample of individuals (ΣN) and the number of species (ΣS) is variable, it is dependent on sample size. If the form of the relationship between S and N is found then the parameter(s) of this allow the characterization of slopes a and b (Fig.

---

* However, I have noted a distinction: whereas the inventors of 'new' indices often roundly condemn their predecessors, the purveyors of 'new' light traps seem completely oblivious of the burgeoning literature that testifies to the parallelism in human thought!

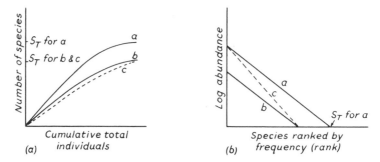

Fig. 13.1 The concepts of species richness and equitability (for evenness) as shown in two simple presentations of diversity data for three habitats *a*, *b* and *c*; *a* and *b* have similar equitability, but different richness, whilst *b* and *c* have similar richness, but different equitability ($S_T=$ 'total' species). *a*. The data presented during the survey showing the rate of accumulation of species against the cumulative total of individuals collected; *b*. The data presented at the end of this 'complete' survey with the log total individuals per species ranked so that the most common species is on the left.

13.1*a*) from any sample size: this line of reasoning led Fisher, Corbet and Williams (1943) to propose the first (and most durable) of the indices (α).

The data may also be considered in their totality and if the log number of individuals (or abundance) of each species is plotted against the rank, the plot will approximate to a straight line (Fig. 13.1*b*) (see also below and Fig. 13.2).

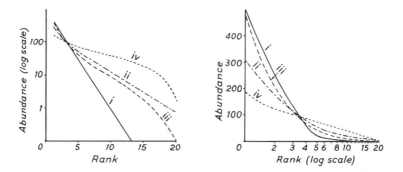

Fig. 13.2 The rank-abundance (or dominance-diversity curves for different underlying distributions when plotted with the log scale on different axes: *a*. log abundance (individuals): rank (species sequence); *b*. abundance: log rank. i = geometric series, ii = log series, iii = log normal, iv = MacArthur broken stick. (after Whittaker, 1972).
Note: the exact relative positions of the curves depend on the assumptions about the parameters. Whilst the broken stick and canonical log normal are defined by one parameter (S_T, here taken as 20), geometric and log series and the general log normal are defined by two parameters: S_T and J, where $J = N_T/m$, *m* being the number of individuals of the rarest species, often in practice one.

Such plots emphasize that the species number (*S*): abundance of individuals (*N*) relationship has two features:
(1) species richness – the total number of species present in the area (S_T)

(2) equitability or evenness – the pattern of distribution of the individuals between the species (a faunal sample of 100 individuals representing 10 species could consist of ten individuals of each (extreme equitability) or, at the other extreme 91 individuals of one species (the dominant) and one each of the other nine).

MODELS FOR THE S:N RELATIONSHIP

The equitability of the species: abundance relationship will be a reflection of the underlying distributions. Four main groups can be recognized: the geometric series, the logarithmic series, the lognormal and MacArthur's 'broken stick' model. The two former will give approximately straight lines when log abundance is plotted against rank (Fig. 13.1 *b* and 13.2*a*), whilst the latter shows up as a straight plot when abundance is plotted against log rank (Fig. 13.2*b*) (Whittaker, 1972; May, 1975*a*).

Geometric series.

If a species occupies a fraction (k) of a habitat, another species the same fraction of the remainder, a third species k of what now remains and so on, the resulting rank-abundance distribution will, as originally shown by I. Motomura, be a geometric series (May, 1975*a*). Termed the 'niche-pre-emption hypothesis', this distribution has often been used to describe floral diversity in temperate regions (Whittaker, 1970, 1972; McNaughton and Wolf, 1970). May (1975*a*) shows that the Odum, Cantlon and Kornicher (1960) formulation (which is expressed in the terms of species and numbers (Fig. 13.1*a*), rather than rank: abundance (Fig. 13.1*b*)) is the equivalent of this series.

Logarithmic (or log) series

Originally suggested as a suitable description of species abundance data by R. A. Fisher (Fisher, Corbet and Williams, 1943) this distribution has been criticized as lacking a theoretical justification at the level of species interaction. Kempton and Taylor (1974) and May (1975*a*, 1976*b*) have provided this. The log series may be regarded as the convenient approximation to gamma model (of which the negative binomial is a well known form – see p. 28) with maximal variance (Kempton and Taylor, 1974). Gamma models often arise as a result of a two stage (compound) process (see p. 34); as May (1975*a*) points out, if we envisage a geometric series type of niche-pre-emption where the fractions of niche pre-empted have arisen from the arrival of successive species at uniform intervals of time, the arrival of the species at random time intervals would lead to the log series (see also Boswell and Patil, 1971). This distribution approximates to a straight line on a rank: log abundance plot (Fig. 13.2*a*) and is described by the parameters S_T and α; the index of diversity:

$$S_T = \alpha \, \log_e (1 + N/\alpha) \qquad (13.1)$$

or in terms of a sampling parameter X, for the expected number of species in a total sample of N individuals:

$$S_{(N)} = \alpha X^N / N \tag{13.2}$$

Lognormal distribution

Preston (1948) suggested that the log normal distribution would give the best description of species-abundance patterns; the assumption being that individuals were distributed between species in accordance with the normal or Gaussian distribution and population growth is geometric (Williams, 1964). May (1975a) enlarges on the biological reasons why the lognormal might be considered to apply to both opportunistic or equilibrium communities. Preston (1948, 1960, 1962) considered the plot of the frequency of species against abundance classes on a log scale (Fig. 13.3). He used logarithms to the

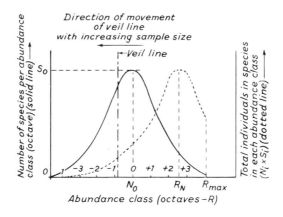

Fig. 13.3 The log normal representation of abundance and species relationships (modified from May, 1975a).

base 2 so that each class, or octave, involved a doubling of the size of the population. (The abundance classes could be on any logarithmic scale). The abundance of the species at the peak (mode) of the distribution is N_0 and the abundance octaves (R) are calculated:

$$R_i = \log_2(N_i / N_0) \tag{13.3}$$

Preston suggested that sample size is usually too small to obtain the species in the lower (rarer) octaves; these species were, he said, hidden behind the 'veil line'. Field data will only represent those species to the right of the veil line and therefore from Fig. 13.3 it will be seen that if the veil line is at N_0 or in any of the positive octaves the plot will approximate a straight line and be of the form shown in Fig. 13.1b (with axes reversed). Preston recognized that the plot of individuals per class against abundance class (dotted line in Fig. 13.3) would not describe the complete 'bell', but be truncated at its crest (the 'canonical condition') so that $R_N = R_{max}$ (Fig. 13.3). This truncated log normal, ending at the crest (unlike Fig. 13.3) is difficult to fit to data, although Bullock (1971a) and Gage and Tett (1973) did so. Kempton and Taylor (1974) and Bulmer (1974) showed that it is more tractable if regarded as a Poisson log normal, described by the parameters S_T and σ (variance). As May (1975a) shows, other parameters may be chosen, commonly:

$$a = (2\sigma^2)^{-1/2} \tag{13.4}$$

and

$$\gamma = R_N/R_{max} = \log_e 2/2a(\log_e S_0)^{1/2} \tag{13.5}$$

Preston (1962) analysed a considerable body of field data and found that $\gamma \sim 1$ was general; likewise Hutchinson (1953) that $a \sim 0.2$. A rational explanation for both has been given by May (1975a); both arise from the mathematics inherent in the log normal given a large value of S_T.

MacArthur's broken stick model
In a theoretical consideration of how niche hyper-volume might be divided MacArthur (1957) postulated three models. One, termed the 'broken stick', assumes niche boundaries are drawn at random and the abundance of the ith most abundant species is given by:

$$N_i = \frac{N_T}{S_T} \sum_{n=i}^{S_T} \frac{1}{n} \tag{13.6}$$

The mathematics of this model have been investigated by Webb (1974); it is more even than the log normal and characterized simply by the parameter S_T. A ratio commonly calculated in faunistic studies is J, defined as the total number of individuals (N_T) divided by the abundance of the rarest species. From the above equation it is clear that for the MacArthur model this will be N_T/S_T^2, and as May (1975a) points out, then $J = S_T^2$.

NON-PARAMETRIC INDICES
The description of the $S:N$ relationship in terms of the parameters of a model implies that the model is at least approximately applicable. It may be argued that non-parametric indices have the advantage that they make no assumptions of this type. There are also relative indices, such as Lloyd and Ghelardi's (1964), that compare the equitability of the distribution with that which would arise if the same sample was distributed according to a certain model, in their case the MacArthur model. Peet (1975) shows that such relative indices possess mathematically undesirable properties. Comparative reviews of these indices are provided by Hurlbert (1971), Whittaker (1972), De Benedictis (1973), De Jong (1975) and Pielou (1975). Three indices, whose relationships to the different models are illuminated by May (1975a), are:

Shannon–Weaver * *function (H)*

A function devised to determine the amount of information in a code, and defined:

$$H = - \sum_{i=1}^{S_T} p_i \log_e p_i \qquad (13.7)$$

where p_i = the proportion of individuals in the ith species and S_T (as before) the total species or in terms of species abundance:

$$H = \log_e N - \frac{1}{N} \sum_{i=1}^{\infty} (p_i \log_e p_i) n_i \qquad (13.8)$$

where n_i = the number of species with i individuals.

Bulmer (1974) and May (1975a) have shown that for large sample size with the log series model:

$$H \simeq 0.577 + \log_e \alpha \qquad (13.9)$$

the constant 0.577 being Euler's constant. Considering the behaviour of H in terms of a changing S_T for all the models, May (1975a) concluded that it was an insensitive measure of the character of the $S{:}N$ relationship and is dominated by the abundant species (Fig. 13.4a).

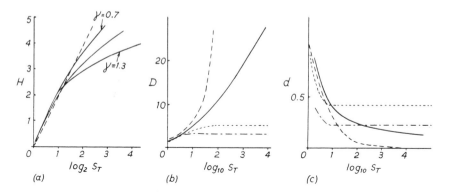

Fig. 13.4 The performance of various diversity indices in relation to the total number of species (S_T) and the underlying model: log normal (solid lines) ($\gamma = 1.0$, unless labelled otherwise), log series (. . . . . . .) $\alpha = 5$, geometric series (- - - - - - -) k = 0.4, and MacArthur model (--------). a. The Shannon–Weaver index; b. The Simpson–Yule index; c. The Dominance index (after May, 1975).

Simpson–Yule index (D)

A diversity index proposed by Simpson (1949) to describe the probability that a second individual drawn from a population should be of the same species as the first. A similar type of index had a few years earlier been proposed by G. Udney Yule to compare an author's characteristic vocabulary (frequency of

*Often, probably correctly, referred to as the Shannon–Wiener function.

different words in his writings). The statistic, C (or Y) is given by:

$$C = \sum_{i}^{S_T} p_i^2 \qquad (13.10)$$

where, strictly, $p_i^2 = \dfrac{N_i(N_i - 1)}{N_T(N_T - 1)}$, but usually:

$$p_i^2 \sim (N_i/N_T)^2 \qquad (13.11)$$

N_i being the number of individuals in the ith species and N_T the total individuals in the sample. The index is:

$$D = 1/C \qquad (13.12)$$

and the larger its value the greater the equitability (range $1 \rightarrow S_T$). In the form $1 - C$ the statistic gives a measure of the probability of the next encounter (by the collector or any animal moving at random) being with another species (Hurlbert, 1971). May's (1975a) comparisons (Fig. 13.4b) show that this index is strongly influenced for values of S_T over ten by the underlying distribution.

Berger–Parker Dominance Index.
This index is simple, both mathematically and conceptually, expressing the proportion of the total catch that is due to the dominant species (whose abundance $= N_{max}$):

$$d = N_{max}/N_T \qquad (13.13)$$

For reasonable values of S_T the index is not influenced by S_T and is relatively independent of the underlying model (Fig. 13.4c). May (1975a) concludes that it seems 'to characterize the distribution as well as any (index), and better than most'.

WHICH MODEL OR INDEX?
The evaluation of the most appropriate model or index may be based on the consideration of their theoretical properties against our knowledge of ecology or by testing them with field data for either their fit or their value in discrimination. Valuable studies in these respects are Bullock (1971a), Hurlbert (1971), Kempton and Taylor (1974), May (1975a), Room (1975), Pielou (1975) and Taylor et al. (1976). There can be no universal 'best-buy', although there are rich opportunities for inappropriate usages! The objective of the analysis must be clearly conceived before the procedure is selected. It is certainly important to distinguish between data on organisms that are part of the trophic structure of the area (e.g. plants that are growing or birds seen foraging) and samples that may contain a number of animals that are outside their trivial ranges (e.g. suction trap catches at several metres above the

vegetation). This is illustrated by Kricher's (1972) studies on the diversity of bird populations at different seasons.

The use of models (e.g. MacArthur's) to gain some insight into resource-apportionment may be justified in areal samples of 'trophically-related' organisms.

From theoretical studies, especially May's (1975a), it can be concluded that the models can be arranged in a series corresponding to maximal niche pre-emption or unevenness and moving through to a more uniform resource apportionment, thus:

(Uneven) Geometric series, log series, log normal, MacArthur (Even)

This appears contrary to Kempton and Taylor's (1974) findings that whereas most populations of macrolepidoptera were well described by the log series, those in unstable habitats with more 'rare' species were better described by the lognormal; it may be that the inclusion of "additional species" has increased the evenness. May (1975a) also concludes that the lognormal pattern may be expected with 'large or heterogeneous assemblies of species'.

Where comparative studies have included the log series the conclusion has generally been that this provides the most 'comprehensible result' (e.g. Bullock, 1971a; Kempton and Taylor, 1974; Room, 1975). In papers that could well establish a new approach to diversity Kempton and Taylor (1974) and Taylor *et al.* (1976) suggest that the 'basic environmental structure' of an area will determine the range of populations it can support and on this assumption they are able to show that the log series is the best model. They point out that the rank: log abundance plot of the log series is almost linear over the mid region (see Fig. 13.2a), so α may be interpreted as the slope of this region 'where the moderately common species reflect most closely the nature of the environment' and fluctuate less violently, from year to year, than the more abundant species. They found that changes in α enabled them to discriminate between diversities in areas where the environment had been known to have changed. For the ecologist the ability of a parameter or index to discriminate between changed conditions may be more relevant than, and is not identical to, the precision of the 'fit' of the underlying model. The richness and evenness parameters of the lognormal (S_1 and σ^2) and the Shannon and Simpson indices were all less useful as indications of environmental change and, even when the fit of the log series was not close, its robustness provided a meaningful α.

A completely novel concept of Taylor *et al.* (1976) is that viewing diversity as a reflection of basic environmental structure, the two meaningful character-istics are not species richness and evenness, but:

(1) diversity as represented by the 'common α', the slope of the line as dominated by the moderately common species.

(2) the fluctuations in numbers, from occasion to occasion (e.g. year to year) as shown by X of the log series (see 13.2).

Their hypothesis, that leads to the conclusion that the log series should be used for diversity data 'unless otherwise proven', is compelling: for the model is robust and biased only with small samples. However one will wish to see it tested against data for other taxa and for groups that are known to be closely related trophically e.g. foraging birds. It will also be noted that this concept of diversity is somewhat different from that of studies concerned with the influence of seasonal or other variations on diversity or with the indications diversity gives on resource allocation between species.

Comparative studies have not given much support to the value of the non-parametric indices, neither the comprehensive and critical reviews of Whittaker (1972) and May (1975a), nor theoretical studies (e.g. Sheldon, 1969; Peet, 1975), nor empirical studies on field data. The Shannon – Weaver Index seems to have been found unsatisfactory (Dickman, 1968; Bullock, 1971a; Hurlbert, 1971; Room, 1975; Taylor *et al.*, 1976) and in spite of some apparently successful uses (Tramer, 1969; Allan, 1975; Muller *et al.*, 1975) should in general be regarded as a distraction rather than an asset in ecological analysis. It is strongly influenced by S_T (species richness) and by the underlying model (Fig. 13.4a).

The Simpson-Yule index is strongly influenced by the few dominant species, although bearing this limitation in mind it could be of value as an indicator of interspecific encounters.

PROCEDURE

In the light of present knowledge the following may be taken as a guide to the analysis of diversity data.

(1) *Draw graph(s) of log abundance on rank (Fig. 13.2)*. It has always been important to biologists actually to examine the form of their data; this point needs even stronger emphasis now that computerized data collection and analysis may tempt the investigator to go straight to the analysis output. Does the data form a straight line? Which are the species that depart most from it? Is there anything unusual in their biology (e.g. they could be migrants *en passage*)? The consideration of these graphs in the light of the earlier discussion in this chapter should indicate whether, exceptionally (*pace* Taylor *et al.*, 1976), other models or indices should be used in addition to those outlined below. These graphs may be an excellent manner of presenting the data for publication (e.g. Sanders, 1969; Bazzaz, 1975; Taylor *et al.*, 1976 (Fig. 5)

(2) *Determine α of the log series*.

This may be done rigorously by extracting the value of α in equation 13.1 by maximum likelihood. For most purposes the value may be read off William's (1947) nomograph (Fig. 13.5). Taylor *et al.* (1976) point out that the true large sample variance is:

$$var\,(\hat{\alpha}) = \frac{\alpha}{-log(1-X)} \qquad (13.14)$$

The standard error contours on Fig. 13.5 exclude variation due to population fluctuations. Which variance is appropriate will clearly depend on the type of comparisons. If there are a series of samples it may be decided to:

(i) calculate the 'common α' (Taylor *et al.*, 1976) by the solution by maximum likelihood of:

$$\sum_{i=1}^{z} S_i = \sum_{i=1}^{z} \hat{\alpha} \log\left[1 + \frac{N_1}{\hat{\alpha}} \right] \tag{13.15}$$

where z = total number of samples and S_i and N_i are the *total* species and individuals in the ith sample.

(ii) calculate the individual sampling factors (X) which indicate the magnitude of population fluctuation:

$$\hat{X}_i = 1 - e^{-S_i/\hat{\alpha}} \tag{13.16}$$

(3) *Calculate the dominance index.* This simple index is, like α free of sample bias and (following Berger and Parker (1970)) is (equation 13.13):

$$d = N_{max}/N_T$$

(4) *Calculate other parameters or statistics.* This will be done where the graphs (1 above) reveal a special need: the appropriate equations are given earlier in the chapter.

13.1.2 Description of β – diversity

β – diversity is concerned with the change in species diversity from habitat to habitat and the comparison of the quantitative and qualitative make-up of different communities: it is a measure of change, of difference (or its obverse, similarity). Many of the methods have been developed and extensively used in work on plant ecology and valuable details will be found in Whittaker (1972), Greig-Smith (1978), Goodall (1973), Mueller-Dombois & Ellenberg (1974), Pielou (1975) and Goldsmith and Harrison (1976). Important examples of the study of β – diversity in bird faunas are those of Bullock (1971b) and Cody (1974; 1975): the former testing many of the existing indices and the latter developing his own approach.

The species composition of two areas may be compared by a suitable indices; if a series of indices are obtained they may be sorted to determine the relationships, but if they are along a habitat gradient they may be related to this directly (Cody, 1975).

INDICES

The general message from the review of α – diversity measurements was that the ecological insights gained were by no means proportional to the mathematical sophistication and complexity of the methods; this lesson will be applied to the evaluation of these indices.

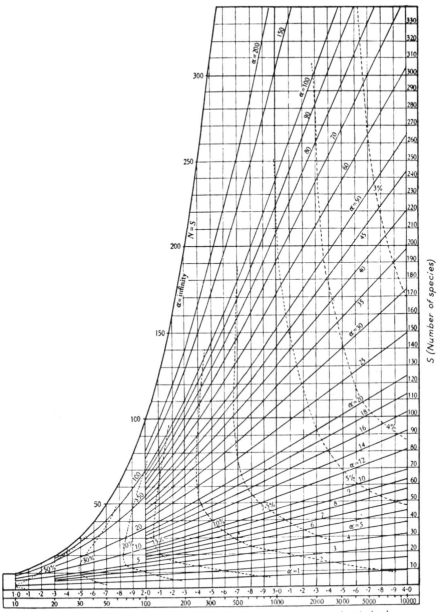

S (Number of species)

N (Number of individuals; log scale above; number below)

Fig. 13.5 Nomograph for determining the Index of Diversity (α) for the number of species (S) and the number of individuals (N) in a random sample of a fauna (after Williams, 1947). Note the standard errors contain only that component of variation due to sampling.

A direct approach to β – diversity would be to compare separate diversity indices, and the Shannon-Weaver Index (or developments of it) has sometimes been used (e.g. Kikkawa, 1968; Hummon, 1974), but in the light of the shortcomings of this index as a measure of α – diversity (see above and Hurlbert, 1971; May, 1975a; Taylor *et al.*, 1976) it seems inappropriate as a general technique. Mountford's (1962) index of similarity is based on the comparison of samples by the log series distribution, but as Bullock (1971b) points out, its assumption that the difference in distributions between different sites will be the converse of the similarities between samples from the same site (and some common α) may not be sound biologically. Bullock's (1971b) tests of the index on actual data tended to support these reservations.

Coefficient of Similarity: This is simple measure of the extent to which two habitats have species (or individuals) in common. It has been formulated in a number of slightly different ways and associated with even more originators:

Jaccard $\hspace{4em}$ $C_J = j/(a+b-j)$ $\hspace{4em}$ (13.17)

Czekanowski or Sørensen $\hspace{2em}$ $C_S = 2j/(a+b)$ $\hspace{4em}$ (13.18)

Kulezynski $\hspace{4em}$ $C_K = \frac{1}{2}C_S$ $\hspace{4em}$ (13.19)

where j is the number of species common (joint) to the two samples and a and b are respectively the total number of species in each sample. It is apparent that $C_S > C_J$ and that the inequality reduces as j approaches the magnitude of $\frac{1}{2}$ $(a+b)$. Applied as defined above, purely in terms of species numbers, these coefficients give equal weight to all species and hence tend to place too much significance on the rare species whose capture will depend heavily on chance. Bray and Curtis (1957) brought abundance into consideration in a modified Sørensen Coefficient and this approach, although criticized (Austin and Orloci, 1966), is widely used in plant ecology (Goldsmith and Harrison, 1976); the coefficient as modified essentially reflects the similarity in individuals between the habitats:

$$C_N = 2jN/(aN+bN) \hspace{3em} (13.20)$$

where aN = the total individuals sampled (N_T) in habitat a, bN = the same in habitat b and jN = the sum of the lesser values for the species common to both habitats (often termed W). This is illustrated from the following set of results; the values marked * are summed for jN ($=88$):

Species	1	2	3	4	5	6	7	8	9	
Habitat a	47*	41	23	15	5	2	1*	0	0	139 = aN
Habitat b	50	28*	5*	6*	1*	0	3	6	1	100 = bN

In habitats where one or a few species have high dominance the modified coefficient (C_N) now underestimates, for diversity studies, the contributions of the moderately common species, which Taylor *et al.* (1976) suggest may be more stable indicators of the characteristic fauna of an area, whilst the rare species have little impact.

Rank correlation – Kendall's tau: Originally suggested by Ghent (1963) as a means of comparing faunas, this measure has not been widely used, but Bullock (1971b) found it particularly valuable, especially if the rarer species were excluded. Specifically as the species lists are made for many samples, some rare species can be absent from both the sublists of the pair of sites for which a χ^2 is being computed. Bullock suggests the elimination of such joint absences; this decreases the number of tied values and increases the sensitivity of the index. Details of its calculation are given by Kendall (1962) and in some other statistical text books.

THE SORTING OF INDICES

Indices may be sorted by classification or ordination: these techniques have been reviewed in the plant ecology texts already cited. Four approaches will be summarized.

Trellis diagram
This old method has been extensively used. An index, such as Sørensen's (1948) (see above), is applied to the data and the sample localities are arranged in an order so that the highest values come on the diagonal, i.e. the samples with the highest similarities are placed together (see Fig. 13.6). A subjective assessment is then made of the groups on the basis of discontinuity.

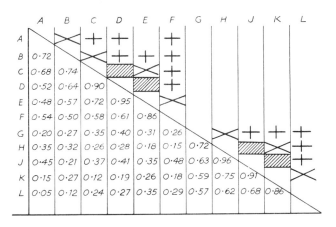

Fig. 13.6 Similarity Indices in a trellis diagram. The shading of the squares ($+ = 0.50 - 0.70$, $\times = 0.70 - 0.90$, cross-hatching $= 0.90 - 1.0$) helps with the interpretation, which in this straightforward case, is that there are two communities ($A-F$ and $G-L$).

Kontkanen (1957) has discussed the value of various indices for this method and Davis (1963) and Macfadyen (1954) give examples of their use.

Dendrogram

A table of indices between pairs of samples (of sites, etc.) is prepared similar to Fig. 13.6. The pair of sites with the highest value of similarity (in Fig. 13.7, H and J) are placed on the apex line and joined at the level of their similarity value, then the site with the highest value to one or other of these is joined at this level (i.e. K which has an index of similarity of 0.91 with J). The process

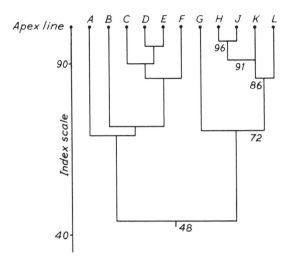

Fig. 13.7 The classification of the sampling sites in Fig. 13.6 by a dendrogram.

continues until one moves to the next group (A–F) where the process is repeated. The link between the groups is at the level of the highest index between any two members; in this case 0.48 between F and J. It should be noted that it is the level of the horizontal linkage lines rather than the order on the apex line that is significant. However it is desirable to arrange sites on the apex line in a sequence of relationship, but there are always elements of arbitrary decision about this. For example F is closest to J, but to arrange them like this (swinging G right round to lie beyond L and reversing H and J) would hide relationships in the G–L group. This is a limitation of two dimensions: in this example one can see that the conflict would be overcome by 'swinging' the A–F group so that it was at right angles to the G–L axes (i.e. coming out of the page). Examples of the use of dendrograms in animal ecology are given in the studies of Kikkawa (1965), Bullock (1971b), Day *et al.* (1971) and Hodkinson (1975). As dendrograms utilize only the highest values, much information is lost (e.g. in the above example (Fig. 13. 7) only 10 of the

55 indices (Fig. 13.6) were used). The following methods do not lead to this information loss.

Cluster Analysis

If the coefficients of similarity (or other indices) are reversed ($1 - C$ or $100 - C$ or $C_{max} - C$) values are obtained that may be taken to correspond with the distances apart of faunas, distances that will be proportional to the $\beta -$ diversity gradient. They may be plotted on an ordination diagram (e.g. Goldsmith and Harrison, 1976). Alternatively a three dimensional model may be constructed with the sites represented by, say, ping-pong balls or other spheres and each of the indices by a piece of wire, glass rod or similar material of the correct length. This is really a form of multivariate technique and has been used by Davis (1963) in a study of soil invertebrates. Bullock (1971b) illustrates how such diagrams may be reduced to two dimensions and the probability values of the links (for an index like χ^2) shown by lines of different thickness.

Multivariate methods

These techniques allow the comparison of a series of samples whose variation is considered to be due to several causes. Cluster analysis, which is really descriptive is referred to above. Principal component analysis has been the method used most frequently in community ecology (e.g. Sinha *et al.*, 1969; Bullock, 1971b), although Buzas (1967) used canonical correlation analysis. These methods are described by Seal (1964) and a computer programme for principal component analysis by Davies (1971). A lucid account of their underlying concepts for the non-mathematician is given by Sinha (1977). Gauch and Whittaker (1972) conclude, however, that these mathematically complex methods are not always the most appropriate methods of ordination and warn that with $\beta -$ diversity gradients in excess of five half-changes neither they, nor simpler methods, perform well.

GRADIENT DESCRIPTION

A different approach to $\beta -$ diversity is to measure diversity along a recognised gradient or at least a linear transect. The slope of the line will measure $\beta -$ diversity and sudden changes will reflect community or higher order boundaries (Odum *et al.*, 1960). Pattern analysis, much used in plant ecology (Greig-Smith, 1978; Kershaw, 1973; Goldsmith and Harrison, 1976) adopts this approach on a small scale.

Whittaker (1972) has summarized many studies on diversity gradients along 'coenoclines' by plotting the mean similarity coefficients for each transect interval: the first part of the resulting curve is extended to zero distance and this allows the 'threshold diversity' (C_0) to be estimated. $\beta -$ diversity, in 'half-changes', is then:

$$\beta_{0 \to i} = \log C_0 - \log C_i / \log 2C_i \qquad (13.21)$$

where C_i = mean similarity coefficient of samples distance i apart.

Another approach has been developed by Cody (summarized in Cody (1975)) in his studies of bird faunas. The techniques are particularly suited for a limited number of species, but probably with the aid of a computer, could be applied to more complex situations.

The data on the density of different species are arranged (Table 13.1) with habitat gradient along one axis and species in an ordered list on the other.

Table 13.1 Species distribution and densities along a gradient, a portion of a table to illustrate Cody's method of calculating species gain and loss rates in faunas along a habitat gradient.

Study site	:	1	2	3	4	 x
Habitat gradient	:	0	5	8.8	14	..... y
Species list ordering	1	15	–	–	–	
	2	–	4	–	–	
	3	20	2	6	–	
	4	10	3	4	2	
	5	5	0	1	3	
	6	–	16	5	22	
	7	–	–	9	11	
	8	–	–	17	25	
	.					
	.					
	.					
	.					
	.					
	i					

Thus this method can only be used when the habitats can clearly be arranged along a quantified gradient. Also the ordering* of the species list is partly subjective allocating the first position to the species whose centre of population density is nearest to the starting point of the gradient; subsequently species are ordered so that the centres of population density run across the table diagonally. In general species can be ordered by the sequence in which they join and are lost from the fauna, as one moves across the gradient, but as ranges are of different 'lengths' this cannot provide a 'rule'. Cody suggests that the table may be envisaged as having 'centres of gravity' along each row, corresponding to the centres of population density of the species, and centres of 'gravity' in each habitat column representing the moments of the species constituting that fauna. The 'moments' are taken as the relative population size or density (n) of the species X its 'distance' measured as its number in the species ordering (1, 2, . . .i,). The sum of this is the 'census moment' of the fauna ($\Sigma (i) . (n_i)$). Cody sets his measure of species

* Cody calls it 'ranking, but this term has a another specific meaning in diversity work, see Fig. 13.1.

lost at the order number of the species reached when 10% of the census moment of a habitat had been accumulated reading down the columns. Thus at site 2 (Table 13.1) the census moments ($\sum_i (i)(n_i)$) are:

$$(2 \times 4) + (3 \times 2) + (4 \times 3) + (6 \times 16) = 122$$

10% of this figure is 12.2 and this is reached by species 3, when the accumulated moments total 14 ($2 \times 4 + 3 \times 2$).

Then the conceptual species loss at habitat gradient 5 is given by the sum of the order numbers of the species below the 10% level + the appropriate fraction of the moment at this level, i.e. $1 + 2 + (12.2/14) = 3.87$. Species gains are calculated in a similar manner from the other ends of the habitat columns. Cody shows that a series of such points across a habitat are usually fitted by the logistic curve and the curve midway between the loss and gain curves is the 'species accumulation curve'; its derivative gives the species turnover rate i.e. β – diversity:

$$\beta(H_j) = \delta\frac{1}{2}\left[g(H) + l(H) \right]/\delta H_j \qquad (13.22)$$

where δH_j is a small change in the habitat gradient ($H_i \to H_k$) and g(H) and l (H) are the species gains and losses respectively over $\delta H(H_i \to H_k)$.

This method could be applied to groups such as butterflies over ranges like those studied by Cody for birds, or on a much smaller scale with smaller organisms, like soil mites. Apart from the subjective aspects of the calculations, it also assumes that one habitat gradient is dominant.

13.2 Species packing

The co-existence of organisms, especially of apparent relatedness, has fascinated naturalists since the time of Alfred Russel Wallace and the extent of interaction between them, the interspecific competition matrix, is the meeting ground of community and population ecology. Field observations on the distribution of organisms against a resource gradient will give a measure of the species packing in relation to this particular component of the species' niche, it is also relatively easy to determine the extent to which species overlap on a particular resource gradient. Thus measures of 'niche breadth' (or width) and 'niche overlap' *on this resource gradient* can be obtained. It is tempting (and indeed desirable) to go beyond this, and having determined similar values on other resource gradients, compound the overlaps in a hopefully appropriate manner to get a true measure of niche overlap. With this 'traveller's cheque' in hand the jump is made across the chiasm between community and population ecology; on arrival the niche overlap measurements are converted to competition coefficients in classical Lotka-Volterra (or Gause) competition equations. (A safer route to these coefficients is by measuring the actual impact of one species on another (e.g. Vandermeer, 1969)). This analogy is not made to cast disparagement on those who have attempted the 'jump' for nothing is achieved (especially in ecology) by pessimistic contemplation of the difficulties, but rather to warn intending travellers in what is one of the most exciting areas of ecology. Functionally the analogy is intended to serve as an explanation as to why the explicit treatment in this section makes but a short part of the 'journey'. At present one can, as Green (1971) pointed out, easily demonstrate from field data that two species do not occupy the same niche, but not, without experiments, that they occupy the same niche. I suspect that in most instances a judicious mixture of the observational and experimental approaches will be necessary before Wallace's observations can be firmly linked to theory (MacArthur and Levins, 1967; May, 1973, 1974, 1975*b*, 1976*b*; Schoener, 1974*a* & *b*; Abrams, 1975; Heck, 1976). Background in this area may be obtained from the references in this section and MacArthur (1972; Ch. 2) and Pianka (1975, 1976*a* and *b*).

In this section the methods of assessing the extent to which species occur together, the usage of a resource spectrum by one species and the comparison of this with other species (species packing) will be described.

13.2.1 Measurement of interspecific association

These measurements may be based either on presence or absence data or on abundance figures; as Hurlbert (1971) points out, presence–absence data is preferable if it is desired to measure the extent to which two species' requirements are similar. Interspecific competition (and other factors) may lead to a 'misleading' lack of association if the measure is based on abundance. However it would seem that whenever possible both types of analysis should be undertaken for a positive association on presence–absence data and a much weaker or negative one on abundance data would suggest (not prove) interspecific competition worthy of further analysis.

THE DEPARTURE OF THE DISTRIBUTION OF PRESENCE OR ABSENCE FROM INDEPENDENCE

These methods measure the departure from independence of the distribution of the two species and *assume that the probability of occurrence of the species is constant for all samples*. Thus if the distribution of two aphid predators were being compared it would only be legitimate to include samples that also contained the prey. A good example of the use of these indices is given by Evans & Freeman (1950), who measured the interspecific association of two species of flea on two different rodents (*Apodemus* and *Clethrionomys*); they found that there was a strong negative association on *Apodemus* and a moderate positive one on *Clethrionomys* and suggested that the coarse and longer fur of *Clethrionomys* allowed the two fleas to avoid competition and exist together on that host. However, when applied to habitats whose uniformity is doubtful, these methods may give results of uncertain value, as was found by Macan (1954) in a study of the associations of various species of corixid bug in different ponds. In such a situation some samples may be from habitats that are outside the environmental ranges of the species; this will have the effect of inflating the value for *d* in the table below and so lead to too many positive associations.

Fager (1957) has demonstrated, by examples, that if two species are rare and therefore both are absent from most of the samples (this could, as just indicated, be due to unsuitable samples being included), a high level of association will be found. Conversely, if two species occur in most of the samples and so are nearly always found together, no association will be shown with these methods. (Although a biological association is obvious, it is equally correct to say that this does not depart significantly from an association that is due to chance and therefore does not necessarily imply any interspecific relationship.)

The contingency table
The basic feature of these methods is the 2×2 contingency table.

Species B	Species A		
	Present	Absent	
Present	a	b	$a+b$
Absent	c	d	$c+d$
	$a+c$	$b+d$	$n = a+b+c+d$

Such a table should always be drawn up so that A is more abundant than B, i.e. $(a+b) < (a+c)$. A number of statistics are available for analysis of such a table, but the corrected chi-square (χ^2) makes fewest assumptions about the type of distribution, and the significance of the value obtained can be determined from tables available in standard statistical textbooks. It is calculated:

$$\chi^2 = \frac{n[\,|ad - bc| - (n/2)]^2}{(a+c)(b+d)(a+b)(c+d)} \qquad (13.23)$$

where the letters are as in the contingency table above and $|ad - bc|$ signifies placing the term in the positive form. If *ad* is greater than *bc* then the association is positive (affinity); if *bc* is greater than *ad* then the association is negative (repulsion). The test in this form is only valid if the expected numbers (if the distribution was random) are not less than 5; there is only one degree of freedom and so the 5% point is 3.84. Therefore, if a χ^2 of less than this is obtained, any apparent association could well be due to chance and further analysis should be abandoned. If the smallest expected number is less than 5 the exact test should be used and the relevant tables are available (e.g. Fisher & Yates, 1960).

Coefficients of association

If χ^2 is significant then one of the coefficients of association may be used to give an actual quantitative value for comparison with other species. They are designed so that the coefficient has the same range as the correlation coefficient (r), i.e. $+1 =$ complete positive association, $-1 =$ complete negative association and $0 =$ no association. Cole (1949) reviews a number of coefficients and points out how, if the above interpretations of values from $+1$ to -1 are to hold for comparative purposes, the plot of the value of the coefficient against the possible number of joint occurrences should be linear; for several of the coefficients it is not, and for those in which it is the plot does not pass through the zero. Some coefficients are given below.

(1) *Coefficient of mean square contingency.* This coefficient makes no assumption about distribution, but it cannot give a value of $+1$ unless $a = d$ and b and $c = 0$. For less extreme forms of association it is useful and easily calculated:

$$C_{AB} = \sqrt{\frac{\chi^2}{n + \chi^2}} \tag{13.24}$$

where $C_{AB} =$ coefficient of association between A and B, $n =$ total number of occurrences, and the χ^2 value is obtained as above.

This coefficient is recommended by Debauche (1962) and was used by Davis (1963).

(2) *Coefficient of interspecific association.* The coefficients originally designed by Cole (1949) have been shown by Hurlbert (1969) to be biased by the species frequencies. He showed this bias was considerably diminished if the coefficient is defined:

$$C^1_{AB} = \frac{ad - bc}{|ad - bc|} \left| \left(\frac{\text{Obs}\chi^2 - \text{Min } \chi^2}{\text{Max}\chi^2 - \text{Min } \chi^2} \right)^{\frac{1}{2}} \right| \tag{13.25}$$

Min χ^2 is the value of χ^2 when the observed a differs from its expected value $(\hat{a})$ by less than 1.0 (except when $a - \hat{a} = 0$ or $= 0.5$, the value of Min χ^2 depends on whether $(ad - bc)$ is positive or negative), formulated:

$$\text{Min } \chi^2 = \frac{n^3(\hat{a} - g[\hat{a}])^2}{(a + b)(a + c)(c + d)(b + d)} \tag{13.26}$$

where $g(\hat{a}) = \hat{a}$, rounded to the next lowest integer when $ad < bc$ or rounded to the next highest integer when $ad \geqslant bc$, if $\hat{a}$ is an integer then $g(\hat{a}) = \hat{a}$.

Max χ^2 is the value of χ^2 when a is as large or small as the marginal totals of the 2×2 table will permit formulated under specified conditions as follows:

Conditions	$ad \geq bc$	$ad < bc$ $a \leq d$	$ad < bc$ $a > d$
Max $\chi^2 =$	$\dfrac{(a+b)(b+d)n}{(a+c)(c+d)}$	$\dfrac{(a+b)(a+c)n}{(a+b)(c+d)}$	$\dfrac{(b+d)(c+d)}{(a+b)(a+c)}$

Obs χ^2 is calculated in the normal manner (see above). Pielou and Pielou (1968) describe a method for species of infrequent occurrence.

Correlation may be utilized if the data can be normalized, otherwise some of the methods of comparing similarity between habitats may be used. For example Sørensen's coefficient, as used by Whittaker and Fairbanks (1958), may be modified, to give the normal range of -1 (no association) to $+1$ (complete association):

$$I_{ai} = 2\left[\frac{J}{A+B} - 0.5\right] \qquad (13.27)$$

where J = no. of individuals of A and B in samples where both species are present and A and B = total of individuals of A and B in all samples.

13.2.2 Measurement of resource utilization
Among the meaningful resources that may be partioned, by apparently co-existing species are: food [defined by quality (type) and size], space [defined in Euclidian terms (e.g. height in vegetation) and physico-chemical characters (e.g. climate, substrate type)] and time (seasonal and diel). This, by no means exhaustive list, gives an idea of the number of 'resource spectra' that contribute to a species' niche (Pianka, 1976b).

In considering the most appropriate way to measure resource utilization it is useful to start from a theoretical basis (May and MacArthur, 1972). The extent of the separation in the resource utilization between potentially competing species is well expressed in terms of the separation measure (d) between the means of their curves and the widths (1 standard deviation in the normal Gaussian curve) (w) (Fig. 13.8a). In this theoretical situation (where

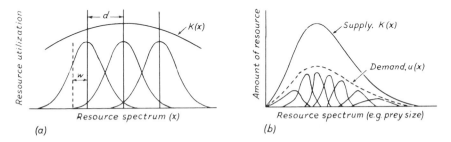

Fig. 13.8 Theoretical resource utilization relationships: *a.* the 'simple case' of three species with similar (and normal) resource utilization curves, d = distance apart of means, w = standard deviation of utilization and d/w = resource separation ratio = 2.5 (in these cases) (modified from May and MacArthur, 1972); *b.* the more typical case with varying resource utilization curves: broadest in the region of fewer resources and less interspecific competition (after Pianka, 1976).

resource utilization functions are Gaussian and equal), then if the resource separation ratio (d/w) is less than 3 there will be some interaction between the species. Theoretically the ratio will have a minimum value below which the competitive exclusion principle operates (the two species cannot co-exist). May's (1973, 1974) studies suggest that one is the minimum value, but the conversion of this ratio to a competition coefficient is, as already indicated, fraught with biological difficulties (May, 1975b; Abrams, 1975; Heck, 1976).

Resource utilization curves are seldom normal; indeed field data suggest that they vary greatly, being broad where resources are scarce, and narrow where they are more plentiful (Fig. 13.8b). The shape of the curve influences the extent of packing as was pointed out by Roughgarden (1974) on mathematical grounds. His point is illustrated simply in Fig. 13.9, which represents two species pairs, the ones above having leptokurtic resource utilization curves and the pair below platykurtic ones. It should be noted that each species takes an identical quantity of resource (the areas under the curves are equal) and the ranges ($x^1 \rightarrow x^{11}$) are equal, but when the lines for w (the

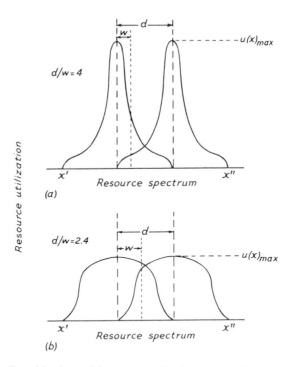

Fig. 13.9 The effect of the shape of the resource utilization curve on the separation measure (d) and hence species packing. Note that the two species with leptokurtic curves (a) and the two with platykurtic curves (b) both utilize the same quantity of resource (the areas under the curves are equal) and same range of resources ($x^1 - x^{11}$), but the extent of utilization, of which $u(x)_{max}$ is one measure, differs.

distance from the mean of 68 % of the resource utilization) are inserted their ratio (leptokurtic: platykurtic) is 3:5. The impact of this on the resource separation ratios (d/w) is shown. The difference implies that, if this was the only resource spectrum, four closely packed ($d/w \sim 1$) 'platykurtic species' could be replaced by seven closely packed 'leptokurtic species'. (These numerical values are, of course, precisely relevant only to Fig. 13.9).

It will be noted that whereas the platykurtic species are not as closely packed as leptokurtic, the latter, for the same resource utilization, have higher $u(x)_{max}$. If we imagine the supply of the resource $K(x)$ to be far above the utilization (demand) level $u(x)$ in Fig. 13.9b, which could be due to predator pressure, then there will be no restrictions on the shapes of the curves that can replace the platykurtic ones and tighter packing, as indicated above, is possible. There are two possible procedures for assessing species packing in respect of a resource dimension:

(1) d/w approach, in which the utilization by each species is considered in turn, the mean and the width (w = 68 % of the utilization on one side of the mean) determined; the differences between the means giving d.

(2) proportional utilization or 'Simpson-Yule' index approach in which the proportion of the species utilizing a particular segment of the resource gradient is determined (p_i), the breadth of utilization ['niche breadth' of Levins (1968)] being equivalent to the Simpson-Yule index.

The former is conceptually simpler as the investigator can visualize the 'areas' of resource occupied by the species, but it assumes a utilization spectrum that approximates to the normal: thus the resource dimension must be approximately ordered. It has seldom been used in field studies. The proportional utilization approach makes no demands on the ordering on the resource dimension and has been widely used. Of course p_i is simply the height of the utilization curve at i, the curve whose width (in terms of one standard deviation) is given by w.

SPECIES PACKING IN TERMS OF MEAN AND WIDTH OF RESOURCE UTILIZATION SPECTRUM ('d/w METHOD')

This method could easily be used when the resource spectrum is continuous: this applies especially to seasonal occurrence, diel occurrence and food items or other resources defined by size. Data on seasonal occurrence is widely available and often shows striking patterns of species packing (e.g. Greenslade, 1965; Shapiro, 1975; Southwood, 1978). The first step is to find the mean and then the total range of time of occurrence can usually be taken as twice the interval between the mean and the earliest occurrence. This method eliminates the extreme, end of season, 'tail': these are probably post-reproductives (this may be checked) and will not affect the fitness of that species (they may however affect the species that occur with them and then a new range will need to be found and used when assessing the impact of this putative competitor on those species). If the frequency of occurrence is approximately normally distributed (see p. 9, and consider the use of probability paper) then the total range will be equivalent to 6 w (three standard deviations on each side of the mean). If the frequency of occurrence shows kurtosis or other irregularity, w may be calculated by a method analogous to the graphical method for estimating the numbers in a stage (see Fig. 10.1). The mean is drawn in and the number of animal/days (or other time period e.g. 5 days) summed visually. Then starting from the mean and working in vertical columns towards an end of the range one sums the 'animal/days' until 68 % of the total that side of the mean have been accumulated. At this point a vertical line can be drawn and the distance of this to the mean $\sim w$. Experience with the graphical method for determining the numbers entering a stage (p. 358) has shown its robustness and as a method of estimating resource utilization it is simple and free of assumptions. Large sets of data could easily be handled on a computer, but it is often necessary to make decisions about stragglers that should be based on biological knowledge about the specific and not on a general rule. This technique allows for different w's either side of the mean and for adjustments, as in the case of post-reproductives mentioned above.

When the values of w are different for the ith and jth species, a 'common w' $[w_{ij}]$ must be calculated:

$$w_{ij} = \left[w_i^2 + w_j^2/2 \right]^{\frac{1}{2}} \tag{13.28}$$

The resource separation ratio is:-

$$\rho_{ij} = d_{ij}/w_{ij} \tag{13.29}$$

where d_{ij} is the difference between the means of resource utilization by the ith and jth species.

Species packing in diel time has engaged the attention of ecologists since Charles Elton utilized it in his discussion of the niche. The data may be treated in the same manner, but with diurnal/nocturnal partitioning (e.g. in Carabids, (Greenslade, 1963) it is necessary to remember the circularity of the resource and the ratio for a pair of species may be in the form $d/2w$, both 'tails' overlapping. Diel peaks of activity within the day (e.g. in bees, Heinrich, 1976) may require the same treatment.

Size may easily be arranged on a linear scale and it governs many aspects of resource utilization. The size (and hence type) of food may show a reasonable relationship to body size (Hespenheide, 1975; Wilson, 1975; Davidson, 1977) or body part (e.g. bees' tongues, (Heinrich, 1976) or birds' bills (Karr and James, 1975); the trivial range is also approximately related to size when account is taken of feeding habits (predator v. herbivore) and habitat type (e.g. desert v. temperate grassland); for ambulatory animals the span between the front and back legs is a better measure of size than body length.

Other resources, such as plant species or types for herbivores, could be arranged as the resource spectrum, but the ordering would require judgement analogous to Cody's (1975) ordering of bird species (see above). Under such circumstances the more versatile proportional utilization method is more objective.

SPECIES PACKING IN TERMS OF PROPORTIONAL UTILIZATION OF DIFFERENT RESOURCE STATES (p_i method)

The breadth of utilization ('niche breadth', Levins, 1968) is given by the Simpson-Yule index applied to the distribution of the individuals between the resource states:

$$B = 1/\sum_i^m p_i^2 = 1/\sum_i^m N_i^2/N_T^2 \qquad (13.30)$$

where N_i = the number of individuals of the species in question in the ith resource state, N_T = the total number of individuals in all (= m) the resource states. In this context the tendency of this index to 'undervalue' rare events is, probably, an advantage.

This method is versatile in that the 'resource states' do not have to be ordered along a continuum; they may, for example, be samples from m different categories of food type found by stomach content analysis or some other method.

The application of the method to a series of random samples from a heterogeneous environment is, however, open to criticism, for the degree of uniformity in distribution could well be influenced by the natural clumping (or otherwise) of the organism (see Chapter 1); one may claim that this is part of the resource, but this is not very illuminating. There is a considerable interpretative advantage if real resource states are sampled separately: parts of the habitat (e.g. different species of plant), that from other knowledge are known to be distinguished by ecologically significant factors. However samples often reveal vagrants, temporary members of the fauna: insects that engage in upper air migration (e.g. aphids, many flies, coccinellids, etc.) regularly occur in places that are not part of *their* niche in any meaningful way. Studies of colonies of Lepidoptera record the movement of individuals across terrain that is again not part of their niche. If these vagrants are relatively few, the Simpson-Yule index's character of emphasizing the high probability occurrences, will minimize their significance. A more general criticism (for the purposes of determining competition is that 'handfuls of animals' taken from their habitat may not represent usage of the same resource; one may live on the surface of the leaf blade, one on the stem; one be active by day, another by night and so on. Thus although this approach is valid for the assessment of the breadth of resources used; its utilization to calculate 'niche overlap' must be handled with caution.

There is a further problem if the different resource states are present (in the samples) at very different frequencies. Colwell & Futuyma (1971) highlighted this and proposed weighting resource states within a resource matrix and expanding the matrix during calculations to a large size so that each resource state is equally represented. The weightings were defined, not in terms of physico-chemical or other measurements, but in terms of the rarity of the fauna. They used the Shannon-Weaver index; comparable manipulations could be made with the Simpson-Yule Index. But Sabath and Jones (1973) used their method (with drosophilids) and found that the weightings did not add to the information gained from unweight matrices and that the expansion of the matrix eventually prevented discrimination between species distributions.

Another detailed study using the Colwell-Futayma weightings was made by Hanski and Koskela (1977) on beetles in dung pats. This would seem to be an ideal 'test' of the concept for the habitat is discrete, its state may also be described by certain physical characters and the beetles collected are a real assemblage, which cannot necessarily be claimed for trap catches, a difficulty of which Sabath and Jones (1973) were well aware. They conclude that weighting by unusual species is unsatisfactory; but they did weight 'resource states' in relation to the total number of individuals in each. However once again they found all the measurements of 'niche breadth' highly correlated and concluded that the main results were fairly insensitive to weighting. It seems therefore that in practice the simple B has much to recommend it as a measure of the width of resource utilization by a species.

The proportional overlap in resource utilization between two species (i and j) is (Schoener, 1968; Colwell & Futuyma, 1971):

$$\Theta_{ij} = 1 - 0.5 \sum_{h=1}^{m} p_{ih} - p_{jh} \qquad (13.31)$$

where p_{ih} = the proportion of species i in resource state h and p_{jh} = the proportion of species j in same resource states and m = the total number of resource states. [Note that Θ measures 'overlap', whilst ρ measures separation). This method has been used with, for example, leaf-hoppers (McClure & Price, 1976) and wolf-spiders (Uetz, 1977).

13.2.3 Niche size and competition coefficients

All the methods discussed above provide information on species packing and their resource utilization. These measures are themselves intrinsically interesting and may be used in studies of community ecology, island biogeography, etc. (e.g. Price, 1971; McClure and Price, 1976; Mühlenberg *et al.*, 1977). The difficulty of equating resource utilization measures with niche breadth have already been outlined. The utilization indices (w and B) pose two problems: how may they be combined and, less easily answered, have all the components of the niche been assessed?

If species packing indices (ρ_{ij} and Θ_{ij}) are to be equated with niche overlap and hence competition attention also has to be paid to the renewability of resources, interactions at other trophic levels and the relative availability of other reources (May, 1973, 1974, Schoener, 1974b, 1976; Levine, 1976; Pianka, 1976b).

Competition coefficients between the *i*th and *j*th species for one resource axis may be calculated (May, 1973, 1975b):

(1) from resource separation ratio:

$$\alpha_{ij} = \exp\left[-d_{ij}^2/4\,w_{ij}^2 \right]$$ (13.32)

(2) from proportional utilization functions:

$$\alpha_{ij} = \sum p_i p_j \, / \, \sqrt{(\sum p_i^2)(\sum p_j^2)}$$ (13.33)

where each term is summed for the array of resource states along this dimension. This expression is due to Pianka (1973); May (1975b) compares it with alternatives.

Weighted competition coefficients for food resources may be calculated (Schoener, 1974a):

$$\alpha_{ij} = \left(\frac{T_j}{T_i}\right)\left[\frac{\sum\limits_{k=1}^{m} (p_{ik}/f_k)(p_{jk}/f_k)\,b_{ik}}{\sqrt{\sum\limits_{k=1}^{m} (p_{ik}/f_k)^2\,(p_{jk}/f_k)^2 b_{ik}}} \right]$$ (13.34)

where T_j/T_i = the ratio of the total number of food items consumed by an individual of the *j*th to that consumed by an individual of the *i*th species, measured over an interval of time that includes all regular fluctuations in consumption for both species; p_{ik} and p_{jk} the frequencies of food type *k* in the diets of *i* and *j* respectively, f_k the frequency of food item *k* in the environment, b_{ik} = the net calories gained by an individual of *i* from one item of *k*, or (more approximately) the calories in an item of *k*, or (still more approximately) the biomass of an item of *k* (see Chapter 14). The summations are taken over all (m) the food items eaten by one (i) of the two species. The equation is modified from Schoener's so as to put it in the same form as the basic expression 13.33 (Pianka's expression). The above expression may also be used to obtain a coefficient in relation to foraging times in the same habitats; the *T*'s are the ratios of total time, b_{ik} the calories obtained from an average item of food in habitat *k*, and the *f*'s are omitted. As T_j/T_i will often approximate to unity, unless the average value of the food items in the different habitats are remarkably different, b_{ik} will not be significant, this will often reduce to the basic (Pianka's) expression. Thus, as indicated above (p. 446), weighting may not in practice always be necessary.

Competition coefficients between species may be expressed as a matrix – the diagonal values representing intraspecific interactions are unity (May, 1973; 1976a).

Species generally interact in more than one resource dimension; as Schoener (1974b) points out, the number of dimensions, their type, and the comparative extent of the interactions, are of intrinsic interest and already certain patterns are emerging, e.g. predators partition a habitat through the diel cycle more often than other types.

The combination of competition coefficients from different resource axes is not straightforward except, as May (1975b) has clearly shown, in the two limiting cases. When the two axes are totally independent, then the coefficients (α's) should be multiplied; when the two axes are completely correlated then the coefficients should be summed. A method for use with field data, allowing for various degrees of independence between resource axes, has not yet been devised.

Nevertheless, as there are so many difficulties over the assumptions of elegant theoretical models (May, 1973, 1974, 1975b; Abrams, 1975; Heck, 1976; Armstrong, 1977), it seems important to determine from field observations which assumptions are justified and which refinements are, in practice, trivial. Field studies should be made over as many resource dimensions as possible, the d/w ratio used to determine the shape and separation of resource utilization and where possible, the impact of perturbations studied [although here too there are difficulties in interpretation (Schoener, 1974b)].

13.3 Habitats

13.3.1 Qualitative
Zoologists frequently delimit their communities by reference to plants or environmental factors. The most universal classification of habits is that of Elton & Miller (1954):

(*i*) Terrestrial system
  Formations: Open-ground type – if any dominant plants, these not more than 15 cm (6 in) high.
  Field type – dominant life form coincides with field layer, usually not more than 2 m in height.
  Scrub type – dominant life form does not exceed a shrub layer, height generally not over 7.6 m (25 ft).
  Woodland type – trees dominant life form.

  Vertical layers: Subsoil and rock.
  Topsoil.
  Ground zone, including low-growing vegetation, less than 15 cm (6 in).

Low canopy – up to about 7.6 m (25 ft).
High canopy.
Air above vegetation.

(*ii*) Aquatic system.
Formation types: these are shown in table below.

	A Very small	B Small	C Medium	D Large	E Very large
1 Still	Tree hole	Small pond 17 m^2	Pond 0.4 hect.	Large pool or Tarn 40 hect.	Lake or Sea
2 Slow	Trickle Gutter	Ditch Field dyke	Canal River bank Water		
3 Medium	Trickle	Lowland brook Small stream	Lowland river	Lowland large river	River estuary
4 Fast	Spring	Upland weir Small torrent stream	Large torrent stream		
5 Vertical or steep	Water drip Pipe outlet	Small weir Water-fall	Large weir Medium waterfall	Large waterfall	

Vertical layer: Bottom, light, dark zones, water mass, light and dark zones – free water not among vegetation
Submerged vegetation.
Water surface – upper and under surface of film of floating leaves.
Emergent vegetation – reed swamp and similar vegetation, the bases of which are in the water.
Air above vegetation.

(*iii*) Aquatic – terrestrial transition system – defined further by body type with which it occurs and by vegetational systems corresponding to the terrestrial system.
(*iv*) Subterranean system – caves and underground waters.
(*v*) Domestic system.
(*vi*) General system: Dying and dead wood.
Macro-fungi.
Dung.
Carrion.
Animal artefacts – nests, etc.
Human artefacts – fence posts, straw stacks, etc.

Further division of the habitat into communities may be made on the type of plant but in soil (Macfadyen, 1952, 1954; Davis, 1963) and freshwater studies (Whittaker & Fairbanks, 1958), divisions have been based on the fauna itself; these techniques (pp. 430–7) have been used in few other situations (Kontkanen 1957), but their use could undoubtedly be extended.

13.3.2 Quantitative

The quantification of the characters of the habitat especially plants, in general terms was pioneered by MacArthur and MacArthur (1961) who assessed foliage height diversity in trees to relate to bird diversity. MacArthur (1972) suggested the best measure of foliage height diversity was D of the Yule index (p. 426), where the various values of p_i are the proportion of the vegetation at various heights or at the different layers in the Eltonian system (opposite). A quick method of assessing fifteen interrelated variables in arboreal vegetation is described by James & Shugart (1970) and utilized by James (1971) to demonstrate the vegetational characteristics of bird habitats. Measures of the patchiness of vegetation (Pielou, 1975; Ch. 5) could also be useful in the study of faunal diversity.

In a review of habitat characteristics Southwood (1977) stressed the need to define spatial heterogeneity in relation to the trivial and migratory ranges of the animal, and temporal heterogeneity in relation to generation time. Stability in space, the length of time the same location remains suitable for the species (durational stability), and stability in time, the seasonal fluctuations in favourableness, should be distinguished; such temporal changes, together with the general level of unfavourableness constitute the 'adversity axes' of Whittaker. It was proposed that most habitats could be characterized against the two axes of adversity and durational stability and that suites of bionomic adaptations were related to the position of the organism's habitat in this chart.

REFERENCES

ABRAMS, P., 1975. Limiting similarity and the form of the competition coefficient. *Theoret. Popul. Biol.* **8**, 356–75.

ALLAN, J. D., 1975. Components of diversity. *Oceologia* [Berl.] **18**, 359–67.

ARMSTRONG, R. A., 1977. Weighting factors and scale effects in the calculation of competition coefficients. *Am. Nat.* **111**, 810–12.

AUSTIN, M. P. and ORLOCI, L., 1966. Geometric models in ecology. II. an evaluation of some ordination techniques. *J. Ecol.* **54**, 217–22.

BAZZAZ, F. A., 1975. Plant species diversity in old field successional ecosystems in Southern Illinois. *Ecology* **56**, 485–8.

BERGER, W. H. and PARKER, F. L., 1970. Diversity of planktonic Foraminifera in deep sea sediments. *Science* **168**, 1345–7.

BOSWELL, M. T. and PATIL, G. P., 1971. Chance mechanisms generating the log arithmetic series distribution used in the analysis of number of species and individuals. *In* Patil, G. P., Pielou, E. C. & Wates, W. E. (eds.). *Statistical Ecology* 3, 99–130. Penn State University Press, Philadelphia.

BRAY, J. R. and CURTIS, C. T., 1957. An ordination of the upland forest communities of southern Wisconsin. *Ecol. Monogr.* **27**, 325–49.

BULLOCK, J. A., 1971a. The investigations of samples containing many species. I. Sample description. *Biol. J. Linn. Soc.* **3**, 1–21.

BULLOCK, J. A., 1971b. The investigations of samples containing many species. II. Sample comparison. *Biol. J. Linn. Soc.* **3**, 23–56.

BULMER, M. G., 1974. On fitting the Poisson log normal distribution to species abundance data. *Biometrics* **30** (1), 101–10.

BUZAS, M. A., 1967. An application of canonical analysis as a method for comparing faunal areas. *J. Anim. Ecol.* **36,** 563–77.

CANCELA da FONSECA, J. P., 1968. L'outil statistique en biologie du sol. IV. *Rev. Ecol. Biol. Sol.* **5,** 41–54.

CANCELA da FONSECA, J. P., 1969. L'outil statistique en biologie du sol. V, VI. *Rev. Ecol. Biol. Sol.* **6,** 1–30, 533–55.

CODY, M. L., 1974. *Competition and the Structure of Bird Communities.* Princeton Univ. Press, Princeton, New Jersey.

CODY, M. L., 1975. Towards a theory of continental species diversity: bird distributions over Mediterranean habitat gradients. *In* Cody, M. L. & Diamond, J. M. (eds.) *Ecology and Evolution of Communities* pp. 215–57. Harvard University Press, Cambridge, Mass.

CODY, M. L. and DIAMOND, J. M., 1975. (eds.) *Ecology and Evolution of Communities,* Harvard University Press, Cambridge, Mass.

COLE, L. C., 1949. The measurement of interspecific association. *Ecology* **30,** 411–24.

COLWELL, R. K. and FUTUYMA, D. J., 1971. On the measurement of niche breadth and overlap. *Ecology* **52,** 567–76.

DAVIDSON, D. W., 1977. Species diversity and community organization in desert seed-eating ants. *Ecology,* **58,** 711–24.

DAVIES, R. G., 1971. *Computer programming in Quantitative Biology.* Academic Press, London.

DAVIS, B. N. K., 1963. A study of micro-arthropod communities in mineral soils near Corby, Northants. *J. Anim. Ecol.* **32,** 49–71.

DAY, J. H., FIELD, J. G. and MONTGOMERY, M. P., 1971. The use of numerical methods to determine the distribution of the benthic fauna across the continental shelf of North Carolina. *J. Anim. Ecol.* **40,** 93–125.

DEBAUCHE, H. R., 1962. The structural analysis of animal communities of the soil. *In* Murphy, P. W. (ed.), *Progress in Soil Zoology,* 10–25. Butterworths, London.

DE BENEDICTIS, P. A., 1973. On the correlations between certain diversity indices. *Am. Natur.* **107,** 295–302.

DE JONG, T. M., 1975. A comparison of three diversity indices based on their components of richness and evenness. *Oikos* **26,** 222–7.

DICKMAN, M., 1968. Some indices of diversity. *Ecology* **49,** 1191–3.·

ELTON, C. S. and MILLER, R. S., 1954. The ecological survey of animal communities with a practical system of classifying habitats by structural characters. *J. Ecol.* **42,** 460–96.

EVANS, F. C. and FREEMAN, R. B., 1950. On the relationship of some mammal fleas to their hosts. *Ann. ent. Soc. Am.* **43,** 320–33.

FAGER, E. W., 1957. Determination and analysis of recurrent groups. *Ecology* **38,** 586–95.

FINDLEY, J. S., 1973. Phenetic packing as a measure of faunal diversity. *Am. Nat.* **107,** 580–4.

FISHER, R. A., CORBET, A. S. and WILLIAMS, C. B., 1943. The relation between the number of species and the number of individuals in a random sample of an animal population. *J. Anim. Ecol.* **12,** 42–58.

FISHER, R. A. and YATES, F., 1960. *Statistical tables for biological, agricultural and medical research.* 138 pp., Longman, Edinburgh and London.

GAGE, J. and TETT, P. B., 1973. The use of log-normal statistics to describe the benthos of Lochs Etive and Creran. *J. Anim. Ecol.* **42,** 373–82.

GAUCH, H. G. and WHITTAKER, R. H., 1972. Comparison of ordination techniques. *Ecology* **53,** 868–75.

GHENT, A. W., 1963. Kendall's 'Tau' coefficient as an Index of similarity in comparisons of plant or animal communities. *Can. Ent.* **95,** 568–75.

GOLDSMITH, F. B. and HARRISON, C. M., 1976. Description and analysis of vegetation. *In* Chapman, S. B. (ed.) *Methods in Plant Ecology*. Blackwells, Oxford, London.

GOODALL, D. W., 1973. Sample similarity and species correlation. *In* Whittaker, R. H. *Handbook of Vegetation Science Part V*, 107–56. Junk, The Hague.

GREEN, R. H., 1971. A multivariate statistical approach to the Hutchinson niche: bivalve molluscs of central Canada. *Ecology* **52**, 543–56.

GREENSLADE, P. J. M., 1963. Daily rhythms of locomotor activity in some Carabidae (Coleoptera). *Entomologia exp. appl.* **6**, 171–80.

GREENSLADE, P. J. M., 1965. On the ecology of some British Carabid beetles, with special reference to life histories. *Trans. Soc. Brit. Ent.* **16** (6), 149–79.

GREIG-SMITH, P., 1978. *Quantitative plant ecology*. 3rd Ed., Butterworths, London.

HANSKI, I. and KOSKELA, H., 1977. Niche relations among dung-inhabiting beetles. *Oecologia* **28**, 203–31.

HECK, K. L., 1976. Some critical considerations of the theory of species packing. *Evol. Theory.* **1**, 247–58.

HEINRICH, B., 1976. Resource partitioning among some unsocial insects: bumblebees. *Ecology* **57**, 874–89.

HENDRICKSON, J. A and EHRLICH, P. R., 1971. An expanded concept of 'species diversity'. *Notulae Naturae Acad. Natur. Sci. Philadelphia.* **439**, 6 pp.

HESPENHEIDE, H. A., 1975. Prey characteristics and predator niche width. *In* Cody, M. L. & Diamond, J. M. (eds.) *Ecology and Evolution of Communities.* pp. 158–80. Harvard University Press, Cambridge, Mass.

HODKINSON, I. D., 1975. A community analysis of the benthic insect fauna of an abandoned beaver pond. *J. Anim. Ecol.* **44**, 533–51.

HUMMON, W. D., 1974. SH: A similarity Index based on shared species diversity used to assess temporal and spatial relations among inter-tidal Marine Gastrotricha. *Oecologia* **17**, 203–20.

HURLBERT, S. H., 1969. A coefficient of interspecific association. *Ecology* **50** (1), 1–9.

HURLBERT, S. H., 1971. The non-concept of species diversity: a critique and alternative parameters. *Ecology* **52** (4), 577–86.

HUTCHINSON, G. E., 1953. The concept of pattern in Ecology. *Proc. Acad. Nat. Sci. Philadelphia* **105**, 1–12.

JAMES, F. C., 1971. Ordination of habitat relationships among breeding birds. *Wilson Bull.* **83**, 215–36.

JAMES, F. C. and SHUGART, H. H., 1970. A quantitative method of habitat description. *Audubon Fld. Notes* **24**(6), 727–36.

KARR, J. R. and JAMES, F. C., 1975. Eco-morphological configurations and convergent evolution in species and communities. *In* Cody, M. L. & Diamond, J. M. (eds.) *Ecology and Evolution of Communities* pp. 258–91. Harvard University Press, Cambridge, Mass.

KENDALL, M. G., 1962. *Rank correlation methods*. 199 pp. Griffin, London.

KEMPTON, R. A. and TAYLOR, L. R., 1974. Log-series and log-normal parameters as diversity discriminants for the *Lepidoptera*. *J. Anim. Ecol.* **43**, 381–99.

KERSHAW, K. A., 1973. *Quantitative and Dynamic Plant Ecology*. 2nd ed. Griffin, London.

KIKKAWA, J., 1968. Ecological association of bird species and habitats in Eastern Australia; similarity analysis. *J. Anim. Ecol.* **37**, 143–65.

KONTKANEN, P., 1957. On the delimitation of communities in research on animal biocoenotics. *In* Demerec, M. (ed.), *Population Studies: Animal Ecology and Demography*. Cold Spring Harb. Sym. quant. Biol. **22**, 373–8.

KRICHER, J. C., 1972. Bird species diversity: the effect of species richness and equitability on the diversity index. *Ecology* **53**, 278–82.

LEVINE, S. H., 1976. Competitive interactions in ecosystems. *Am. Nat.* **110**, 903–10.

LEVINS, R., 1968. *Evolution in Changing Environments*. Princeton University Press, Princeton, New Jersey.

LLOYD, M. and GHELARDI, R. J., 1964. A table for calculating the 'equitability' component of species diversity. *J. Anim. Ecol.* **33**, 217–25.

MACAN, T. T., 1954. A contribution to the study of the ecology of the Corixidae (Hemipt.). *J. Anim. Ecol.* **23**, 115–41.

MACARTHUR, R. H., 1957. On the relative abundance of bird species. *Proc. Nat. Acad. Sci.* **43**, 293–5.

MACARTHUR, R. H., 1972. *Geographical Ecology: Patterns in the Distribution of Species.* 269 pp. Harper & Row, New York and London.

MACARTHUR, R. H. and LEVINS, R., 1967. The limiting similarity, convergence and divergence of coexisting species. *Am. Nat.* **101**, 377–85.

MACARTHUR, R. H. and MACARTHUR, J. W., 1961. On bird species diversity. *Ecology* **42**, 594–8.

MACARTHUR, R. H. and WILSON, E. O., 1967. *The Theory of Island Biogeography*. Princeton University Press, Princeton, New Jersey.

McCLURE, M. S. and PRICE, P. W., 1976. Ecotype characteristics of co-existing *Erythoneura* Leafhoppers (Homoptera: Cicadellidae) on sycamore. *Ecology* **57**, 928–40.

MACFADYEN, A., 1952. The small Arthropods of a *Molinia* fen at Cothill. *J. Anim. Ecol.* **21**, 87–117.

MACFADYEN, A., 1954. The invertebrate fauna of Jan Meyen Island (East Greenland). *J. Anim. Ecol.* **23**, 261–98.

McNAUGHTON, S. J. and WOLF, L. L., 1970. Dominance and the niche in ecological systems. *Science* **167**, 131–6.

MAY, R. M., 1973. *Stability and Complexity in Model Ecosystems*. Princeton University Press, Princeton, New Jersey.

MAY, R. M., 1974. On the theory of niche overlap. *Theoret. Popul. Biol.* **5**, 297–332.

MAY, R. M., 1975a. Patterns of species abundance and diversity. *In* Cody, M. L. & Diamond, J. M. (eds.) *Ecology and Evolution of Communities*, Harvard University Press, Cambridge, Mass.

MAY, R. M., 1975b. Some notes on estimating the competition matrix α. *Ecology* **56**, 737–41.

MAY, R. M., 1976a. Models for two interacting populations. *In* May, R. M. (ed.) *Theoretical Ecology*, Blackwells, Oxford. pp. 49–70.

MAY, R. M., 1976b. Patterns in multi-species communities. *In* May, R. M. (ed.) *Theoretical Ecology*, Blackwells, Oxford. pp. 142–62.

MAY, R. M. and MacARTHUR, R. H., 1972. Niche overlap as a function of Environmental variability. *Proc. Nat. Acad. Sci.* **69**, 1109–13.

MOUND, L. A. and WALOFF, N., 1978. The diversity of insect faunas. *Symp. R. ent. Soc. Lond.* **9** (in press).

MOUNTFORD, M. D., 1962. An index of similarity and its application to classificatory problems. *In* Murphy, P. W. (ed.), *Progress in Soil Zoology*, 43–50. Butterworths, London.

MUELLER-DOMBOIS, D. and ELLENBERG, H., 1974. *Aims and Methods of Vegetation Ecology*. 547 pp. John Wiley & Sons, New York, London, Sydney, Toronto.

MÜHLENBERG, M., LEIPOLD, D., MADER, H. J. and STEINHAUER, B., 1977. Island ecology of arthropods I & II. *Oecologia* **29**, 117–44.

MULLER, P., KLOMANN, U., NAGEL, P., REIS, H. and SCHAFER, A., 1975. Indikaturwert unterschiedlicher biotischer diversität im verdichtungsraum von Saarbrucken. *Verh. Ges. Okologie* **1974**, 113–28.

ODUM, H. T., CANTLON, J. E. and KORNICKER, L. S., 1960. An organizational hierarchy postulate for the interpretation of species–individual distributions, species entropy, ecosystem evolution and the meaning of a species–variety index. *Ecology* **41**, 395–9.

PEET, R. K., 1974. The measurement of species diversity. *Ann. Rev. Ecol. Syst.* **5**, 285–307.

PEET, R. K., 1975. Relative diversity indices. *Ecology* **56**, 496–8.

PIANKA, E. R., 1973. The structure of lizard communities. *Rev. Ecol. Syst.* **4**, 53–74.

PIANKA, E. R., 1975. Niche relations of desert lizards. *In* Cody, M. L. & Diamond, J. M. (eds.) *Ecology and Evolution of Communities.* Harvard University Press, Cambridge, Mass.

PIANKA, E. R., 1976a. *Evolutionary Ecology.* Harper and Row, New York, 356 pp.

PIANKA, E. R., 1976b. Competition and Niche Theory. *In* May, R. M. (ed.) *Theoretical Ecology,* Blackwells, Oxford.

PIELOU, E. C., 1975. *Ecological Diversity.* 165 pp. Wiley, New York.

PIELOU, D. P. and PIELOU, E. C., 1968. Association among species of infrequent occurrence: the insect and spider fauna of *Polyporus betulinus* (Bulliard) Fries. *J. Theoret. Biol.* **21**, 202–16.

PRESTON, F. W., 1948. The commonness and rarity of species. *Ecology* **29**, 254–83.

PRESTON, F. W., 1960. Time and space and the variation of species. *Ecology* **41**, 611–27.

PRESTON, F. W., 1962. The canonical distribution of commonness and rarity. Part I. *Ecology* **39**, 185–215. Part II. *ibid.* **39**, 410–32.

PRICE, P. W., 1971. Niche breadth and dominance of parasitic insects sharing the same host species. *Ecology* **52**, 587–96.

ROOM, P. M., 1975. Diversity and organisation of the ground foraging ant faunas of forest, grassland and tree crops in Papua, New Guinea. *Aust. J. Zool.* **23**, 71–89.

ROUGHGARDEN, J., 1974. Species packing and the competition function with illustrations from coral reef fish. *Theor. Pop. Biol.* **5**, 163–86.

SABATH, M. D. and JONES, J. M., 1973. Measurement of niche breadth and overlap the Colwell-Futuyma method. *Ecology* **54**, 1143–7.

SANDERS, H. L., 1969. Benthic marine diversity and the stability-time hypothesis. *In Diversity and Stability in Ecological Systems, Brookhaven Symp. Biol.* **22**, 71–81.

SCHOENER, T. W., 1968. The Anolis lizards of Bimini: resource partitioning in a complex fauna. *Ecology* **49**, 704–26.

SCHOENER, T. W., 1974a. Some methods for calculating competition coefficients from resource-utilization spectra. *Am. Nat.* **108**, 332–40.

SCHOENER, T. W., 1974b. Resource partitioning in ecological communities. *Science* **185**, 27–39.

SEAL, H. L., 1964. *Multivariate statistical analysis for biologists.* 207 pp. Methuen, London.

SHAPIRO, A. M., 1975. The temporal component of butterfly species diversity. *In* Cody, M. L. & Diamond, J. M. (eds) *Ecology and Evolution of Communities.* p. 181–95, Harvard University Press, Cambridge, Mass.

SHELDON, A. L., 1969. Equitability indices: dependence on the species count. *Ecology* **50**, 466–7.

SIMPSON, E. H., 1949. Measurement of diversity. *Nature* **163**, 688.

SINHA, R. A., 1977. Uses of multivariate methods in the study of stored-grain ecosystems. *Environ. Ent.* **6** (2), 185–92.

SINHA, R. N., WALLACE, H. A. H. and CHEBIB, F. S., 1969. Principal component analysis of interrelations among fungi, mites and insects in grain bulk ecosystems. *Ecology* **50**, 536–47.

SØRENSEN, T., 1948. A method of establishing groups of equal amplitude in plant sociology based on similarity of species content and its application to analyses of the vegetation on Danish commons. *Biol. Skr.* (*K. danske vidensk. Selsk.* N. S.) **5**, 1–34.

SOUTHWOOD, T. R. E., 1978. The components of diversity. *In* Mound, L. A. & Waloff, N. (eds.) *The Diversity of Insect Faunas, Symp. R. ent. Soc. Lond.* **9** (in press).

SOUTHWOOD, T. R. E., 1977. Habitat, the Template for Ecological Strategies. *BES Presidential Address. J. Anim. Ecol.* **46**, 337–65.

TAYLOR, L. R., KEMPTON, R. A. and WOIWOOD, I. P., 1976. Diversity statistics and the log-series model. *J. Anim. Ecol.* **45**, 255–72.

TRAMER, E. J., 1969. Bird species diversity: components of Shannon's formula. *Ecology* **50**, 927–9.

UETZ, G. W., 1977. Coexistence in a guild of wandering spiders. *J. Anim. Ecol.* **46**, 531–41.

VANDERMEER, J. H., 1969. The community matrix and the number of species in a community. *Ecology* **50**, 362–71.

VAN VALEN, L., 1965. Morphological variation and width of ecological niche. *Am. Nat.* **99**, 377–90.

WEBB, D. J., 1974. The statistics of relative abundance and diversity. *J. Theor. Biol.* **43**, 277–92.

WHITTAKER, R. H., 1970. *Communities and Ecosystems.* 162 pp. Macmillan, London.

WHITTAKER, R. H., 1972. Evolution and measurement of species diversity. *Taxon* **21**, 213–51.

WHITTAKER, R. H. and FAIRBANKS, C. W., 1958. A study of plankton copepod communities in the Columbia basin, south eastern Washington. *Ecology* **39**, 46–65.

WILLIAMS, C. B., 1947. The logarithmic series and the comparison of island floras. *Proc. Linn. Soc. Lond.* **158**, 104–8.

WILLIAMS, C. B., 1964. *Patterns in the balance of Nature and related problems in quantitative ecology.* 324 pp. Academic Press, London and New York.

WILSON, D. S., 1975. The adequacy of body size as a niche difference. *Amer. Nat.* **109**, 769–84.

WILSON, E. O., 1969. The species equilibrium. *In Diversity and Stability in Ecological Systems. Brookhaven Symp. Biol.* **22**, 38–47.

14

The Estimation of Productivity and the Construction of Energy Budgets

The size of a population and the interactions between populations within an ecosystem may be expressed in terms of biomass (weight of living material) or energy content, as well as in numbers. Biomass and energy are useful to ecologists in that they provide a common unit for the description of populations of animals and plants of different sizes. Descriptions of the predator in these terms are often essential in studies on the effect of insect predators of varying ages on a prey population (Szalay-Marzsó, 1958), and the prey consumed by general predators, e.g. insectivorous birds, is best expressed as biomass or calories. Conversely if the energy requirements are known from metabolic measurements, they may be used to predict the food requirements in the field (Pearson, 1954; Stiven, 1961), although the quality of the food in terms of specific amino-acids, vitamins and other constituents will also be important (Boyd & Goodyear, 1971; Iversen, 1974; Schroeder, 1977; Onuf *et al.*, 1977; McNeill & Southwood, 1978).

The concept of energy is fundamental in the consideration of the functioning of the ecosystem; Ivlev's (1939, 1945) and Lindeman's (1942) now classical papers pointed the way to this approach, reviewed by Petrusewicz (1967); Bourliere & Lamotte (1967, 1968); Gallucci (1973); and Wiegert (1976). Various trophic levels can be distinguished in the ecosystem; firstly there are the primary producers (i.e. plants) that build up complex substances from simple inorganic substances utilizing the energy from sunlight. The total energy retained by the plant as fixed carbon is usually referred to as the *gross production* and the change in energy level (standing crop or stock) plus any losses due to grazing or mortality during the year, but less imports (e.g. migrants) is termed the *net production*. The amount of living material present at any given time is the *standing crop* or *biomass*. Unfortunately the use of these terms has been confused, but Macfadyen (1963) has given a useful table of synonymy and Petrusewicz & Macfadyen (1970) have established standard definitions and notation. Information on primary productivity is given in Newbould (1967), Milner & Hughes (1968), Vollenweider (1969) and Chapman (1976). Other trophic levels are occupied by consumers and decomposers, i.e. animals and certain plants (e.g. fungi, bacteria), often in the soil (Phillipson, 1971; Parkinson *et al.* 1971).

456

The energy equations for an individual may be expressed:

Gross energy =	Digestible energy	=	Metabolizable energy	=	Resting energy
or	+		+		+
energy intake	Faecal waste →		Urinary waste →		Activity
or					+
ingestion					Growth
or					+
consumption					Reproduction

(Note – only the upper term partakes in the equation to its right.)

The energy budget of a population or a trophic level can be expressed in several equations (Wiegert, 1964a, 1968; Petrusewicz & Macfadyen, 1970). These may be related to the equations above, noting that the energy used up in resting activity and the 'work' of growth is equated with respiration, so that:

$$C = D + F = A + U + F = R + P + (F + U) \tag{14.1}$$

where C = consumption (= ingestion), D = digestable energy, F = faecal waste, A = assimilation (metabolizable energy), U = urinary waste, R = respiration and P = productivity (= growth + reproduction).

This basic equation may be divided into a number of simpler equations for parts of the process, which are of value in the construction of energy budgets; for if all but one of the terms are known the other may be found. Thus:

$$P = A - R \tag{14.2}$$

$$A = C - (F + U) \tag{14.3}$$

Productivity is the increase (Δ) in the biomass of the standing crop (B), plus production eliminated (E), through death, emigration, exuviae or other products (e.g. feathers of birds at moult, spiders' silk, byssus of mussels); then:

$$P = \Delta B + E \tag{14.4}$$

One will also note from the composite equation that

$$P = Pg + Pv + E \tag{14.5}$$

where Pg = productivity as growth and Pr = productivity through the reproduction (the biomass of offspring).

These equations demonstrate the usefulness of energy, as opposed to simply biomass, in the description of this aspect of ecosystem or population dynamics, for calories provide the common unit of expression. The studies of Odum & Odum (1955) on a coral reef, of Odum (1957) on a hot spring, of Teal (1957) on a cold spring and of Whittaker *et al.* (1974) on a deciduous forest exemplify the value of measurements of biomass and energy in describing and analysing the trophic structure of ecosystems. These were the major themes in the International Biological Programme (IBP) (van Dobben & Lowe-McConnell, 1975), which has also published a series of handbooks (e.g. Ricker

(1968), Golley & Buechner (1968), Petrusewicz & Macfadyen, (1970), Edmondson & Winberg (1971), and Grodzinski *et al.* (1975). These and Paine's (1971) review should be consulted for further information on the measurement of productivity and energy flow.

14.1 Estimation of standing crop

If a detailed life-table or budget has been constructed (see Chapter 10) then the standing crop of the population at any given time can be determined by converting numbers to biomass (dry weight is usually the most appropriate measure), and converting this to energy in terms of the caloric value.

14.1.1 Measurement of biomass

Biomass is generally expressed in terms of dry weight; a known number of animals are dried until their weight is constant and this value is taken as the dry weight. Material should be dried at a low temperature (freeze-drying or at maximum of 60°C) to avoid the loss of volatile – especially lipoid – constituents, for although their contribution to the total dry weight may not be great, they may contribute significantly to the caloric value which is usually determined on the same material.

With small animals, especially plankton, it may be more convenient to work with the volume of the animals rather than the actual numbers (e.g. in Neess & Dugdale's method, see below). A number of devices for measuring the volume of small animals have been developed (e.g. Gnanamuthu, 1952; Andrássy, 1956; Yentsch & Hebard, 1957; Frolander, 1957; Stanford, 1973).

14.1.2 Determination of caloric value

The energy content of a material may be determined directly by oxidation either by potassium dichromate in sulphuric acid (Ivlev, 1934; Teal, 1957) or by burning in oxygen and determining the amount of heat liberated. The latter method – bomb calorimetry – is most convenient and widely used in ecology; however, it involves drying the material, and even when precautions are taken to avoid loss of volatile substances the energy content of dry plant material has been found to be about 8% less than that of the fresh (Komor, 1940).

Phillipson (1964) has described a miniature non-adiabatic bomb (or ballistic) calorimeter that is simple and suitable for handling materials in the order of 5–100 mg dry weight. It is operated by placing the material in a platinum pan which is enclosed in the 'bomb'. The bomb is filled with oxygen until a pressure of 30 atmospheres is reached, cooled until the temperature, as indicated by a thermocouple, is steady, and then enclosed in a polystyrene insulating jacket. The sample is ignited by passing an electric current through a thin wire which presses against it. The temperature rises rapidly, and this is recorded by the deflection of a potentiometer attached to the thermocouple.

The maximum deflection in a firing is a measure of the caloric content of the sample. The instrument is calibrated by burning given weights of a substance of known caloric value. The results from the Phillipson calorimeter have been found to be virtually identical with those obtained by the more complex instrument used by Slobodkin & Richman (1960, 1961). Commercial versions of ballistic microcalorimeters are available.*

The amount of ash present in a substance can influence its caloric value if measured in this way; magnesium and calcium carbonates may decompose at the high temperatures momentarily reached and this reaction will absorb some of the heat produced. If the weight of ash is subtracted from the dry weight, the ash-free dry weight (AFDW) is obtained: these values are more consistent throughout the season (in plants) (Caspers, 1977) or between stages in insects (Wiegert, 1965). If these values are used then biomass must be corrected to its ash-free value (Reiners & Reiners, 1972).

Naturally not all the energy in an animal or plant is available to an animal that eats it, and quite frequently only a proportion of the potentially digestible energy is assimilated. (An extreme and iconoclastic view on availability is presented by Verdun (1972). The assimilation rates for various types and quantities of food need to be determined. The total energy is, however, available to the ecosystem as micro-organisms can break down almost all organic compounds.

Values for the energy content of various plants have been published by Ovington & Heitkamp (1960) and Golley (1960) and of animals by Golley (1960), Slobodkin & Richman (1961), Comita & Schindler (1963) and others†; some of these values, together with others for animals, are given in Table 14.1. There is, however, considerable variation in caloric value which, unless recognized, could be a major source of error in energy budgets. Both plants (Boyd, 1969; Singh & Yadava, 1973; Caspers, 1977) and animals (Wiegert, 1965; Hinton, 1971; Wiegert & Coleman, 1970; Wissing & Hasler, 1971) vary from season to season. This variation is due to variations in the levels of fat (which account for much of the interspecific variation in caloric value in animals (Slobodkin & Richman, 1961; Griffiths, 1977) and of mineral composition, which changes markedly in plants (Boyd & Blackburn, 1970; Boyd, 1970; Caspers, 1977), insects (Wiegert, 1965) and crustaceans (Griffiths, 1977). Different stages of the same insect also vary markedly (Chlodny *et al.* 1967; Gyllenberg, 1969; McNeill, 1971; Wiegert, 1965; Hinton, 1971; Holter, 1974; Hagvar, 1975) (Table 14.2): in the froghoppers, *Philaenus* and *Neophilaenus* the immature stages contain more ash than the adults, the same appears to be true, to a lesser extent, of the willow leaf beetle *Melosoma*, which showed a fall in caloric value throughout development: in *Aphodius* in contrast it rose markedly during the last larval instar.

* From Gentry Instruments Incorp., Whiskey Rd., Aiken, S. Carolina, U.S.A. and Newham Electronics Ltd., Plashet Rd., London E13. U.K.
† See also Cumins & Wuycheck (1971) *Mitt. Int. ver. Theor. Angew. Limnol.* **18**, 1–158.

Table 14.1 Energy content of various substances and living materials (from Golley, 1961; Westlake, 1963; Slobodkin & Richman, 1961; Wiegert, 1964a; Wissing & Hasler, 1971; Singh & Yadava, 1973; Caspers, 1977, and others). Ranges represent seasonal or site-to-site variation.

	Kcal/g dry wt.		Kcal/g dry wt.
Monosaccharides, e.g. glucose	3.7	Grass, *Poa annua*	4.13–4.21
Disaccharides, e.g. sucrose	4.0	Herb, *Stellaria media*	3.71–3.78
Polysaccharides, e.g. starch,		Herb, *Euphorbia peplus*	4.29
cellulose	4.2		
Protein: crude	5.7	Mollusca: *Modiolus*	4.6
Fat: ether extract	9.5	*Succinea*	5.4
Crude fibre	4.5–4.7	Crustacea: *Daphnia*	4.17–5.12
Leaves of plants	4.2	*Hyalella azteca*	3.68–4.23
Stems of plants	4.3	*Calanus hyperboreus*	7.4
Roots of plants	4.7	Dictyoptera: *Mantis*	5.21
Seeds	5.7	Coleoptera: *Tenebrio molitor* la.	6.58
Litter	4.3	*T. molitor* adult	5.00
Tomato xylem sap	4.5	*Cantharis tenuicollis*	7.54
Philaenus spumarius – exuvia	5.2	*Mylabris pustulata*	6.04
Philaenus spumarius – spittle	4.7	*Diapromorpha turcica*	5.81

Table 14.2 Caloric values (Kcal/g dry wt. or ash free dry wt.*) of different stages of the same insect.

Insect	Data source	Egg	1	2	3	4	5	Pupa	Adult	♂	♀
Homoptera											
*Neophilaenus**	Hinton, 1971		5.44	5.47	5.79	5.80	5.91			5.46–6.00	5.56–
Philaenus	Wiegert, 1965	6.31	4.58	5.27	5.45	5.53	5.71			5.49–5.95	5.16–6.08
*Philaenus**	Wiegert, 1965	6.50	4.98	5.67	5.80	5.78	5.90			5.58–5.95	5.50–6.08
Heteroptera											
Leptopterna	McNeill, 1971	6.30	5.70	5.81		5.86				6.05	6.06
Coleoptera											
Melosoma	Hagvar, 1975	6.16		5.28	5.24			5.71			
Aphodius	Holter, 1975			4.67	4.66–5.80				5.68		
Orthoptera											
*Encoptoloptius**	Bailey & Riegert, 1972		5.01	5.23	5.30	5.40	5.16			5.25	5.34
*Chorthippus**	Gyllenberg, 1969			5.05	5.46	5.80				5.61	5.85

14.2 Estimation of energy flow

Energy flow in a population is usually estimated using one or more of the equations listed in the introduction to this chapter, more especially (14.2 and 14.3):

A(ssimilation) $= R$(espiration) $+ P$(roduction)
A(ssimilation) $= C$(onsumption) $- (F$(aecal) $+ U$(rinary waste)$)$*

Frequently two of the terms in the equations will be found directly and the third by calculation. For example, with a particular animal it might be difficult to determine the respiratory rate in the field, but ingestion and consumption and waste excretion (egestion + excretion) per individual could be measured and these values would be converted to calories; knowing the numbers of individuals in the field, the total energy assimilation (the energy income) of the population could be calcuated. Assuming the numbers of individuals of different ages throughout the season is known from a budget and the calorific value per individual of each age group has been found, the (net) production can be calculated. The difference between the assimilation and production will be a measure of respiration.

The construction of an energy budget requires the initial development of a numerical budget. Normally it is necessary to calculate the standing crop and its caloric value (described above) and from this the production, together with the independent measurement of either consumption and waste excretion rates or respiration. Seasonal variations will affect these values (e.g. Jonasson & Kristiansen, 1967); the energy budget will, perforce, contain a number of approximations and the more variables that can be determined the more reliance can be placed upon it; clearly if all the terms in any of the equations are determined (e.g. Schroeder, 1973) and their magnitudes prove compatible, when they are substituted in the equations, this provides valuable evidence as to their reliability. Thus, as with a numerical budget (p. 357) a detailed energy budget will expose its own errors. Wiegert's (1964a); McNeill's (1971); Lawton's (1971), Bailey & Riegert's (1972) and Holter's (1974, 1975) studies may be taken as model examples of energy budget construction for insects; energy budgets for crustacea are described by Richman (1958), for nematodes by Marchant & Nicholas (1974), and for a rotifer by Doohan (1973).

14.2.1 The measurement of production

Production consists of the total amount of material synthesized by the animal and is obtained by summing:

   (*i*) The increase in the standing crop during the season (or other time unit).
   (*ii*) The biomass of all individuals that died or were eaten during the season.

---

* Because faecal and urinary waste are mixed in insects digestible energy (D) cannot be obtained (Reichle, 1969).

(*iii*) The biomass of the total number of exuviae shed by all the individuals (those alive and those now dead) and of any other product, e.g. the spittle of cercopids (Wiegert, 1964a), the byssus of mussels (Kuenzler, 1961), the silk of many insects, mites and spiders.

(*iv*) The biomass of any reproductive products (sperms, eggs), young individuals or adults that have left the area (i.e. emigration).

Strictly the amount of nitrogenous waste excreted should also be measured. The biomass of any individuals that immigrated into the population must be subtracted from the total obtained above.

The calculation of production from a budget, as was done, for example, by Smalley (1960), Wiegert (1964a), Golley & Gentry (1964), McNeill (1971) and Karo (1973), demands that the form of the mortality curve, the actual population at various stages and immigration and emigration are accurately known. The weights of the various products can be determined in the laboratory, but caution should be exercised in assuming that laboratory-determined rates hold under field conditions.

A somewhat different approach was developed by Allen (1951) and Neess & Dugdale (1959) for the study of the production of chironomid larvae in a lake. Instead of separating the numbers into different age categories (as in a budget) and then multiplying the age categories by their respective biomasses, Neess & Dugdale use the product of these variables as expressed in the field in terms of weight.

The net production of a developing larval population is the outcome of two opposing processes:

increase in weight of the individual: $k_g Q_t = dQ_t/d_t$ (14.6)

fall in numbers due to mortality $= -k_m N_t = dN_t/d_t$ (14.7)

where Q_t = weight of the individual at time t, N_t = number of individuals at time t, k_g = growth rate and k_m = mortality rate. These equations (14.6 and 7) can be combined by the elimination of t and integrated to give:

$$\log_e Q_t^{1/k_g} = \log_e C N_t^{-1/k_m} \text{ or } Q_t^{-k_m} = C N_t^{k_g}$$ (14.8)

C may be found in powers of Q_t and N_t; if this is expressed at $t = 0$, when Q_0 = weight of a newly hatched larva and N_0 = total hatch, and substituted for C in 14.8:

$$Q_t = Q_0/(N_0/N_t)^{-k_g/k_m}$$ (14.9)

For given values of N_0 and Q_0 the shape of the curves describing the relationship between Q_t and N_t is determined solely by the ratio k_g/k_m; if the ratio remains constant, the growth-survivorship curve will trace out the same path. Neess & Dugdale noted that within the range of values of the ratio up to 2, large changes occur in the area under the curve for small variations in the actual ratio. Now, the area under the curve corresponds to the actual net production. Therefore, if it can be shown that k_g/k_m is a constant, a smooth curve may be drawn of numbers on mean weight and the area beneath it found. As equation 14.9 can be rearranged to:

$$\log Q_t = \log Q_0 + (k_g/k_m) \log (N_0/N_t) \tag{14.10}$$

the constancy of the ratio may be tested by plotting $\log N_t$ against $\log Q_t$. A straight line will indicate that the ratio is constant.

The numbers per unit area can then be plotted against the mean weight, a smooth curve (an 'Allen curve') drawn and the area below it (ACDE in Fig. 14.1) found with a planimeter. This is the actual net production in biomass per unit area (the units corresponding of course to those of Q_t and N_t).

A number of terms may be defined by reference to Fig. 14.1.

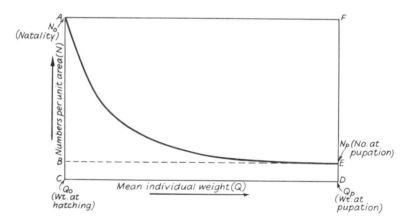

Fig. 14.1 The relationship between numbers and mean individual weight in an insect population during larval development; the area below the curve may be measured to give actual net production. The letters indicate areas representing other productivity terms (see text) (modified from Neess & Dugdale, 1959).

ACDE   Actual net production.
BCED   Actual net recruitment or increase in standing crop due to this cohort = $Q_p N_p$.
ABE   Directly recycled production.
ACDF   Potential net production = $N_0 Q_p$.
AEF   Lost potential net production.
ABEF   Lost potential recruitment.

This method is most appropriate when the development of the individuals of a generation is closely synchronized or where a particular cohort can be recognized and distinguished from others, as in dragonflies (Benke, 1976).

14.2.2 The measurement of feeding and assimilation

THE QUALITY OF THE FOOD EATEN

In some instances, such as with many arboreal lepidopterous larvae, the nature of an animal's food may be determined by observation in the field; observations in the laboratory may also be used, but with caution, for many animals will eat unnatural foods under artificial conditions. Another method, frequently used by ornithologists, is to examine the crop contents (p. 312). Such methods may be used with chewing insects and the remains of different types of plant or animal identified under the microscope (e.g. Hanna, 1957; Mulkern & Anderson, 1959). This approach can be quantified using a sedimentation technique (Brown, 1961). In plant-feeding sucking insects the actual tissue – phloem, xylem, mesophyll cells – from which the nutriment is obtained should be determined. The salivary sheath (Miles, 1972) or the position of the stylets (Wiegert, 1964b; Pollard, 1973) that may sometimes be kept in place by a high voltage shock (Ledbetter & Flemion, 1954) are precise indications, but examination of a freshly damaged leaf may sometimes be satisfactory (McNeill, 1971).

The identification of the meals of predators by serological techniques is discussed in Chapter 9 (p. 312), where reference is also made to the use of radio-isotope tagged prey. Following a pioneer study by Pendleton & Grundmann (1954), radio-isotopes have often been used to trace out food webs, (e.g. Odum & Kuenzler, 1963; Paris & Sikora, 1965; Marples, 1966; Wiegert, Odum & Schnell, 1967; Reichle, 1967; de la Cruz & Wiegert, 1967; Coleman, 1968; Van Hook, 1971; Shure, 1973). The food plant or primary food source is tagged and development of radioactivity (assessed quantitatively by counting) in the animals demonstrates a trophic link; the rate of build-up of radioactivity in the animal has been taken as proportional to the trophic distance e.g. a peak in activity soon after tagging would indicate a herbivore, a later and lower peak, a predator or saphrophagous species. However, as Shure & Pearson (1969) and Shure (1970) point out, such interpretations should be made with care because herbivores may change their host plant and those moving to the labelled plant during the experiment would show accumulation curves resembling a predator's. Appropriate field observation can obviate such misinterpretations. Trophic links in soil invertebrates may be investigated using pesticides for the selective removal of animal groups (Edwards *et al.*, 1969), a method analagous to the 'insecticidal check-method' (p 320).

The forage-ratio and methods of assessing food preferences are discussed in Chapter 9 (p. 326).

FEEDING AND ASSIMILATION RATES

There are four basic approaches to the measurement of these rates and related parameters:

(1) Radiotracer – measuring the passage of radio-isotopes from the food to the animal and their subsequent loss.

(2) Gravimetric – by the direct weighing of the food, the faeces and the animal.

(3) Indicator – by marking the food with an inert non-absorbed indicator, the increase in its concentration as it passes through the alimentary canal measures the amount of matter absorbed.

(4) Faecal – by measuring faecal output and relating this to consumption.

Radiotracer techniques

Radiotracers* may be used singly or as a double marking method. When used singly there are two approaches: the amount of radio-isotope used may be small relative to equilibrium body burden of the animal, then the rate of increase of assimilation of the isotope is linearly related to the rate of assimilation of the food, as for example in the studies of Engelmann (1961), Strong & Landes (1965), Hubbell *et al.* (1965), Malone & Nelson (1969) and Moore *et al.* (1974). Alternatively the body burden may be increased so as to reach equilibrium when the following relationship holds:

$$r = \frac{KQ_e}{a} \tag{14.11}$$

where r = feeding rate (measured in μc/day), a = proportion of ingested nuclide assimilated, K = the elimination constant[†] $= 0.693$ – biological half-life (in days) and Q_e = the whole body radioactivity (in μc) in a steady state equilibrium. Therefore if the whole body radioactivity is measured and the assimilation rate and the biological half-life are known, the feeding rate may be calculated. This approach has been developed by Crossley (1963*a* and *b*, 1966), who worked in an environment heavily contaminated with [137]cesium This isotope is almost completely assimilated so that the term 'a' can be dropped from the above equation. Crossley (1963*b* and *c*) found that the biological half-life appeared to be linearly related to body weight, except that in the pupal stage, of course, there was no elimination. However, as Odum & Golley (1963) point out and Hubbell *et al.* (1965) confirm, the biological half-life of an isotope in a given animal is variable and will be influenced by temperature, activity, food and other factors, so that the half-life in the field is likely to be different from that determined in the laboratory.

Working on the isopod, *Armadillidium*, Hubbell *et al.* (1965) also show that the percentage of the isotope ingested that is actually assimilated will vary with the feeding rate; they used [85]Sr and, with this biologically significant

---

* Further information on radiotracer techniques is given on pp. 78–86 and 317–319.

† This formula only applies to isotopes with long half-lives, so that the effective half-life = the biological half-life. Where radioactive decay is significant then this elimination rate needs to be added to that due to the biological half-life.

element (it is utilized in the exoskeleton), its assimilation rate paralleled the actual assimilation of nutriments from the food. The biological half-life of an isotope may be different in the two sexes, as Williams & Reichle (1968) found in a chrysomelid beetle.

In conclusion therefore, before radio-isotopes can be used to estimate feeding and assimilation rates in field populations, it is necessary to determine:

(1) The variability of the rate of assimilation of the isotope under field conditions.

(2) The variability of the biological half-life of the isotope under field conditions.

(3) Whether the level of isotope in the organism is low relative to the equilibrium level or whether it is high, a steady state having been reached. This is done from the equation

$$ra = KQ_e \qquad (14.12)$$

already given, but some measure of R and a must be obtained experimentally so that the expected equilibrium body burden of radio nuclide (Q_e) can be compared with that actually observed.

(4) That if the isotope is likely to be generally present in the media, as in studies with aquatic predators and labelled prey, it is only taken-up by feeding. Dragonfly larvae appear to absorb ^{65}Zn on their surface (Kormondy, 1965).

For biologically indeterminate (i.e. with no specific metabolic function) trace substances, whether radioactive or not, Fagerström (1977) considers that the biological half-life will be proportional to body weight raised to $(1 - b)$ where b is the exponent relating body weight to metabolic rate – normally taken as 0.8.

Feeding (or ingestion) and assimilation can usually be separated, for when an animal is fed on radio-isotope tagged food and the fall-off of radioactivity with time is measured, it will be found that at first the curve is steep, reflecting the loss of the ingested but non-assimilated component; later the curve becomes shallow and this represents the elimination rate of the assimilated isotope (Odum & Golley, 1963) (Fig. 14.2). Furthermore, Odum & Golley point out that the assimilated isotope is eliminated in different ways depending on the element, e.g. iodine mainly through the exuviae, zinc mainly by excretion, especially after moulting, or with the female, by the eggs.

A double-tracer method was developed by Calow & Fletcher (1972) in which the isotopes are used as a modification of the indicator technique (see below). They point out that the indicator and gravimetric methods assume that all material in the faeces is derived from the food, although strictly this is not true. By using two isotopes – one, carbon (^{14}C) that is biologically active, and the other chromium (^{51}Cr) that is not absorbed from the gut – they were able to allow for errors arising from this assumption. These two isotopes may

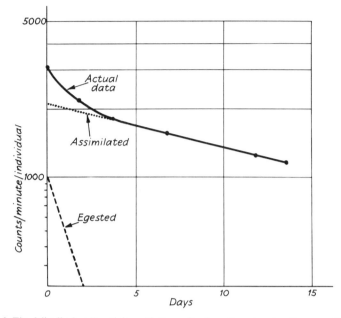

Fig. 14.2 The fall-off of radioactivity with time after ingestion showing the separation of the component that is assimilated from the component that is egested (from Odum & Golley, 1963; data based on marine isopod, *Idothea*, ingesting seaweed tagged with [65]zinc).

be distinguished by their different radiations, β from [14]C (detected by liquid scintillation counting) and α from [51]Cr (detected by crystal scintillation counting). If their levels (counts/min (ct/min)) are determined in the food and faeces, then:

$$\% \ A = 1 - \frac{\text{ct/min}\ ^{51}\text{Cr (food)}}{\text{ct/min}\ ^{14}\text{C (food)}} \times \frac{\text{ct/min}\ ^{14}\text{C (faeces)}}{\text{ct/min}\ ^{51}\text{Cr (faeces)}} \times 100 \quad (14.13)$$

[14]C could be replaced by another absorbed isotope (e.g. [3]H) and ^{51}Cr by a non-absorbed isotope.

The proper homogenization of material for accurate scintillation counting is sometimes difficult; with [14]C – labelled material this may be overcome by combustion. Burnison & Perez (1974) describe a simple apparatus for this.

Gravimetric techniques

The simplest approach to the measurement of ingestion and one that is frequently applicable is to weigh the food before and after the animal has fed (e.g. Smith, 1959; Phillipson, 1960; Fewkes 1960; Evans, 1962; Strong & Landes, 1965; Hubbell *et al.*, 1965; Lawton, 1970); if all values are reduced to dry weight, errors due to the loss of moisture during feeding will be eliminated, otherwise this must be determined from controls.

Aquatic predators may 'lose' part of the body contents of their prey during feeding. This would be most significant when the prey is chewed, and large relative to the predator; Dagg (1974) found an amphipod could lose nearly 40 % of its 'apparent' meal in this way. The quantity of leaf litter destroyed by soil arthropods may be determined by exposing a known quantity in mesh (e.g. nylon) bags (Crossley & Hoglund, 1962; Wiegert, 1974).

In the laboratory the faeces may usually be collected and their dry weight and caloric value found; the latter will of course, differ from that of the food (Gere, 1956; Hubbell *et al.*, 1965) and the difference in energy content between ingestion and egestion may be more striking than the actual weight differences. The collection in the field of the total amount of faeces produced by a population is impossible (p. 289), but a measure could be obtained for a small number of arboreal larvae under semi-experimental conditions.

Hubbell *et al.* (1965) have compared the results of gravimetric and radiotracer methods using [85]strontium for determining the feeding and assimilation rates of the woodlouse, *Armadillidium*, and have found close agreement.

The only other weight that can be conveniently measured is the increase in weight of the animal (Fewkes, 1960; Johnson, 1960; Evans, 1962); this is a measure of the amount of growth. When converted to dry weights two equations (14.2 and 14.4) hold:

Assimilation = Production + Respiration
Production = Growth + Exuviae and other Products

Clearly, then, if respiration can be measured, growth + the weight of the products may be used to give an estimate of assimilation. If wet weights are used allowance must also be made for the loss of weight due to the evaporation of water from the insect (Strong & Landes, 1965).

Just as the rate of assimilation of radio-isotopes has been found to vary under different conditions (p. 466), so does the nutritional assimilation rate and the relation between assimilation and production. For example, Sush-chenya (1962) found with the brine shrimp (*Artemia salina*) that the greater the intake of food the smaller the percentage of it that is assimilated, but a slightly higher percentage of the assimilated energy was used for growth at the higher levels of food intake. Likewise in the grasshopper, *Poecelocerus*, when food availability was lowered, a lower proportion of the assimilated energy was used for growth (Muthukrishnan & Delvi, 1974). Assimilation efficiencies are also influenced by temperature conditions (in poikilotherms) and even relatively small differences in food e.g. slugs on lettuces and potatoes (D. H. Davidson, 1977). Thus caution needs to be exercised in the extrapolation of laboratory-determined values to field conditions.

Indicator methods
Assimilation can also be measured if the food can be marked with an indicator

which is easily measured quantitatively, is non-toxic at the concentrations used and is not absorbed by the gut. The percentage assimilation of the food is given by:

$$\% \ A = 1 - \left(\frac{\text{conc. indicator in food/unit dry wt}}{\text{conc. indicator in faeces/unit dry wt}} \right) \times 100 \quad (14.14)$$

Chromic oxide has been widely used as an indicator in studies on vertebrate nutrition (Corbett *et al.*, 1960) and McGinnis & Kasting (1964*a* and *b*) have shown how the chemical, or more conveniently, chromic oxide paper, may be used with insects as an indicator to measure assimilation from finely divided food. As the indicator must be mixed homogeneously throughout the food, this method will be inappropriate in many instances, especially with natural diets, but Holter (1973) found it very suitable for measuring food consumption in the larvae of the dung beetle, *Aphodius*.

Measurement of faecal output

Provided the assimilation rate is not too sensitive to variations in food quantity, for which there is some evidence (Muthukrishnan & Delvi, 1974), faecal output should be related to consumption, and once standardized for the particular food, could be used for this purpose. Mathavan & Pandian's (1974) studies with various lepidopterous larvae suggest it is reliable in this group (see also p. 289). Lawton (1971) used gut clearance times to give an independent estimate of consumption in the larva of a damsel fly and Humphreys (1975) obtained a measure of the food consumed in the field by wolf spiders in the fourteen days prior to capture from the guanine content of the excreta during the following seven days.

14.2.3 The measurement of the energy loss due to respiration and metabolic processes

CALORIMETRIC

The amount of energy lost through respiration may be determined directly by the measurement of the heat given out by the animal, but as some energy will have been used to vaporize water from the animal this will be less than the total energy loss and a correction must be applied. Furthermore, it is difficult to work with small animals; few studies have been made with insects using this approach (Pratt, 1954; Heinrich, 1972*b*).

THE EXCHANGE OF RESPIRATORY GASES

The energy equivalents of oxygen and carbon dioxide

During metabolic processes, in which energy is liberated, oxygen is utilized and carbon dioxide produced; the proportion of carbon dioxide evolved, to oxygen used, depends on the actual metabolic process and is referred to as the respiratory quotient expressed:

$$RQ = \frac{\text{Carbon dioxide produced}}{\text{Oxygen utilized}} \qquad (14.15)$$

Thus if two of the three values are known the third may be found; the gases may be measured by a number of different techniques and the respiratory quotient by noting the change in volume at constant pressure and temperature when the animal respires in an enclosed space.

The amount of energy liberated in the process can be determined from the quantity of oxygen (or carbon dioxide) involved, if the respiratory quotient is known and certain assumptions are made as to the actual metabolic processes.

If carbohydrates alone are being utilized for energy $RQ = 1$ and 0.198 cm^3 (at S.T.P.) of oxygen will be consumed and the same volume of carbon dioxide produced during the liberation of one calorie of energy; with fats alone $RQ = 0.707$ and then the consumption of 0.218 cm^3 of oxygen and the production of 0.154 cm^3 of carbon dioxide will result from the liberation of one calorie. Tables are widely available of the calorific equivalents of unit volumes of oxygen and carbon dioxide for different RQ's assuming that carbohydrate and fat utilization are the only metabolic processes involved (Brody, 1945; White *et al.*, 1959; Table 14.3) and ecologists have frequently approximated by using the RQ obtained from experiments as equivalent to an 'N-free RQ', disregarding the effect of protein and other metabolisms.

Table 14.3 The caloric equivalents of oxygen and carbon dioxide for various values of RQ, due to utilization of different proportions of carbohydrates and fats (modified from an extensive table in Brody, 1945; original data from Zuntz & Schumberg)

RQ	*Oxygen* kcal* /litre	*Carbon dioxide* kcal/litre	% *oxygen consumed by carbohydrate component*
0.70	4.686	6.694	0.0
0.75	4.729	6.319	14.7
0.80	4.801	6.001	31.7
0.85	4.683	5.721	48.8
0.90	4.924	5.471	65.9
0.95	4.985	5.247	82.9
1.00	5.047	5.047	100.0

If greater accuracy is to be obtained the quantity of nitrogen excreted must be measured (Shaw & Beadle, 1949) enabling the component of the gases due to protein katabolism to be established and the 'N-free RQ' determined. The respiratory exchange involved in the production of 1 g of nitrogen from the katabolism of a protein will depend on the proportions by weight of the

* $1 \text{ kcal} = 4.18 \text{ kJ}$

various elements in its molecule; the necessary calculations are explained by Kleiber (1961).

However, even this approach assumes that fat, carbohydrate and protein katabolism are the only processes contributing to the respiratory quotient, but many others may theoretically influence the RQ and as these have RQ values outside the 0.7–1.0 range small amounts may modify the composite RQ disproportionately; e.g. alcohol breakdown has an $RQ = 0.67$, the synthesis of fat from carbohydrate an $RQ = 8.0$ (Kleiber, 1961).

It is discouraging for the ecologist to note that Cahn (1956) and Kleiber (1961) report studies that show that the estimate of energy expended, obtained indirectly (from the respiratory quotient), may be found to deviate by as much as 25 % from the true value calculated by direct calorimetry. As Macfadyen (1963) has pointed out, when doubt exists as to the exact katabolic process, but assuming that these involve only fats, carbohydrates and proteins, errors will be reduced if calculations are based on the oxygen uptake rather than on the carbon dioxide produced; for the minimum volume of oxygen that is required to produce 1 calorie is 90.8 % of the maximum volume, but the minimum volume of carbon dioxide that results from the production of 1 calorie is as little as 77.7 % of the maximum.

The respiratory rate
Starved warm-blooded animals exhibit a level of respiration referred to as the basic metabolism, but there is no comparable standard available for poikilothermic animals. It is important to remember, therefore, that although respiration rate and body weight show in general a linear relationship (Brody, 1945; Berg & Ockelmann, 1959; Engelmann, 1961), with poikilotherms the rate may be varied by many factors: temperature, age, season, oxygen concentration and even the type of respirometer used for the measurement (Berg, 1953; Berg *et al.,* 1958; Allen, 1959; Berg & Ockelmann, 1959; Edwards & Learner, 1960; Keister & Buck, 1964; Golley & Gentry, 1964). When respiration rate is being related to the weight of an invertebrate it is necessary to eliminate individual variations in weight due to the contents of the gut (Allen, 1959).

Gas analysis
The precise analysis of the respiratory gas to determine the proportions of oxygen and carbon dioxide is described in detail in textbooks such as Conway (1947), Calvert & Pratt (1956), Kleiber (1961) and Kay (1964), and certain of the methods are discussed in special works: Dixon (1951) and Umbreit *et al.* (1957) describe manometric techniques and Kolthoff & Lingane (1952) polarography. The methods are extensively reviewed in Petrusewicz & Macfadyen (1970), Edmondson & Winberg (1971) and Grodzinski *et al.* (1975); only a general outline will be given here; normally it is necessary to modify the precise form of the instrument to suit the particular animal and

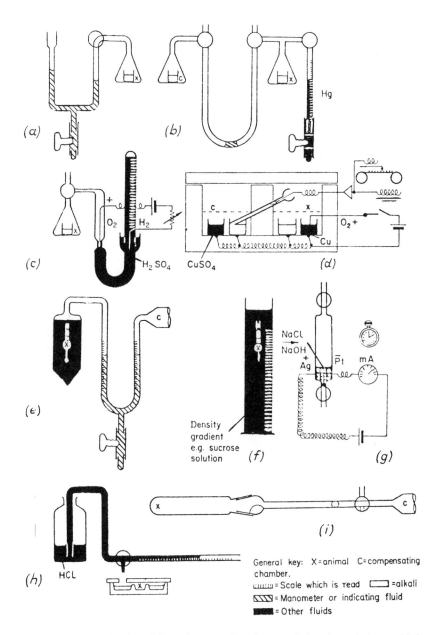

Fig. 14.3 Summary sketches of the main types of respirometer (other than calorimeters) (after Petruzewicz and Macfadyen, 1970): *a.* constant volume Warburg – open; *b.* compensating with mechanical restoration; *c.* simple electrolytic – open; *d.* compensating automatic; *e.* cartesian diver; *f.* gradient diver; *d.* Kopf's method of CO_2 determination; *h.* Conway microdiffusions method; *i.* constant pressure.

investigation. The main types of respirometer, for both the gaseous phase and dissolved gaseous are illustrated in Fig. 14.3; some comparisons of their performance are given in Table 14.4; detailed comparative data is given by Lawton & Richards (1970).

(1) *Isotopes.* Although radioactive and stable isotopes of carbon, hydrogen and oxygen have been used by physiologists to investigate the details of respiratory metabolism (Kay, 1964), they have not been used by ecologists. McClintock & Lifson (1958) showed that the total carbon dioxide production of an animal could be determined by using 'double-labelled' water. The heavy isotope of hydrogen is eliminated only as water, but that of oxygen (^{18}O) is eliminated both as carbon dioxide and as water; the difference between the two elimination rates will give a measure of the carbon dioxide evolved. This method is of great value to the ecologist for the determination of respiratory rates under natural conditions; the animals can be fed with double-labelled water, released in the field, recaptured at some later time and the levels of the isotopes determined by mass spectrometry (neutron activation, p. 77) and scintillation counting. It has been used with, for example, birds (Le Febvre, 1964; Utter & Le Febvre, 1973) and lizards (Nagy, 1975; enett & Nagy, 1977).

(2) *Analysis in the gaseous phase for air-breathing animals.* The animal is enclosed in a chamber and the changes caused by its respiration in the composition of the air in the chamber determind in some way. Several techniques are based on the absorption of the evolved carbon dioxide by an alkali solution (e.g. sodium hydroxide) so that the pressure or volume of the air will change proportionally to the consumed oxygen. In constant-volume respirometers, e.g. the Warburg (Dixon, 1951; Umbreit *et al.*, 1957), the reduction in pressure is measured with a manometer.

Alternatively the pressure can be kept constant and the volume allowed to change. This change can be measured, for example, by the movement of an oil droplet in a capillary tube; Smith & Douglas (1949) developed an apparatus of this type that was subsequently modified by Engelmann (1961), Wiegert (1964a) and Davies (1966). The change in volume can also be utilized in ultra-microrespirometers that operate on the principle of the Cartesian diver (Holter, 1943; Zeuthen, 1950; Nielsen, 1961; Kay, 1964; Gregg & Lints, 1967; Wood & Lawton, 1973).

The amount of carbon dioxide evolved may be determined by the titration of the alkali (Itô, 1964). Methods of construction are described by Arlian (1973). An experiment using these types of respirometer cannot be continued for a long period, as the respiration may become abnormal as an increasing proportion of the available oxygen is utilized*.

If the respiratory rate is to be recorded over a period of time the oxygen utilized by the animal must be replaced. A number of electrolytic respirometers have been developed in which the change in volume resulting from the

---

* Wightman, J. A., 1977 (*N.Z.J. Zool.* **4**, 453–69) considers that closed-vessel respiration rates need correction by × 2.5.

Table 14.4 Comparison of the performance of Respirometers (from Petrusewicz & Macfadyen, 1970)

(1) Respirometer	(2) Measures O₂, CO₂, cals	(3) Container volume ml	(4) Normal rates µl/hr	(5) Normal accuracy	(6) Automation	(7) Sensitivity: smallest detectable change (µl)
WARBURG Normal	O_2 and CO_2	25	10–500	$2\mu l/hr$	N	0.5
Small vessels	O_2 and CO_2	5	2–50	$1\mu l/hr$	N	0.5
GILSON Normal	O_2 and CO_2	16	5–500	$2\mu l/hr$	P	0.5
DIVER Holter	O_2 and CO_2	0.005–0.05	0.01–1.0	5%	P	0.0001
Gregg and Lints	O_2 and CO_2	0.005–0.05	0.01–1.0	5%	Y	0.005
Gradient	O_2 and CO_2	0.001–0.05	0.01–1.0	3%	Y	0.0001
PHILLIPSON automatic	O_2 and CO_2	50	50–300	8%	Y	10
MACFADYEN electrolytic	O_2 and CO_2	7	1–100	$1\mu l/hr$	Y	0.01
Capillary const. vol.	O_2	1.0	2.0 ca	5%	P	0.05
Const. pressure	O_2	1–10	0.1–5	2%	N	0.05
CONWAY microdiffusion	CO_2	20	50–1000+	$20\mu l/hr$	N	20
JENSEN CO_2 diffusion	CO_2	10	1.0 ca	$0.05\mu l/hr$	Y	–
CALVET microcalorimeter	heat	15–100	20–1000+	1%	Y	0.2
WINKLER (dissolved oxygen)	O_2	10	100 ca	2%	N	30

Notes: All quantities are converted to equivalents of oxygen volumes, regardless of the parameter actually measured.
(5) Accuracy is expressed as the expected standard error of a series of readings over the normal range or, where accuracy declines at low ranges, as the probable minimum detectable rate.
(6) The column 'Automation' indicates whether application of automatic recording is possible (Y), inherently impossible (N), or potentially possible (P).

consumption of oxygen switches on an electrolysis apparatus that generates oxygen; when the previous volume has been restored the current is switched off. The activity of the compensating oxygen generator is recorded and after suitable calibration this may be converted to give the volume of oxygen utilized (Winteringham, 1959; Macfadyen, 1961: Phillipson, 1962; Hayward *et al.*, 1963; Annis & Nicol, 1975; Arnold & Keith, 1976). In all these volumetric or manometric methods constant temperature must be maintained throughout the experiment; this is usually done with a water bath. Under certain circumstances the production of ozone is a hazard.

The quantities of carbon dioxide and oxygen in the air may also be assessed by methods depending on differences in the thermal conductivity, viscosity, magnetic susceptibility and other properties of the gases (Kleiber, 1961; Kay, 1964).

(3) *Analysis of dissolved gases for completely aquatic species.* The animal must be enclosed in a limited volume of water out of contact with the atmospheric air; Wohlschag (1957) describes a convenient 'Plexiglass' container for moderately sized animals. The gases dissolved in the water after a given time can be determined by a number of methods. Chemical methods, especially the 'Winkler', have been widely used (e.g. Teal, 1957; Richman, 1958; Beyers & Smith, 1971); the procedures of the Winkler and another titration technique, the phenosafranine, are described by Dowdeswell (1959). The accuracy of these can, however, be impaired by the presence of ferrous iron, nitrites and other 'impurities' in the water (Allee & Oesting, 1934).

Polarometry or polarography in which the concentration of dissolved oxygen is measured electrolytically has also been used in many ecological studies (Berg, 1953; Mann, 1956; Heywood & Edwards, 1961) and is described by Kolthoff & Lingane (1952) and Kay (1964). Oxygen electrodes that are particularly suitable for ecological work have been developed by Carey & Teal (1965) and their construction is described in Edmondson & Winberg (1971); they may also be used to measure the oxygen concentration in aquatic environments.

A tonometric method has been developed by Jones (1959) for use with very small volumes of fluid, and is thus more applicable to samples of fluid from within animals than from without! A small bubble of air is enclosed with the fluid and the carbon dioxide, oxygen and nitrogen in it allowed to come into diffusion equilibrium with these gases in the fluid. The composition of the bubble can then be analysed (Krogh, 1908).

Gasometric methods involve the expulsion of the gases from the fluid and their subsequent analysis. Van Slyke & Neill (1924) extracted the gases under vacuum and the quantities of both oxygen and carbon dioxide can be determined (Milburn & Beadle, 1960; Elkan & Moore, 1962; Kay, 1964). Only the oxygen and nitrogen can be measured in the method of Scholander *et al.* (1955), in which the gases are expelled from the sample by the addition of acid and a carbonate.

14.3 The energy budget, efficiencies and transfer coefficients

The final energy budget may be presented in various forms: emphasis may be placed on the population studied (e.g. Lawton, 1970, 1971; McNeill, 1971), trophic links (e.g. Gyllenberg, 1970; Shure, 1973) or on biomass turnover and features of the ecosystem (e.g. Teal, 1962; Jonasson & Kristiansen, 1967; Burky, 1971). Budgets are generally expressed in kilocalories (kcal) or joules (1 kcal = 4186 joules) or carbon (1 gm C ∼ 0.5357 kcal). For most purposes kcal is convenient.

14.3.1 The energy budget of a population (or trophic level)

A graphic representation of the energy budget of a caddis fly population is given by Otto (1975) (Fig. 14.4). This budget is interesting too in the way it emphasizes that a system is seldom without 'imports' (e.g. imagines from downstream and other streams) and 'exports' (e.g. predation of the adults by terrestrial animals that move away from the area), and that trophic links may cross habitat barriers, e.g. the predation of the larvae by crows.

Energy budgets can usefully be summarized and compared if the efficiencies of various processes are calculated. There have been many terms applied to these (Kozlovsky, 1968); but the following, that largely follow Wiegert (1964a), are now recognised as standard (Petrusewicz & Macfadyen, 1970) (the notation follows equation 14.1 at the start of this chapter):

Assimilation/consumption	A/C
Production/assimilation or 'net' production efficiency	P/A
Production/consumption efficiency	P/C
Production/respiration ratio	P/R

As Wiegert (1964a) emphasized, the terms (numerator and denominator) in these ratios should be precisely defined and the ratios described by the symbols (e.g. P/R ratio) or by a name that indicates these. The efficiencies, which represent energy transfers within organisms, are usually expressed as percentages.

Wiegert & Evans (1967) published an interesting table of P/C efficiencies; some values for other invertebrates are given in Table 14.5. The year to year variations (Gyllenberg, 1970; McNeill, 1971; and Lawton, 1971) in the same population, which give an indication of the 'condition' of the population, its food supply and environment, show that too much stress must not be placed on small differences. However, not surprisingly, herbivores have lower P/C values than predators and, from Wiegert & Evans (1967), homiotherms than poikilotherms. The distinction between the two latter groups is clearly shown when known P/R ratios are plotted; there are two linear relationships, the homiotherms having a higher respiratory expenditure for the same production (McNeill & Lawton, 1970; Wiegert, 1976).

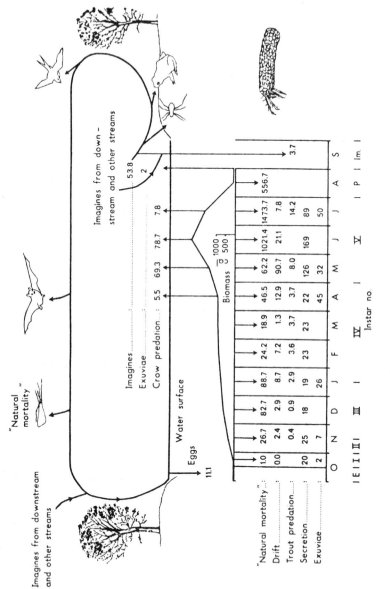

Fig. 14.4 The energy budget of a population of the caddis fly, *Potamophylax cingulatus*, calculated on a monthly basis in terms of cal m⁻² (after Otto, 1975).

Table 14.5 Some energy efficiencies for populations of different invertebrates showing inter- and intra-specific variations

Organism		Efficiencies			Source
		A/C %	P/A %	P/C %	
Predators					
Pyrrhosoma (Odonata)	Year 1	90	52	47	Lawton, 1971
	Year 2	88	48	42	,, ,,
Phonoctonus (Het.)				c.45	Evans, 1962
Herbivores					
Leptopterna (Het.)	Year 1	29	58	17	McNeill, 1971
	Year 2	36	51	18	,, ,,
	Year 3	36	52	19	,, ,,
	Year 4	34	59	20	,, ,,
	Year 5	28	55	16	,, ,,
Rhynchaenus (Col.)	Larva I	24		20	Grimm, 1973
	Larva II	24		17	,, ,,
	Larva III	20		14	,, ,,
Melasoma (Col.) la.		c.72	35	24	Hågvar, 1975
Danaus (Lep.)			45	21	Schroeder, 1976
Euchaetias (Lep.)			49	20	Schroeder, 1977
Potamophylax (Trich.)		19	45	8	Otto, 1975
Saprophytic/microbial feeders					
Aphodius (Col.)		8	51	4	Holter, 1975
Pelodera (Nematoda)		60	38	22	Marchant & Nicholas, 1974
Brachionus (Rotifer) adult		19	57	11	Doohan, 1973

14.3.2 Energy transfer across trophic links

The pathways of energy transfer in an ecosystem are logically considered against a flow model of the type proposed by Wiegert & Owen (1971) (Fig. 14.5). The energy flow within the 'boxes' is described by the efficiencies or ratios listed above.

The rate at which energy is leaving one 'box' for other 'boxes' is best described as:

$$\text{Production turnover rate} \quad — \quad P_t/B = \Theta_P \qquad (14.16)$$

$$\text{Elimination turnover rate} \quad — \quad E_t/B = \Theta_E \qquad (14.17)$$

where B = the average biomass over time unit t and P_t and E_t are the total production and elimination values over time. These two rates (which are, of course, related; see Equation 14.4) express the rate at which the production of the population (or trophic level) is being passed-on to the rest of the ecosystem. The transfer of energy across particular links between boxes [e.g. between the autotroph and the first order biophage (or herbivore)] may be expressed:

$$\text{Harvest/Production transfer coefficient} = E_t''/P_t' \qquad (14.18)$$

$$\text{Yield/Production transfer coefficient} \quad = C''/P'_t \quad (14.19)$$

where $E''_t =$ the total biomass of the 'donor' (prey) eliminated by the next level biophage during time t, $P'_t =$ total production of the donor during this time period and $C''_t =$ total consumption by the biophage or saprophage of the donor during period t. It will be noted that for the links to the biophages (the boxes on the right of Fig.14.5), there are two transfer coefficients. The 'harvest transfer coefficient' represents the removal of production to the biophage, i.e. the kill or cull by a predator. However, as Dagg (1974) found, a predator, especially if its prey is large relative to it, may only consume about half the energy it harvests. Its actual consumption, the yield to it, the flow across one of the links between the boxes in Fig. 14.5, is given by the 'yield transfer coefficient'.

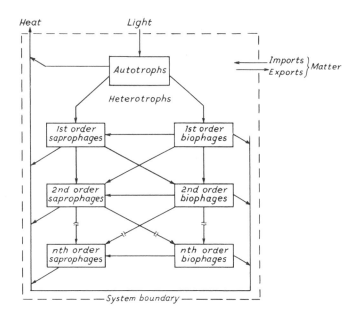

Fig. 14.5 Energy transfer in an ecosystem: dioristic model of Wiegert and Owen (1971).

It is important to make these distinctions between the *efficiencies* within organisms, the *rates* at which production is passed on through trophic links, the impact (in energy terms) of one trophic level on that below it (*Harvest transfer coefficient*) and the actual energy flow along a particular link (*Yield transfer coefficient*). Ratios, other than those given here, may be useful in a particular context, but it should always be made clear whether the energy flow described is within a trophic level or between levels, or spans both levels and links.

14.4 Assessment of energy and time costs of strategies

Ecologists and evolutionary biologists have become increasingly interested in the quantitative comparison of different strategies e.g. different forms of foraging strategies. The problem is of interest, not only for between species comparisons, but within species that show polymorphism in these characters. The ultimate criterion is, of course, reproductive success, but valuable indications of the relative costs and benefits can be obtained from the consideration of time and energy budgets (Emlen, 1966; Schoener, 1969, 1971; VanValen, 1976; Norberg, 1977). The methods utilized will often be similar to those described above, e.g. measurement of respiratory rate, caloric value of 'harvest', but these will have to be related to particular activities. In essence the following are usually required:

(1) A time budget of the various activities*.

(2) The rate of energy expenditure characteristic of each type of activity

(3) The rate of energy gain food harvested or usually yield (food ingested) for each type of activity.

Studies of this type have been made on territoriality and resources (e.g. McNab, 1963; Smith, 1968; Gill & Wolf, 1975*a* and *b*) on speed and pattern of movement in relation to resource availability (e.g. Calow, 1974) and foraging strategies (e.g. Mukerji & Le Roux, 1969; Heinrich, 1972*a* 1975; Charnov, 1976; D. W. Davidson, 1977).

REFERENCES

ALLEE, W. C. and OESTING, R., 1934. A critical examination of Winkler's method for determining dissolved oxygen in respiration studies with aquatic animals. *Physiol. Zool.* **7**, 509–41.

ALLEN, K. R., 1951. The Horokiwi Stream, a study of a trout population. *N.Z. Mar. Dep. Fish. Rep. DIV Bull.* **10**, 1–238.

ALLEN, M. D., 1959. Respiration rates of worker honeybees of different ages and at different temperatures. *J. exp. Biol.* **36**, 92–101.

ANDRÁSSY, I., 1956. Die Rauminhalts- und Gewichtsbestimmung der Fadenwürmer (Nematoden) *Acta Zool. Budapest* **2**, 1–15.

ANNIS, P. C. and NICOL, G. R., 1975. Respirometry system for small biological samples. *J. appl. Ecol.* **12**, 137–41.

ARLIAN, L. G., 1973. Methods for making a cartesian diver for use with small arthropods. *Ann. Ent. Soc. Am.* **66**, 694–5.

ARNOLD, D. J. and KEITH, D. E., 1976. A simple continuous-flow respirometer for comparative respirometry changes in medium-sized aquatic organisms. *Water Research* **10**, 261–4.

BAILEY, C. G. and RIEGERT, P. W., 1972. Energy dynamics of *Encoptolophus sordidus cortalis* (Scudder) (Orthoptera: Acrididae) in a grassland ecosystem. *Can. J. Zool.* **51**, 91–100.

BENKE, A. C., 1976. Dragonfly production and prey turnover. *Ecology* **57**, 915–27.

* See Hassell, M. P. and Southwood, T. R. E. (1979), *Ann. Rev. Ecol. Syst.* **9**, 75–98 (especially Table 2).

BENNETT, A. F. and NAGY, K. A., 1977. Energy expenditure in free-ranging lizards. *Ecology* **58**, 697–700.

BERG, K., 1953. The problem of respiratory acclimatization. *Hydrobiologia* **5**, 331–50.

BERG, K., LUMBYE, J. and OCKELMANN, K. W., 1958. Seasonal and experimental variations of the oxygen consumption of the limpet *Ancylus fluviatilis* (O. F. Müller). *J. exp. Biol.* **35**, 43–73.

BERG, K. and OCKELMANN, K. W., 1959. The respiration of freshwater snails. *J. exp. Biol.* **36**, 690–708.

BEYERS, R. J. and SMITH, M. H., 1971. A calorimetric method for determining oxygen concentration in terrestrial situations. *Ecology* **52**, 374–5.

BOURLIÈRE, F. and LAMOTTE, M., 1967, 1968. Les fondements physiologiques et démographiques des notions de production et de rendements bioénergétiques. *Rev. Quest. Scient.* **138**(4), 509–41 and **139**(1), 5–24.

BOYD, C. E., 1969. The nutritive value of three species of water weeds. *Economic Botany* **23**, 123–7.

BOYD, C. E., 1970. Chemical analyses of some vascular aquatic plants. *Arch. Hydrobiol.* **67**(1), 78–85.

BOYD, C. E. and BLACKBURN, R. D., 1970. Seasonal changes in the proximate composition of some common aquatic weeds. *Hyacinth control J.* **8**, 42–144.

BOYD, C. E. and GOODYEAR, C. P., 1971. Nutritive quality of food in ecological systems. *Arch. Hydrobiol.* **69**, 256–70.

BRODY, S., 1945. *Bioenergetics and growth.* 1023 pp. Hafner, New York.

BROWN, D. S., 1961. The food of the larvae of *Chloëon dipterum* L. and *Baëtis rhodani* (Pictet) (Insecta, Ephemeroptera). *J. Anim. Ecol.* **30**, 55–75.

BURKY, A. J., 1971. Biomass turnover, respiration and interpopulation variation in the stream limpet *Ferrissia rivularis* (Say). *Ecol. Monogr.* **41**(3), 235–51.

BURNISON, B. K. and PEREZ, K. T., 1974. A simple method for the dry combustion of ^{14}C-labelled materials. *Ecology* **55**, 899–902.

CAHN, T., 1956. *La régulation des processus métaboloques dans l'organism.* Hermann et Cie, Paris.

CALOW, P., 1974. Some observations on locomotory strategies and their metabolic effects in two species of freshwater gastropods, *Anylus fluviatilis* and *Planorbis contortus* Linn. *Oecologia* **16**, 149–61.

CALOW, P. and FLETCHER, C. R., 1972. A new radio tracer technique involving ^{14}C and ^{51}Cr for estimating the assimilation efficiencies of aquatic primary consumers. *Oecologia* **9**, 155–70.

CALVERT, E. and PRATT, H., 1956. *Microcalorimétrie.* 396 pp. Masson, Paris.

CAREY, F. G. and TEAL, J. M., 1965. Responses of oxygen electrodes to variables in construction, assembly and use. *J. Appl. Physiol.* **20**, 1074–7.

CASPERS, N., 1977. Seasonal variations of caloric values in herbaceous plants. *Oecologia* **26**, 379–83.

CHAPMAN, S. B., 1976. Production Ecology and nutrient budgets. In Chapman, S. B. (ed.) *Methods in Plant Ecology*, pp. 157–228, Blackwells, Oxford.

CHARNOV, E. L., 1976. Optimal foraging: attack strategy of a mantid. *Am. Nat.* **110**, 141–51.

CHLODNY, J., GROMADZKA, J. and TOJAN, P., 1967. Energetic budget of development of the Colorado beetle – *Leptinotarsa decemlineata* Say. *Bull. Acad. pol. Sci.* Ch. 11 Ser. Sci. Biol. *XV* (12): 743–747

COLEMAN, D. C., 1968. Food webs of small arthropods of a broomsedge field studied with radio-isotope-labelled fungi. *Proc. IBP. Tech. Meeting on Methods of Study in Soil Zoology* 203–207, UNESCO, Paris.

COMITA, G. W. and SCHINDLER, D. W., 1963. Calorific values of microcrustacea. *Science* **140**, 1394–6.

CONWAY, E. J., 1947. *Microdiffusion Analysis and Volumetric Error* 2nd ed. 357 pp. Lockwood, London.

CORBETT, J. L., GREENHALGH, F. D., MCDONALD, I. and FLORENCE, E., 1960. Excretion of chromium sesquioxide administered as a component of paper to sheep. *Brit. J. Nutr.* **14**, 289–99.

CROSSLEY, D. A., 1963*a*. Movement and accumulation of radiostrontium and radio-cesium in insects. *In* Schultz, V. & Klement, A. W. (eds.), *Radioecology* 103–5. Rheinhold, New York.

CROSSLEY, D. A., 1963*b*. Consumption of vegetation by insects. *In* Schultz, V. & Klement, A. W. (eds.), *Radioecology* 427–30. Rheinhold, New York.

CROSSLEY, D. A., 1963*c*. Use of radioactive tracers in the study of insect-plant relationships. *Radiation and radioisotopes applied to insects of agricultural importance* (Int. Atom. Energy Ag.) STI/PUB 74, 43–54.

CROSSLEY, D. A., 1966. Radio-isotope measurement of food consumption by a leaf beetle species, *Chrysomela knabi* Brown. *Ecology* **47**(1), 1–8.

CROSSLEY, D. A. and HOGLUND, M. P., 1962. A litter-bag method for the study of micro-arthropods inhabiting leaf litter. *Ecology* **43**, 571–3.

CRUZ, A. A. DE LA and WIEGERT, R. G., 1967. 32-Phosphorus tracer studies of a horseweed-aphid-ant food chain. *Am. Midl. Nat.* **77**(2), 501–9.

DAGG, M. J., 1974. Loss of prey body contents during feeding by an aquatic predator. *Ecology* **55**, 903–6.

DAVIDSON, D. H., 1977. Assimilation efficiencies of slugs on different food materials. *Oecologia* **26**: 267–73.

DAVIDSON, D. W., 1977. Foraging ecology and community organization in desert-seed-eating ants. *Ecology* **58**, 725–37.

DAVIES, P., 1966. A constant pressure respirometer for medium-sized animals. *Oikos* **17**, 108–12.

DIXON, M., 1951. *Manometric methods as applied to the measurement of cell respiration and other processes.* (3rd ed.) Cambridge University Press, Cambridge.

DOBBEN, W. H. VAN and LOWE-MCCONNELL, R. H. 1975. *Unifying concepts in Ecology.* The Hague, 302 pp.

DOOHAN, M., 1973. An Energy Budget for adult *Brachionus plicatilis* Muller (Rotatoria) *Oecologia* (Berl.) **13**, 351–62.

DOWDESWELL, W. H., 1959. *Practical animal ecology.* 315 pp. Methuen, London.

EDMONDSON, W. T. and WINBERG, G. G. (eds.), 1971. *A manual on methods for the assessment of secondary productivity in fresh waters.* (I.B.P. Handbook 17), 358 pp. Blackwells, Oxford.

EDWARDS, C. A., REICHLE, D. E. and CROSSLEY, D. A. (Jr.). 1969. Experimental manipulation of soil invertebrate populations for trophic studies. *Ecology* **50**(3), 495–8.

EDWARDS, R. W. and LEARNER, M. A., 1960. Some factors affecting the oxygen consumption of *Asellus. J. exp. Biol.* **37**, 706–18.

ELKAN, G. H. and MOORE, W. E. C., 1962. A rapid method for measurement of CO_2 evolution by soil microorganisms, *Ecology* **43**, 775–6.

EMLEN, J. M., 1966. The role of time and energy in food preferences. *Am. Nat.* **100**, 611–17.

ENGELMANN, M. D., 1961. The role of soil arthropods in the energetics of an old field community. *Ecol. Monogr.* **31**, 221–38.

EVANS, D. E., 1962. The food requirements of *Phonoctonus nigrofasciatus* Stål (Hemiptera, Reduviidae). *Entomologia exp. appl.* **5**, 33–9.

FAGERSTRÖM, T., 1977. Bodyweight, metabolic rate and trace substance turnover in animals. *Oecologia* **29**(2), 99–116.

FEWKES, D. W., 1960. The food requirements by weight of some British Nabidae (Heteroptera). *Entomologia exp. appl.* **3**, 231–7.

FROLANDER, H. F., 1957. A plankton volume indicator. *J. Cons. perm. int. Explor. Mer.* **22**, 278–83.

GALLUCCI, V. F., 1973. On the principles of thermodynamics and ecology. *Ann. Rev. Ecol. Syst.* **4**, 329–57.

GERE, G., 1956. Investigations concerning the energy turnover of the *Hyphantria cunea* Drury caterpillars. *Opusc. Zool., Budapest* **1**, 29–32.

GILL, F. B. and WOLF, L. L., 1975a. Economics of feeding territoriality in the Golden Winged Sunbird. *Ecology* **56**, 333–45.

GILL, F. B. and WOLF, L. L., 1975b. Foraging strategies and energetics of east African sunbirds at mistletoe flowers. *Am. Nat.* **109**, 491–510.

GILSON, W. E., 1963. Differential respirometer of simplified and improved design. *Science* **141**, 531–2.

GNANAMUTHU, C. P., 1952. A simple device for measuring the volume of an aquatic animal. *Nature* **170**, 587.

GOLLEY, F. B., 1960. Energy dynamics of a food chain of an old-field community. *Ecol. Monogr.* **30**, 187–206.

GOLLEY, F. B. and BUECHNER, H. K. (Eds.), 1968. *A practical guide to the study of the productivity of large Herbivores.* (I.B.P. Handbook 7) 308 pp., Blackwells, Oxford.

GOLLEY, F. B. and GENTRY, J. B., 1964. Bioenergetics of the southern Harvester ant, *Pogonomyrmex badius, Ecology* **45**, 217–25.

GREGG, J. H. and LINTS, F. A., 1967. A constant-volume respirometer for *Drosophila* imagos. *C.R. Lab. Carlsberg* **36**, 25–34.

GRIFFITHS, D., 1977. Caloric variation in Crustacea and other animals. *J. Anim. Ecol.* **46**, 593–605.

GRIMM, R., 1973. Zum Energieumsatz phytophager Insekten im Buchenwald I. *Oecologia* **11**, 187–262.

GRODZINSKI, W., KLEKOWSKI, R. Z. and DUNCAN, A. (Eds.). 1975. *Methods for Ecological Bioenergetics.* (I.B.P. Handbook 24), Blackwells, Oxford.

GYLLENBERG, G., 1969. The energy flow through a *Chorthippus parallelus* (Zett.) (Orthoptera) population on a meadow in Tyärminne, Finland. *Acta. Zool. Fenn.* **123**, 1–74.

GYLLENBERG, G., 1970. Energy flow through a simple food chain of a meadow ecosystem in four years. *Ann. Zool. Fennici.* **7**, 283–9.

HAGVAR, S., 1975. Energy budget and growth during the development of *Melasoma collaris* (Coleoptera) *Oikos* **26**, 140–6.

HANNA, H. M., 1957. A study of the growth and feeding habits of the larvae of four species of caddis flies. *Proc. R. ent. Soc. Lond. A* **32**, 139–46.

HAYWARD, J. S., NÖRDAN, H. C. and WOOD, A. J., 1963. A simple electrolytic respirometer for small animals. *Can. J. Zool.* **41**, 63–8.

HEINRICH, B., 1972. Energetics of temperature regulation and foraging in a bumblebee, *Bombus terricola* Kirby. *J. Comp. Physiol.* **77**, 48–64.

HEINRICH, B., 1975. The role of energetics in bumblebee-flower interrelationships. *In* Gilbert, L. E. & Raven, P. H. (eds.). *Co-evolution of animals and plants.* 141–58. Univ. of Texas Press, Austin.

HEYWOOD, J. and EDWARDS, R. W., 1961. Some aspects of the ecology of *Potamopyrgus jenkinsi* Smith. *J. Anim. Ecol.* **31**, 239–50.

HINTON, J. M., 1971. Energy flow in a natural population of *Neophilaenus lineatus* (Homoptera) *Oikos* **22**, 155–71.

HOLTER, H., 1943. Technique of the cartesian diver. *C. R. Lab. Carlsberg (Ser. Chin.)* **24**, 400–78.

HOLTER, P., 1973. A chromic oxide method for measuring consumption in dung-eating *Aphodius* larvae. *Oikos* **24**, 117–22.

HOLTER, P., 1974. Food utilization of dung-eating *Aphodius* larvae (Scarabaeidae). *Oikos* **25**, 71–9.

HOLTER, P., 1975. Energy budget of a natural population of *Aphodius rufipes* larvae (Scarabaeidae). *Oikos* **26**, 177–86.

HUBBELL, S. P., SIKORA, A. and PARIS, O. H., 1966. Radiotracer, gravimetric and calorimetric studies of ingestion and assimilation rates of an Isopod. *Health Physics* **11**(12), 1485–1501.

HUMPHREYS, W. F., 1975. The food consumption of a Wolf Spider, *Geolycosa godeffroyi* (Aracridae: Lycoridae), in the Australian Capital Territory. *Oecologia (Berl.)* **18**, 343–58.

ITÔ, Y., 1964. Preliminary studies on the respiratory energy loss of a spider, *Lycosa pseudoannulata*. *Res. Popul. Ecol.* **6**, 13–21.

IVERSEN, T. M., 1974. Ingestion and growth in *Sericostoma personatum* (Trichoptera) in relation to the nitrogen content of the ingested leaves. *Oikos* **25**, 278–82.

IVLEV, V. S., 1934. Eine Mikromethode zur Bestimmung des Kaloriengehalts von Nährstoffen. *Biochem. Z.* **275**, 49–55.

IVLEV, V. S., 1939. Transformation of energy by aquatic animals. *Int. Revue ges. Hydrobiol. Hydrogr.* **38**, 449–58.

IVLEV, V. S., 1945. The biological productivity of waters. [In Russian.] *Usp. sovrem. Biol.* **19**, 98–120.

JENSEN, C. R., VAN. GUNDY, S. D. and STOLZY, L. H. 1966. Diffusion exchange respiratometer using the CO_2 electrode. *Nature* **211**, 608–610.

JOHNSON, C. G., 1960. The relation of weight of food ingested to increase in body-weight during growth in the bed-bug, *Cimex lectularius* L. (Hemiptera). *Ent. exp. appl.* **3**, 238–40.

JÓNASSON, P. M. and KRISTIANSEN, J., 1967. Primary and secondary production in Lake Esrom. Growth of *Chironomus anthracinus* in relation to seasonal cycles of phytoplankton and dissolved oxygen. *Int. Rev. ges. Hydrobiol.* **52**, 163–217.

JONES, J. D., 1959. A new tonometric method for the determination of dissolved oxygen and carbon dioxide in small samples. *J. exp. Biol.* **36**, 177–90.

KARO, J., 1973. An attempt to estimate the energy flow through the population of Colorado Beetle (*Leptinotarsa decemlineata* Say) *Ekologia Polska* **21**(1), 239–50.

KAY, R. H., 1964. *Experimental biology. Measurement and analysis.* 416 pp., Chapman and Hall, London.

KEISTER, M. and BUCK, J., 1964. Respiration: some exogenous and endogenous effects on the rate of respiration. *In* Rockstein, M. (ed.), *The Physiology of Insecta* **3**, 617–58. Academic Press, London and New York.

KLEIBER, M., 1961. *The fire of life. An introduction to animal energetics.* Wiley, New York.

KOLTHOFF, J. M. and LINGANE, J. J., 1952. *Polarography.* Wiley, New York.

KOMOR, J., 1940. Über die Ausnützung des Sonnenlichtes beim Wachstum der grünen Pflanzen, *Biochem. Z.* **305**, 381–95.

KORMONDY, E. J., 1965. Uptake and loss of zinc-65 in the dragonfly *Plathemis lydia*. *Limnol. Oceanogr.* **10**, 427–33.

KOZLOVSKY, D. G., 1968. A critical evaluation of the trophic level concept. 1. Ecological efficiencies. *Ecology* **49**(1), 48–60.

KROGH, A., 1908. On micro-analysis of gases. *Skand. Arch. Physiol.* **20**, 279–88.

KUENZLER, E. J., 1961. Structure and energy flow of a mussel population in a Georgia salt marsh. *Limnol. Oceanogr.* **6**, 191–204.

LAWTON, J. H., 1970. Feeding and food energy assimilation in larvae of the damselfly *Pyrrhosoma nymphula* (Sulz.) (Odonata: Zygoptera). *J. Anim. Ecol.* **39**, 669–89.

LAWTON, J. H., 1971. Ecological energetics studies on larvae of the damselfly *Pyrrhosoma nymphula* (Sulz.) (Odonata: Zygoptera). *J. Anim. Ecol.* **40**, 385–419.

LAWTON, J. H. and RICHARDS, J., 1970. Comparibility of Cartesian diver, Gilson, Warburg and Winkler methods of measuring the respiratory rates of aquatic invertebrates in ecological studies. *Oecologia (Berl.)* **4**, 319–24.

LEDBETTER, M. C. and FLEMION, F., 1954. A method for obtaining piercing-sucking mouth parts in host tissues from the tarnished plant bug by high voltage shock. *Contrib. Boyce Thompson Inst.* **17**(6), 343–6.

LE FEBVRE, E. A., 1964. The use of D_2O^{18} for measuring energy metabolism in *Columba livia* at rest and in flight. *Auk* **81**, 403–16.

LINDEMAN, R. L., 1942. The trophic-dynamic aspect of ecology. *Ecology* **23**, 399–418.

MCCLINTOCK, R. and LIFSON, N., 1958. Determination of the total carbon dioxide output of rats by the D^2O^{18} method. *Am. J. Physiol.* **192**, 76–8.

MACFADYEN, A., 1961. A new system for continuous respirometry of small air-breathing invertebrates under near-natural conditions. *J. exp. Biol.* **38**, 323–43.

MACFADYEN, A., 1963. *Animal ecology. Aims and methods.* 2nd ed. 344 pp. Pitman, London and New York.

MCGINNIS, A. J. and KASTING, R., 1964a. Chromic oxide indicator method for measuring food utilization in a plant-feeding insect. *Science* **144**, 1464–5.

MCGINNIS, A. J. and KASTING, R., 1964b. Digestion in insects, colorimetric analysis of chromic oxide used to study food utilization by phytophagous insects. *J. agric. Fd. Chem.* **12**, 259–62.

MCNAB, B. K., 1963. Bioenergetics and the determination of home-range size. *Am. Nat.* **97**, 133–40.

MCNEILL, S., 1971. The energetics of a population of *Leptopterna dolabrata* (Heteroptera: Miridae). *J. Anim. Ecol.* **40**, 127–40.

MCNEILL, S. and LAWTON, J. H., 1970. Annual production and respiration in animal populations. *Nature* **225**, 472–4.

MCNEILL, S. and SOUTHWOOD, T. R. E., 1978. Role of nitrogen in the development of insect plant relationhips. In Harborne, J. (ed.), *Biochemical Aspects of Plant and Animal Coevolution.* Academic Press, London (in press).

MALONE, C. R. and NELSON, D. J., 1969. Feeding rates of freshwater snails (*Goniobasis clavaeformis*) determined with Cobalt[60]. *Ecology* **50**(4), 728–30.

MANN, K. H., 1956. A study of the oxygen consumption of five species of leech. *J. exp. Biol.* **33**, 615–26.

MARCHANT, R. and NICHOLAS, W. L., 1974. An energy budget for the free-living Nematode *Pelodera* (Rhabditidae). *Oecologia (Berl.)* **16**, 237–52.

MARPLES, T. G., 1966. A radionuclide tracer study of arthropod food chains in a Spartina salt marsh ecosystem. *Ecology* **47**(2), 270–77.

MATHAVAN, S. and PANDIAN, T. J., 1974. Use of faecal weight as an indicator of food consumption in some lepidopterans. *Oecologia (Berl.)* **15**, 177–85.

MILBURN, T. R. and BEADLE, L. C., 1960. The determination of total carbon dioxide in water. *J. exp. Biol.* **37**, 444–60.

MILES, P. W., 1972. The saliva of hemiptera. *Adv. Insect. Physiol.* **9**, 183–225.

MILNER, C. and HUGHES, R. E., 1968. *Methods for the measurement of the primary production of Grasslands.* (I.B.P. Handbook 6) 82 pp. Blackwells, Oxford.

MOORE, S. T., SCHUSTER, M. F. and HARRIS, F. A., 1974. Radioisotope technique for estimating lady beetle consumption of tobacco budworm eggs and larvae. *J. econ. Ent.* **67**(6), 703–5.

MUKERJI, M. K. and LE ROUX, E. J., 1969. A study of energetics of *Podisus maculiventris* (Hemiptera: Pentatomidae). *Can. Ent.* **101**, 449–460.

MULKERN, G. B. and ANDERSON, J. F., 1959. A technique for studying the food habits and preferences of grasshoppers, *J. econ. Ent.* **52**, 342.

MUTHUKRISHNAN, J. and DELVI, M. R., 1974. Effect of ration levels on food utilisation in the grasshopper *Poecilacerus pictus*. *Oecologica (Berl.)* **16**, 227–36.

NAGY, K. A., 1975. Nitrogen requirement and its relation to dietary water and potassium content in the lizard *Sauromalus obesus*. *J. comp. Physiol.* **104**, 49–58.

NEESS, J. and DUGDALE, C., 1959. Computation of production for populations of aquatic midge larvae. *Ecology* **40**, 425–30.

NEWBOULD, P. J., 1967. *Methods for estimating the primary production of forests.* (I.B.P. Handbook 2) 72 pp. Blackwells, Oxford.

NIELSEN, C. O., 1961. Respiratory metabolism of some populations of Enchytraeid worms and free living Nematodes. *Oikos* **12**, 17–35.

NORBERG, R. A., 1977. An ecological theory on foraging time and energetics and choice of optimal food-searching method. *J. Anim. Ecol.* **46**, 511–29.

ODUM, E. P. and GOLLEY, F. B., 1963. Radioactive tracers as an aid to the measurement of energy flow at the population level in nature. *In* Schultz, V. & Klement, A. W. (eds.), *Radioecology* 403–10.

ODUM, E. P. and KUENZLER, E. J., 1963. Experimental isolation of food chains in an old-field ecosystem with the use of phosphorus-32. *In* Schultz, V. & Klement, A. W. (eds.), *Radioecology* 113–20.

ODUM, H. T., 1957. Trophic structure and productivity of Silver springs, Florida. *Ecol. Monogr.* **27**, 55–112.

ODUM, H. T. and ODUM, E. P., 1955. Trophic structure and productivity of a Windward coral reef community on Eniwetok Atoll. *Ecol. Monogr.* **25**, 291–320.

ONUF, C. P., TEAL, J. M. and VALIELA, I., 1977. Interactions of nutrients, plant growth and herbivores in a mangrove ecosystem. *Ecology* **58**, 514–26.

OTTO, C., 1975. Energetic relationships of the larval population of *Potamophylax cingulatus* (Trichoptera) in a South Swedish stream. *Oikos* **26**, 159–69.

OVINGTON, J. D. and HEITKAMP, D., 1960. The accumulation of energy in forest plantations in Britain. *J. Ecol.* **48**, 639–46.

PAINE, R. T., 1971. The measurement and application of the calorie to ecological problems. *Ann. Rev. Ecol. Syst.* **2**, 145–64.

PARIS, O. H. and SIKORA, A., 1965. Radiotracer demonstration of Isopod herbivory. *Ecology* **46**, 729–34.

PARKINSON, D., GRAY, T. R. G. and WILLIAMS, S. T., 1971. *Methods for studying the ecology of soil microorganisms.* (I.B.P. Handbook 19), 128 pp. Blackwells, Oxford.

PEARSON, O. P., 1954. The daily energy requirements of a wild anna hummingbird. *Condor* **56**, 317–22.

PENDLETON, R. C. and GRUNDMANN, A. W., 1954. Use of phosphorus-32 in tracing some insect-plant relationships of the thistle, *Cirsium undulatum*. *Ecology* **35**, 187–91.

PETRUSEWICZ, K. (Ed.). 1967. *Secondary productivity of terrestrial ecosystems (Principles and Methods).* Warzawa–Kraków.

PETRUSEWICZ, K. and MACFADYEN, A., 1970. *Productivity of terrestrial animals: principles and methods.* (I.B.P. Handbook 13), 190 pp. Blackwells, Oxford.

PHILLIPSON, J., 1960. The food consumption of different instars of *Mitopus morio* (F.) (Phalangiida) under natural conditions. *J. Anim. Ecol.* **29**, 299–307.

PHILLIPSON, J., 1962. Respirometry and the study of energy turnover in natural systems with particular reference to harvest spiders (Phalangiida). *Oikos* **13**, 311–22.

PHILLIPSON, J., 1964. A miniature bomb calorimeter for small biological samples. *Oikos* **15**, 130–9.

PHILLIPSON, J. (Ed.)., 1971. *Methods of study in quantitative soil ecology.* (I.B.P. Handbook 18), 308 pp. Blackwells Oxford.

POLLARD, D. G., 1973. Plant penetration by feeding aphids (Hemiptera: Aphoidea): a review. *Bull. ent. Res.* **62**, 631–714.

PRATT, H., 1954. Analyse microcalorimetrique des variations de la thermogenèse chez divers insectes. *Can. J. Zool.* **32**, 172–94.

REICHLE, D. E., 1967. Radioisotope turnover and energy flow in terrestrial isopod populations. *Ecology* **48**(3), 351–66.

REICHLE, D. E., 1969. Measurement of elemental assimilation by animals from radioisotope retention patterns. *Ecology* **50**(6), 1102–4.

REINERS, W. A. and REINERS, N. M., 1972. Comparison of oxygen-bomb combustion with standard ignition techniques for determining total ash. *Ecology* **53**, 132–6.

RICHMAN, S., 1958. The transformation of energy by *Daphnia pulex*. *Ecol. Monogr.* **28**, 273–91.

RICKER, W. E. (Ed.), 1968. *Methods for assessment of Fish Production in fresh waters.* (I.B.P. Handbook 3) 313 pp., Blackwells, Oxford.

SCHOENER, T. W., 1969. Optimal size and specialization in constant and fluctuating environments: An energy-time approach. *In Diversity and Stability in Ecological Systems. Brookhaven Symp. Biol.* **22**, 103–14.

SCHOENER, T. W., 1971. Theory of feeding strategies. *Ann. Rev. Ecol. Syst.* **2**, 369–404.

SCHOLANDER, P. F., VANDAM, L., CLAFF, C. L. and KANWISHER, J. W., 1955. Microgasometric determination of dissolved oxygen and nitrogen. *Biol. Bull., Woods Hole* **109**, 328–34.

SCHROEDER, A., 1973. Energy budget of the larvae of the moth *Pachysphinx modesta.* *Oikos* **24**, 278–81.

SCHROEDER, L. A., 1976. Energy, matter and nitrogen utilization by larvae of the monarch butterfly *Danaus plexippus* (Danaidae: Lepidoptera) *Oikos* **27**.

SCHROEDER, L. A., 1977. Energy, matter and nitrogen utilization by larvae of the milkweed tiger moth *Euchretias egle*. *Oikos* **28**, 27–31.

SHAW, J. and BEADLE, L. C., 1949. A simplified ultra-micro Kjeldahl method for the estimation of protein and total nitrogen in fluid samples of less than 1.0μ. *J. exp. Biol.* **26**, 15–23.

SHURE, D. J., 1970. Limitations in radio-tracer determination of consumer trophic positions. *Ecology* **51**(5), 899–901.

SHURE, D. J., 1973. Radionuclide tracer analysis of trophic relationships in an old-field ecosystem. *Ecol. Monogr.* **43**(1), 1–19.

SHURE, D. J. and PEARSON, P. G., 1969. Distribution of P^{32} in *Ambrosia artemisiifolix*; its implication for trophic transfer studies. *Ecology* **50**(4), 724–26.

SINGH, J. S. and YADAVA, P. S., 1973. Caloric values of plant and insect species of a tropical grassland. *Oikos* **24**, 186–94.

SLOBODKIN, L. B. and RICHMAN, S., 1960. The availability of a miniature bomb calorimeter for ecology. *Ecology* **41**, 784.

SLOBODKIN, L. B. and RICHMAN, S., 1961. Calories/gm in species of animals. *Nature* **191**, 299.

SLYKE, D. D. VAN and NEILL, J. M., 1924. The determination of gases in blood and other solutions by extraction and manometric measurement. *J. Biol. Chem.* **61**, 523–73.

SMALLEY, A. E., 1960. Energy flow of a salt marsh grasshopper population. *Ecology* **41**, 672–7.

SMITH, A. H. and DOUGLAS, J. R., 1949. An insect respirometer. *Ann. ent. Soc. Am.* **42**, 14–18.

SMITH, C. C., 1968. The adaptive nature of social organization in the genus of three squirrels *Tamiasciurus*. *Ecol. Monogr.* **38**, 31–63.

SMITH, D. S., 1959. Utilization of food plants by the migratory grasshopper, *Melanoplus bilituratus* (Walker) (Orthoptera: Acrididae) with some observations on the nutritional value of the plants. *Ann. ent. Soc. Am.* **52**, 674–80.

STANFORD, J. A., 1973. A centrifuge method for determining live weights of aquatic insect larvae, with a note on weight loss in preservative. *Ecology* **54**, 449–451.

STIVEN, A. E., 1961. Food energy available for and required by the blue grouse chick. *Ecology* **42**, 547–53.

STRONG, F. E. and LANDES, D. A., 1965. Feeding and nutrition of *Lygus hesperus*

(Hemiptera: Miridae). II. An estimation of normal feeding rates. *Ann. ent. Soc. Am.* **58**, 309–14.

SUSHCHENYA, L. M., 1962. [Quantitative data on nutrition and energy balance in *Artemia salina* (L.)] [In Russian.] *Doklady Adad. Nauk S.S.S.R.* **143** (5), 1205–7.

SZALAY-MARZSÓ, L., 1958. Populationsdynamische Untersuchungen an Beständen der Rübenblattlaus (*Aphis (Dorsalis) fabae* Scop.) in Ungarn, in den Jahren 1955 und 56. *Acta agron.* **8**, 187–211.

TEAL, J. M., 1957. Community metabolism in a temperate cold spring. *Ecol. Monogr.* **27**, 283–302.

TEAL, J. M., 1962. Energy flow in the salt marsh ecosystems of Georgia. *Ecology* **43**, 614–624.

UMBREIT, W. W., BURRIS, R. H. and STAUFFER, J. F., 1957. *Manometric techniques.* (3rd ed.) Burgess, Minneapolis.

UTTER, J. M. and LE FEBVRE, E. A., 1973. Daily energy expenditure of purple martins (*Progne subis*) during the breeding season: estimates using D_2O^{18} and time budget methods. *Ecology* **54**, 597–604.

VAN HOOK, R. I., 1971. Energy and nutrient dynamics of spiders and orthopteran populations in a grassland ecosystem. *Ecological Monographs* **41**(1), 1–26.

VAN VALEN, L., 1976. Energy and evolution. *Evolut. Theory* **1**, 179–229.

VERDUN, J., 1972. Caloric content and available energy in plant matter. *Ecology* **53**, 982.

VOLLENWEIDER, R. A., 1969. *A manual on methods of measuring primary production in aquatic environments.* (I.B.P. Handbook 12) 224 pp. Blackwells, Oxford.

WESTLAKE, D. F., 1963. Comparisons of plant productivity. *Biol. Rev.* **38**, 85–425.

WHITE, A., HANDLER, P. and SMITH, E. L., 1959. *Principles of biochemistry.* McGraw-Hill, New York.

WHITTAKER, R. H., BORMANN, F. H., LIKENS, G. E. and SICCAMA, T. G., 1974. The Hubbard Brook ecosystem study: Forest biomass and production. *Ecol. Monogr.* **44**(2), 233–52.

WIEGERT, R. G., 1964*a*. Population energetics of meadow spittle bugs (*Philaenus spumarius* L.) as affected by migration and habitat. *Ecol. Monogr.* **34**(2), 225–41.

WIEGERT, R. G., 1964*b*. The ingestion of xylem sap by meadow spittle bugs, *Philaenus spumarius* (L.), *Am. Midl. Nat.* **71**, 422–8.

WIEGERT, R. G., 1965. Intraspecific variation in calories/g of meadow spittle bugs (*Philaenus spumarius* (L.)) *BioScience* **15**, 543–5.

WIEGERT, R. G., 1968. Thermodynamic considerations in animal nutrition. *Am. Zool.* **8**, 71–81.

WIEGERT, R. G., 1974. Litterbug studies of microarthropod populations in three South Carolina old fields. *Ecology* **55**, 94–102.

WIEGERT, R. G., (Ed.). 1976. *Ecological Energetics. Benchmark Papers in Ecology* **4**, 457 pp. Dowden, Hutchinson & Ross, Pennsylvania.

WIEGERT, R. G. and COLEMAN, D. C., 1970. Ecological significance of low oxygen consumption and high fat accumulation by *Nasutitermes costalis* (Isoptera: Termitidae). *BioScience* **20**, 663–5.

WIEGERT, R. G. and EVANS, F. C., 1967. Investigations of secondary productivity in grasslands. *In* Petrusewicz, K. (ed.). *Secondary Productivity of Terrestrial Ecosystems,* 161–176, Warzawa-Kraków.

WIEGERT, R. G., ODUM, E. P. and SCHNELL, J. H., 1967. Forb-arthropod food chains in a one-year experimental field. *Ecology* **48** (1), 75–83.

WIEGERT, R. G. and OWEN, D. F., 1971. Trophic structure, available resources and population density in terrestrial vs. aquatic ecosystems. *J. theor. Biol.* **30**, 69–81.

WILLIAMS, E. C. and REICHLE, D. E., 1968. Radioactive tracers in the study of energy turnover by a grazing insect (*Chrysochus amatus* Fab.: Coleoptera Chrysomelidae) *Oikos* **19,** 10–8.

WINTERINGHAM, F. P. W., 1959. An electrolytic respirometer for insects. *Lab. Practice* **8,** 372–5.

WISSING, T. E. and HASLER, A. D., 1971. Intraseasonal change in caloric content of some freshwater invertebrates. *Ecology* **52**(2), 371–373.

WOHLSCHAG, D. E., 1957. Differences in metabolic rates of migratory and resident freshwater forms of an arctic whitefish, *Ecology* **38,** 502–10.

WOOD, T. G. and LAWTON, J. H., 1973. Experimental studies on the respiratory rates of mites (Acari) from beech-woodland leaf litter. *Oecologia (Berl.)* **12,** 169–91.

YENTSCH, C. S. and HEBART. J. F., 1957. A gauge for determining plankton volume by the mercury immersion method. *J. Cons. perm. int. Explor. Mer.* **22,** 184–90.

ZEUTHEN, E., 1950. Cartesian diver respirometer. *Biol. Bull. Mar. Lab. Woods Hole* **98,** 139–143.

Author Index

Bibliographical citations are in *italics*.

General Index

The titles of papers cited in the bibliographies are not indexed.